ANIMALS, PLANTS AND AFTERIMAGES

ANIMALS, PLANTS AND AFTERIMAGES

The Art and Science of Representing Extinction

Edited by Valérie Bienvenue and Nicholas Chare

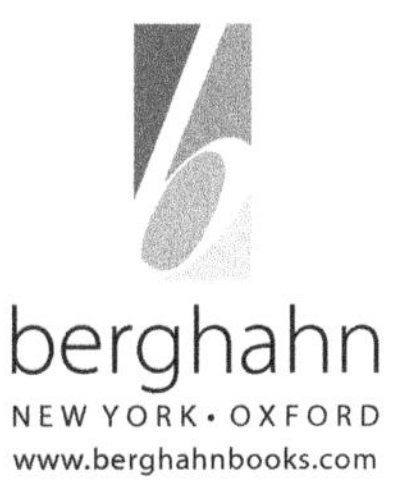

berghahn
NEW YORK · OXFORD
www.berghahnbooks.com

First published in 2022 by
Berghahn Books
www.berghahnbooks.com

Library of Congress Cataloging-in-Publication Data

Names: Chare, Nicholas, editor. | Bienvenue, Valérie, editor.
Title: Animals, plants and afterimages : the art and science of
representing extinction / [edited by] Nicholas Chare and Valérie
Bienvenue.
Description: New York : Berghahn, 2022. | Includes bibliographical
references and index.
Identifiers: LCCN 2021042546 (print) | LCCN 2021042547 (ebook) | ISBN
9781800734258 (hardback) | ISBN 9781805390671 (open access ebook)
Subjects: LCSH: Extinct animals in art. | Extinct plants in art. | Extinct
animals--Exhibitions. | Extinct plants--Exhibitions. | Art and biology.
| Museum techniques--Moral and ethical aspects.
Classification: LCC N7663 .A55 2022 (print) | LCC N7663 (ebook) | DDC
704.9/43--dc23/eng/20220210
LC record available at https://lccn.loc.gov/2021042546
LC ebook record available at https://lccn.loc.gov/2021042547

British Library Cataloguing in Publication Data
A catalogue record for this book is available from the British Library

ISBN 978-1-80073-425-8 hardback
ISBN 978-1-80539-332-0 paperback
ISBN 978-1-80073-426-5 epub
ISBN 978-1-80539-067-1 web pdf

https://doi.org/10.3167/9781800734258

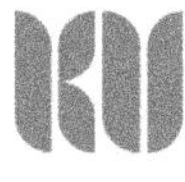

An electronic version of this book is freely available thanks to the support of
libraries working with Knowledge Unlatched. KU is a collaborative initiative
designed to make high-quality books Open Access for the public good. More
information about the initiative and links to the Open Access version can be
found at knowledgeunlatched.org.

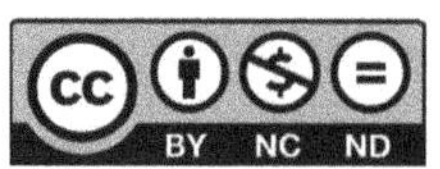

Contents

Part III. Representing Avian and Insect Extinctions

Part IV. Representing Extinct Plants and Fungi

Part V. Representing Extinct Mammals

Part VI. Exhibiting Extinction

Illustrations, Figures and Tables

Figure

Tables

Acknowledgements

The outlines of this project emerged from a conversation with Darrin Lunde, the collections manager of the Mammals Collection at the National Museum of Natural History, Washington DC. We are very grateful to him for sharing his insights regarding mammalian extinctions and natural history more broadly. Belinda Bauer of the Zoology Unit at the Tasmanian Museum and Art Gallery, and Trish Koh, the collections manager of the Tiegs Museum in Melbourne, also gave generously of their time and expertise during research visits. Their understandings of museum access and display helped us to shape this project. We thank Elizabeth 'Biz' Bell and Robert Prŷs-Jones for providing valuable information during the genesis of the volume. Additionally, we are indebted to Anne Bruneau of the Département de sciences biologiques at the Université de Montréal for her invaluable advice relating to plants. We are also grateful to Ursula K. Heise for suggesting potential contributors.

Much of the work on this manuscript has been carried out during the COVID-19 pandemic. As we write these acknowledgements many parts of the world, including Quebec, are still heavily impacted by the outbreak. Throughout this time, which has been very anxious and demanding for many, the staff at the Bibliothèque des lettres et sciences humaines have continued to respond to our queries and source materials. We are obliged, in particular, to the library staff Marie-Ève Ménard, Vincent Perrault and Nancy Théberge. We also wish to acknowledge the help given to us by Maryna Beaulieu of the Bibliothèque des livres rares et collections spéciales at the Université de Montréal. Elsewhere, Rosalind Grooms, the Cambridge University Press archivist, lent us invaluable assistance.

We are beholden to the Social Sciences and Humanities Research Council of Canada for funding this project. We had the invaluable aid of three research assistants as we prepared the volume for press. Many thanks therefore go to Camila de Oliveira Savoi, and also Jade Brais-Dussault and I. Milan Cail. Additionally, we are particularly grateful to Mitchell B. Frank for his perceptive readings of parts of the manuscript and his helpful comments. Along the way, we also benefited from assistance from Patricia Correa-Romero, Marisa Flores, Marie-Janine Georges, Michelle

Gewurtz, Peter Krausz, Laurence Labor Palaniaye, Sylviane Latour, Vincent Lavoie, Michel Ledoux, Sébastien Lévesque, Silvestra Mariniello, Milena Marinkova, Suzanne Paquet, Julie Pelletier, David Sume, Kristine Tanton, Tom Tyler, Jeff Tshibaka and Dominic Williams. More generally, we wish to express our gratitude to our students on the courses *Introduction aux cultures visuelles* and *L'art du témoignage* who shared their thoughts and perceptions relating to issues concerning the representation of extinct species in art, cinema and literature.

We are also very grateful to Marion Berghahn and to the commissioning and production teams at Berghahn Books, particularly our commissioning editor Amanda Horn, our copyeditor Nigel Smith, our production editor Caroline Kuhtz and the editorial assistant Sulaiman Ahmad. Finally, we are thankful to the anonymous reviewers of the initial proposal and of the manuscript for their valuable advice and comments.

Introduction

Representing Extinction

Art, Science and Afterimages

Valérie Bienvenue and Nicholas Chare

Exordium: Afterimages

In his remarkable study of the Upper Palaeolithic parietal art of Western Europe, *The Mind in the Cave*, David Lewis-Williams suggests that some imagery present in caves such as Chauvet and Lascaux was inspired by experiences of altered states of consciousness.[1] Emergence from such altered states can be accompanied by the appearance of afterimages, mental pictures that hang suspended in the field of vision for a minute or more. These images gradually lessen in intensity and clarity, slowly blending with the background of the surrounding visual field before ultimately disappearing. Lewis-Williams suggests that through their drawings and paintings, prehistoric people sought to 'fix' such fleeting images, granting them a measure of permanence. Many images in the caves are of animals, including bears, bison, deer, horses, ibex and mammoths.[2]

Some cave imagery reveals characteristics about these animals that cannot be gleaned from the fossil record, such as the likely belt patterning on Eurasian rhinoceroses or the partial striping of horses in the Late Pleistocene.[3] Species such as the cave bear (*Ursus spelaeus*) and the woolly mammoth (*Mammuthus primigenius*) that appear in parietal art are now extinct. Prehistoric paintings and engravings tell us much about the appearance and distribution of these animals.[4] Insights about extinct animals provided by ancient rock art are not restricted to Europe. In mainland Australia, for instance, there are numerous depictions of thylacines (*Thylacinus cynocephalus*), an animal that probably became extinct there roughly three thousand years ago.[5] A petroglyph in Murujuga in Western Australia has been interpreted by Ken Mulvaney as registering Aboriginal recognition of a local decline in the species and incorporating efforts to remedy it.[6]

The ways in which afterimages have been understood has changed through time. In modernity, as Jonathan Crary examines, they became associated with autonomous vision, with sensory perception 'cut from any necessary link with the external referent'.[7] Afterimages, as durational, were subject to quantitative study, and efforts were made to formally classify them based on their appearance. Pioneered by Johann (Jan) Purkinje, these efforts at classification, which necessitated making drawings of afterimages, also involved efforts to 'fix' transient optical phenomena in the present. In both prehistory and modern times, afterimages have been linked to the desire to keep something that is transitory from disappearing. Writing in 1819, Purkinje describes the effort that must be expended to keep an afterimage in the field of vision, as it 'disappears as soon as the will slackens'.[8] His sketches, abstract forms, provide artful records of the persistence of vision while simultaneously indexing his own tenacity, his scientific resolve. The study of afterimages in the nineteenth century occurred at the intersection of art and science.

Both Crary and Lewis-Williams refer to afterimages as physiological phenomena, but, as is the case here, the word is also often used figuratively.[9] One of the reasons that Laura Mulvey, for example, employs the term 'afterimages' is that she hears echoes of the term 'afterwardsness' (as Sigmund Freud's term *Nachträglichkeit* is often translated into English) in it.[10] For Freud, 'afterwardsness' describes a belated grasp of the significance of an event, particularly a traumatic one.[11] Extinction is linked etymologically to the Latin verb *exstinguo*, which means to put out, quench, extinguish, kill or destroy. It is therefore unsurprising that extinction, the dying out of a particular species of organism, is frequently conceived as a trauma. In *Imagining Extinction*, for example, Ursula Heise explores how the disappearance of the ivory-billed woodpecker (*Campephilus principalis*) 'points to a traumatic past, the history of large-scale ecological exploitation and deforestation of the American South'.[12] Heise also reads the conclusion to Lydia Millet's novel *Magnificence* – in which a hidden collection of endangered and extinct species is revealed – through the prism of ideas about trauma and genocide.[13]

Trauma is frequently conceived as an unassimilated, unsymbolized experience.[14] It refers to occurrences the significance of which are only retrospectively *realized*. Much trauma theory relates to individuals with pathological conditions that prevent them coming to terms with past events from their personal histories. Relief emerges through therapy that encourages the articulation of the experience, its representation through words or images. Thinkers such as Heise, however, conceive of trauma in broader, cultural terms as impacting group rather than individual consciousness. When comprehended as a cultural phenomenon, efforts to alleviate trauma become bound up with 'public acts of commemoration, cultural representation, and public political struggle'.[15] Whether related to the individual or the collective, practices of representation therefore have a key role to play in expressing traumatic events and coming to know them.[16] It is for this reason that Griselda Pollock links artistic responses to personal or cultural traumatic experiences with afterimages. For her, art as afterimage is potentially transformative, opening a space for aesthetic encounter with 'that which, by definition, is not yet in the grasp of representation'.[17] Art, in this conception, helps something of a traumatic event to be processed.

While not making direct reference to trauma, several of the chapters in this volume reflect on extinction in ways that encourage making links between species disappearance and traumatic experience. The chapters by David Maynard and Kathryn Medlock, for instance, which explore museum exhibits devoted to the thylacine and its destruction, recognize that the marsupial's fate was bound up with colonial conquest. European settler colonists decimated lutruwita's (Tasmania's) Indigenous population. They also eradicated native species including the Tasmanian emu (*Dromaius novaehollandaie diemensis*) and the thylacine.[18] The emu and the thylacine formed part of Aboriginal Country.[19] Their indiscriminate killing by settler colonists would have gone against the sustainable use of resources associated with 'caring for Country'.[20] Colonialism's after-effects also feature prominently in *hagwil hayetsk*'s chapter, which examines the impact of overharvesting of *bilhaa* or Northern abalone (*Haliotis kamtschatkana*) by Canadian settler colonists upon traditional cultural practices of the Gitxaała people of Laxyuup Gitxaała.

Mulvey's use of the term 'afterimage' is not solely motivated by its association with trauma. For Mulvey, with her particular interest in cinema, afterimages also evoke a kind of visual afterlife that can be granted to things by motion picture technology: 'the medium preserves the living presence of human figures, often long dead, through the film machine'.[21] Film is therefore often haunted by its images, its capacity to revive the past in the present.[22] Murray Leeder describes cinema as having become '[d]eliberately or accidentally . . . a storehouse for our dead'.[23] Mulvey uses 'the figure of the ghost and haunting to evoke the complex implication of a past persisting into its future'.[24] This implication as it relates to representations of extinction is one that extends well beyond film. Photographs and sound recordings of many recently vanished species endure that grant them a phantom presence in the present. Physical remains of numerous extinct organisms also persist, including taxidermy mounts which are sometimes presented as if alive, a phenomenon Barbara Creed discusses in her chapter in this volume. Taxidermy mounts, highly illusionistic forms of representation which Rachel Poliquin understands as always bound up with remembrance, are also discussed in several other chapters.[25]

Meditating on a stuffed parrot, Poliquin notes that taxidermy forms a haunting spectacle, affording an animal a kind of diminished yet enduring afterlife.[26] Mark Barrow's book on American efforts to recognize and legislate against human-caused extinction is titled *Nature's Ghosts*.[27] Much of the book is dedicated to examining how the threat of extinction haunted American naturalists, with haunting understood as a troubling or discomfiting.[28] Clearly, however, 'nature's ghosts' are species that have disappeared, such as the taxidermied Carolina parakeet (*Conuropsis carolinensis*) which graces the book's cover. In *British Animals Extinct Within Historic Times*, James Edmund Harting writes of extinction as 'disappearance beyond recall'.[29] The idea of absence also underpins Tim Flannery and Peter Schouten's titling of their book on extinct animals: *A Gap in Nature*.[30] Despite the emphasis on loss and vanishment, these books all attest to an extensive capacity to remember extinct species through drawing on sources such as physical remains and textual accounts. The disappearance of a known species is never total. There is always some trace, some afterimage, even if it is no more than a name and the picture it conjures, that endures, that haunts us. Avery Gordon observes of apparitions in general that they are felt, sensed, rather than known.[31] Conceived in this way, extinction is affectively charged, amorphous and elusive.

Sight Unseen

Disappearance, with its etymological roots in the Latin *parere* (to come into view), and vanishment, which has its origins in the Latin *evanescere* (fading from sight), are both visual metaphors used regularly to describe extinction. The disappeared and the vanished can no longer be *seen*. Through these synonyms, extinction is therefore conceived as something rendered visibly absent. The artist Lucienne Rickard, whose work is examined in Jeanette Hoorn's chapter for this volume, gave literal expression to this idea. Her durational performance, *Extinction Studies* (2019–21), involved a palimpsestic process of meticulously drawing and then erasing examples of recently extinct plant and animal species. Two such images, *Half-erased Camballerus alvarezi* and *Erasing Madhuca insignis*, are reproduced on the cover of this volume. The traces of earlier erased species persist as wraithlike afterimages within these later drawings. The title of Rickard's performance was intended to connote both art and science: a study is a technical term in art for a preparatory work, yet it can also refer to a practice of knowledge acquisition.[32] Although an exploration of vanishment, Rickard's erasures leave visible traces. If they are to be known, all extinctions must leave some remnant, some kind of remainder acting as a reminder.

The reminder may be a physical specimen or a part thereof, sometimes it is a visual or aural record, and, occasionally, it is only a pictorial or textual reference. These last forms of visibility are referred to by Alexander Lees and Stuart Pimm as 'anecdotal evidence' of the existence of a species – an eyewitness (or, more rarely, earwitness) account.[33] Samuel Turvey has analysed paintings, for instance, to substantiate the presence of a now extinct species of macaw in Jamaica.[34] The *Anapurú* parrot, which is referred to by the Portuguese Jesuit administrator Fernão Cardim in his *Treatise on the Land and Climate of Brazil*, offers another example.[35] Cardim, who was in Brazil from 1583 to 1590, describes the 'very beautiful' bird as reportedly having a body 'splashed and bespeckled' [*salpicado e espargido*] with 'red, green, yellow, black, blue, brown [and] lavender [*côr de rosmaninho*]'.[36] As Luciano Moreira-Lima notes, the identity of this 'mysterious parrot' remains unknown today.[37] Cardim also refers to an example mentioned by Lees and Pimm, a black macaw, the *Ararúna*, which seems distinct from Brazil's known macaws.[38] The bird was already identified as rare by Cardim.[39] The *Anapurú* and the *Araréna* have a shadowy ornithological existence, as no holotypes (type specimens) exist for the birds. They persist only as words.

The textual descriptions nonetheless grant a kind of afterlife to the birds. Something, no matter how slight and/or vague, survives of them. Many organisms exit this world unnoted. Elizabeth Kolbert notes this in relation to the present plight of amphibians, a number of which have disappeared before their existence was scientifically recorded.[40] A comparable situation exists for plants. In an article discussing how many plant species potentially exist, Stuart Pimm and Lucas Joppa suggest that some 'missing species', plants not named and catalogued, 'went extinct before we could even estimate that they were missing from the taxonomic catalogue'.[41] They use orchids from lowland areas of the Atlantic Coast as an example, noting that 'many species could have lived in areas completely destroyed before taxonomists explored them'.[42] Recent research on Malagasy grasses suggests 'at least a 50% rate of unrecorded extinctions'.[43] This figure is region specific, but it gives an indication of the potential scale of unnoted extinctions in range-restricted areas with high endemic plant populations. Given that they have never been noted, to speak of these hypothetical species as 'disappeared' or 'vanished' is disjunctive.

For a species to 'exist', a necessary precursor to it becoming extinct, it must first be described. Description is a key dimension of taxonomy: the study of naming, defining and classifying organisms. Usually a holotype, a physical example of an organism, forms the basis for such a description.[44] Taxonomy as a practice has changed considerably through the ages. Contemporary taxonomic methods are usually traced back to Carl Linnaeus (1707–1778), particularly to the publication of *Species Plantarum* [The species of plants] in 1753, and the 10th edition of *Systema Naturæ* [System of nature] in two volumes in 1758 and 1759. Linnaean innovations such as the consistent use of Latin binomials and of divisions such as class, order, genus and species, continue to inform taxonomic practices today.[45] As Staffan Müller-Wille summarizes, Linnaeus's descriptions were arrived at through 'a straightforward inductive process that involved the careful comparison of individual species'.[46] The botanist's notion of description was subsequently understood in varied ways by different people. Harriet Ritvo cites an unpublished manuscript by an English devotee of Linnaeus, Leonard Chappelow, which 'equated taxonomy with works of art' and characterized the Linnaean system as 'a series of descriptive pictures'.[47] Chappelow's account suggests an artful aspect to scientific description. It is in the context of art that Michael Baxandall notes that the description of a picture 'is a representation of thinking about a picture more than a representation of a picture'.[48] It might similarly be said that description of a specimen is a representation of thinking about that specimen rather than a representation of the specimen.

As a kind of representation, description shapes our perceptions of what it refers to. Pollock defines representation as 'something refashioned, coded in rhetorical, textual or pictorial terms'.[49] Representations of species (extinct or otherwise), as coding, are related to, yet distinct from, the physical organisms they describe.[50] The gap between a species and the ways in which it is represented is one that many of the chapters in this volume examine. This consideration is not motivated by the aim of establishing a 'true' picture of a given organism and assessing how specific representations deviate from it, but rather addressing the kinds of ideas and beliefs that underpin particular examples.[51] The way a species is represented raises important ethical issues. Depictions that portray a species as potentially dangerous to humans and human interests, for instance, can reduce sympathy for the plight of that species.[52] Even descriptions that aim to be precise and unambiguous, including those linked to morphology, are equivocal. Morphology, the study of the form and structure of organisms, necessarily involves processes of explanation.[53] Morphological form is not pregiven in any simple sense. It must be described by way of text and/or images. These descriptions then circulate as data about a given organism.

Becoming an Afterimage

Morphological data, as description, is linguistic in a broad sense. Language, for example, moulds the way shape is used as a descriptor of an organism. Yet shape, as Norman MacLeod and Peter Forey have noted, involves employing terms that possess ambiguity. Using the example of 'leaf shape: oval, round', they ask: 'where in the context of any particular systematic comparison does "round" stop and "oval" begin?'.[54] The question draws attention to an instance of arbitrary decision-making in relation to the description of shape. The shift from observation of phenomena to their description, from things to their conceptualization, is complex and necessarily transformative. Language 'puts pressure on us to discriminate in its way and in this sense every language is tendentious'.[55] Valérie Bienvenue's chapter in this volume reflects on how a language can be developed that refuses to generalize and to 'overwrite' the particularities of an individual example of an extinct species, respecting instead something of its singularity.

The tendentiousness of language is well demonstrated by William Harvey's summary of the general characteristics of the (now extinct) protist Bennett's seaweed (*Vanvoorstia bennettiana* also known as *Claudea bennettiana*):

> *Frond* stipiate; stipes filiform, merging in the marginal rib of a flat, unilateral, open network, formed of several series of anastomosing, slender leaflets. *Fructification*: 1, *ceramidia* containing within a membranaceous pericarp a tuft of pear-shaped spores; 2, *stichidia* formed from the bars of the network, and studded with triangularly parted tetraspores in transverse rows. –CLAUDEA (*Lamour.*), in honour of Claude Lamouroux, father of the botanist of that name.[56]

Harvey's description of the alga is scrupulous but not dispassionate, manifesting a poetic rather than prosaic precision. By accident or design, the account is lyrical, including considerable alliteration and assonance (notably more so than in other of Harvey's descriptions). We would suggest that this probably stems from Harvey's sense of the organism's exquisiteness and uniqueness. He calls it a 'beautiful and curious species' that he finds 'very remarkable'.[57] Something of this splendour and extraordinary significance registers into the exceptionally crafted description he accords it. The *feel* Harvey has for the frond inflects the textual depiction. In the guise of a quest for precise elaboration, he is able to provide a paean. The idea that emotional investment in a species influences morphological description would be anathema to many scientists, but Harvey's short sketch gestures towards just such a possibility.[58]

Some words in the description also manifest ambiguity of the kind that troubles MacLeod and Forey. Calling the spores 'pear-shaped', for instance, evidently suggests an image of a spore that tapers towards the top. To arrive at this image, however, requires working back from another image, that of a common or garden pear. Pears come in many shapes and sizes, some of which, such as the Asian pear (*Pyrus pyrifolia*), are not 'pear-shaped'. The analogy functions only for a specific audience, one that, when they hear 'pear-shaped', form a mental image of the European pear (*Pyrus communis*).[59] The reference to a tuft is also ambiguous. Tufts are bunches or clumps or clusters of small things. The physical appearance of some alga has encouraged visual analogies with tufts or tresses of hair.[60] Clearly, for Bennett's seaweed and other species (such as *Euchema speciosum*), Harvey also perceives this resemblance in their spores. The tuft as a descriptor is nonetheless vague, implying something held together at its base but also having loose ends.

To supplement his text, Harvey also included a plate, a lithograph by Vincent Brooks. The lithograph features three illustrations of the alga, one showing it at natural size (which is small) and two providing magnifications.[61] The illustrations seem designed to reflect the description, to show the alga as network. Preserved examples of *Vanvoorstia bennettiana* do not give such a neat demonstration of their own structure, despite Harvey's claim that the life-size illustration 'is an exact facsimile as to form and size'.[62] Harvey clearly believes that combining images and text enables him to augment the accuracy of his description. For morphological purposes, images can never describe organisms in themselves, they require textual supplementation. Images do nevertheless 'serve as visual support and provide empirical substantiation for a given descriptive statement by documenting the observational basis for this description'.[63] Harvey employs the image in this way, showing the importance of visual representation for taxonomy and, by extension, of art for science.[64]

Although Harvey's focus on morphology is not absolute (he also provides limited information regarding the alga's geographical distribution, which he lists as New South Wales), form is clearly of paramount importance. Yet form is only one mode of understanding an organism and its significance – a mode that privileges external and internal structure over, for example, behaviour and that organism's role in the broader ecosystem.[65] In the plate showing *Vanvoorstia bennettiana* the background is the blank of the page, the alga has become free-floating, ecologically unmoored. There is no sense of a marine environment. The importance of form to taxonomic practice comes at the expense of an acknowledgement of the entanglement of Bennett's seaweed with other organisms. The alga was a source of nutrition and of shelter for other marine life. Brooks's lithograph artificially disentangles the protist from its coastal community and all the dynamic interactions with other organisms that accompanied it. It is a mode of representation that visibly negates interrelationality among species, concealing their often co-dependence. Brooks relocates the seaweed to an abstract realm the better to communicate taxonomic knowledge. This raises ethical issues about privileging morphology over ecological integratedness as a mode of (visually) knowing the protist. The lithograph produces a particular kind of seaweed, self-contained and self-sufficient, shaping how the alga is perceived and understood.[66] In this sense, the representation manifests agency, acting on the world rather than simply reflecting it.

The previously mentioned petroglyph from Murujuga was also conceived as agential. Mulvaney suggests the carving, which is of a thylacine, was used as part of thalu or increase ceremonies. These ceremonies are employed to encourage the regeneration of plants or animals.[67] The site located at what is now known as Patterson Valley must therefore have been strongly associated with the Dreaming power of the thylacine.[68] In Australian Aboriginal culture (which is not homogeneous), the ancestor spirits of specific animals, plants and insects remain at particular sites. Patterson Valley was a sacred site linked to the thylacine. Mulvaney indicates that the ceremony at Murujuga involved pounding the interior of the thylacine motif, a process which caused large cupules to be produced. The patination of the cupules suggests that they are of a similar age to the motif. In addition, 'pecked and scored lines radiate out from the quadruped, and several of these lines continue over the surface of adjacent boulders'.[69] The weathering is less pronounced for the lines, implying that they are of a more recent date. Additionally, surrounding the petroglyph in an area of 200 square metres are twenty-three carvings of macropods.[70]

Mulvaney observes that '[t]here appears to be a spatial and arguably a symbolic association between the macropod and the quadruped motifs'.[71] For him, the relationship between the thylacine and the macropods is distinctive:

> There is something unique about this combination and treatment of images. With the evident demise of the thylacine 4000–3000 years ago, those charged with its ritual maintenance would have been inevitably challenged. For the custodians of the site, the sacramental practitioners, altering the usual may have constituted a final and desperate attempt to ensure the continued existence of *Thylacinus cynocephalus*. What we may have documented is a continuity of ritual practice which has an antiquity spanning an extinction event.[72]

The group of petroglyphs, as imagery that foregrounds the fundamental material entanglement of species (here of thylacine and macropod), differ radically from Brooks's lithograph.[73] In the petroglyphs (which form part of the very habitat of the animals they portray), predator and prey are shown as mutually dependent. Mulvaney reads the lines as radiating out from the thylacine, yet they should, perhaps, be read as bidirectional.[74] Prey sustain predators and predators often perform an important role in maintaining the health of prey.[75] Additionally, Mulvaney gestures towards the symbolic entanglement of the custodians of the site with the animals. The desperation he believes the incised lines index might be linked to the totemic importance of the thylacine for the custodians.[76] Another kind of enmeshing, one also running counter to views of species as discrete, is explored by *hagwil hayetsk* in this volume.[77] He explains that the symbolic entanglement of the Gitaaxła people with *bilhaa* potentially generates respect for the latter.[78]

Indigenous Knowledge and Extinction

Indigenous peoples such as the First Nations Gitaaxła frequently stress respect for nature and call attention to the need for ecological balance. Too often, however, Indigenous voices are still marginalized in discussions about extinction. The protest coalition Wretched of the Earth (which includes the activist group Indigenous Environmental Network) has drawn attention to how movements such as the Extinction Rebellion view ecological and environmental issues from a position of White privilege, and sideline Indigenous voices and expertise.[79] Considerable efforts are now being made to foster pan-Indigenous solidarity regarding environmental and other issues, while also recognizing each community's singularity.[80] A delicate balancing act is underway aimed towards alliance that is respectful of difference. Such efforts must continually guard against linguistic and other forms of exclusion.[81] Decolonization is a long-standing and key shared concern. Thohahoken Michael Doxtater notes that Indigenous scholarship has been confronting colonial-power-knowledge since the 1960s.[82] There are, however, some Indigenous communities that have not been subject to colonization yet are also at risk of losing traditional knowledge. A Chinese minority community such as the Chuanqing, for example, possess considerable expertise relating to the flora of the mountainous regions of central Guizhou. Until recently, no efforts had been made outside the community to record that understanding.[83] Samuel Turvey's chapter in this volume, which focuses on the Yangtze ecosystem and Hainan in China, foregrounds the speed with which Indigenous knowledge about extinct species can be lost.

For a long time, the value of Indigenous knowledge for sustainable development has been recognized. Writing thirty years ago, André Lalonde listed numerous kinds of knowledge held by Indigenous African societies that could contribute to conservation efforts, including social taxonomy, pest management, agronomic practices and approaches to anti-desertification.[84] Yet Abayneh Unasho Gandile, Solomon Mengitsu Tessema and Fisha Mesfine Nake suggest that Indigenous knowledge continues to be overlooked, is inadequately recorded and is itself at risk of extinction.[85] They do not view Indigenous knowledge as antithetical to Western scientific knowledge and think that bringing different belief systems into dialogue can be mutually enriching. A similar perspective is advanced in the context of Abya Yala (South America) by C. Dustin Becker and Kabita Ghimire, who examine how synergy between traditional knowledge and Western knowledge has aided forest preservation in Ecuador.[86]

Many Indigenous peoples, however, are circumspect about how their knowledge has been sourced and used by non-Indigenous conservationists. In a report on the topic of Indigenous knowledge as it links to extinction-related research, Audra Mitchell, Zoe Todd and Pitseolak Pfeifer draw attention to the way Western secular scientific logics continue to restrictively shape responses to the contemporary extinction crisis through entrenching divisions between human and 'nature', framing non-humans as resources for instrumental use and privileging techno-scientific and economic management solutions to ecological crises.[87] Focusing on Canada's First Nations, Inuit and Métis communities, Mitchell, Todd and Pfeifer note that the turn to Indigenous knowledge to aid conservation initiatives is usually accompanied by a failure to recognize the distinct ontologies and epistemologies of the communities from which the insights are derived. Indigenous communities are mined for 'data' and then sidelined. Our decision to place the section on 'Indigenous Peoples and Extinction' early in the volume is motivated by a wish to avoid this kind of marginalization and foreground the importance of traditional knowledges for ongoing efforts to address today's biodiversity crisis. The issues raised in the section by *hagwil hayetsk* regarding interspecies relations, kinship and the toxic legacy of colonialism, have profound implications in terms of the uneven power relations that characterize many discussions and representations of extinction and possess broad relevance.

In the context of representation, natural history museums often maintain colonial values and unreflectively display artefacts that are of cultural significance to Indigenous peoples. Natural history is always also cultural history, with many valuable collections built on colonial exploitation and violence. Scholarship in this area has tended to focus on how ethnographic displays perpetuate racist worldviews.[88] The entire classificatory system (Linnaean taxonomy) used in most natural history museums is a form of descriptive domination. Many species with Latin binomials were historically well known to Indigenous peoples yet by other names. Subhadra Das and Miranda Lowe discuss the presence of a panel in the Hintze Hall of the Natural History Museum in London that portrays the plant *Quassia amara*, the binomial chosen by Linnaeus to commemorate the Ghanaian slave Kwasimukamba (Graman Quassi) who brought the medicinal qualities of the Surinamese shrub to the attention of Europeans. Das and Lowe lament the failure of the museum to commemorate Kwasimukamba, an important figure in Black history, or the plant that now bears his name. Their discussion, however, obscures the reality that the medicinal properties of the shrub were already recognized in Surinam.[89] The plant is widespread through the Caribbean and Abya Yala (Central America). Most of its current regional names are derived from the languages of the colonizers (Dutch, English, French, Portuguese and Spanish) but some, such as the Ulwa name of *battaka di basta*, are Indigenous.[90] Continuing to solely use the standardizing Latin binomial, even if it harbours a hidden dimension of Afro-Surinamese history and foregrounds the role of extra-European expertise in botanical discovery, erases historical Indigenous recognition of the plant. This example shows in microcosm some of the difficulties that accompany the ongoing process of decolonizing natural history museums.

Das and Lowe rightly stress the value of narratives as part of decolonization efforts. Anna Guasco similarly argues that extinction storytelling in the museum 'may allow us to pay closer attention to the ways in which political and economic inequity, racism, (neo)colonialism, imperialism and ecological debt amongst nations intersect with issues of extinction and biodiversity loss'.[91] Storytelling, for Guasco, seems a reflexive form of narrative that embraces interdisciplinarity and is attentive to social justice issues. There should also be space for Indigenous storytelling in contemporary natural history museums. In many Indigenous cultures, storytelling is an embodied mode of knowledge sharing. The storyteller, their status and the language they use, is often inseparable from the 'content' of a given story. Elements such as facial expression, gesture and vocal intonation are crucial.[92]

Mitchell, Todd and Pfeifer suggest that researchers concerned with Indigenous conceptions of extinction need to look beyond scientific narratives and engage with oral history and other cultural forms such as art, film and poetry. Music might be added to the list. The Inuk singer Tanya Tagaq has powerfully demonstrated how music can embody Indigenous activism, cultural revitalization and political critique. We want briefly to examine her music here, as it ably demonstrates the kinds of insights into

Indigenous understandings of the land and its ecology that a non-scientific account can potentially offer. Tagaq is inspired by *katajjaq*, an Inuit throat-singing game of stamina usually played by two women in which each mimics aspects of their regional soundscape, such as the elements, fauna and human activities. On her albums, Tagaq throat-sings solo, accompanied by Western classical instruments and electronica. As Alexa Woloshyn has noted, Tagaq's 'musical practice demonstrates the limits of the easy binaries of traditional/modern and past/present'.[93] Her vocalizations run the gamut from aggression and pain, to the highly erotic, the breathy and the ecstatic.[94] The songs are strongly affective, achieving their political force through the conjuring of moods, of atmospheres that refuse to coalesce into clearly defined messages. Although lyrics are often present, there is a studied refusal to offer slogans or platitudes. In her thoughtful and nuanced engagement with Tagaq's music, Kate Galloway suggests that the singer gives voice to 'ecological trauma' and also invites a reconnection of human bodies to the land, fostering an ethics of kinship with the non-human.[95]

The album *Animism* (2014) includes songs that highlight human and non-human animal connectedness in Inuit culture. In 'Tulugak', Tagaq voices the *tulugak* or raven (*Corvus corax*), her larynx transformed to syrinx. Through her vocal inhabiting of the bird, Tagaq embodies her connection with it. What she accomplishes using her voice is frequently signalled in Inuit culture by other means such as through clothing. As Heather Igloliorte explains, clothing can symbolize 'the correlation and affinity between humans and animals, and is a form of transformation iconography'.[96] Clothing that resembles a given animal, such as a caribou, transfers something of the qualities of that animal to the wearer. In this context, Tagaq's singing, her replication of non-human animal communication, should not be understood as mimicry where she simply 'sounds like' a raven or other animal. In Inuit cosmology, an extended sense of personhood exists and, as such, Tagaq's vocal practice can be heard to perform the unity of human and non-human animals.[97]

The short film *Tungijuq* (Dir. Félix Lajeunesse and Paul Raphaël, Canada, 2009), starring Tagaq, is about the value of hunting to the Arctic ecosystem and gives the reality of human and non-human interconnection visual expression through employing CGI and prosthetics. The latter are used to give Tagaq lupine eyes and a tail. In the film, she also embodies a dying caribou and a seal. One scene shows Tagaq naked against the arctic landscape, a cut of meat nestled between her breasts and on her abdomen. She caresses the bloody flesh, joying in the meat.[98] Later Tagaq eats raw seal, a smile flickering across her face. Like *Angry Inuk* (Dir. Alethea Arnaquq-Baril, Canada, 2016), *Tungijuq* foregrounds the importance of seal hunting to Inuit culture. Canadian Inuit have faced social opprobrium for their continued hunting of *nattiq* or ringed seal (*Pusa hispida*) despite protests against sealing historically being prompted by settler colonial killing of harp seal (*Pagophilus groenlandicus*). Like the Gitxaała Nation discussed by *hagwil hayetsk*, who must now endure considerable restrictions on their harvesting of *bilhaa* because of settler colonial over exploitation of the marine snails, the Inuit pay the price for the actions and practices of others.

One of Tagaq's best-known songs, 'Fracking' (from the album *Animism*) is a vocal condemnation of hydraulic fracturing, a process used to extract fossil fuels such as natural gas and petroleum from rock formations. This technique has been condemned because of its environmental impact, which includes ground- and surface-water contamination, noise pollution, and seismic activity. Nunavut possesses shale that is rich in natural gas and might be mined using fracking. In addition to resource exploitation, the Arctic has been heavily impacted by climate change, with declining insect numbers noted and populations of some species of shorebird also diminishing.[99] *Tuktu* or Peary caribou (*Rangifer tarandus*) have decreased dramatically because of recent severe winters, making 'anthropogenic climate change … the caribou's worst enemy'.[100] Overhunting by Europeans historically also led to the extinction of the *isarukitsok* or great auk (*Pinguinus impennis*) and the *akpingak* or Eskimo curlew (*Numenius borealis*).[101] Galloway suggests that 'Fracking' 'gives voice to the trauma inflicted on the non-human environment', a reading that presumes humans exist as separate from the environment rather than being on a continuum with it. We believe that through compositions such as 'Fracking', Tagaq is signalling that the land is a part of her, not set apart from her. The cover version of Nirvana's 'Rape Me', from Tagaq's album *Retribution* (2016), is interpreted by Galloway as alluding 'to the actual and metaphorical violation of Indigenous lands'.[102] 'Fracking' is clearly also a song about assault, about fucking (with) the earth without its consent. Tagaq has, in fact, said fracking is like 'earth-rape'.[103]

In 'Fracking', Tagaq's vocal energy transmits a powerful sense of a land in pain, one that makes the hairs stand on end. Her voice physically affects the listener, registering corporeally as horripilation. VK Preston, writing of a live performance by Tagaq, describes the singer as communicating a 'felt politics'.[104] Olivia Michiko Gagnon similarly calls Tagaq's music a 'sensate politics'.[105] Tagaq does not offer a representation of earth violence but rather provides an affective enactment of it. In this sense, her music expresses 'active agencies that reach beyond representational logic and any anthropocentric perspective'.[106] D. Ferrett's notion of 'dark sound' as potentially articulating both the darkness of humanity's violent effects on the environment and the usually unperceived inaudible frequencies of nature, has considerable relevance in the context of albums such as *Animism* and *Retribution*.[107] For Ferrett eco-activist music that embraces 'dark sound' can shift understandings, alter human perception and introduce 'the possibility of change in behaviour'.[108] Tagaq's music exhibits agency of this kind, not simply representing the Arctic environment but embodying it and working to change perceptions of it.

How to Do Things with Pictures

Both the petroglyph and the lithograph discussed previously are agential kinds of image. The petroglyph, like Tagaq's music, was knowingly conceived to effect change in the world. It is unlikely Brooks and Harvey thought of the lithograph and its accompanying text in this way. Yet images are commonly accorded vitality in Western culture. In *What do Pictures Want?*, motivated by ideas of animism, W.J.T. Mitchell describes images as like living organisms. One example he uses for what he calls 'living images' or 'animated icons' is Dolly the sheep, the first mammal cloned from an adult somatic cell. For Mitchell, Dolly foregrounds the reality that living things are also images. Dolly conjures wonder and dread as *idea*, as 'icon of cloning and biotechnology'.[109] In a similar sense, even before Harvey embarked on naming and describing Bennett's seaweed, as a marine botanist he saw the alga through the prism of phycology, he had an *idea* of what it was. He knew what he was 'doing' when he named and described the seaweed. Naming a taxon also creates that taxon.[110] It is as much a doing as a describing. The lithograph of the seaweed contributed to this process of species realization. It produced rather than simply reflected *Vanvoorstia bennettiana*.

The idea that pictures possess agency and can act upon the world, potentially changing it, also inspires many contemporary artists. Mark Dion's artworks, for instance, which are examined by Anne-Sophie Miclo in her chapter for this volume, are intended to contribute to the building of 'a culture of nature that features regeneration over destruction, sustainability over depletion [and] nurturing over domination'.[111] Dion hopes his art practice will encourage change in relation to attitudes towards conservation and the environment. Rickard's semi-erasure of the critically endangered swift parrot (*Lathamus discolor*) for her *Extinction Studies* has also been interpreted as 'a moving call to action' regarding the fate of the bird.[112] There is a varied relationship between images and agency across time and geographical contexts.[113] Although Harvey's nineteenth-century description and the accompanying lithograph can be conceived as performative, their symbolic dimension renders them readily assimilable to ideas of representation. This cannot be said for the thalu site discussed earlier. It includes renderings of animals that might be praised for their mimetic competence, yet it was a place of process rather than of the re-presentation of things in the world, a reflection of the pre-existent. The action of striking was intended to provoke a response that would positively impact the physical animal. Striking the image was to strike a thylacine Dreaming. The image is not representational, it is the Dreaming. Although this volume is subtitled 'The Art and Science of Representing Extinction', it is clear that extinction as it appears in art and visual culture is not always representational, nor is it wholly explicable through theories of representation.

Images that feature extinct organisms can, for example, often generate a strong affective response. Affect is conceived in varied ways. In psychoanalysis, it is understood as an accumulation of excitation that is, by its very nature, resistant to identification and formulation. It refers to responses to stimuli that are 'not as specific as the emotions' being 'more diffused and shapeless'.[114] Melancholy provides a good example of such a response. Melancholia is experienced as a feeling of loss that bypasses signification.[115] It is registered intensely but cannot be put into words. Often extinction involves knowing we have lost something, a species, but not knowing what it is that we have lost. In many instances, the nature of this loss necessarily remains unresolved. Our knowledge of most extinct species is fragmentary and unlikely to increase significantly. To look at an image of a vanished organism is never to see the whole picture.

Works such as Rickard's *Extinction Studies*, in which loss is given powerful visual expression, are not melancholic in Freud's understanding of the term because what is being lost, the artist's detailed drawing, is readily knowable and recorded. Bill Hammond's 1995 acrylic work *Living Large 6*, which references the extinct Aotearoa/New Zealand bird the huia (*Heteralocha acutirostris*), gives a better sense of melancholy as an affective state. The sombre, surreal scene, painted in varying shades of blue and grey, shows a hippocephalic humanoid in evening dress on what may be a dais. The figure has a cello and seems about to give a performance. Their audience is a flock of bird's heads, specifically huia heads. Hammond leaves drips of paint across the picture that Cameron Boyle equates with tears. Boyle describes the painting as a 'melancholic scene'.[116] He rightly suggests it has the air of a requiem. As Boyle also notes, Hammond eschews straightforward imitations: the huia in the painting are spectral, elusive. The drips of paint draw the viewer earthwards, downwards, signalling the gravity of the scene. These drips also imply fluidity, a refusal of form. Suffused by affect, the work gives off a downbeat air. It is allusive, registering as a mood rather than communicating a meaning and coalescing into a specific emotion.

Sarah Bezan's chapter in this volume also attends to the affective dimension of artworks, specifically Jakob Kudsk Steensen's recent Virtual Reality (VR) installation, *Re-Animated*. Bezan's chapter additionally draws attention to how technological innovations such as VR potentially open new directions for the portrayal of extinction. This potential is also affirmed in Jeffrey Benca's chapter, which explains how he used measuring data from a fossil of a lycopsid (a form of lycophyte, a spore-bearing vascular plant) and a vector software program to reconstruct the plant's structural intricacies.

The Fossil Record

The lycopsid branch of the tree of life is still extant today, having survived several mass extinction events. These events form part of background or natural extinctions, which are to be differentiated from anthropogenic extinctions.[117] In his description of the tree of life, Charles Darwin noted that: '[f]rom the first growth of the tree, many a limb and branch has decayed and dropped off; and those lost branches of various sizes may represent the whole orders, families, and genera which have now no living representatives, and which are now known to us only from having been found in a fossil state'.[118] Darwin's description of limbs and branches dropping off the tree through decay suggests a kind of gradual failure. Extinction, so figured, is a slow process; yet some extinctions in the deep past, those associated with mass extinction events, occurred relatively abruptly. Darwin's metaphor also forecloses tree branches being sawn off, extinctions being caused by human actions. The mass extinctions that have been identified across geological time all occurred prior to the emergence of humankind. Five mass extinctions are generally accepted to have happened but there is a strong argument to also acknowledge a sixth.[119] Any human-made representations of these extinctions have been produced retrospectively, millions or billions of years after the disappearance of the organisms they depict.

We know of these organisms because some have left fossil traces. Perhaps the earliest such traces, from the Archean era, exist as biogenic graphite. Using electron microscopy, it has been established that graphite in western Greenland is probably composed of carbonate sediments from marine organisms. The graphite, dating from 3.7 billion years ago, therefore indexes early ocean life. The scientific paper that discussed the graphite sample was accompanied by several figures, including a graph, bar charts and transmission electron microscopy images.[120] The microscopy images show polygonal grains of graphite, their shape indicating they derive from biological material, from once living matter. The graphite is not, however, identifiable as a specific life form. The earliest *identifiable* life forms are fossil stromat-olites. The fossil record from the Archean era is meagre, meaning that reconstructing the nature of life at that time is difficult. Fossils from later in geological time are more numerous.

W.J.T. Mitchell describes the fossil record as 'a material and pictorial record, a vast iconic and indexical archive of species, most of them ex-tinct'.[121] He also notes their allegorical potency, seeing them as *memento mori*.[122] Fossils are often conceived as solid, as durable, despite many being incredibly fragile. Their perceived hardness seems to lend them substantial evidential value: fossils offer 'rock solid' data concerning extinct organisms. In reality, the organisms preserved as fossils have often been compressed and deformed. The process of fossilization, which is varied, transforms the organism. Soft tissue decays and, in a process known as diagenesis, hard tissue becomes modified geochemically and physically. Most ancient species are only known through this distorted record, a geological archive which, as with any archive, is partial. The majority of organisms that die do not fossilize.

The organisms that are most likely to fossilize are those with tissues re-sistant to decay. There is also a greater likelihood of finding more common organisms in the fossil record. Organisms living in low-energy environ-ments are more likely to be preserved as their environment is less abra-sive. Yet organisms in high-energy environments may develop protective coverings, dermises or shells, that increase their chances of fossilization.[123] Organisms from marine environments have more prospect of becoming fossilized than those on land. Norman MacLeod notes that fossil assem-blages where a group of organisms overcome by a sudden catastrophe are conserved together 'often [preserve] important aspects of the spatial sys-tems, ecological systems, behavioural systems, developmental systems and in some cases even the social systems of which the living organisms were part'.[124] As our picture of the natural world of the past is built from out of the fossil record, it is a highly incomplete and uncertain one. The way we represent ancient biota is influenced by what life became fossilized and by how those fossils are now interpreted.

It was through the study of fossils that the reality of extinction was first recognized. In the sixteenth century, Bernard Palissy correctly surmised that fossils embodied once living organisms.[125] In the early nineteenth century, through analyses of the fossil record, Georges Cuvier would argue for the possibility of species becoming extinct. Responding to critics of his ideas, Cuvier suggested that the considerable morphological differences between superficially similar living species of quadrupeds and those in the fossil record could not be explained away through gradual modification, as no traces of such modifications were visible in the record: 'the bowels of the earth have not preserved monuments of this strange genealogy'.[126] The absence of such data implied a catastrophic event had caused the sudden disappearance of some species. Much of Cuvier's work involved the Elephantidae family and the Mammutidae family. These fossil materials he worked from (of mammoths and mastodons) were relatively familiar and identifiable. Most remains of ancient organisms are unrecognizable for what they once were.

Fossil fuels, for example, are formed through the decomposition of dead organisms. These organisms are now unknowable but the species they belonged to will, almost invariably, be extinct. Fossil fuels such as coal, natural gas and crude oil are side effects of death and extinction. These side effects are now contributing to the current Holocene (or Anthropocene) extinction. Most anthropogenic carbon dioxide emissions are caused by the burning of fossil fuels. These emissions massively impact global warming. The extinction of the Bramble Cays melomys (*Melomys rubicola*), a rodent endemic to a coral cay in the Torres Strait, was directly attributed to climate change. Rising sea levels and extreme weather events had caused erosion and significant vegetation loss on the cay. A report on the demise of the rodent suggested: 'repeated ocean inundation driven by anthropogenic climate change posed the most severe, immediate and all-pervasive threat to this rodent population, ultimately sealing its fate'.[127] Through human intervention, the remains of past extinction events have generated a new one.

Crude oil is formed principally from algae and zooplankton that have degraded and gradually broken down into their chemical constituents, chemicals that have subsequently recombined and transformed. The chemical terms used to describe the composition of crude oil – hydrocarbon compounds (including alkanes and naphthenes), non-hydrocarbon compounds and organometallic compounds – gives only an abstract and limited sense of its living origins.[128] During the refining of crude oil, petrochemicals such as aromatics and olefins are obtained which are used in many everyday products. Olefins such as ethylene and propylene contribute to the production of numerous plastics. Plastics used in bags, films and bottles are composed of polyethylene (which utilizes ethylene) and those used to manufacture many bottle tops and containers feature polypropylene (which has propene as a foundation). Benzene, an aromatic, contributes to the production of polystyrene (among many others uses, employed to make petri dishes and test tubes) and some nylons (often present in clothing). The housing and the keyboard of the computer being used to type these lines is probably derived from petrochemicals, the fingers of our hands touching material that once formed living organisms, life forms which are now extinct species. Traces of extinction, in this sense, are literally beneath our fingertips.

The sheer quantity of materials surrounding us that derive from fossil fuels provides insight into the underlying omnipresence of extinction in the history of life on earth. As Norman MacLeod's chapter in this volume discusses, these events have involved a massive loss of biodiversity while also enabling new organisms to emerge. There is sometimes a productive dimension to extinction. Mass extinctions are unusual, but background extinctions are common. The majority of extinctions that have occurred – 95 per cent of all species – are background extinctions.[129] Most species that have lived are now extinct. These extinctions are so numerous as to resist conceptualization. This may explain why extinction is usually pictured in relation to very recent examples. The current Wikipedia page devoted to 'Extinction', for example, includes images of the thylacine, the golden toad (*Incilius periglenes*), the dodo (*Raphus cucullatus*), Haast's eagle or pouakai (*Hieraaetus moorei*), the moa (possibly *Diornis novaezelandiae*), the passenger pigeon (*Ectopistes migratorius*) and the critically endangered great hammerhead (*Sphyrna mokarran*). These animals all became extinct or endangered in the last six hundred years. At the time of writing, there are no images of recently extinct plants included in the encyclopedia entry. Older examples of extinction are provided by a composite image of skeletons of six extinct dinosaurs, a drawing of a tyrannosaurus, a plate from Cuvier's 1799 essay 'Mémoire sur les espèces d'éléphans vivantes et fossiles' that compares mammoth and elephant jaws, and a photograph of an external mould (the imprint of the outside) of a prehistoric plant, *Lepidodendron*, from the Upper Carboniferous period (roughly 320 million to 298 million years ago). The majority of images on Wikipedia therefore relate to recent extinctions. If we look beyond modern times, it is invariably to the age of the dinosaurs that we turn. Excepting the ancestors of modern birds, most dinosaurs lived between 240 million and 66 million years ago.

Given life on earth has existed for approximately 2.4 billion years, the visible emphasis on organisms from only a span of 174 million years is highly restrictive. This volume similarly focuses predominantly on recent extinctions. The chapters by Jeffrey Benca, Norman MacLeod, W.J.T. Mitchell and Jingmai O'Connor are, however, intended to give an indication of how the rich research currently being undertaken regarding ancient extinct species intersects with issues of representation. MacLeod's chapter considers how museums should exhibit extinction, including the natural mass extinctions of prehistory. Mitchell revisits ideas from the *The Last Dinosaur Book*, in which he analysed the modern and contemporary cultural significance of dinosaurs.[130] Benca and O'Connor use case studies from specific clades, lycophytes and enantiornithines respectively, to discuss issues of reconstruction in relation to extinct species. Benca details the reconstruction of a lycopsid or clubmoss from the Middle Devonian period (roughly 385 million years ago). O'Connor examines the many insights that the fossil record provides about enantiornithines (a group of extinct avialans) from the Cretaceous period. Her chapter demonstrates that sometimes, if the fossil record is relatively substantial, a remarkable amount of data can be deduced from fossils regarding issues such as behaviour and colouration. Techniques of representation, including palaeo-art, form a vital means of communicating such insights.

A major technology of representation used to impart knowledge gleaned from the fossil record is photography. Joanna Zylinska has argued that photographs can themselves be conceived as a kind of fossil: 'a light-induced process of fossilization'.[131] Using Louis-Jacques-Mandé Daguerre's 1839 daguerreotype *Shells and Fossils [Coquillages]* as an example, Zylinska also draws attention to how fossils feature as a subject almost from photography's inception. The image *Shells and Fossils* was taken in the Muséum national d'Histoire naturelle in Paris and is carefully framed so that an ammonite is at the centre of the composition. For Zylinska, *Shells and Fossils* 'showing deep-time artefacts carefully arranged into a sculptural grid reflecting light at various angles, placed photography in its very nascence between science and art, while also hinting at its geological entanglement'.[132] Nicholas Mirzoeff has also recognized that the choice of subject-matter is not coincidental, given debates about extinction in the period. For him, the photograph denotes human technological potency, embodying the ability to 'fossilize' things or events as images almost instantaneously, in contrast to the slow process of preservation associated with geological time that produces traditional fossils.[133]

Zylinska is attentive to how the arrangement of shells and fossils is illuminated. Photographing fossils is often difficult because of a lack of contrast or relief between the fossil and its surrounds.[134] There is therefore considerable discussion in palaeontology about how best to photograph them as photographs of fossils are a key means of presenting data. For this reason, Hans Kerp and Benjamin Bomfleur suggest that '[w]hen illustrating fossils, the same care should be taken as when describing and interpreting them'.[135] Illustrating here seems to be understood as 'explanation through pictures', with pre-existing descriptions exemplified by way of images. This supports Kerp and Bomfleur's claim that photography 'forms the least subjective means of presenting . . . fossils'.[136] Their qualified endorsement indicates that feelings and judgements still potentially inflect photographic images, including those of extinct species. Kerp and Bomfleur focus on fossil plant photography, examining its history and the impact of new technology on its practice.[137] We, however, would like to briefly examine photographs that record more recent extinct species. Photographs of extinct species have a significant power as representations, because of the truth status accorded to them. They are little discussed in other chapters in the volume, W.J.T. Mitchell's being the exception, so we now devote some attention to them here.

On Photography

On Tuesday 1 September 1914, the last-known passenger pigeon, a female called Martha, was found dead on the floor of her cage. She is often identified as an endling, as the last surviving individual of her species. After her death, Martha was skinned and mounted. Her internal organs were also studied and preserved. In addition to these physical remains, several black-and-white photographs exist of her that were taken in life. These include two that are included in Errol Fuller's remarkable volume of photographs of extinct bird and mammal species, *Lost Animals: Extinction and the Photographic Record*. Another bird featured in Fuller's book is the Aldabra brush warbler (*Nesillas aldabranus*), an avian that was endemic to the coral atoll of Aldabra in the Seychelles.[138] The warbler was discovered in 1967 when a Royal Society expedition to Aldabra sourced two adult specimens and three eggs.[139] The holotype is an adult female caught on 11 December 1967.[140] Her nest, which contained the three eggs, was collected on the same day.[141] The other specimen is an adult male caught on 29 January 1968. Unlike the passenger pigeon, which was first described in the eighteenth century by Mark Catesby and once numbered in the millions, the warbler population was always small, estimated in the 1970s at not more than twenty-five individuals.[142] The warbler is described by Fuller as 'virtually unknown', whereas the pigeon has become a poster bird for anthropogenic extinction.[143] Fuller reproduces two colour photographs of the warbler (Illustrations 0.1 and 0.2), both taken by Robert Prŷs-Jones in 1975. These images show solitary birds perched on plant limbs against a backdrop of tropical vegetation.

Similarities and differences between the photographs of Martha and the unnamed warblers are revealing about photography as a record, a form of evidence about the past. In the first photograph Fuller reproduces of Martha, she is photographed from below, alighted on a metal pole, the wire mesh of her aviary visible above her. She is clearly a captive bird, held in a humanmade environment. The second photograph grants Martha an illusory freedom.[144] The pigeon is perched on what is either a branch or twined wire, framed against a backdrop that could be interpreted as open sky. Both photographs are in black and white, a medium associated with agedness and artfulness. In the context of Martha, the black and white images date her, distancing her from the present. The colour photographs

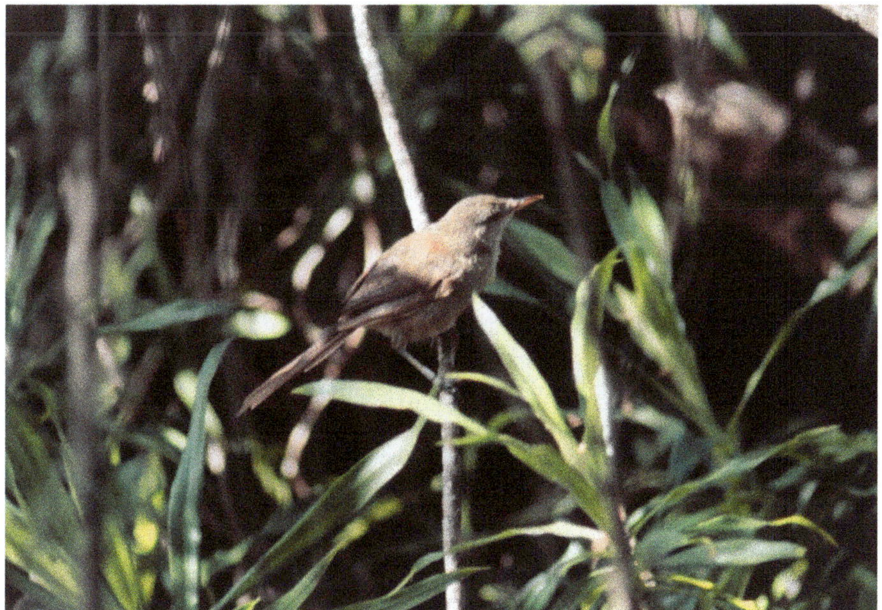

Illustration 0.1 Aldabra brush warbler (*Nesillas aldabranus*). Photo: Robert Prŷs-Jones.

of the warblers seem at less of a temporal remove. The warblers are also pictured in their natural habitat, tropical understory.[145] One bird is shown amid pandanus and the other, photographed at head height, is surrounded by what look like the leaves of *Cissampelos pareira* (also known as velvetleaf).[146] Both birds have been ringed. The bird bounded by pandanus has a silver metal British Trust for Ornithology (BTO) ring visible on their right leg. The ring gave the bird a unique identification number. The other bird has a yellow ring visible on his left leg. This identifies him as a pairbonded male, first observed by Prŷs-Jones on 17 December 1974.[147]

In the photographs of the pigeon and the warblers, the birds are all positioned at the centre of the composition.[148] This is intuitively how most people photograph a specific thing. Centring helps to affirm what constitutes the main subject. It also gives a sense of balance. In the black and white photographs, when coupled with the colourless surrounds, it also contributes to a sense of isolation. The colour photographs, with their verdant, sun-dappled backgrounds, do not share this seclusion. Our using the term composition in the context of the four images foregrounds that, even for those pictures that were taken in the wild, a process of framing has occurred. In each instance, there was a modicum of reflection about how best to 'capture' or represent the bird.[149] The way the photographs are set up reveals that they do not provide straightforward mechanically produced objective depictions but index subjectively informed perspectives regarding their avian subject matter. The set-up also shows recognition of the technical constraints of the medium. The birds are all 'flightless' as any pictures taken of a bird on the wing would be out of focus and unsatisfactory. In *Lost Animals*, the only bird show in flight is the Imperial woodpecker (*Campephilus imperialis*), and the image is a film still rather than a photograph.[150] The historical limitations imposed by the camera therefore shape the kind of bird that can be represented.

The second photograph (Illustration 0.2) from Aldabra shows the greatest spontaneity. Foliage in the right foreground is out of focus and lens flare obscures the bottom left of the image. This phenomenon, caused by bright light scattering in the lens system of a camera, is sometimes perceived as a sign of authenticity, the imperfection foregrounding the reality that a given photograph was taken *en plein air*. Flare, however, is also viewed negatively as it obscures aspects of the image and draws attention to the mediating presence of the camera.[151] The blur and the flare suggest this was a snapshot, a quickly taken picture, perhaps a reaction to the sound of the bird's call. This warbler has his beak open, seemingly in song, chirruping. Recordings of warbler vocalizations were used by Prŷs-Jones to try to attract individuals.[152] Bird calls are linked to communicating location, demarcating territory and attracting a mate. In this image, the bird

Illustration 0.2 Aldabra brush warbler (*Nesillas aldabranus*). Photo: Robert Prŷs-Jones.

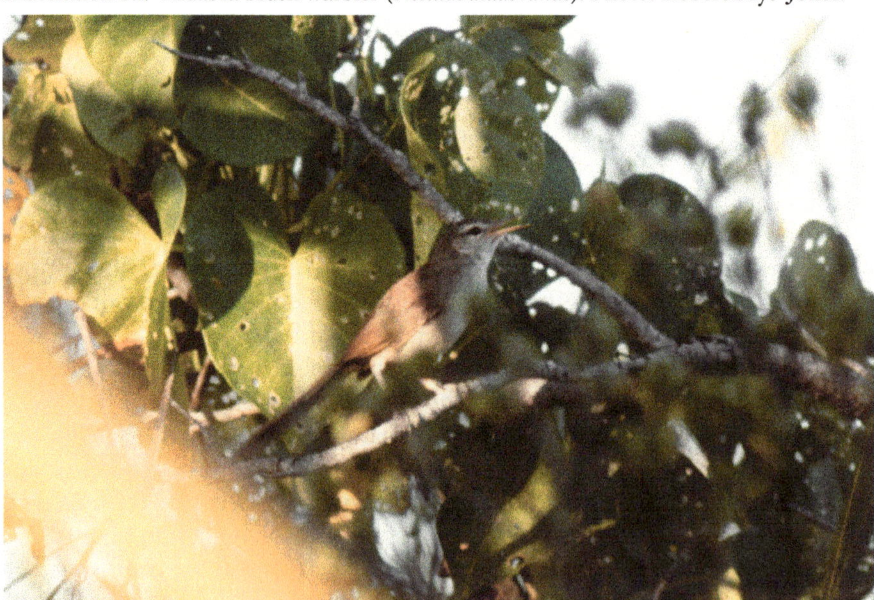

may be responding to one of Prŷs-Jones's recordings, he may potentially be answering a call from himself. The calls of the warbler have been described by David Pearson as a short 'chickachoy' or 'chak-chir', a repeated 'tak' and a high-pitched rattle of varying rhythm.[153] These calls were recorded in the wild. Most of the information about the vocalizations of passenger pigeons was derived from Wallace Craig listening to non-breeding birds held in captivity and is therefore, by his own admission, partial.[154] There are no audio recordings of passenger pigeons. Their voice is now purely textual.

The photographs in *Lost Animals*, mechanically produced images, embody varied codes of representation. Prŷs-Jones was carrying out fieldwork, his gaze, at least in part, that of a scientist. This shaped his actions behind the camera. The photographs, recording rare instances of encounter, were probably intended primarily as data. A desire to showcase the morphology of the bird likely contributed to the structuring of some of the shots or to retrospective judgements about which images were deemed a success. If fidelity to form was an underlying concern, then Prŷs-Jones would have privileged compositions (and resultant photographs) that most resembled his sense of the 'ideal form' of the Aldabra brush warbler. A preening warbler, in such a context, would fall foul of his formal concerns as the bird would likely have their head at an 'odd' angle and other parts of their body twisted. In *Lost Animals*, none of the birds seem to be engaged in preening, this despite the fact preening 'takes up a considerable proportion of the time budget of birds'.[155] The photographic record of these extinct species as presented by Fuller is therefore lacking when it comes to cataloguing common behaviours (few of the birds are shown feeding either). This could be because the photographers concerned were focused on recording appearance rather than behaviour, enacting a kind of morphological objectification. Their image-taking is influenced by concerns about often scientifically predefined visible criteria of form and structure.

The first photograph of Martha in *Lost Animals* might have been taken by the animal painter Enno Meyer (who worked as a photographer at the Cincinnati Zoological Gardens) or the pharmacologist and amateur ornithologist William C. Herman (Fuller suggests both men as the possible photographer).[156] The second photograph is regularly attributed to Meyer although Fuller notes that the circumstances of its taking are unknown.[157] These photographs were more likely produced to memorialize Martha, created in the knowledge that when she died, her species would die with her. Martha is burdened with representing the last of her kind. Prŷs-Jones knew the warbler was endangered but not that the extinction of the species was imminent. He could not foresee that in just over a decade the bird would disappear, joining the passenger pigeon and other species of bird, such as the Laysan rail (*Zapornia palmeri*) and the laughing owl (*Sceloglaux albifacies*), that died out in the twentieth century and also feature in *Lost Animals*. Even if the photographers we are discussing fantasized a particular reception for their images, the responses of real readers to verbal or visual texts are 'idiosyncratic, unpredictable and/or undisciplined'.[158]

In the cases of the Laysan rail and the laughing owl, Fuller additionally includes reproductions of artworks of the birds, an acrylic painting featuring the rail and a hand-coloured lithograph of the owl. There is also an acrylic painting of a passel of passenger pigeons. Fuller clarifies that these images feature as aids to a better visualization of the birds because the relevant photographs are either blurred or in black and white. The other visual media are therefore called upon to supplement the book's lens-based images, the 'photographic record'. Mechanically reproduced images are sometimes found visually wanting, inadequate to the task of bearing witness to an animal that has been 'lost'. Fuller explains that there are, however, no drawings or paintings of the Guam flycatcher (*Myiagra freycineti*), because it is 'clearly depicted in the existing photos'.[159] He openly shares his rationale for what renders a photograph adequately intelligible for him: colour and clarity. There are three photographs of the flycatcher, two in black and white of a nestling and an adult, both at liberty, and one in colour of a male adult being gently held in profile. It may be the photographer, the field biologist Anne Maben, whose right hand is visible clasping the bird. The male was captured in an effort to establish a breeding programme, but unfortunately no female could be found. The quantity and quality of images of a given animal do not, however, seem to fully explain Fuller's decision-making. In a similar way to how Harvey's refined description of Bennett's seaweed betrays his enthusiasm for that particular protist, Fuller's species preferences can be seen to manifest in *Lost Animals*.

Icons of Extinction

Most animals and birds in *Lost Animals* are represented by four or five photographs and four of five pages of accompanying text. Some species, such as the ivory-billed woodpecker (*Campephilus principalis*), the passenger pigeon and the thylacine are granted more extended analyses. Of these, the thylacine is the most disproportionately represented. The cover image to the book is of a 'yawning' thylacine.[160] An image of them also features in the Introduction (as does one of the ivory-billed woodpecker). There is a ten-page entry dedicated to the carnivorous marsupial, which includes seven photographs. A hand-coloured lithograph of the species features in the appendix. A photograph of the last captive thylacine, 'Benjamin', is reproduced across two pages after the Further Reading section.[161] A different photograph of 'Benjamin' is repeated six times across the two pages that precede the Index, each time the image is more bleached (a practice comparable to Rickard's erasures) causing the animal to fade away. Finally, on the last numbered page of the book, an 1864 photograph by Frank Haes of a thylacine in London Zoo is reproduced. The sheer number of images of the thylacine indicate that Fuller seems to remark the loss of the animal more acutely than any of the other species featured in *Lost Animals*. The iconic status of the thylacine probably informed the decision to use a photograph of the species as a cover image: the vanished marsupial circulates widely as 'a symbol of extinction, or a reminder of the fragile state of the environment; a figure that evokes feelings of sadness, curiosity and concern'.[162]

Fuller's varied emotional investments in the species he includes also manifests through the minimal attention accorded another mammal, the greater short-tailed bat (*Mystacina robusta*) or pekapeka. A single photograph of the bat is included in *Lost Animals*, yet despite it offering a restricted view (only the head and a wing is shown), Fuller includes no further illustrations.[163] Drawings and other photographs of the bat do, however, exist. In 1962, Peter Dwyer, for instance, sketched a specimen from the collection of what was then the Dominion Museum (now the Museum of New Zealand/Te Papa Tongarewa) which features in his article 'Studies on the Two New Zealand Bats'.[164] His drawing includes two views of the bat's head (lateral and ventral) and one of his body and his left wing, as seen from above. The drawing also includes lateral and ventral views of the head of a subspecies (*Mystacina tuberculata tuberculata*) of the lesser short-tailed bat. Fuller is not averse to including works that feature more than one species in *Lost Animals*, the painting of the Laysan rail also includes an albatross (*Phoebastria immutabilis*). If he was familiar with Dwyer's drawing, it probably did not appeal because it is a monochrome scientific illustration that is openly instrumental, too obviously designed to support an argument (about morphological differences). The bat also looks lifeless.

Dwyer's drawing can productively be compared with the scientific illustrations that do appear in *Lost Animals*, most notably several hand-coloured lithographs by John Gerrard Keulemans. These highly detailed works were created to accompany scientific texts.[165] Dwyer's work is also meticulous, noting minutiae. Additionally, he extracts details: the head of the bat is accorded particular significance through being separated from the body, with the placement of the bat's ears and the form of his nostrils carefully foregrounded. The dorsal view of the bat also omits his right wing. The mammal is shown as parts or only in part. Keulemans usually portrays birds in their entirety, as is the case with the lithograph of the Hawai'i mamos (*Drepanis pacifica*) that Fuller sourced from *Avifauna of*

Laysan. In this desire for wholeness, Keulemans is similar to the artist John James Audubon, whose works are examined in Gordon Sayre's chapter for this volume.

In *Avifauna of Laysan* there are details of the mamo included on a separate plate alongside those of other birds, namely drawings of their bill (with a close-up), tongue (at life-size, and magnified by 10) and sternum. These were not produced by Keulemans but by Frederick William Frohawk. The tongue and sternum could not be drawn from life, and call attention to the reality that the works were produced post-mortem. Both the birds that Keulemans portrays as alive, perching on foliage, were based on collected specimens.[166] Dwyer's subject is far more obviously dead. As Fuller's narrative of loss relies on pictures of living (or seemingly living) animals, drawings such as Dwyer's or photographs of dead extinct animals have no place. The emotiveness of Fuller's images derives from their being open to reading at the same time as foreshadowing extinction and following on from it: '*This will be* and *this has been*'.[167] To paraphrase Roland Barthes in *Camera Lucida*, we observe with horror an anterior future of which extinction is the stake.[168]

More than this, however, we believe Fuller accords less attention to the bat because it is a bat. Bats have traditionally been viewed negatively in Western culture as carriers of disease.[169] Their association with vampirism has also contributed to their stigmatization. In fact, only three extant species of bat (out of the approximately one thousand known species) feed exclusively on blood. The positive contributions which bats make to human well-being are not popular knowledge. They destroy agricultural 'pests' thereby contributing to increased crop yields. They also aid the pollination of many plants. Since 1900, at least five species of bat have become extinct. In addition to the greater short-tailed bat, the Philippine bare-backed fruit bat (*Dobsonia chapmani*), the Nendo tube-nosed bat (*Nyctimene sanctacrucis*), Sturdee's pipistrelle (*Pipistrellus sturdeei*) and the Guam flying fox (*Pteropus tokudae*) have vanished. None of these mammal extinctions, however, has a high profile. Bats, like rodents, fall foul of what might be called 'species aesthetics'.[170] They are not seen as beautiful or striking and do not possess 'cuteness'.[171] Steve Baker has drawn attention to the need to combat harmful stereotypes that are embodied in representations of some species.[172] The bat requires a representational rehabilitation in order to counteract the half-truths and untruths that circulate about the mammal.

Fuller wrongly affirms that the photograph he reproduces is '[t]he only known photo of a greater short-tailed bat'.[173] In fact, there are at least two others. The first of these photographs (Illustration 0.3) was taken in what was referred to as the 'bat cave' at Taukihepa/Big South Cape Island in April 1961.[174] It shows a colony of bats at liberty in part of their natural

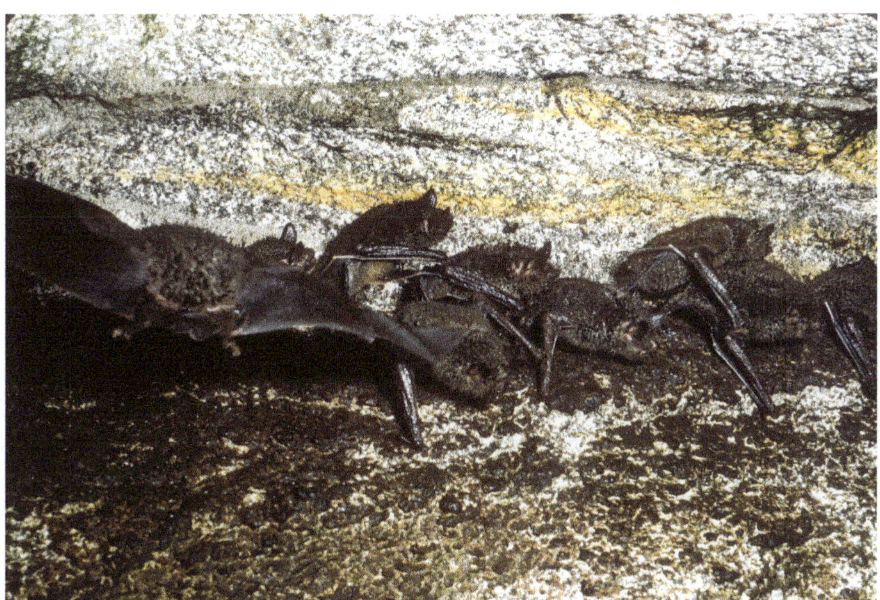

Illustration 0.3 Greater short-tailed bats (*Mystacina robusta*) in the 'bat cave', Taukihepa/ Big South Cape Island, April 1961. Credit: Brian D. Bell.

environment. Their guano has visibly accreted on the rock surface beneath them. This situated-in-the-environment depiction is much less able to communicate form than Dwyer's out-of-the-environment post-mortem portrayal. The photograph, however, foregrounds bat sociality. As a form of representation, it draws attention to the reality that the extinction of the greater short-tailed bat involved the loss of a specific bat society.[175] The second photograph (Illustration 0.4), which was taken in the same month, shows the conservationist Don Merton holding a bat that was caught in a mist net.[176] The ornithologist Brian Bell took both the photographs. Although greater short-tailed bats were unusually terrestrial in their habits, Merton carefully holds the captured bat's wings open to show the span and structure. The bat population on Taukihepa/Big South Cape Island was decimated after the inadvertent introduction of rats to the isle in 1964. The greater short-tailed bat was last sighted in 1967. The species is significant enough to merit mention in *Lost Animals* but is accorded none of the attention granted the thylacine or the ivory-billed woodpecker.

The issue of iconic status is of great importance in relation to issues of conservation and memorialization. Endangered species perceived as aesthetically appealing are likely to secure public sympathy for their plight (often accompanied by financial contributions to help safeguard the future of the species) to the potential detriment of other flora and fauna. The Save the Panda campaign of the 1970s and 1980s provides a good example. A Save the Central Rock Rat campaign would be unlikely to generate comparable publicity or support. The vulnerable giant panda (*Ailuropoda melanoleuca*) is seen as huggable and adorable and thus high profile; the central rock rat (*Zyzomys penduculatus*), a critically endangered Australian rodent, has no public profile, but as a rodent the species would probably generate negative reactions. Birds also have both iconic and ignored extinct and endangered species. The dodo and the passenger pigeon are household names whereas the Aldabra brush warbler and the Ryukyu wood pigeon (*Columba jouyi*) are little known.[177] As Mark Halsey has foregrounded, however, there are also risks in singling out species as endangered or vulnerable. Halsey notes that 'in naming certain species as rare, more precious, more vital than others, there . . . marches the problems attending the hierarchization of the earth'.[178] The endangered births the expendable.

Our aim in this volume has been to engage with some iconic extinct species and how they are represented, such as the thylacine, the quagga (*Equus quagga quagga*) and the great auk, and also to consider relatively neglected examples. Almost the entire kingdom of plants, for instance, is still largely overlooked in extinction debates, despite the many species that are disappearing. Naming a specific species of plant that has gone extinct in the past hundred years would probably be difficult for many people. Unsurprisingly, the issue of plant blindness or plant awareness disparity is therefore referenced in all the chapters in this volume relating to plants. Jeffrey Benca's chapter is on prehistoric plants, yet plant blindness is still an issue in that context. He foregrounds how such blindness impacts on the ways the palaeontological record is approached and interpreted. Plant blindness, a failure to notice plants and/or an inability to distinguish between plant species and/or to recognize the general importance of plants to the biosphere, is not culturally uniform. Some cultures demonstrate a highly nuanced understanding and appreciation of plant life.[179] In Western culture, however, plants, if they are perceived, are often seen as boring.[180] They do not generally tend to generate strong emotions, with trees forming the most notable exception. Historically, trees have held considerable cultural significance in Western and other cultures.[181] Darwin's tree of life metaphor, which we discussed earlier, reflects this notability.[182] Exploiting arboreal symbolic capital, the practice of tree sitting (building and occupying a platform in a tree to stall its felling) features in many contemporary environmental protests.[183] Because of the status accorded to trees, silviculture is also an area where Indigenous knowledge is increasingly being sought, with insights regarding sustainability incorporated into forest management.[184]

Debates about plant rights are still in their infancy.[185] There is no Plant Liberation Front equivalent to the Animal Liberation Front. In 1972, Christopher D. Stone published his classic essay (subsequently developed into a book), 'Should Trees Have Standing? – Toward Legal Rights for Natural Objects', in which he used forestry as one example among several to argue for environmental rights, ultimately calling for a shift in perception such that we 'get away from the view that Nature is a collection of useful senseless objects'.[186] Stone emphasizes a need to acknowledge human entwinement with the environment and to develop humility about humanity's role within it. His example to illustrate such a change in consciousness is a short story by Carson McCullers, thereby foregrounding how an art form such as literature can contribute to altering outlooks. In the specific context of flora, early education has been identified as crucial to shifting opinions.[187] The potential role of textual or visual representation in this process is emphasized by Howard Thomas, Helen Ougham and Dawn Sanders in their article 'Plant Blindness and Sustainability', which mentions art-based research practice and literary creativity as important ways of raising awareness of plants.[188]

In relation to plant extinction, initiatives such as the art installation *Resurrecting the Sublime* have helped to draw attention to specific extinctions.[189] The installation was the result of a collaboration between the artists Daisy Ginsberg and Sissel Tolaas, and the scientist Christina Agapakis, and involved drawing on palaeogenomic expertise to recreate the smells of extinct plants. Three plants featured in the project, all from the United States. One, Maui hau kuahiwi (*Hibiscadelphus wilderianus*), was known from a single specimen discovered circa 1910. A fragment of this shrub is held in the Harvard herbarium. There, a branch is carefully attached to a white card backing, its brown leaves, dry and creased, stark against an artificial sky that is itself made from cellulose fibres.

Besides the species of *hibiscadelphus* featured in the project, many other plant species from Oceania have disappeared in modern times, including the Daintree banana (*Musa fitzalanii*) in Queensland, Australia, and the Philip Island glory pea (*Streblorrhiza speciosa*) formerly endemic to Philip Island in the Norfolk Island group. In his entry dedicated to the shrub, which he calls the flesh-coloured glory pea and seeks to transfer to the Clianthus genus as *Clianthus carneus*, John Lindley hopes that the figure he includes illustrating the plant will convince his readers of its relative beauty when in flower.[190] The plate in question is by the British artist Sarah Anne Drake. Drake was a botanical illustrator who contributed to important botanical publications in the 1830s and 1840s, producing drawings and paintings of approximately 1,300 plants.[191] Women such as Drake have historically played a crucial, yet under-appreciated, role in botanical research. Jeanette Hoorn's chapter in this volume examines the work of an Australian botanical illustrator, Ellis Rowan. Rowan's work is noteworthy for eschewing a purely morphological approach in favour of emphasizing a plant's surroundings or habitat. Her choice of subject, however, was restrictive. She had a definite hierarchy, preferring to depict big, orchidaceous species rather than small, unshowy ones. To her, tiny-flowered, herbaceous plants seemingly lacked sufficient visual interest to merit transforming into art.[192]

The Lens of Compassion

As well as plant extinctions, those of amphibians, reptiles and invertebrates are also overshadowed by the attention accorded to birds and mammals. Among invertebrates, insects are ecologically of immense importance, yet their high contemporary extinction rate is not common knowledge.[193] The behaviours by insects that directly or indirectly benefit humans include preying on 'pest' insects, pollinating plants and consuming waste. Pedro Cardoso's chapter in this volume considers ways in which representational strategies in museums might help to draw attention to the plight of arachnids and insects. A key issue for Cardoso is to foster an increase in empathy towards insects. Cary Wolfe has proposed the idea of 'trans-species empathy', an empathy that foregrounds similarities rather than differences across species.[194] Cardoso's suggestion is to create artificial similarities. In the case of insects, he notes that issues of scale need to be considered when seeking to generate fellow-feeling towards insects. As adult humans are much bigger than all insects, using representations that magnify insects can help to overcome size differentials that may impede identification and empathy.[195] Insects, as anthropods, as creatures that have jointed appendages, already bear some similarity to humans and may therefore lend themselves to recognition in ways that non-anthropod invertebrates may not. The sediment-dwelling marine invertebrates the Loricifera, for instance, appear far more 'alien'.[196] Encouraging sympathy for other species may consequently require different strategies dependent upon issues of morphology and, perhaps, behaviour.

The role of feelings in relation to animal ethics has long been debated. In 1996, for instance, Josephine Donovan argued the merits for an ethics grounded in sympathy and compassion rather than abstract reasoning.[197] Sympathy in Donovan's terms involves thoughtful compassion. She frames her call for a sympathetic considerateness towards other species as grounded in an expanded visual field, a capacity to see the bigger political picture: 'people exercising attentive love *see* the tree; but they also see the logging industry'.[198] Some visual representations gesture towards such an ethic of care. One example is the second photograph (Illustration 0.4) of the greater short-tailed bat, which shows Don Merton gently, if firmly, holding the mammal. The night-time photograph merits sustained analysis. Seemingly illuminated by torch light, Merton's blue-grey-black check shirt visually mirrors the blue-black of the bat's wing membranes. The conservationist's mud-caked hands and dirt-ingrained fingernails attest to the fieldwork he has been engaged in. This is a particular type of nature photograph, comparable in some ways to the archaeological 'hand-held artefact shot' in which a 'still-dirty artefact' is shown in the excavator's hands.[199] As Kelsie Martinez has discussed, such archaeological photographs are imbued with 'the romantic conditions of discovery'. Dirt signals that the artefact is 'fresh from the trench', granting it an imprimatur of authenticity. The soil adhering to Merton's hands signifies differently. It draws attention to him being out in the field, unafraid to get his hands dirty in the interests of conservation. The grime derives from the bat's natural habitat. Something of this habitat now adheres to the outsider.

The visibility of Merton's hands also foregrounds human presence in a way many wildlife photographs do not.[200] In the photograph of the 'bat cave', for instance, Brian Bell's being there is only implied. He remains invisible. This enables the shot of the cave to offer the illusion of an 'untouched

Illustration 0.4 Greater short-tailed bat (*Mystacina robusta*) held by D.V. Merton, Taukihepa/Big South Cape Island, April 1961. Credit: Brian D. Bell.

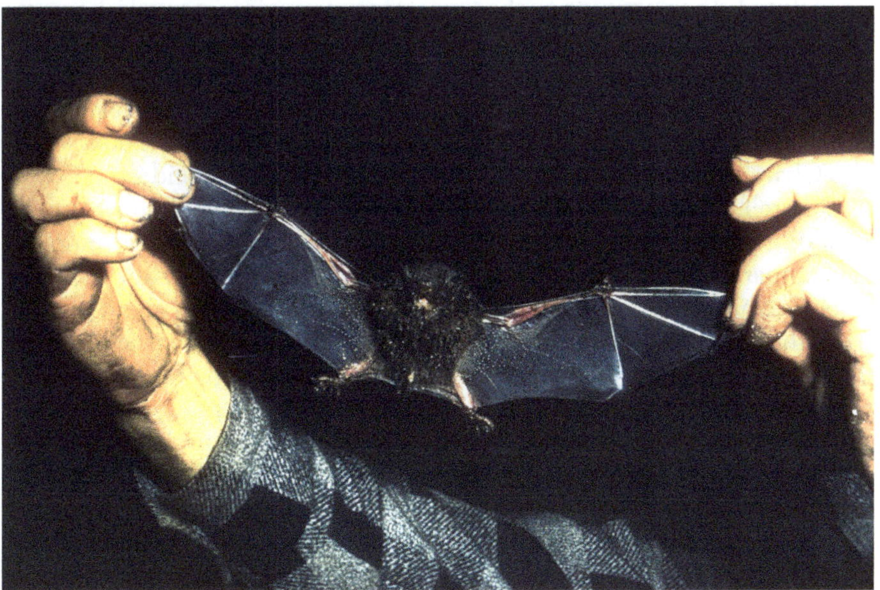

wilderness'. Bell's photograph featuring Merton, by contrast, foregrounds human interaction with the environment. That Merton has touched the bat's island habitat is clear from his hands. Human entanglement with the natural world is affirmed rather than suppressed. Other photographs attest to this entanglement in a markedly different way. A picture of the 'bullocky' Albert Quarrell with a dead thylacine resting across his lap – his right hand cradling the marsupial's muzzle, his left holding his work dog against his chest – also evinces entanglement.[201]

Quarrell shot the thylacine in late 1911 or in 1912 (accounts vary), close to Fitzgerald near Maydena in the Derwent Valley of Tasmania. He had initially hoped to capture the animal alive to sell it to a zoo and therefore viewed the thylacine as an economic opportunity, a commodity. The trees that form the backdrop to the photograph were also commodities. Quarrell's team of bullocks were used to haul timber. The township of Fitzgerald benefited from logging and the harvesting of trees such as eucalyptus and myrtle.[202] When Quarrell's efforts to catch the thylacine by the tail ended in failure, he killed the animal with his shotgun. Two other men are visible in the photograph: a man usually identified as the trapper D. Pearce is seated behind and to the left of Quarrell, and the hand of an anonymous figure can be seen to his right holding the stiff tail of the thylacine aloft.[203] A fourth figure, the photographer, is implied but invisible. In this photograph the thylacine is displayed as a trophy. The death of the animal is celebrated and memorialized. Quarrell, with his muddy boots, and Merton, with his dirty hands, are both marked by the environment in which they are pictured. Their attitude to the fauna of those environments is, however, qualitatively different. Merton's taking care not to injure the bat is palpable. Quarrell, by contrast, revels in the thylacine's destruction. Barbara Creed's chapter in this volume offers an explanation of where celebratory destructivity such as that exhibited by Quarrell comes from, as does Nicholas Chare's chapter, with its consideration of animal genocide.

Quarrell supports the animal's deadweight the better to showcase his achievement: the killing of the 'pest'. Both photographs are posed, employing comparable logics of display. The marsupial carnivore is shot in profile to best exhibit the animal's size and banding. The bat is photographed from the front and below to display body size and wing structure. That these animal subjects were seen as unusual is attested by the very taking of the photographs. Quarrell knew the thylacine was rare yet he shot it anyway. Merton was aware the bat was uncommon, but probably not how critically endangered it was. Had he been, he might not have released the mammal but considered his or her capture and relocation.

The bat looks towards the camera, opening the possibility of an exchange of gazes between species. Being netted and manhandled was probably traumatic. This is not an image of a non-dominative interspecies relationship. The bat does not give consent to be photographed. Their perspective on events remains opaque, yet the framing of the image affords the bat a potential subjectivity. In the photograph of the thylacine, the marsupial's lifeless eyes are closed. Quarrell's dog, however, looks at the photographer. The inclusion of the domesticated animal is noteworthy. The dog, perhaps a kelpie, is clearly valued. Quarrell also makes his living from using bullocks as draught animals. He perceives merit in non-human animals but only those that have been rendered dependent and offer labour power. Merton's sensitive grasp of the bat suggests care towards other species is not contingent upon their usefulness (in any direct sense). The photographs therefore both represent extinct species, vanished mammals, but the interspecies relations they embody are highly dissimilar. Bell and Merton provide a vision of compassion towards another species, one communicated by way of a medium, photography, often associated with mechanical detachment and inhumanity.[204] Images have the capacity to reinforce or reshape our existing perceptions of extinct animals. At times, this power, while related to species that have disappeared, has considerable importance for influencing our understanding and behaviour in the present.

Beyond Species Extinction

Extinction is usually thought of as linked to the disappearance of a species, making it of potentially limited value when addressing major ecosystem changes. As one of the peer reviewers for this volume emphasized, however, there are potentially extinct landscapes.[205] Landscape is often considered a visual phenomenon and thought of in aesthetic terms.[206] Here we understand the term more broadly, equating it with an ecosystem. This is how the wildlife biologist David E. Brown understood the word when, writing in 1979, he noted that due to arroyo-cutting, river rechannelling and groundwater pumping, the beaver-friendly riverine marshlands of the Turtle Island (North American) Southwest have become 'an almost extinct landscape'.[207] Brown's example of an almost vanished ecosystem is modern. Many extinct landscapes are ancient; Antarctica, for instance, used to be temperate rainforest.[208] Similarly, the Arctic was once warm and forested.[209] Most landmasses have undergone substantial ecological change across geological periods. For example, the province of Quebec in Canada, where we write these words, was once part of a far warmer region made up of temperate forests rather than the boreal forests and tundra of today. In the Cretaceous period, the landmass forming Quebec and Ontario was separated from the landmass that is British Columbia not by the Canadian Prairies but by the Western Interior Seaway, by water not grassland.

Envisioning such vastly different, vanished topographies is immensely difficult. Maps and drawings, for instance, offer a flat earth, cultivating a detached engagement. Computer modelling provides useful insights but findings are usually presented as graphs, the lost world rendered only in abstract terms.[210] Fossils, as small fragments of a complex assemblage of plants and animals that inhabited a landscape, cannot usually provide the lay of an ancient land either. Only occasionally do they give a sense of terrain. Petrified forests, for example, particularly where some tree trunks still stand upright, seem like meeting points of different geological times and places. Ichnites are also sometimes highly evocative. Frequently derived from ancient shorelines, when they are excavated the tracks are feted for the information they hold about the organisms that made them. Dinosaur prints, for instance, can potentially reveal data about foot shape, skin texture, behaviour and locomotion, among other things.[211] Tracks capture the imagination because they were made in life and index vitality. More than that, while now sometimes preserved in bedrock, the impressions also suggest regolith, sand or soft mud, something malleable and changeable. They summon images of a dynamic, lived-in landscape.

Beyond the imaginings prompted by evocative fossils of this kind, ancient landscapes are potentially best conveyed through immersive media such as film and video games. Immersion fosters the 'feeling of being enveloped' and 'of being transported into a non-immediate reality'.[212] It is associated with what Carl Therrien identifies as a credible and vivid representation.[213] Sarah Bezan's chapter in this volume discusses immersive VR technology in the context of a modern avian extinction, the Kaua'i 'ō'ō (*Moho braccatus*). In the context of the remote past, video games such as the dinosaur simulator *Saurian* (2017) set in the Upper Cretaceous, and the action-adventure *Far Cry Primal* (2016) set in the Upper Palaeolithic, conjure simulated prehistoric worlds, visually reviving extinct landscapes.[214] *Saurian* is particularly noteworthy as it strives for a scientifically accurate depiction of an Upper Cretaceous ecosystem, drawing inspiration from fossil finds from the Hell Creek Formation in the United States. A survival game set 66 million years ago, *Saurian* invites players to choose one of six species of dinosaur and then seek to nurture the animal from birth to maturity. The game designers consulted palaeontologists as they endeavoured to produce plausible reconstructions of flora and fauna from six distinct biomes.

Saurian gives a sense of the sights, sounds and textures of the Upper Cretaceous, of an extinct landscape, in a way a still image cannot. The game, however, also gives the appearance of a seamless reconstruction. In reality, there are gaps in our knowledge of the Hell Creek ecosystems. Insect fossils, for example, are extremely rare, although the discovery of dipteran and zygopteran insects in amber this century has augmented our knowledge.[215] Some vertebrate species are also only known from microfossils such as teeth, as their small body size, fragile skeletons and/or small populations make them rare as macrofossils.[216] Any vision of the ecosystems of the Upper Cretaceous, traces of which are now preserved at Hell Creek, is therefore partial. In the same way that ethical approaches to the restoration of cultural artefacts (rather than their mere conservation) frequently call for transparency about repair work, an argument can be made that gaps in the reconstruction of an extinct landscape should be foregrounded. In the case of *Saurian*, this might, perhaps, take the form of an intentional, intermittent ethical pixilation or the incorporation of blank spaces.[217]

The success of projects such as *Saurian* is also not haphazard. The simulator secured considerable funding through Kickstarter because dinosaurs possess popular appeal. The idea of a game called *Ostracoderm* in which the player is an Astrapsis from the Ordovician seems far-fetched. Market forces dictate the kinds of extinct ecosystems that are likely to be reconstructed as video games. The Upper Cretaceous has a financial power that the Ordovician does not. Museums, however, have slightly more flexibility and can potentially design immersive reconstructions of less 'exciting' extinct land and seascapes.

Landscape change continues to occur today but now the causes are predominantly anthropogenic. Various human activities such as damming, farming, logging and mining have impacted the environment, harming and even destroying ecosystems.[218] Habitat loss and fragmentation accompany landscape change and are a well-known extinction driver. The ecosystem within which habitats are found is less commonly recognized as subject to extinction. There has historically been clear recognition that, with colonization and modernization, landscapes and ways of life have disappeared, often rapidly. Nancy C. Shour reads James Fenimore Cooper's novel *The Pathfinder; or, The Inland Sea* (1840) as a re-envisioning of 'the extinct landscape of the western prairies'.[219] This echoes Cooper's explicit aim in his tale centred on Lake Ontario to give a 'passing glimpse ... of what that vast region so lately was'.[220] Cooper is referring not solely to the landscape but to its First Nations inhabitants. Sometimes it is not the landscape that vanishes but a way of experiencing it. Augustin Berque uses the example of the Anangu, Aboriginal peoples of Australia's Western Desert, to illustrate his idea of '*cosmoctonie*' (cosmocide), the destruction not of the land but of a mode of conceiving it.[221] Berque goes on to note that the Anangu are now reaffirming their understanding and connection to Country.

Today, certain landscapes, while not extinct, also take on the appearance of 'museum pieces'. Abayneh Unasho Gandile, Solomon Mengitsu Tessema and Fisha Mesfine Nake write of tropical dry forest reserves in Tanzania that have benefited from protection through Indigenous belief systems (including taboos linked to some species of tree) that they now 'stand out as ecological museums of local vegetation'.[222] In this understanding, patches of landscape become something like display cases, giving us a peek into an ecosystem that once was. The forests, living museums, are viewed as relics. Representing extinction is therefore not necessarily restricted to single species but can encompass broader ecological loss.

Conclusion: Thinking across Art and Science

This volume includes chapters by scholars working in the humanities and natural sciences, or across those fields (as is particularly the case for some of the museum curators who have contributed). Scientific modes of enquiry are usually distinct from humanistic approaches to the study of nature and culture, each field employing different languages and epistemologies. Differences at the level of language certainly register in the chapters included here. When contributors use terms from their respective disciplines that might be unfamiliar to some readers, we have asked them to provide a gloss. There is also a greater emphasis on empirical knowledge in some of the scientific chapters, with ethical and political issues more to the fore in several of the humanities chapters. All the contributors share a commitment to thinking through how art and science intersect or combine in efforts to represent extinct species.

Each contributor agreed to participate in the project in the knowledge that the aim was to create a space for research from scholars in both the humanities and natural sciences to appear together. There have been previous efforts to discuss extinction in the humanities drawing on scientific insights but the involvement of members of the scientific community in these projects has been limited.[223] Here, however, the work of biologists, ecologists, palaeontologists and zoologists appears alongside that of art historians, comparative literature scholars, historians and screen studies specialists. Additionally, museum curators who work at the interface between science and visual culture have contributed. Although humanities and science scholars often display methodological divergences, there are shared areas of concern. These chapters are therefore each in dialogue at some level. Issues such as fidelity regarding representations to a given species, questions of empathy or sympathy, and problems associated with hierarchies of perceived interest or importance (such as the undue emphasis given to mammals in extinction discourse) form common points of reference. Extant publications concerning representations of multiple extinct species often simply reproduce such representations without critically engaging with them.[224] This is the case, for example, with the previously discussed *A Gap in Nature* (paintings) or *Lost Animals* (photographs). Similar volumes include David Day's poetic meditation *Nevermore* and his more encyclopedic *Vanished Species*. Our book differs in that each contributor reflects on the effects of images of extinction, considering the potentials and pitfalls of varied representational strategies for shaping understanding of a given disappeared species and of extinction events more broadly.

Part I, 'Dialogues about Extinction', comprises three conversations. This part sets the tone for the rest of the volume, which we also view as an expanded conversation regarding extinction and its representation. In the first chapter, 'The Dinosaur as Cultural Symbol and Totem', W.J.T. Mitchell returns to his ground-breaking research on the cultural symbolism and function of dinosaurs in modern and contemporary society. In a rich, wide-ranging discussion, Mitchell reflects at length on what images of extinction, including fossils, reveal about the nature of images more broadly. He also discusses gender issues in relation to depictions of dinosaurs and the discipline of palaeontology.[225] Contemporary portrayals of endangered species, such as Joel Sartore's 'Photo Ark', are also addressed. The next chapter, 'Visualizing Extinction' by Harriet Ritvo, centres upon extinct species that have accrued 'significant human interest' such as the dodo, the mammoth and the thylacine. Ritvo provides a compelling meditation on the ways these species have historically been represented, with some co-opted for nationalist or political ends. The chapter also mulls the specificities of animal history, including the need for a broad-minded approach that draws extensively on scientific sources. The final chapter in this part, Stuart Pimm's 'Putting Nature Back Together Again', discusses several key issues that recur in later chapters, including the pitfalls of the tendency to hierarchize species (to the detriment of, for instance, plants), the problem of generating interest in non-aesthetically pleasing or visually interesting fauna and flora, and the role of museums in engaging public interest in biodiversity and related conservation issues. Additionally, Pimm provides fascinating insights into extinct species known only anecdotally, particularly avifauna.

Part II, 'Indigenous Peoples and Extinction', features chapters engaging with Canadian and Chinese Indigenous perspectives on extinction. The first chapter in the part, Nicholas Chare's chapter 'The Beothuk, the Greak Auk and the Newfoundland Wolf', focuses on animal and human genocide in Newfoundland. Chare examines negative representations of the Beothuk, a First Nations people of Newfoundland, and of the Newfoundland wolf (*Canis lupus beothucus*) produced by settler colonists. The next chapter, 'Cultural Memory of Recent Extinction' by Samuel Turvey, addresses the relative neglect accorded to Indigenous knowledge of local biodiversity and extinctions. Using China as a case study, Turvey presents a careful examination of the way representations and understandings of recently extinct species articulated by local peoples provide important insights into those species. Using extinct species of apes, cetaceans and paddlefish as examples, Turvey also forcefully demonstrates how quickly such knowledge is lost. The final chapter in the part, *hagwil hayetsk*'s (Charles Menzies's) 'Grief, Extinction and *Bilhaa* (Abalone)' is a moving reflection regarding the impact of protective legislation designed to safeguard the *bilhaa* upon the First Nations Gitxaała people of Laxyuup Gitxaała, a national territory located in British Columbia. The legislation was enacted in response to overharvesting of the marine snails by settler colonists, yet it has had a profound impact on Gitxaała culture because of the profound symbolic significance of *bilhaa*. *hagwil hayetsk* also draws attention to the value that Indigenous knowledge of sustainability offers to efforts to safeguard against future extinctions.

Part III, 'Representing Avian and Insect Extinctions', contains four chapters, three relating to avian extinctions and one to insects. It begins with Jingmai O'Connor's remarkable study of efforts to represent enantiornithines, 'Sparrows with Teeth and Claws?' Enantiornithines are a group of extinct avialans, similar in some respects to what are today called birds. Although they are known only from the fossil record, O'Connor provides a highly textured picture of these avialans, including their form and behaviour. The wealth of available fossil data feeds into the work of palaeoartists, some of whose art is reproduced in the chapter. Collaborative portrayals of this kind, in which professional palaeontologists offer advice, are invaluable for counteracting speculative depictions of enantiornithines and other prehistoric species. These circulate widely on the internet, creating false impressions of extinct life. The next chapter, Gordon Sayre's 'Rare Birds and Rare Books', is a beautiful meditation on the portrayal of the greak auk and the Californian condor (*Gymnogyps californianus*) in John James Audubon's *Birds of America*. Sayre examines tensions also identified by Pimm between the drive to collect and the need to preserve species. He also invites reflection on the modern practice of initiating breeding programmes for endangered species in zoos, a process that transforms the species into a work of art that, like a painting or sculpture, is dependent upon humans for its preservation and value. The third chapter, Sarah Bezan's 'The Species Revivalist Sublime', considers the use of virtual reality (VR) technologies to stage encounters with extinct species. Bezan uses Jakob Kudsk Steensen's installation *Re-Animated,* which features the Kaua'i 'ō'ō bird (*Moho braccatus*) as a case study. Bezan argues that the highly affective sublime experience engendered by Steensen's staging of virtual encounters with the extinct bird moves away from the elegiac responses commonly associated with the contemplation of disappeared wildlife. The part concludes with Pedro Cardoso's 'Insects, Spiders, Snails and Empathy', which considers negative public perceptions of invertebrates such as insects and assesses ways in which natural history museum exhibitions might help to change such opinions. Like Bezan, Cardoso sees considerable value in VR technology because of the immersive, multisensory experiences they can provide.

Part IV, 'Representing Extinct Plants and Fungi', contains three chapters about plants and one about yeasts. It begins with a chapter considering reconstructions of extinct plants from the fossil record, Jeffrey Benca's 'Reconstructing Lycopsids Lost to the Deep Past'. Benca is a palaeontologist and a palaeoartist, and his chapter includes an example of his extraordinary, incredibly detailed work. In the chapter, he draws attention to how contemporary perceptions of plants inflect palaeontological practice, with the portrayal of plants in reconstructions of prehistoric ecosystems often unrealistic. The next chapter, 'Ellis Rowan, Extinction and the Politics of Flower Painting' by Jeanette Hoorn, presents exquisite readings of some of Rowan's botanical illustrations and salon paintings. Hoorn suggests the artist-scientist accords plants a subjectivity that has important implications for today, given that plant extinction is a pressing contemporary issue. Dawn Sanders's powerful chapter 'Towards Extinction' offers close analyses of three art installations by Snæbjörnsdóttir/Wilson. Sanders suggests that, by way of artworks such as these, contact zones between plant and human can be generated. Discussions of scale and immersivity in the chapter resonate with issues raised in earlier chapters by Cardoso and Bezan. For Sanders, contact zones encourage greater recognition and understanding of plants, extinct or otherwise. The final chapter by Robert Dunn, Monica Sanchez and Matthew Booker explores yeast extinctions. Yeasts, a kind of fungi, are often overlooked despite their importance and their omnipresence through history. Dunn, Sanchez and Booker seek to grant yeasts a new measure of visibility through the history they provide. They draw attention to the mutually beneficial relationship that exists between many insects and yeasts and to the key role insects play in yeast ecology. Insects are umbrella species, so protecting them would also safeguard yeast diversity. The authors also reflect on the need to think about yeasts beyond optical representation, as they are visually uninteresting. Their remarkableness resides in what they enable, not what they look like.

Part V, 'Representing Extinct Mammals', includes a broad discussion of mammal extinction and then two species-specific case studies. It begins with Barbara Creed's 'Animal Extinction, Film and the Death Drive', a persuasive, psychoanalytically informed meditation on extinction drivers in the Anthropocene. Creed suggests that the death drive, the drive towards death and destruction identified by Sabrina Spielrein and developed by Sigmund Freud, has played a notable but little-understood role in many recent extinctions. Her exploration of the drive is structured around a consideration of three animal documentaries by Nicolas Philibert, and also the art practice of Janet Laurence. The next chapter by David Maynard, 'Tasmanian Tiger: Precious Little Remains', provides a fascinating account of the rationale underpinning an ongoing exhibition dedicated to the thylacine. The exhibition draws on both art and science to trace the animal's fate and to reflect on its continuing importance within Tasmanian society. The final chapter in the part, 'From the General to the Particular', focuses on an individual quagga located at the Muséum national d'Histoire naturelle in Paris. By weaving together insights derived from the physical, textual and visual archives relating to this particular quagga's life and afterlife, Valérie Bienvenue enables something of the singularity of this specific individual, an animal once owned by Louis XVI, to emerge. Now labelled simply as A544, the quagga appears abstract, a 'specimen' rather than an individual. Bienvenue's aim in reconstructing aspects of his life is to grant him a personality, to render him noteworthy, so that we have a tangible sense of what we lost when he died and can therefore begin to grieve that loss.

The final part of the volume, Part VI, 'Exhibiting Extinction', contains three chapters that examine the significance of museums as spaces through which extinction is represented.[226] Museum exhibitions, as Serge Chaumier notes, provide narratives that express particular ideas and draw attention to pressing issues.[227] Thom van Dooren's emphasis on the value of storytelling is of note in this context. He suggests that stories about extinction can 'add flesh to the bones of the dead and dying', breathing life into specific species disappearances through sharing the 'intimate particularities' of a given organism.[228] Writing of textual representations, he understands his 'thick descriptions' as ethically purposive, potentially giving rise 'to genuine care and concern'.[229] For him, stories act upon the world rather than simply reflect dimensions of it.[230] These three chapters on museum exhibitions embody a similar perspective.

The first chapter in this part, 'Three Variations on the Theme of Extinction' by Anne-Sophie Miclo, provides an extended consideration of the art practice of Mark Dion. Dion's art, which frequently explores extinction issues, is always informed by scientific research. His works have been displayed in contemporary art galleries and in natural history museums, bringing art into the domain of science and science into the sphere of art. Miclo persuasively argues that the playful seriousness associated with Dion's earlier works has now been replaced by an urgent gravitas. These later works, which invite critical reflection on contemporary ecological issues, form a call to action. The middle chapter, Norman MacLeod's 'The Exhibition of Extinct Species', offers a measured critique of contemporary museum approaches to the exhibition of extinction. MacLeod suggests many museums have adopted an overly sentimental vision of extinction that fails to adequately acknowledge the complexities of past extinction events. For example, the mass extinctions of deep time led to the disappearance of some species but also ultimately promoted biodiversity. In this context, the melancholic discourse associated with extinctions in general is misplaced. Modern human-driven extinctions should not be conflated with the natural mass extinctions of the past. The final chapter, 'Exhibiting Extinction' by Kathryn Medlock, offers an absorbing exhibition history of the thylacine at the Tasmanian Museum and Art Gallery (TMAG) from the 1850s to the present. Medlock traces the shifting ideologies underpinning how the thylacine has been represented by the museum over time, culminating with the present exhibition in which the thylacine display functions both to share knowledge and elicit feelings of 'sadness and guilt' at this striking example of anthropogenic extinction. Today, the thylacine exhibition at TMAG provides a powerful warning of the dangers humans sometimes pose to the world they are a part of and, thus, to themselves.

Valérie Bienvenue is a doctoral candidate in the Department of History of Art and Film Studies at the Université de Montréal. Her thesis critically examines human–equine relations through the prism of modern art and visual culture. Prior to her academic career, she worked for ten years in equestrian circles, including teaching bareback riding and rehabilitating horses suffering from physical and psychological trauma. She is the author of several articles and book chapters.

Nicholas Chare is professor of art history in the Department of History of Art and Film Studies at the Université de Montréal. He is the author of *After Francis Bacon* (2012). In 2017, with Sébastien Lévesque and Silvestra Mariniello, he founded the baccalaureate (BACCAP) in visual cultures at the Université de Montréal.

Notes

1. Lewis-Williams focuses on afterimages caused by sensory deprivation or psychotropic experiences. Images that persist in the field of vision can also be caused in other ways, such as by looking into bright light and then turning away.
2. Prehistoric cave art elsewhere in the world also includes depictions of animals. The oldest known cave painting, which was made at least 43,900 years ago in a limestone cave in South Sulawesi, is of a warty pig (*Sus celebensis*). See Brumm et al., 'Oldest Cave Art Found'.
3. Guthrie, *The Nature of Paleolithic Art*, 79–84, 109.
4. For an overview of mammoth imagery, see Braun and Palombo, '*Mammuthus primigenius*'. The majority of works depicting bears probably show the brown bear (*Ursus arctos*) but cave bears are also depicted. See Stiner, 'Cave Bear Ecology', 53.
5. A relict population survived on the island of lutruwita/Tasmania into the twentieth century. For an exploration of the mainland extinction, see Johnson, *Australia's Mammal Extinctions*, 154–55. Johnson suggests overhunting as an explanation for the extinction of the mainland thylacine. Robert Paddle argues the thylacine also survived in mainland Australia into the nineteenth century. See Paddle, *The Last Tasmanian Tiger*, 23–24.
6. Mulvaney, 'Dating the Dreaming'. Mulvaney refers to the image simply as a 'quadruped' for much of his essay before finally suggesting it is a thylacine (46). His attribution is lent plausibility by the probable age of the image (it cannot be a dingo, for instance). For a general discussion of the petroglyphs at Murujuga, see Chare, 'Vision and Indifference'. For a reading of the petroglyph that builds on Mulvaney's ideas, see Chare, 'After the Thylacine', 139–42. Mulvaney provides further discussion of thylacine imagery in *Murujuga Marni*, 225–35.
7. Crary, 'Techniques of the Observer', 9.
8. Our translation. All translations are our own unless otherwise stated. Purkinje, *Beiträge zur Kenntniſs des Sehens*, 167.
9. As part of a summary of Jonathan Crary's work, W.J.T Mitchell refers to 'afterimages' as psychophysiological phenomena and associates their emergence in the nineteenth century as something to be studied with a shift to more bodily conceptions of visual experience. See Mitchell, *Picture Theory*, 19–20.
10. Mulvey, *Afterimages*, 10.
11. The ready association between afterimages and afterwardness as a belated recognition of trauma also registers in the title of Hirsch's *Afterimage*, a book that examines ethical and political issues related to representations of the Shoah. For a brief but insightful overview of Freud's idea of *Nachträglichkeit*, see De Lauretis, *Freud's Drive*, 118. The performance artwork *Becoming an Image* by Cassils has been interpreted as exploiting the artificial encouragement of afterimages through the use of flash photography as a means to explore the trauma-inducing violence perpetrated against trans* people. See Steinbock, 'Photographic Flashes', 261.
12. Heise, *Imagining Extinction*, 42. There are periodic claims that the woodpecker is not extinct. See, for instance, Hill et al. 'Evidence'.
13. Heise, *Imagining Extinction*, 60–61.
14. See, for instance, Caruth, *Unclaimed Experience*. For a problematizing of this understanding of trauma, see Chare and Williams, *Matters of Testimony*. For an incisive overview of trauma theory generally, see Ball, 'Introduction'.
15. Alexander, 'Toward a Theory of Cultural Trauma', 7.

16. Communicating a traumatic experience and grasping its significance does not equate with recovering from it. For a discussion of debates about trauma and healing, see Ionescu and Margaroni, 'Introduction'.
17. Pollock, *After-affects/After-images*, xxvi.
18. Fat from the emu was mixed with pulverized metal (the metal, referred to as 'lateenar', was pounded using a tool referred to as a 'patener') and used as body paint by Tasmanian Aborigines. See Anonymous, 'A Relic from the Past', *The Mercury*, 29 January 1874, p. 2. The importance of the emu to the Aboriginal population is also signalled by the recorded existence of a now lost charcoal drawing of the bird made on bark. The drawing of the bird was located at an abandoned settlement in an area known by the Europeans as Painter's Plains (a name inspired by the presence of the artworks). See Burn, 'Narrative of the Overland Journey', 373.
19. The importance of the emu to Aboriginal ceremony in lutruwita (Tasmania) is attested by Adolphus Schayer. See McFarlane, 'Adolphus Schayer', 112–13.
20. Country is all encompassing and attacks on part of it therefore injure the whole rather than a part. For a discussion of trauma theory in relation to Indigenous trauma in a Canadian context (focusing on the Beothuk), see Chare, 'In Her Hands'.
21. Mulvey, *Afterimages*, 10.
22. For further discussion of the idea of 'the past in the present' as it relates to film, see Chare, 'Once More with Feeling'.
23. Leeder, 'Introduction', 3.
24. Mulvey, *Afterimages*, 10.
25. Poliquin, *The Breathless Zoo*.
26. Ibid., 203.
27. Barrow, *Nature's Ghosts*.
28. See, for example, ibid., 100, 107, 139.
29. Harting, *British Animals Extinct*, 3. Harting's aim in writing his book was to bridge a gap in British scholarship between the study of the fossil record and research into the natural world in the present day. This gap related to 'such animals as have become extinct in Britain *within historic times* (v)'; emphasis in the original. Harting is essentially offering animals a history to supplement the prehistory already accorded them.
30. For Flannery and Schouten, however, their collaboration is conceived as filling in gaps, resurrecting vanished species through creating life-sized artwork of them (Flannery and Schouten, *A Gap in Nature*, xii–xiii).
31. Gordon, *Ghostly Matters*, 18. In this sense, their social and political effects might be said to register affectively.
32. Bird specimens in natural history collections are also referred to as 'study skins'. For further information on Rickard's project, see the Tasmanian Museum and Art Gallery website: https://www.tmag.tas.gov.au/whats_on/exhibitions/current_upcoming/info/extinction_studies (last accessed 4 February 2021).
33. Lees and Pimm, 'Species, Extinct Before We Know Them?', R177.
34. Turvey, 'New Historical Record'.
35. William Dean suggests that both 'now unidentifiable' birds mentioned by Cardim might constitute 'the first extinctions brought about in the Atlantic Forest by the European invasion'. Dean, *With Broadax and Firebrand*, 49.
36. Cardim, *Tratados*, 50–51. The Portuguese word *rosmaninho* used by Cardim to describe the parrots colouring is often translated as 'rosemary' in English but actually refers to lavender, and probably indicates a mauve colour. It is presumably because of this confusion with rosemary, which can have pink flowers, that the first English

translation of Cardim's text in 1625 referred to 'the colour of the Bulfinch [*sic*]'. Bull-finches are pink-breasted. See Purchas, *Hakluytus Posthumus*, 461.

37. Moreira-Lima, 'Aves de Mata Atlântica', 9.
38. Cardim, *Tratados*, 119.
39. The feathers were used to fabricate diadems and colourings (*esmaltes*) by Indigenous peoples. Cardim, *Tratados*, 51.
40. Kolbert, *The Sixth Extinction*, 10.
41. Pimm and Joppa, 'How Many Plant Species', 174. See also Pimm, 'What We Need to Know', 10.
42. Pimm and Joppa, 'How Many Plant Species', 174.
43. Vorontsova et al., 'Inequality in Plant Diversity Knowledge', 56.
44. For a discussion of possible exceptions, see Stuart Pimm's chapter in this volume.
45. For an extended discussion of binomials and species naming more generally, see Chapter 2 of Ritvo, *The Platypus and the Mermaid*.
46. Müller-Wille, 'Systems and How Linnaeus Looked', 314. Susanne Renner detects an important diagnostic dimension to Linnaeus's approach. See Renner, 'A Return to Linnaeus's Focus on Diagnosis'.
47. Ritvo, 'New Presbyter or Old Priest?', 265.
48. Baxandall, *Patters of Intention*, 5.
49. Pollock, *Vision and Difference*, 6.
50. We understand coding here to refer to language (verbal, visual or of another kind) as a social system that substitutes symbols for things and experiences. These symbols, which are subject to convention and open to variation, are used as a means to intentional communication. The choice that exists of which symbols to use demonstrates that language is both coded and behavioural in character.
51. Steve Baker rightly cautions against the idea of 'better or more true' images of animals that escape any misconceptions about a given species. See Baker, *Picturing the Beast*, 189.
52. See Carol Freeman's extended analysis in *Paper Tiger* of the demonization of the thylacine, and how this may have hindered efforts to protect the species.
53. Genetic information about species gleaned from tissue samples also involves an element of description. Species identifications based on morphology and molecular genetics sometimes diverge, with physical appearance and genetic make-up failing to generate the same outcome. The existence of cryptic species – ones that look the same but are different – demonstrates the limits of appearance or form as a guide to distinctiveness.
54. MacLeod and Forey, 'Introduction', 1.
55. Baxandall, *Giotto and the Orators*, 9.
56. Harvey, *Phycologia Australica*, unpaginated. The seaweed was first collected in the vicinity of Spectacle Island in 1855 and finally near Shark Island in 1886. See Millar, 'Bennett's Seaweed'.
57. Harvey, *Phycologia Australica*, unpaginated.
58. Harvey is writing in the nineteenth century. Nowadays, taxonomic descriptions are more standardized, so the opportunity for feelings to manifest through data representation is significantly reduced but not, we would suggest, entirely extinguished.
59. The influence of European flora and fauna on perceptions of Australian species is also registered in the plant species now known as the woody pear (*Xylomelum pyriforme*). The non-pear-related Aboriginal name for the plant, which went unnoted when it was classified, is Meridja couroo. See Maiden, *The Useful Native Plants of Australia*, 615.

60. In his entry for the red alga *Plocamium preissianum*, Harvey explains that etymologically the Plocamium derive their name for the Greek for 'tuft of hair'. See Harvey, *Phycologia Australica*, unpaginated.
61. Harvey acknowledges that the example he is using is 'only in a young state, and probably the fully developed frond would be of different shape and considerably larger'. He is, however, unconcerned by this as he feels he has been able to establish the essential character of the organism.
62. Harvey, *Phycologia Australica*, unpaginated. Both Harvey's Plate (number 61) and a photograph of a preserved specimen are reproduced as part of the entry for *Vanvoorstia bennettiana* at AlgaeBase. https://www.algaebase.org/search/species/detail/?species_id=23738&-session=abv4:AC1F22890327930DD6wQ6E2349F5 (last accessed 29 January 2021). The vivid red colour of the alga in the Plate probably reflects how *Vanvoorstia bennettiana* appeared in life. The dry specimen is a dull maroon.
63. Vogt, Bartolomaeus and Giribet, 'The Linguistic Problem of Morphology', 308.
64. In the context of morphology, Johann Wolfgang von Goethe's aesthetically informed morphological studies offer a particularly complex historical affirmation of how art and science can combine in the study of nature. See Steigerwald, 'Goethe's Morphology'.
65. Australian Aboriginal notions of species difference, for example, include attention to form but also to colour, behaviour and habitat. See Rose, *Dingo Makes Us Human*, 45.
66. In its natural environment, seaweed is reliant upon microbial communities for nutrients and well-being. See Singh and Reddy, 'Seaweed–Microbial Interactions'.
67. David Daniel discusses thalu sites for birds, fish, insects, mammals and plants in *Thalu Sites of the West Pilbara*.
68. For an overview of the Dreaming, see Chapter 3 of Rose, *Dingo Makes Us Human*.
69. Mulvaney, 'Dating the Dreaming', 43.
70. Mulvaney identifies these as fat-tailed and argues they may portray an extinct species of macropod. Although the petroglyph identified as a quadruped looks like one, it is not impossible that the largely obliterated image could be the same macropod found engraved on surrounding rocks. Mulvaney's argument about the ritual significance of the site would not change (as the macropod also appears to be extinct) but there would no longer be a predator–prey dynamic present. Katie Glaskin has recently critiqued Mulvaney's reading and questioned whether the disappearance of the thylacine would have been understood in terms of extinction. Glaskin notes that increase rituals are usually seasonal rather than 'specifically reactive to apparently scare supply' (9). This, however, ignores Mulvaney's identification of the Patterson Valley thalu as qualitatively different to other increase sites in the region. See Glaskin, 'Extinction, Inscription and the Dreaming'. In *Thalu Sites*, Daniel notes of a whale thalu in the vicinity of Murujuga which was formerly used by the Yaburara, that how it functioned is now unknown. Each site requires the use of specific traditional knowledge.
71. Mulvaney, 'Dating the Dreaming', 44. He also notes that petroglyphs embodying a spatial relation of predator–prey are found elsewhere in Murujuga, at Gum Tree Valley.
72. Mulvaney, 'Dating the Dreaming', 46. Mulvaney's use of the Latin binomial is, of course, anachronistic. The name given by the Aboriginal custodians to the animal now known as *Thylacinus cynocephalus* is lost. It was not the cynocephalic or 'dog-headed' thylacine they strove to save but an animal viewed very differently.
73. Writing specifically of rock engravings of fat-tailed macropods (but also with a knowledge of Mulvaney's discussion of the thalu site featuring the thylacine), Steve Brown, in 'Tales of a Fat-Tailed Macropod', suggests the production of such images 'solidifies the relationship between human and non-human' (253). He also argues the petro-

glyphs should not be viewed as representations but instead as 'emergent relational entanglements across human/non-human/more-than-human domains' (254).

74. Reading the lines as radiating outwards institutes a hierarchy and conceives of species difference in oppositional terms. Aboriginal understandings of difference, however, usually emphasize complementarity. See Rose, 'Common Property Regimes', 138.

75. Macropods such as kangaroos and some wallabies live in mobs (groups). When faced with a group of prey animals, predators often do not kill indiscriminately but select prey they perceive as weaker. See Allen, *Wolves of Minong*, 88–112.

76. For a discussion of totemism in relation to Aboriginal Australia, see Rose, 'Common Property Regimes'; Merlan, 'Ghost Twitter'.

77. The idea of species as discrete entities forms part of a metaphysics of individualism. See Vint, 'Entangled Posthumanism', 315. Summarizing Karen Barad's ideas, Vint suggests terms such as 'interaction' should be replaced by 'intra-action' to foreground the ontological inseparability of phenomena.

78. See also Menzies, 'Dm sibilhaa'nm da laxyuubm Gitxaała', 216.

79. We are grateful to one of the anonymous peer reviewers of this volume for helping us to refine our thinking regarding Indigenous knowledge and extinction.

80. The Creature Collective formed in 2016, for instance, unites scholars with ties to Australia, Canada, Malaysian Borneo, the Philippines and the United States of America. See Hernández et al., 'The Creature Collective'.

81. Bagele Chilisa notes the overemphasis on English in research and stresses the need for bilingual research. Indigenous cultures without numerous English-speakers risk being sidelined from contemporary collaborations and debates. See Chilisa, *Indigenous Research Methodologies*, 205–6.

82. Doxtater, 'Indigenous Knowledge in the Decolonial Era', 626.

83. A recent study has started to rectify this problem. See Wang et al., 'Ethnobotanical Study'.

84. Lalonde, 'African Indigenous Knowledge', 56. Luc Hens discusses the value of Indigenous knowledge in the specific context of Ghana in his article 'Indigenous Knowledge and Biodiversity'.

85. Gandile, Tessema and Nake, 'Biodiversity Conservation'. It is noteworthy that of the fifty-five members of the Zeyse, Zergula and Ganta communities interviewed for the study, fifty-three were men. As women in rural areas provide the majority of agricultural labour, their lack of representation in the study is regrettable.

86. Becker and Ghimire, 'Synergy'. We use the Guna name Abya Yala for South America in deference to the region's Indigenous communities but recognize there are alternative terms such as the Nahuatl name Ixachilan.

87. Mitchell, Todd and Pfeifer, 'Aboriginal Knowledge Systems', 6.

88. See, for example, the important article by Das and Lowe, 'Nature Read in Black and White'.

89. See Londa Schiebinger's discussion of the plant, where she notes its discoverer (displaced African or Indigenous Surinamese) is actually unknown. Schiebinger, 'Naming and Knowing', 99.

90. The Ulwa are a people of the eastern part of what is now called Nicaragua. For a discussion of various names ascribed to the plant, see Duke, *Duke's Handbook*, 589–90. *Battaka di basta* is referenced in Coe and Anderson, 'Ethnobotany', 383.

91. Guasco, "As Dead as a Dodo".

92. For a discussion of Indigenous storytelling (*récits*), see Brais-Dussault, 'Art du tatouage'. See also Chare, 'Memory'.

93. Woloshyn, "Welcome to the Tundra", 4.

94. The song 'Hunger' on the album *Auk/Blood* is particularly sensual. Tagaq's tongue becomes an erotic instrument, articulating a softly spoke desire that simultaneously voices a hard-edged refusal of binary logic through lyrics such as 'I want to be your slave and your master'.
95. Galloway, 'The Aurality of Pipeline Politics', 122.
96. Igloliorte, 'Nunatsiavummi Sananguagusigisimajangit', 206.
97. Boerchers, "To Bring a Little Bit of the Land", 31.
98. We believe that through its bloody eroticism, the scene forms a deliberate nod to Carolee Schneemann's performance *Meat Joy* (1964). It is interpreted differently by Boerchers, being seen as a continuation of the preceding scene in which Tagaq is shown embodying a dying caribou, clad in reindeer fur and bleeding. See Boerchers, "To Bring a Little Bit of the Land", 33.
99. For references to literature on the decline of Arctic insects, see Wagner, 'Insect Declines'.
100. Aniskowicz-Fowler, 'Terrestrial Arctic Fauna', 113.
101. *Isarukitsok* or *isarokitsok* is the name used by the Kalaallit or Greenland Inuit for the auk. See Buffon, *Histoire naturelle*, 409n2. The Inuit name *akpingak* is noted by Austin in *The Birds of Newfoundland Labrador*, 90. Writing in 1932, Austin already believed the *akpingak*/curlew extinct.
102. Galloway, 'The Aurality of Pipeline Politics', 140.
103. Cited in Drew Nelles, 'The Rise of Tanya Tagaq', *The Walrus* (2015). https://thewalrus.ca/howl/ (last accessed 14 October 2021)
104. Preston, 'Tanya Tagaq in Concert', 649.
105. Gagnon, 'Singing with *Nanook of the North*', 48.
106. Ferrett, *Dark Sound*, 146.
107. One of Ferrett's case studies is Jana Winderen's composition *The Listener* (2016) which records the sounds of underwater insects. See Chapter 5 of *Dark Sound*.
108. Ferrett, *Dark Sound*, 146–47.
109. Mitchell, *What do Pictures Want?*, 15. For a discussion of cloning in relation to species de-extinction, see Martinelli, Oksanen and Siipi, 'De-extinction'.
110. Cambefort, 'Zoological Nomenclature', 145.
111. See 'Q+A with Mark Dion' (2017) https://www.icaboston.org/articles/qa-mark-dion (last accessed 4 February 2021). In this interview, Dion stresses that he does not view himself an activist, but his work clearly lends itself to environmental activism.
112. Stephanie Eslake, '"I almost cracked": 16-month artistic performance of mass extinction comes to a close', *The Guardian*, 25 January 2021. https://www.theguardian.com/artanddesign/2021/jan/25/i-almost-cracked-16-month-artistic-performance-of-mass-extinction-comes-to-a-close (last accessed 4 February 2021).
113. For an in-depth examination of this topic, see Gell, *Art and Agency*.
114. Pollock, *Art in the Time-Space*, 54.
115. Pollock, *After-affects/After-images*, 193.
116. Boyle, 'Remembering the Huia', 81.
117. Samuel Turvey identifies the Holocene as a particularly important period for studying how anthropogenic factors contribute to extinctions, this because of the epoch's relatively stable climatic conditions in contrast to the Pleistocene. See Turvey, 'Preface', ix. Cary Wolfe cautions against uncritical uses of the term 'anthropogenic' that fail to acknowledge the varied implication of human groups and societies in extinction events. See Wolfe, 'Foreword'.
118. Darwin, *Origin of Species*, 129–30.
119. In 1982, David M. Raup and J. John Seposki identified five clearly defined mass

extinctions in the Late Ordovician, Late Devonian, Late Permian, Late Triassic and Late Cretaceous. See Raup and Seposki, 'Mass Extinctions', 1502. The Great Oxygenation Event, which occurred roughly 2.4 billion years ago, was the first mass extinction and, if added to Raup and Seposki's list, means we have already experienced six mass extinctions. The current mass extinction would therefore be the seventh rather than, as it is commonly called, the sixth.

120. Ohtomo et al., 'Evidence for Biogenic Graphite'.
121. Mitchell, *Image Science*, 35.
122. Ibid., 36.
123. For an incisive discussion of biases in the fossil record, see Milsom and Rigby, *Fossils at a Glance*, 7.
124. MacLeod, *The Great Extinctions*, 40.
125. For a discussion of Palissy's understanding of fossils, see Delord, *L'extinction d'espèce*, 93–136; Shell, 'Casting Life, Recasting Experience', 24–29.
126. Cuvier, *Theory of the Earth*, 115.
127. Gynther, Waller and Leung, 'Confirmation of the Extinction', 26.
128. For a brief summary of the composition of crude oil, see Matar and Hatch, *Chemistry of Petrochemical Processes*, 12–19.
129. MacLeod, *The Great Extinctions*, 45.
130. Mitchell, *The Last Dinosaur Book*.
131. Zylinska, *Nonhuman Photography*, 104.
132. Ibid., 111.
133. Mirzoeff, *How to See the World*, 23.
134. This issue is evident from Daguerre's photograph. Several of the fossils are backed by white card which helps them to stand out.
135. Kerp and Bomfleur, 'Photography of Plant Fossils', 118.
136. Ibid.
137. Probably the first fossil plant photograph, a salt print, was produced in 1857 by Auguste-Adolphe Bertsch, and shows a magnification of a section of fossilized wood. The image was labelled 'Bois fossile du Kingia'. Kerp and Bomfleur, 'Photography of Plant Fossils', 119.
138. The warbler probably disappeared in the 1980s. Its extinction is attributed to the introduction of rats to the island. Writing in 1893, W.L. Abbott described the rats on Aldabra as swarming everywhere and as 'very destructive'. See Abbott, 'Notes on the Natural History of Aldabra', 762.
139. Given the small population of the bird, this act of collecting was potentially devastating. A photograph by Malcolm Penny of the nest in situ is reproduced in Benson and Penny, 'A New Species of Warbler', 104. Benson and Penny describe the warbler as a 'dingy coloured species' (102). A colour hierarchy is clearly in operation here, with brown plumage being viewed as dull and drab.
140. The bird is catalogued at the Natural History Museum as 1968.43.1.
141. Wherever possible in this chapter, we have adopted his or her pronouns to reflect the known gender of specific animals. When we do not know the gender of a given animal, we have opted for they, their or them. These decisions impose human conceptions of gender on non-human animals but aid us to resist objectifying the animals we are referring to. We have not extended this approach to our writing about plants, but perhaps we should have. Plants are more resistant to having human conceptions of gender mapped onto them (often they are 'hermaphroditic' but sometimes they are 'dioecious') but, in the context of flora, we could have adopted they, their or them pronouns.

142. Catesby describes the passenger pigeon (referred to by him as *Palumbus migratorius*, the pigeon of passage) and includes an illustration of it in *The Natural History of Carolina, Florida and the Bahama Islands*. He attributes some of his knowledge of the bird to an unnamed 'Canada Indian' (23).
143. Fuller, *Lost Animals*, 130.
144. Our reading of these images is inspired by Carol Freeman's remarkable work on photographs featuring the thylacine in captivity. See chapters 5 and 6 of Freeman's *Paper Tiger*.
145. Prŷs-Jones used Ektachrome 400 slide film when photographing the birds (Robert Prŷs-Jones, email message to Nicholas Chare, 18 January 2021). This kind of film is well suited to for reproducing colours faithfully and with saturation, even in dim light.
146. We are indebted to Anne Bruneau for proposing a possible identification for this plant. That *Cissampelos pareira* is native to Aldabra is confirmed by William Hemsley in his study of the island's flora. Hemsley also cites a report by Rivalz Dupont that draws on local perspectives, which refers to the flowering season on Aldabra as unlike anywhere else in the Seychelles, such is the 'display of wild flowers with its accompaniment of insects and birds' (110). It seems probable that the as yet unnamed Aldabra brush warbler was one of the species of birds included in this description. See Hemsley, 'Flora of Aldabra'.
147. We are grateful to Robert Prŷs-Jones for clarifying information about the rings (Robert Prŷs-Jones, email message to Nicholas Chare, 10 February 2021). Prŷs-Jones stated to Fuller that he took all his photographs of warblers in 1975.
148. Fuller notes that the photographs he reproduces have not been tampered with or enhanced, but in the case of the second picture of Martha he appears to have reproduced a version that had been cropped in the past. Fuller, *Lost Animals*, 10.
149. This reflection extends to which photographs of the birds to circulate. Exercising his aesthetic judgement, Prŷs-Jones has only digitized two of his slides of the warbler because he believes the others are probably not of acceptable quality for reproduction (Robert Prŷs-Jones, email message to Nicholas Chare, 10 February 2021). A blurred picture of the warbler, for instance, may appear to hold little informational worth, although it attests to the difficulties involved in photographing a small, energetic bird and therefore possesses a significant story with a different kind of value.
150. Films of birds do not always provide clear depictions. The April 2004 video footage of what is identified by some as an ivory-billed woodpecker is a case in point. The attribution has been questioned. Stills from the video are reproduced in Fitzpatrick et al., 'Ivory-billed Woodpecker.'
151. For a useful discussion of lens flare, see Campbell, *Science, Entertainment*, 68.
152. Whether the vocalizations were calls or songs and whether of male or female warblers (or of warblers of unknown sex) is not noted by Prŷs-Jones in 'The Ecology and Conservation', 212.
153. Pearson, 'Aldabra Brush-Warbler'.
154. Craig, 'Expressions of Emotion'.
155. Delius, 'Preening', 40.
156. For a discussion of aspects of Meyer's career, see Greenland, 'Connections in the Collections'. A month and a half after Martha's death, Herman wrote a letter to the editor of the *Journal of the American Medical Association* arguing for the importance of ducks as destroyers of insects and suggesting their usefulness could be exploited to guard against their extinction. Herman, 'The Duck as an Insect

Destroyer', 1410. Herman wrote a well-known article about Martha for *The Auk* in which he compared her to the 'Last of the Mohicans'. Herman, 'The Last Passenger Pigeon', 80.

157. Fuller, *Lost Animals*, 71.
158. Willis, *Reception*, 74.
159. Fuller, *Lost Animals*, 9.
160. The 'yawn' is actually a sign of distress. Fuller's book is therefore fronted by an animal signalling immense anxiety in response to the act of being photographed. For further discussion of the 'yawn', see Chare, 'After the Thylacine', 132.
161. For an account of 'Benjamin', see Sleightholme, 'Confirmation of the Gender'.
162. Freeman, *Paper Tiger*, 180.
163. See Bell, Bell and Merton, 'The Legacy of Big South Cape', 215.
164. Dwyer identifies the specimen he sketched as 1083. This seems to be the bat now catalogued as LM001893, which is the holotype for *Mystacina robusta*. The bat was collected by L. Bell on 29 April 1955 from Pukeotakeo (Big South Cape Island). See Dwyer, 'Studies on the Two New Zealand Bats', 4.
165. Like Dwyer, Keulemans worked in monochrome. The colouring of the lithographs in his name was outsourced to others.
166. Rothschild, *The Avifauna of Laysan*, 162.
167. Barthes, *Camera Lucida*, 96.
168. Ibid.
169. Some bats do carry disease. Rabies, for instance, likely exists in the United Kingdom bat populations. Not all bats, however, carry diseases in equal measure. See Hoffmaster, Vonk and Mies, 'Education to Action'.
170. For further discussion of this phenomenon, see our conversation with Stuart Pimm in this volume.
171. For an extended discussion of cuteness in the context of felines, see Lavoie, *Trop mignon*. Lavoie interprets cuteness as an ambiguous value judgement that holds painful as well as pleasant connotations.
172. Baker provides a subtle and sophisticated analysis of how stereotypical representations of animals can encourage conceptual closure. For him, stereotypes are simplistic and prejudicial (Baker, *Picturing the Beast*, 28–29).
173. Fuller, *Lost Animals*, 180.
174. The name provides a good literal description of the location but also seems intended to connote Batman's Batcave, which was first mentioned in 1943 (it was initially populated by bats as well as by the eponymous superhero). The affectionate name given to the bat roost in Aotearoa/New Zealand demonstrates the influence that popular culture can have upon species and habitat perception. See also the discussion of the 'Skywalker' hoolock gibbon in Stuart Pimm's chapter in this volume.
175. For a general overview of bat sociality, see Kerth, 'Causes and Consequences'.
176. We are grateful to Elizabeth Bell for providing information about the circumstances in which the two photographs were taken (Elizabeth Bell, email message to Nicholas Chare, 21 January 2021).
177. The Ryukyu wood pigeon was described as 'of little importance zoogeographically' in an article on Ryukyu Island zoogeography. This demonstrates that the attention accorded a species by naturalists also varies significantly dependent upon the kinds of issues that interest them. See Short, 'Notes on Okinawan Birds', 265.
178. Halsey, *Deleuze and Environmental Damage*, 235.
179. See Balding and Williams, 'Plant Blindness'.

180. Plants that produce 'impressive' flowers, however, are cherished. Some orchid species are much admired. In botanical science, a disproportionate number of researchers seem to work on orchids.

181. For a discussion of trees as cultural symbols, see Rival, 'Trees'. For insights into the significance that timber might have held in the Neolithic period, see Chare and Price, 'The Dagenham Idol', 30–35.

182. Darwin also countenanced the 'coral of life'. See Hellström, 'Darwin and the Tree of Life'.

183. For a summary of a specific case of tree sitting ('Hector the Forest Protector') and its exploitation for media impact, see Lister and Hutchins, 'Power Games', 586–87. Most trees used for tree sitting are subsequently felled. An exception is the coast redwood (*Sequoia sempervirens*), Luna, that Julia Butterfly Hill occupied for 738 days in 1997–99.

184. For an exploration of this in relation to First Nations in Canada, see Wyatt, 'First Nations, Forest Lands'.

185. The relative lack of debate about plant rights is linked to a failure to countenance ideas of plant cognition and plant sentience. As recognition of the complexity of plant behaviour increases, calls for plant rights may become more energetic. For a recent investigation into plant cognition, see Parise, Gagliano and Souza, 'Extended Cognition in Plants'.

186. Stone, 'Should Trees Have Standing?', 496. Stone employs the term 'standing' to refer to legal standing and to general status. The term might also indicate why a tree is his natural object of choice, as uprightness is associated with being human and, as Adriana Cavarero notes, trees are often linked with the phallus and, by extension, with masculinity. See Chapter 5 of Cavarero's *Inclinations*.

187. See Jose, Wu and Kamoun, 'Overcoming Plant Blindness', 171.

188. Thomas, Ougham and Sanders, 'Plant Blindness and Sustainability'.

189. For a discussion of this project, see https://www.ginkgobioworks.com/2019/05/03/reviving-the-smell-of-extinct-plants/ (last accessed 12 February 2021).

190. Entry 51 in Lindley, *Edwards's Botanical Register*, unpaginated.

191. Pardoe and Lazarus, 'Images of Botany', 554.

192. See McKay, 'Ellis Rowan', 93.

193. See, for example, Robert Dunn, 'Modern Insect Extinctions'.

194. For an excellent summary of Wolfe's position, see Chiew, 'Posthuman Ethics'.

195. Some adult stick insects (*Phasmatodea*) such as *Ctenomorpha gargantua* exceed the average length of a human newborn. Another option for fostering empathy would be to scale down, as has occurred in wildlife television series such as *Bellamy's Backyard Safari* (1981), in which the naturalist David Bellamy was filmed to appear smaller than the insects he was studying. Maria Fernanda Cardoso blends art and science in her installation artwork exploring the genitalia of Harvestman spiders, 'It Is Not Size That Matters, It Is Shape', to critique the hierarchy regarding sexual anatomy in which size is privileged over form but also to demonstrate the need for scientific technologies of visual enlargement to realize the sculptural beauty of arachnid penises. See Buiani and Genosko, 'Putting Penises under the Microscope'.

196. No Loricifera of the roughly forty species identified since 1983 (when the phylum was discovered) has become extinct.

197. More recently, yet in a similar vein, ethologist and activist Jane Goodall delivered a powerful speech about the need for empathy in scientific research, and for the head and the heart to work together. 'Journey from the Jungle', 2019 Beatty Lecture, McGill University, 26 September 2019.

198. Donovan, 'Attention to Suffering', 98.
199. See Martinez, 'Excavating Experience', 69–70.
200. Photographs produced during fieldwork often do feature the hands of researchers, but these are not commonly reproduced outside scientific papers unless no alternative images exist. See, for instance, the small number of photographs of birds in the hand in Brewer's *Birds New to Science*.
201. For an extended discussion of this image, see Bailey, 'Tales of the Old Tasmanian Bushmen', 3. A second photograph exists of Quarrell standing and holding the thylacine.
202. For an early examination of the Tasmanian timber industry, see Penny, *Tasmanian Forestry*. The cover image for Penny's book is of a bullock team of the kind driven by Quarrell.
203. David Pearce was born in Hamilton to John Pearce and Sarah Pearce (née Jones) on 24 January 1856. He died on 23 March 1898 so he cannot be the man in the photograph. See Scott Cook, 'David Pearce (1856–1898).' https://www.wikitree.com/wiki/Pearce-4736 (last accessed 15 October 2021). Many of the Pearce clan lived in the vicinity of Fitzgerald so the man could be another member of the family.
204. Pollock, *Art in the Time-Space*, 14.
205. The term 'landscape', with its emphasis on land that has been shaped, suggests human intervention; yet we use it here to describe all topography, including that which existed prior to humankind.
206. Landscape (*paysage*), as Augustin Berque affirms, is also a modern Western concept. He cautions that it implies a subject/object dichotomy and stresses that landscapes must not be conflated with environments. See Chapter 4 of Berque, *Médiance*.
207. Brown, 'Southwestern Wetlands', 280.
208. See Klages et al., 'Temperate Rainforests'.
209. See Greenwood and Basinger, 'Paleoecology of High-Latitude Eocene Swamp Forests'.
210. See, for instance, Winder, 'Dynamic Modelling of an Extinct Ecosystem'.
211. For an overview of methods of studying dinosaur tracks and the kind and quality of information that can be derived from them, see Falkingham, Marty and Richter, *Dinosaur Tracks*.
212. Therrien, 'Immersion', 451.
213. Ibid.
214. *Far Cry Primal* features several distinct ecosystems including coniferous forest and tundra. Megaherbivores and extinct carnivores such as the dire wolf (*Aenocyon dirus*) appear in the game's open world.
215. See DePalma et al., 'Preliminary Notes'.
216. See Gates, Zanno and Makovicky, 'Theropod Teeth'.
217. The issue of whether mends to ceramics or paintings should be visible or concealed, of whether intervention should be acknowledged or instead hidden for reasons of aesthetic unity, are ongoing. For a brief discussion of the topic, see Chare, 'Material Witness'.
218. Sometimes these activities cause damage indirectly. Roads built for logging in West Africa, for example, have also served to facilitate trade in bushmeat and may have caused the extinction of Miss Waldron's red colobus monkey (*Procolobus badius waldroni*). A relict population of the monkey might survive in Côte d'Ivoire. See Laverty and Gibbs, 'Ecosystem Loss and Fragmentation', 80.
219. Shour, *The Prospect of a Nation*, 255.
220. Cooper, *The Pathfinder or the Inland Sea*, viii.

221. Berque, *Écoumène*, 41n27.
222. Gandile, Tessema and Nake, 'Biodiversity Conservation', 169.
223. Although it does not focus on extinction, for its efforts to bring together voices from the humanities and sciences, see Tsing et al., *Arts of Living on a Damaged Planet*.
224. Books dedicated to the exploration of the extinction of a single species often provide far more nuanced explorations of representation. This is particularly the case for Freeman's ground-breaking book *Paper Tiger* about portrayals of the thylacine.
225. The phenomenon of ArtActivistBarbie is notable in this context. The brainchild of artist-academic Sarah Williamson, ArtActivistBarbie has drawn attention to sexism in the nineteenth century and to how the discoveries of Mary Anning were often used by male scientists without her being acknowledged. In a series of photographs, a bonneted Barbie in Victorian dress poses with a bucket beside fossils on the shoreline of Lyme Regis. Williamson was inspired to stage the photograph by the film *Ammonite* (Dir. Francis Lee, UK, 2020).
226. The role of the museum as a site of public engagement with extinction is of increasing interest in museology. See, for example, the forthcoming special issue of *Museum and Society* (edited by Dolly Jørgensen) on the theme of 'Exhibiting Extinction'.
227. Chaumier, 'Writing and the Museum'.
228. Van Dooren's focus is on birds.
229. Van Dooren, *Flight Ways*, 9.
230. Ibid., 10.

Bibliography

Abbott, W.L. 'Notes on the Natural History of Aldabra, Assumption and Glorioso Islands, Indian Ocean'. *Proceedings of the United States National Museum* 16 (1893): 759–64.

Alexander, Jeffrey C. 'Toward a Theory of Cultural Trauma', in Jeffrey C. Alexander et al. (eds), *Cultural Trauma and Collective Identity* (Berkeley: University of California Press, 2004), 1–30.

Allen, Durward L. *Wolves of Minong: Their Vital Role in a Wild Community*. Boston, MA: Houghton Mifflin, 1979.

Aniskowicz-Fowler, Theresa. 'Terrestrial Arctic Fauna', in Antoni G. Lewkowicz (ed.), *Poles Apart: A Study in Contrasts* (Ottawa: University of Ottawa Press, 1997), 105–16.

Austin, Oliver L. *The Birds of Newfoundland Labrador*. Cambridge, MA: Nuttall Ornithological Club, 1932.

Bailey, Col. 'Tales of the Old Tasmanian Bushmen', *Natural Worlds*, n.d. Retrieved 13 February 2021 from http://www.naturalworlds.org/thylacine/history/bushmen/bushmen_3.htm.

Baker, Steve. *Picturing the Beast: Animals, Identity, and Representation*. 2nd edition. Urbana: University of Illinois Press, 2001.

Balding, Mung, and Kathryn J.H. Williams. 'Plant Blindness and the Implications for Plant Conservation'. *Conservation Biology* 30(6) (2016): 1192–99.

Ball, Karyn. 'Introduction: Traumatizing Theory', in Karyn Ball (ed.), *Traumatizing Theory: The Cultural Politics of Affect In and Beyond Psychoanalysis* (New York: Other Press, 2007), xvii–li.

Barrow, Mark. *Nature's Ghosts: Confronting Extinction from the Age of Jefferson to the Age of Ecology*. Chicago: University of Chicago Press, 2009.

Barthes, Roland. *Camera Lucida*. Translated by Richard Howard. London: Vintage, (1980) 1993.

Baxandall, Michael. *Giotto and the Orators: Humanist Observers of Painting in Italy and the Discovery of Pictorial Composition, 1350–1450*. Oxford: Clarendon Press, 1971.

———. *Patterns of Intention: On the Historical Explanation of Pictures*. New Haven, CT: Yale University Press, 1985.

Becker, C. Dustin, and Kabita Ghimire. 'Synergy between Traditional Ecological Knowledge and Conservation Science Supports Forest Preservation in Ecuador'. *Conservation Ecology* 8(1) (2003): unpaginated.

Bell, Elizabeth A., Brian D. Bell and Don V. Merton. 'The Legacy of Big South Cape'. *New Zealand Journal of Ecology* 40(2) (2016): 212–18.

Benson, Constantine Walter, and Malcolm J. Penny, 'A New Species of Warbler from the Aldabra Atoll'. *Bulletin of the British Ornithologists Club* 88 (1968): 102–8.

Berque, Augustin. *Écoumène: Introduction à l'étude des milieux humains*. Paris: Éditions Belin, 2010.

———. *Médiance: De milieux en paysages*. Paris: Éditions Belin, 2000.

Boerchers, Meredith. '"To Bring a Little Bit of the Land": Tanya Tagaq Performing at the Intersection of Decolonization and Ecocriticism'. Unpublished MA thesis, Carleton University, Ottawa, 2019.

Boyle, Cameron. 'Remembering the Huia: Extinction and Nostalgia in a Bird World'. *Animal Studies Journal* 8(1) (2019): 66–91.

Brais-Dussault, Jade. 'Art du tatouage autochtone contemporain au Canada: Empuissancement, résurgence culturelle et affirmation identitaire'. Unpublished MA thesis, Université de Montréal, July 2020.

Braun, Ingmar M., and Maria Rita Palombo. '*Mammuthus primigenius* in the Cave and Portable Art: An overview with a short account of the elephant fossil record in Southern Europe during the last glacial'. *Quaternary International* 276–77 (2012): 61–76.

Brewer, David. *Birds New to Science: Fifty Years of Avian Discoveries*. London: Christopher Helm, 2018.

Brown, David E. 'Southwestern Wetlands: Their Classification and Characteristics', in R. Roy Johnson and J. Frank McCormick (eds), *Floodplain Wetlands and Other Riparian Ecosystems* (Washington DC: US Department of Agriculture, 1979), 269–82.

Brown, Steve. 'Tales of a Fat-Tailed Macropod', in Michelle Langley et al. (eds), *The Archaeology of Portable Art: Southeast Asian, Pacific, and Australian Perspectives* (London: Routledge, 2018), 241–57.

Brumm, Adam, et al. 'Oldest Cave Art Found in Sulawesi'. *Science Advances* 7 (2021): 1–12.

Buffon, Georges Louis Leclerc, Comte de. *Histoire naturelle, générale et particulière: Volume 62*. Paris: F. Dufart.

Buiani, Robert and Gary Genosko, 'Putting Penises Under the Microscope'. *InVisible Culture* 20 (Spring 2014). Retrieved 12 October 2021 from https://ivc.lib.rochester.edu/putting-penises-under-the-microscope/

Burn, David. 'Narrative of the Overland Journey of Sir John and Lady Franklin and Party from Hobart Town to Macquarie Harbour', in *Colburn's United Service Magazine II* (London: Henry Colburn, 1843), 208–17, 372–83, 564–74.

Cambefort, Yves. 'Zoological Nomenclature and Speech Act Theory', in Karine Chemla and Jacques Virbel (eds), *Texts, Textual Acts and the History of Science* (Cham: Springer, 2015), 143–82.

Campbell, Vincent. *Science, Entertainment and Television Documentary*. Houndmills: Palgrave Macmillan, 2016.

Cardim, Fernão. *Tratados de terra e gente do Brasil*. Rio de Janeiro: J. Leite & Co., 1925.

Caruth, Cathy. *Unclaimed Experience: Trauma, Narrative and History*. Baltimore, MD: Johns Hopkins University Press, 1996.

Catesby, Mark. *The Natural History of Carolina, Florida and the Bahama Islands*. London: Self Published, 1731.

Cavarero, Adriana. *Inclinations: A Critique of Rectitude*. Translated by Amanda Minervini and Adam Sitze. Stanford, CA: Stanford University Press, 2016.

Chare, Nicholas. 'After the Thylacine: In Pursuit of Cinematic and Literary Improvised Encounters with the Extinct'. *Liminalities* 14(1) (2018): 124–68.

———. 'In Her Hands: Affect, Encounter and Gestures of Wit(h)nessing in Shanawdithit's Drawings'. *parallax* 26(3) (2020): 286–302.

———. 'Material Witness: Conservation Ethics and the Scrolls of Auschwitz'. *symplokē* 24(1/2) (2016): 81–97.

———. 'Memory', in Jeffrey R. Di Leo (ed.), *The Bloomsbury Handbook of Literary and Cultural Theory* (London: Bloomsbury Academic, 2019), 571–572.

———. 'Once More with Feeling: Re-Investigating the Smuttynose Island Murders'. *Intermédialités* 28–29 (Fall 2016, Spring 2017): unpaginated.

———. 'Vision and Indifference: Writing Perceptions of the Petroglyphs of Murujuga'. *Images, Imagini, Images: Journal of Visual and Cultural Studies* 17 (2017): 97–123.

Chare, Nicholas, and Nathanael Price. 'The Dagenham Idol: A Neolithic Wooden Figurine Denuded', in Nicholas Chare and Ersy Contogouris (eds), *On the Nude: Looking Anew at the Naked Body in Art* (London: Routledge, 2022), 23–40.

Chare, Nicholas, and Dominic Williams. *Matters of Testimony: Interpreting the Scrolls of Auschwitz*. New York: Berghahn Books, 2016.

Chaumier, Serge. 'Writing and the Museum: The Structure of Museum Revolutions'. Translated by Nicholas Chare. *RACAR* 45(1) (2020): 22–35.

Chiew, Florence. 'Posthuman Ethics with Cary Wolfe and Karen Barad: Animal Compassion as Trans-Species Entanglement'. *Theory, Culture & Society* 31(4) (2014): 51–69.

Chilisa, Bagele. *Indigenous Research Methodologies*, 2nd edition. Thousand Oaks, CA: Sage, 2020.

Coe, Felix G., and Gregory J. Anderson. 'Ethnobotany of the Sumu (Ulwa) of Southeastern Nicaragua and Comparisons with Miskitu Plant Lore'. *Economic Botany* 53(4) (1999): 363–86.

Cooper, James Fenimore. *The Pathfinder; or, The Inland Sea*. New York: Hurd & Houghton, (1840) 1877.

Craig, Wallace. 'The Expressions of Emotion in the Pigeons: III The Passenger Pigeon (*Ectopistes migratorius* Linn.)'. *The Auk* 28(4) (1911): 408–27.

Crary, Jonathan. 'Techniques of the Observer'. *October* 45 (1988): 3–35.

Cuvier, Georges. *Essay on the Theory of the Earth*. Translated by Robert Kerr. Edinburgh: W. Blackwood, (1813) 1815.

Daniel, David. *Thalu Sites of the West Pilbara*. Perth: Department of Aboriginal Sites, 1990.

Darwin, Charles. *On the Origin of Species by Means of Natural Selection*. London: John Murray, 1859.

Das, Subhadra, and Miranda Lowe. 'Nature Read in Black and White: Decolonial Approaches to Interpreting Natural History Collections'. *Journal of Natural Science Collections* 6 (2018): 4–14.

Day, David. *Nevermore: A Book of Hours*. Toronto: Fourfront, 2012.

———. *Vanished Species*. Bloomington, IN: Gallery Books, 1989.

Dean, Warren. *With Broadax and Firebrand: Destruction of the Brazilian Atlantic Forest*. Berkeley: University of California Press, 1995.

De Lauretis, Teresa. *Freud's Drive: Psychoanalysis, Literature and Film*. New York: Palgrave Macmillan, 2008.

Delius, Juan D. 'Preening and Associated Comfort Behavior in Birds'. *Annals of the New York Academy of Sciences* 525(1) (1988): 40–55.

Delord, Julien. *L'extinction d'espèce: Histoire d'un concept et enjeux éthiques*. Paris: Publications Scientifiques du Muséum, 2010.

DePalma, Robert, et al. 'Preliminary Notes on the First Recorded Amber Insects from the Hell Creek Formation'. *The Journal of Palaeontological Sciences* 2 (2010): 1–7.

Donovan, Josephine. 'Attention to Suffering: A Feminist Caring Ethic for the Treatment of Animals'. *Journal of Social Philosophy* 27(1) (1996): 81–102.

Doxtater, Michael G. 'Indigenous Knowledge in the Decolonial Era'. *American Indian Quarterly* 28(3/4) (2004): 618–33.

Duke, James A. *Duke's Handbook of Medicinal Plants of Latin America*. Boca Raton, FL: CRC Press, 2009.

Dunn, Robert. 'Modern Insect Extinctions, the Neglected Majority'. *Conservation Biology* 19(4) (2005): 1030–36.

Dwyer, P.D. 'Studies on the Two New Zealand Bats'. *Zoology Publications from Victoria University of Wellington* 28 (1962): 2–28.

Falkingham, Peter L., Daniel Marty and Annette Richter (eds), *Dinosaur Tracks: The Next Steps*. Bloomington: Indiana University Press, 2016.

Ferrett, D. *Dark Sound: Feminine Voices in Sonic Shadow*. New York: Bloomsbury, 2020.

Fitzpatrick, John W., et al. 'Ivory-billed Woodpecker (*Campephilus principalis*) persists in North America'. *Science* 308 (2005): 1460–62.

Flannery, Tim, and Peter Schouten. *A Gap in Nature: Discovering the World's Extinct Animals*. New York: Atlantic Monthly, 2001.

Freeman, Carol. *Paper Tiger: How Pictures Shaped the Thylacine*. Hobart: Forty South Publishing, 2014.

Fuller, Errol. *Lost Animals: Extinction and the Photographic Record*. London: Bloomsbury, 2013.

Gagnon, Olivia Michiko. 'Singing with *Nanook of the North*: On Tanya Tagaq, Feeling Entangled, and Colonial Archives of Indigineity'. *ASAP* 5(1) (2020): 45–78.

Galloway, Kate. 'The Aurality of Pipeline Politics and Listening for Nacreous Clouds: Voicing Indigenous Ecological Knowledge in Tanya Tagaq's *Animism* and *Retribution*'. *Popular Music* 39(1) (2020): 121–44.

Gandile, Abayneh Unasho, Solomon Mengitsu Tessema and Fisha Mesfine Nake. 'Biodiversity Conservation Using the Indigenous Knowledge System: The Priority Agenda in the Case of Zeyse, Zergula and Ganta Communities in Gamo Gofa Zone (Southern Ethiopia)'. *International Journal of Biodiversity and Conservation* 9(6) (2017): 167–82.

Gates, Terry A., Lindsay E. Zanno and Peter J. Makovicky. 'Theropod Teeth from the Upper Maastrichtian Hell Creek Formation "Sue" Quarry: New Morphotypes and Faunal Comparisons'. *Acta Palaeontologica Polonica* 60(1) (2015): 131–39.

Gell, Alfred. *Art and Agency: An Anthropological Theory*. Oxford: Clarendon Press, 1998.

Glaskin, Katie. 'Extinction, Inscription and the Dreaming: Exploring a Thylacine Connection'. *Anthropological Forum* (2021). doi: 10.1080/00664677.2021.1937513.

Gordon, Avery. *Ghostly Matters*. Minneapolis: University of Minnesota Press, 1997.

Greenland, Lory. 'Connections in the Collections: Cincinnati Museum Center's Enno Meyer Collection'. *Ohio Valley History* 17(3) (2017): 63–66.

Greenwood, David R., and James F. Basinger. 'The Paleoecology of High-Latitude Eocene Swamp Forests from Axel Heiberg Island, Canadian High Arctic'. *Review of Palaeobotany and Palynology* 81 (1994): 83–97.

Guasco, Anna. "'As Dead as a Dodo": Extinction Narratives and Multispecies Justice in the Museum'. *Environment and Planning E: Nature and Space* (August 2020). doi: 10.1177/2514848620945310

Guthrie, R. Dale. *The Nature of Paleolithic Art*. Chicago: University of Chicago Press, 2005.

Gynther, Ian, Natalie Waller and Luke K.-P. Leung. 'Confirmation of the Extinction of the Bramble Cay Melomys (*Melomys rubicola*) on Bramble Cay, Torres Strait: Results and Conclusions from a Comprehensive Survey in August–September 2014'. Unpublished report to the Department of Environment and Heritage Protection, Queensland Government, Brisbane.

Halsey, Mark. *Deleuze and Environmental Damage: Violence of the Text*. Aldershot: Ashgate, 2006.

Harting, James Edmund. *British Animals Extinct Within Historic Times*. Boston: J.R. Osgood & Co., 1880.

Harvey, William Henry. *Phycologia Australica: Volume II*. London: Lovell Reeve, 1859.

Heise, Ursula K. *Imagining Extinction: The Cultural Meaning of Endangered Species*. Chicago: University of Chicago Press, 2016.

Hellström, Nils Peter. 'Darwin and the Tree of Life: The Roots of the Evolutionary Tree'. *Archives of Natural History* 39(2) (2012): 234–52.

Hemsley, William Botting. 'Flora of Aldabra: With Notes on the Flora of Neighbouring Islands'. *Bulletin of Miscellaneous Information* 3 (1919): 108–53.

Hens, Luc. 'Indigenous Knowledge and Biodiversity Conservation and Management in Ghana'. *Journal of Human Ecology* 20(1) (2006): 21–30.

Herman, William C. 'The Duck as an Insect Destroyer'. *Journal AMA* (17 October 1914), 1410.

———. 'The Last Passenger Pigeon'. *The Auk* 65(1) (1948): 77–80.

Hernández, K.J., et al. 'The Creature Collective: Manifestings'. *Environment and Planning E: Nature and Space* (July 2020). doi: 10.1177/2514848620938316

Hill, G.E., et al. 'Evidence Suggesting the Ivory-billed Woodpeckers (*Campephilus principalis*) Exist in Florida'. *Avian Conservation and Ecology* 1(3) (2006). Retrieved 3 February 2021 from http://www.ace-eco.org/vol1/iss3/art2/.

Hirsch, Joshua. *Afterimage: Film, Trauma, and the Holocaust*. Philadelphia: Temple University Press, 2004.

Hoffmaster, Eric, Jennifer Vonk and Rob Mies. 'Education to Action: Improving Public Perception of Bats'. *Animals* 6(6) (2016). Retrieved 12 February 2021 from https://www.mdpi.com/2076-2615/6/1/6.

Igloliorte, Heather. 'Nunatsiavummi Sananguagusigisimajangit / Nunatsiavut Art History: Continuity, Resilience, and Transformation in Inuit Art'. Unpublished PhD thesis, Carleton University, Ottawa, 2013.

Ionescu, Arleen, and Maria Margaroni. 'Introduction', in Arleen Ionescu and Maria Margaroni (eds), *Arts of Healing: Cultural Narratives of Trauma* (London: Rowman & Littlefield, 2020), ix–xxxviii.

Johnson, Chris. *Australia's Mammal Extinctions: A 50,000-Year History*. Cambridge: Cambridge University Press, 2006.

Jose, Sarah B., Chih-Hang Wu and Sophien Kamoun. 'Overcoming Plant Blindness in Science, Education, and Society'. *Plants, People, Planet* 1(3) (2019): 169–72.

Kerp, Hans, and Benjamin Bomfleur. 'Photography of Plant Fossils – New Techniques, Old Tricks'. *Review of Paleobotany and Palynology* 166 (2011): 117–51.

Kerth, Gerald. 'Causes and Consequences of Sociality in Bats'. *BioScience* 58(8) (2008): 737–46.

Klages, Johann P., et al. 'Temperate Rainforests near the South Pole during Peak Cretaceous Warmth'. *Nature* 580 (2020): 81–86.

Kolbert, Elizabeth. *The Sixth Extinction: An Unnatural History*. New York: Henry Holt, 2014.

Lalonde, Andre. 'African Indigenous Knowledge and its Relevance to Sustainable Development', in Julian T. Inglis (ed.), *Traditional Ecological Knowledge: Concepts and Cases* (Ottawa: International Program on Traditional Ecological Knowledge, 1993), 55–62.

Laverty, Melissa F., and James P. Gibbs. 'Ecosystem Loss and Fragmentation: A Synthesis'. *Lessons in Conservation* 1 (2007): 72–96.

Lavoie, Vincent. *Trop mignon: Mythologies du cute*. Paris: PUF, 2020.

Leeder, Murray. 'Introduction', in Murray Leeder (ed.), *Cinematic Ghosts: Haunting and Spectrality from Silent Cinema to the Digital Era* (New York: Bloomsbury, 2015), 1–14.

Lees, Alexander C., and Stuart Pimm. 'Species, Extinct Before We Know Them?' *Current Biology* 25(5) (2015): R177–R180.

Lindley, John. *Edwards's Botanical Register*, Vol. IV. London: James Ridgeway & Sons, 1841.

Lister, Libby, and Brett Hutchins. 'Power Games: Environmental Protest, News Media and the Internet'. *Media, Culture & Society* 31(4) (2009): 579–95.

MacLeod, Norman. *The Great Extinctions: What Causes Them and How They Shape Life*. London: Firefly, 2013.

MacLeod, Norman, and Peter L. Forey. 'Introduction: Morphology, Shape and Phylogenetics', in Norman MacLeod and Peter L. Forey (eds), *Morphology, Shape and Phylogeny* (London: Taylor & Francis, 2002), 1–7.

Maiden, Joseph Henry. *The Useful Native Plants of Australia*. Sydney: Turner & Henderson, 1889.

Martinez, Kelsie A. 'Excavating Experience: An Examination of Archaeological Field Schools as Communities of Practice'. Unpublished MA dissertation, Binghamton University, NY, 2014.

Matar, Sami, and Lewis F. Hatch. *Chemistry of Petrochemical Processes*. 2nd edition. Boston, MA: Gulf Professional Publishing, 2001.

McFarlane, Ian. 'Adolphus Schayer: Van Diemen's Land and the Berlin Papers'. *Tasmanian Historical Research Association: Papers and Proceedings* 57(2) (2010): 105–18.

McKay, Judith. 'Ellis Rowan: Flower-hunting in the Tropics'. *Queensland Review* 10(2) (2003): 89–97.

Menzies, Charles. 'Dm sibilhaa'nm da laxyuubm Gitxaała: Picking Abalone in Gitxaała Territory'. *Human Organization* 69(3) (2010): 213–20.

Merlan, Francesca. 'Ghost Twitter in Indigenous Australia: Sentience, Agency and Ontological Difference'. *HAU: Journal of Ethnographic Theory* 10(1) (2020): 209–35.

Millar, Alan J.K. 'Bennett's Seaweed', in Jonathan E.M. Baillie, Craig Hilton-Taylor and Simon N. Stuart (eds), *2004 IUCN Red List of Threatened Species: A Global Species Assessment* (Gland: IUCN, 2004), 38.

Milsom, Clare, and Sue Rigby. *Fossils at a Glance*. 2nd edition. Oxford: Wiley Blackwell, 2010.

Mirzoeff, Nicholas. *How to See the World*. St Ives: Pelican, 2015.

Mitchell, Audra, Zoe Todd and Pitseolak Pfeifer. 'How Can Aboriginal Knowledge Systems in Canada Contribute to Interdisciplinary Research on the Global Extinction Crisis?' SSHRC Knowledge Synthesis Research Report, 2017.

Mitchell, W.J.T. *Image Science*. Chicago: University of Chicago Press, 2015.

———. *The Last Dinosaur Book*. Chicago: University of Chicago Press, 1998.

———. *Picture Theory: Essays on Verbal and Visual Representation*. Chicago: University of Chicago Press, 1994.

————. *What Do Pictures Want?: The Lives and Loves of Images*. Chicago: University of Chicago Press, 2005.

Moreira-Lima, Luciano. 'Aves de Mata Atlântica: riqueza, composição, status, endemismos e conservação'. MA dissertation, University of São Paulo, 2013.

Müller-Wille, Staffan. 'Systems and How Linnaeus Looked at them in Retrospect'. *Annals of Science* 70(3) (2013): 305–17.

Mulvaney, Ken. 'Dating the Dreaming: Extinct Fauna in the Petroglyphs of the Pilbara Region, Western Australia'. *Archaeology in Oceania* 44 Supplement (2009): 40–48.

————. *Murujuga Marni: Rock Art of the Macropod Hunters and Mollusc Harvesters*. Crawley: UWA Publishing, 2015.

Mulvey, Laura. *Afterimages: On Cinema, Women and Changing Times*. London: Reaktion, 2019.

Martinelli, Lucia, and Markku Oksanen and Helena Siipi, 'De-extinction: A Novel and Remarkable Case of Bio-Objectification'. *Croatian Medical Journal* 55(4) (2014): 423–427.

Ohtomo, Yoko, et al. 'Evidence for Biogenic Graphite in Early Archean Isua Metasedimentary Rocks'. *Nature Geoscience* 7 (2014): 25–28.

Paddle, Robert. *The Last Tasmanian Tiger: The History and Extinction of the Thylacine*. Cambridge: Cambridge University Press, 2000.

Pardoe, Heather, and Maureen Lazarus. 'Images of Botany: Celebrating the Contribution of Women to the History of Botanical Illustration'. *Collections* 14(4) (2018): 547–67.

Parise, André Geremia, Monica Gagliano and Gustavo Maia Souza. 'Extended Cognition in Plants: Is it Possible?'. *Plant Signaling and Behavior* 15(2) (2020). Retrieved 12 February 2021 from https://www.tandfonline.com/doi/full/10.1080/15592324.2019.1710661.

Pearson, D. 'Aldabra Brush-Warbler (*Nesillas aldabrana*), version 1.0', in J. del Hoyo et al., *Birds of the World* (Ithaca, NY: Cornell Lab of Ornithology, 2020). Retrieved 15 January 2021 from https://doi.org/10.2173/bow.albwar1.01.

Penny, J. Compton. *Tasmanian Forestry: Timber Products and Sawmilling Industry*. 2nd edition. Hobart: John Vail, 1910.

Pimm, Stuart L. 'What We Need to Know to Prevent a Mass Extinction of Plant Species'. *Plants, People, Planet* 3 (2021): 7–15.

Pimm, Stuart L., and Lucas Joppa. 'How Many Plants Species Are There, Where Are They, and at What Rate Are They Going Extinct'. *Annals of the Missouri Botanical Garden* 100(3) (2015): 170–76.

Poliquin, Rachel. *The Breathless Zoo: Taxidermy and Cultures of Longing*. University Park: Pennsylvania State University Press, 2012.

Pollock, Griselda. *After-affects/After-images: Trauma and Aesthetic Transformation in the Virtual Feminist Museum*. Manchester: Manchester University Press, 2013.

————. *Art in the Time-Space of Memory and Migration: Sigmund Freud, Anna Freud and Bracha L. Ettinger in the Freud Museum*. Leeds: Wild Pansy Press, 2013.

————. *Vision and Difference: Femininity, Feminism, and Histories of Art*. London: Routledge, 1988.

Preston, VK. 'Tanya Tagaq in Concert with *Nanook of the North*'. *Theatre Journal* 68 (2017): 649–50.

Prŷs-Jones, R.P. 'The Ecology and Conservation of the Aldabran Brush Warbler *Nesillas aldabranus*'. *Philosophical Transactions of the Royal Society of London. Series B, Biological Sciences* 286(1011) (1979): 211–24.

Purchas, Samuel. *Hakluytus Posthumus or, Purchas his Pilgrimes*. Volume 16. Cambridge: Cambridge University Press, (1625) 2014.

Purkinje, Johann. *Beiträge zur Kenntnifs des Sehens in subjectiver Hinsicht*. Prag: Gedruckt bei Fr. Vetterl Edlen von Wildenbrunn, 1819.

Raup, David M., and J. John Seposki. 'Mass Extinctions in the Marine Fossil Record'. *Science* 215(4539) (1982): 1501–3.

Renner, Susanne S. 'A Return to Linnaeus's Focus on Diagnosis, not Description: The Use of DNA Characters in Formal Naming of Species'. *Systematic Biology* 65(6) (2016): 1085–95.

Ritvo, Harriet. 'New Presbyter or Old Priest? Reconsidering Zoological Taxonomy in Britain'. *History of the Human Sciences* 3(2) (1990): 259–76.

———. *The Platypus and the Mermaid and Other Figments of the Classifying Imagination*. Cambridge, MA: Harvard University Press, 1997.

Rival, Laura. 'Trees, from Symbols of Life and Regeneration to Cultural Artefacts', in Laura Rival (ed.), *The Social Life of Trees: Anthropological Perspectives on Tree Symbolism* (Abingdon: Routledge, [1998] 2020), 1–36.

Rose, Deborah Bird. 'Common Property Regimes in Aboriginal Australia: Totemism Revisited', in Peter Larmour (ed.), *The Governance of Common Property in the Pacific Region* (Canberra: ANU Press, 2013), 127–43.

———. *Dingo Makes Us Human: Life and Land in an Australian Aboriginal Culture*. Cambridge: Cambridge University Press, 2000.

Rothschild, Walter. *The Avifauna of Laysan and the Neighbouring Islands*. London: R.H. Porter, 1893–1900.

Schiebinger, Londa. 'Naming and Knowing: The Global Politics of Eighteenth-Century Botanical Nomenclatures', in Pamela H. Smith and Benjamin Schmidt (eds), *Making Knowledge in Early Modern Europe: Practices, Objects, and Texts, 1400–1800* (Chicago: University of Chicago Press, 2007), 90–105.

Shell, Hanna Rose. 'Casting Life, Recasting Experience: Bernard Palissy's Occupation between Maker and Nature'. *Configurations* 12 (2004): 1–40.

Short, L.L. 'Notes on Okinawan Birds and Ryukyu Island Zoogeography'. *Ibis* 115 (1973): 264–67.

Shour, Nancy C. *The Prospect of a Nation: Expansionist Landscape and American Identity, 1770–1850*. Berkeley: University of California Press, 1990.

Singh, Ravindra Pal, and C.R.K. Reddy. 'Seaweed–Microbial Interactions: Key Functions of Seaweed Associated Bacteria'. *FEMS Microbiology Ecology* 88(2) (2014): 213–30.

Sleightholme, Stephen R. 'Confirmation of the Gender of the Last Captive Thylacine'. *Australian Zoologist* 35(4) (2011): 953–56.

Steigerwald, Joan. 'Goethe's Morphology: Urphänomene and Aesthetic Appraisal'. *Journal of the History of Biology* 35(2) (2002): 291–328.

Steinbock, Eliza. 'Photographic Flashes: On Imaging Trans Violence in Heather Cassils' Durational Art'. *Photography and Culture* 7(3) (2014): 253–68.

Stiner, Mary C. 'Cave Bear Ecology and Interactions with Pleistocene Humans'. *Ursus* 11 (1999): 41–58.

Stone, Christopher D. 'Should Trees Have Standing? Toward Legal Rights for Natural Objects'. *Southern California Law Review* 45 (1972): 450–501.

Therrien, Carl. 'Immersion', in Mark Wolf and Bernard Perron (eds), *The Routledge Companion to Video Games Studies* (New York: Routledge, 2014), 451–58.

Thomas, Howard, Helen Ougham and Dawn Sanders. 'Plant Blindness and Sustainability'. *International Journal of Sustainability in Higher Education* (early cite). Retrieved 17 February 2021 from https://www.emerald.com/insight/content/doi/10.1108/IJSHE-09-2020-0335/full/html.

Tsing, Anna, et al. (eds). *Arts of Living on a Damaged Planet*. Minneapolis: University of Minnesota Press, 2017.

Turvey, Samuel. 'A New Historical Record of Macaws on Jamaica'. *Archives of Natural History* 37(2) (2010): 348–51.

———. 'Preface', in Samuel Turvey (ed.), *Holocene Extinctions* (Oxford: Oxford University Press, 2009), ix–x.

Van Dooren, Thom. *Flight Ways: Life and Loss at the Edge of Extinction*. New York: Columbia University Press, 2014.

Vint, Sherryl. 'Entangled Posthumanism'. *Science Fiction Studies* 35(2) (2008): 313–19.

Vogt, Lars, Thomas Bartolomaeus and Gonzalo Giribet. 'The Linguistic Problem of Morphology: Structure versus Homology, and the Standardization of Morphological Data'. *Cladistics* 26 (2010): 301–25.

Vorontsova, Maria S., et al. 'Inequality in Plant Diversity Knowledge and Unrecorded Plant Extinctions: An Example from the Grasses of Madagascar'. *Plants, People, Planet* 3(1) (2021): 45–60.

Wagner, David L. 'Insect Declines in the Anthropocene'. *Annual Review of Entomology* 65 (2020): 457–80.

Wang, Qinghe, et al. 'Ethnobotanical Study on Herbal Market at the Dragon Boat Festival of Chuanqing People in China'. *Journal of Ethnobiology and Ethnomedicine* 17(19) (2021). doi.org/10.1186/s13002-021-00447-y

Willis, Ika. *Reception*. London: Routledge, 2018.

Winder, Nick. 'Dynamic Modelling of an Extinct Ecosystem: Refugia, Resilience and the Overkill Hypothesis in Palaeolithic Epirus', in Geoff Bailey (ed.), *Klithi: Palaeolithic Settlement and Quaternary Landscapes in Northwest Greece* (Cambridge: McDonald Institute for Archaeological Research, 1997), 615–36.

Wolfe, Cary. 'Foreword', in Deborah Bird Rose, Thom van Dooren and Matthew Chrulew (eds), *Extinction Studies: Stories of Time, Death, and Generations* (New York: Columbia University Press, 2017), vii–xvi.

Woloshyn, Alexa. '"Welcome to the Tundra": Tanya Tagaq's Creative and Communicative Agency as Political Strategy'. *Journal of Popular Music Studies* 29 (2017): 1–14.

Wyatt, Stephen. 'First Nations, Forest Lands, and "Aboriginal Forestry" in Canada: From Exclusion to Comanagement and Beyond'. *Canadian Journal of Forestry Research* 38 (2008): 171–80.

Zylinska, Joanna. *Nonhuman Photography*. Cambridge, MA: MIT Press, 2017.

Part I

Dialogues about Extinction

Chapter 1

The Dinosaur as Cultural Symbol and Totem

W.J.T. Mitchell in Conversation

W.J.T. Mitchell

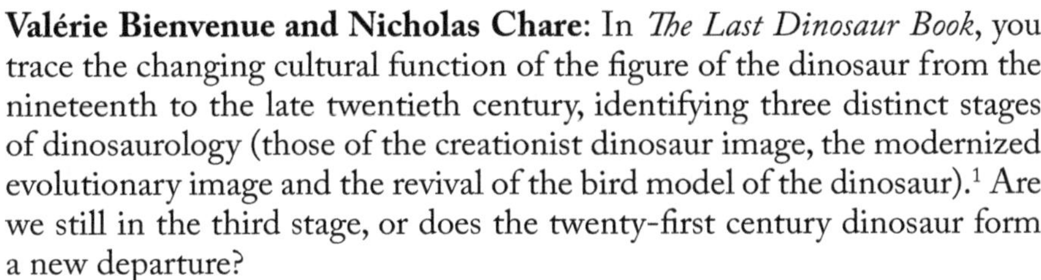

Valérie Bienvenue and Nicholas Chare: In *The Last Dinosaur Book*, you trace the changing cultural function of the figure of the dinosaur from the nineteenth to the late twentieth century, identifying three distinct stages of dinosaurology (those of the creationist dinosaur image, the modernized evolutionary image and the revival of the bird model of the dinosaur).[1] Are we still in the third stage, or does the twenty-first century dinosaur form a new departure?

W.J.T. Mitchell: I think most palaeontologists are still in the third stage, seeing dinosaurs mainly as flightless birds rather than walking reptiles. But the 'stages' operate at more than one level, depending on whether we are talking about narratives of extinction or cultural symbolism. The Victorian story of extinction was centred on air pollution, understandable in the unbreathable air of nineteenth-century London. The modern story was a hodgepodge of stories: entropy (loss of energy); stupidity and failure to adapt; drought and the Dustbowl of the 1930s, the contemporary frame for Disney's *Fantasia*. The third, 'postmodern' stage was probably launched by the onset of the nuclear age, which links dinosaur extinction to the atomic bomb via the bombing of the earth by a giant meteorite – in short, a 'catastrophic' narrative blaming extinction on a single event. Contemporary extinction stories rely on a fourth phase that emphasizes the gradual disappearance of the dinosaurs as a result of multiple factors: climate change and disruptions to the food chain (low calcium intake makes dinosaur eggs vulnerable to hungry rodents), coupled with catastrophic events. The fact is that not all the dinosaurs were wiped out at the same time. And of course, the bird model suggests that they never really died out – they adapted and evolved.

I am most interested in the cultural symbolism of the dinosaur, both in its continuities and deviations. The stage we are in was inaugurated by *Jurassic Park*'s portrayal of the dinosaur as revivable by means of genetic engineering, the onset of cloning technologies. The old dream of reviving the dinosaurs with sculptural, pictorial and cinematic representations was replaced by a new fantasy of literally bringing them back to life. I think it is no accident that this dream of resurrection is accompanied by an increasingly widespread awareness that climate change in our time is no longer a distant danger, but an imminent threat. Hundreds of species are disappearing from the planet every day, and it is no longer possible to deny that human beings may be the first species in the cosmos to bring on its own extinction. The dinosaurs died out through no fault of their own. We have no such excuse.

VB & NC: A powerful dimension of *The Last Dinosaur Book* is your willingness to engage with how the dinosaur has been employed and understood across a variety of forms of cultural expression, including art and science. At one point you affirm that 'Nature *is* culture; science *is* art', openly challenging efforts to set the domains of culture and science in opposition to each other.[2] Dialogue across the humanities and the natural sciences is, however, unfortunately still uncommon. Collaborations of the kind you have engaged in with Norman MacLeod, for example, remain rare. Why do you think that is? Should it be a cause for concern?

WJTM: The old 'two culture split' (C.P. Snow) is far down the list of contemporary concerns in my view.[3] To the extent that both science and the humanities are today under assault by forms of know-nothing populism, climate change denial, and hatred of intellectual elites, we humanists have found a common cause (and a common enemy) in solidarity with scientists. The 'post-truth' era of 'alternative facts', and the gaslighting of large populations, is the bigger concern that we share with all intellectuals engaged in the production of useful, reliable knowledge. There is still a danger that the preference of capital for the STEM disciplines (Science, Technology, Engineering, Mathematics) is leading to a decline of the humanities and social sciences, but this is a quite different issue from the traditional two culture divide.

VB & NC: Yes, the emphasis placed on STEM is certainly a concern. In some countries, however, women are still not encouraged to pursue tertiary education in STEM disciplines because of gender inequality. Could we turn to gender issues as they link with the dinosaur as metaphor, a theme that forms a common thread running through *The Last Dinosaur Book*. You compellingly tease out how an interest in dinosaurs has traditionally been aligned with a particular kind of (phallic) masculinity, a form of masculinity buttressed by an interest in finding and possessing big bones. The 1995 artwork *Brontosaurus* by Sam Taylor-Johnson (née Taylor-Wood), in which a naked man dances to techno music (the sound of which the artist has removed) while a soft toy stegosaurus perches on one of the stereo speakers, seems to expose and undermine connections between male potency and dinosaurs. What is the relationship between dinosaurs and masculinity today?

WJTM: I think they are still primarily coded as male/female=predator/ prey, in a taxonomy dominated by *T. rex*. But the women are gaining, as the 'clever girls' of *Jurassic Park* showed long ago. The female velociraptors of Spielberg's dinosaur park are an apt expression of male hysteria about the 'new women' invading the old political and economic preserves of male dominance. Spielberg succeeded in adding a new 'folk taxon' to the bestiary of extinction, so that the city of Toronto could adopt the raptor as the totem animal of its NBA championship team. The raptors are fast-moving, clever with their hands (they open doors in *Jurassic Park*'s control room), and they are highly adaptable, hunting in packs, and 'figuring things out' fast enough to take control of their own reproductive processes.

VB & NC: Staying with your description of the velociraptors as 'clever girls', in *The Last Dinosaur Book* you read this gendering as symptomatic of a patriarchal backlash against third-wave feminism, against 'the arrival of career women and "clever girls" in multinational corporations'.[4] In a profession such as palaeontology, however, women still lack career visibility despite a long history of contributing to the field. The film *Ammonite* about Mary Anning (who you mention in *The Last Dinosaur Book*) may help to raise visibility in this context. In the United States, Annie Montague Alexander (who, like Anning, contributed to our knowledge of ichthyosaurs) merits mention. Are there alternative histories of palaeontology to the one of the macho male dino-hunter?

WJTM: Most definitely. One symptom of this is the emphasis on dinosaur reproduction, with the 'good mother' Maiasaurus portrayed as a protectively nesting herbivore. A feminist history of palaeontology needs to be written, and the dinosaur's function as a totem animal will no doubt begin to reflect the contemporary deconstruction of gender binaries, and replace them with a host of 'trans' figures of indeterminate or mutable gender. The crazy female scientist in the B-movie *Carnosaur* (1993) is one indication that the dominance of the patriarchal *T. rex* is in serious danger. Even in all-male corporate boardrooms, the CEO is urging his colleagues to 'trade in the brute force of the *T. rex* for the cunning of the raptor'.[5]

VB & NC: What is now referred to as the Sixth Extinction has become one of the most pressing concerns of our time. Writing twenty years ago, you already draw attention to fears about contemporary extinction. More recently, in 'Planetary Madness', you invoked the modern cult of the dinosaur as part of a discussion of current anxieties about extinction.[6] Now that the alarming rate at which species are disappearing has become more broadly known, what role does the figure of the dinosaur have to play in contemporary extinction debates?

WJTM: I think it remains central as the totemic cultural icon. But from a scientific standpoint, we learn more from micropalaeontology and the evolutionary development of the small critters. The fossil record is more continuous with clamshells, and gives us much more insight into evolution.

VB & NC: Continuing with the theme of extinction, you discuss the greater degree of imaginative activity required to represent dinosaurs because of the absence of what might be referred to as 'life models', the fact they are extinct, and the need to work from 'fragmentary traces and fossil remains'.[7] Do you see the representation of more recent extinct species as qualitatively different in nature from that of dinosaurs?

WJTM: Yes. The first thing that comes to mind is Joel Sartore's *National Geographic* 'Photo Ark', which emphasizes the thousands of endangered species that will soon only be known by way of their photographic effigies.

VB & NC: That's a remarkable project. What do you make of the images that are used to form the 'Photo Ark'? As you know, Sartore always photographs the animals against black or white backgrounds, decontextualizing them. He also selects a particular shot to showcase each species. What kind of effigies does Sartore leave us with?

WJTM: I think Sartore is very faithful to the concept of the *ark* and the *archive*. The images of the animals are preserved in amber, as it were, deliberately removed from context, from a living environment. There is no attempt to put them back into their world, into the habitat that made them possible. It is a rescue operation only of images; a melancholy reminder that these living creatures represented are vanishing or vanished. There is no attempt to fool the beholder with a simulation of the habitat, as in the dioramas of natural history museums. Sartore's genius is to insist on the ghostly presence of an absence. Every photograph, of course, removes its subject from its context. As Roland Barthes notes, family photographs are always associated with the departed, with the vanishing of grandparents into albums. But Sartore's photos are even more radical in their insistence on absence. No offspring of these gorgeous creatures will be around. Their biological lineage has given way to a purely iconological survival.

VB & NC: Coming back to fossils, in *What Do Pictures Want?* you describe them as 'melancholy figures'.[8] We can't help hearing Freud in that description; fossils register a loss but not one we can come to fully know. In that sense, they leave us *wanting*. You also describe fossils as 'natural images'.[9] Could you tell us more about the kinds of insights that fossils as things and as metaphors permit regarding other forms of image?

WJTM: In semiotics, the theory of signs, images have always been contrasted with words as 'natural' signs, with words playing the role of arbitrary, conventional symbols. This contrast is a bit reductive, however, and it begs the question of what we mean by 'natural' in the first place. How is it that we speak of 'natural languages' like English, French, Chinese, etc. if language is 'artificial'? To claim something is natural is often only a way of saying it is conventional, customary, or 'normal'. I follow Charles Sanders Peirce's semiotics, which divides sign-functions into the iconic, indexical and symbolic (signs by similarity or analogy; signs by pointing, tracing, proximity, or cause and effect; signs by convention). All signs are mixtures of these functions, so that fossils, for instance, are clearly the result of a long process of petrification of some part of a plant or animal. The fossil is a trace or index (e.g. a footprint), but it is also an icon in that it resembles a foot. And it is easy for a fossil to move into the realm of the symbolic, as it becomes associated with cultural values. Of all the images we deal with, fossils seem to be the most deeply embedded in non-human, natural processes – life, death, and the species death known as extinction.

VB & NC: In *Image Science*, you suggest that it 'is no accident that most palaeontologists have highly developed visual acuity and . . . many of them are artists and image processors'.[10] Could you expand on this a little for us? Why are visual skills and image sensibility so crucial for palaeontology?

WJTM: My remark was based mostly on my conversations with palaeontologist Paul Sereno, who teaches biological drawing at the University of Chicago. Palaeontologists strike me as imaginative detectives, watching for evidence, and assembling their images out of fragments. Ever since Waterhouse Hawkins brought Richard Owen's terrible lizards to life in reinforced brick and concrete sculptures, and Charles Knight's paintings unveiled their worlds to audiences in New York and Chicago, palaeontology has been in a love affair with the visual arts.

VB & NC: And, as *The Last Dinosaur Book* ably demonstrates, artists have been in a love affair with palaeontology. Could we ask you about what is, perhaps, one of your own loves, a particular scene in *Jurassic Park* in which a velociraptor inadvertently starts a film projector that screens a film describing how dinosaurs in the titular park were resurrected from ancient DNA? In *Image Science*, you read this scene as 'a nexus point for [your] speculations on the science of images' and as an allegory for the end of the odyssey of the image.[11] In *What Do Pictures Want?* you also discuss the scene, this time as part of a reflection on the relation of analogical and digital codes.[12] You include a still from the scene in *The Last Dinosaur Book* as part of an observation regarding the dinosaur as a cyborg 'in both the story [of *Jurassic Park*] and the medium in which the story is represented'.[13] These varied readings, each persuasive, show how aspects of *Jurassic Park* have continued to resonate with you over time. Why do you feel the film has been so intellectually stimulating for you over the years?

WJTM: I think *Jurassic Park* was a brilliant synthesis of themes and technologies that captured the imagination of an entire epoch. The fact that its biocybernetic premise (cloning of dinosaurs) was accompanied by a momentous shift in animation techniques from animatronics and robotics to digital animation is one symptom of this timeliness. Its clever linking of these themes to the modern phenomenon of the theme park, and the spectacular revival of a 'cinema of attractions' organized around speculative capital, made the film a global sensation. The funniest moment in the film is the scene of the corporate lawyer who promises to make billions from the park being snatched off the toilet by the *T. rex* and devoured in a single gulp. As a metapicture, a story about image-making, and the realization of the ancient dream of making images come to life, it played a central role in my thinking about images as such. I could not have written a book like *What Do Pictures Want?*, an exploration of the survival of animist and vitalist mythologies around image-making, without passing through this fantasy of the resurrection of extinct animals.

VB & NC: We wanted to ask you about your understanding of extinction in *What Do Pictures Want?* It's there that you pose the fascinating question as to whether images can become extinct (p. 91). As you know, our volume is mainly concerned with images of extinct species, but can you tell us a little about how images themselves can sometimes be thought to (figuratively) die out?

WJTM: I am still puzzling over this question. On the one hand, I want to say that images can never become extinct. They may disappear, be buried, destroyed. But that only means they may come to life later. This is why I distinguish images from 'pictures' – the concrete, material vessels in which images appear. You can destroy a picture, tear it to pieces. But that does not destroy the image that appeared in it if (for instance) you *remember* the image that was in that picture. Or if there is some other picture that contains the same or a similar image. On the other hand, I suppose that a thorough purging of all images of something ('down the memory hole', as Orwell puts it in *1984*) is conceivable. It's the sort of thing that totalitarian regimes always try, but usually fail, to achieve. Let me just compromise and say that images are extremely difficult to kill. They have a way of coming back to life, and palaeontology is a science devoted to exactly that task. It is a way of bringing dead images back to life.

VB & NC: In *What Do Pictures Want?* you also compare the iconologist to a natural historian, suggesting that both professions primarily seek to 'explain why things are the way they are' rather than engaging in value judgements.[14] In that context, you suggest 'a species is neither good nor bad: it simply *is*' (p. 86). Is a natural historian's encounter with a species ever unmediated though? Aren't they beset with the same issues an iconologist faces? Naturalists have, historically, often unconsciously engaged in value judgements about species that have indirectly contributed to their extinction through influencing perceptions of them. To give an example, George Prideaux Robert Harris compared the thylacine [referred to by him as *Didelphis cynocephala*] to a wolf and described them as 'stupid'.[15] That was in 1807 but, more recently, tremendous sums of money have been invested in science to try and clone this now extinct mammal. Such efforts also don't seem value-free. Why clone the thylacine and not another extinct species – a rodent, for example? Isn't natural history, and natural science more broadly, also value-laden? Aren't aesthetic judgements, for instance, potentially made about species within scientific contexts? Perhaps some scientists could learn from the self reflexivity and/or critical reflexivity of iconologists?

WJTM: This is a very good point. Of course palaeontologists, like other scientists, are constantly making value judgements. You rightly point out that we need to specify exactly what kinds of values are at stake. The dinosaur, from the standpoint of studying the complete, unbroken record of extinctions, is a relatively 'bad' species. Micropalaeontologists like Norman MacLeod know that a much more precise and full record of evolution is contained in the archive of small creatures (shellfish, for instance). They don't leave so many gaps in the record. But dinosaurs are a very good species from the standpoint of public interest and fund-raising. And the further question of what specific form of value they contain is always of interest: are they 'totemic' (as dinosaurs certainly are) or aesthetic – trilobites appreciated for their formal beauty.

VB & NC: It would be remiss of us here in Montreal not to mention the cultural phenomenon that is the television series *Dino Dan* (and its more recent spin off series *Dino Dana*) produced in Canada and syndicated worldwide. Reception of the show at first involved considerable speculation about whether Dan, who is passionate about dinosaurs and sees them alive in his everyday suburban life, had autism or Asperger's. His great interest in dinosaurs was interpreted as indexing a developmental disorder, as autism is sometimes viewed. The popularity of the show nevertheless demonstrates that dinosaurs continue to form a rite of passage for children. In *The Last Dinosaur Book* you offer substantial reflections on the role of dinosaurs in education. Has your thinking changed or developed at all in this regard?

WJTM: *Dino Dan* is strong evidence that dinosaurs are still great for activating a lively imagination. I think the TV series flirts with treating it as a mild disorder, but not nearly as serious (or funny) as *Calvin & Hobbes*, where the big lizards activate a childish megalomania. I wish *Dino Dan* were not quite so nice. Although it is better than the syrupy love calls of Barney, it drains almost all the violence from the dino myth and replaces it with a suburban fantasy of good mothers and nuclear families. The only touch of darkness I could find is in the character of a rather stupid and incompetent elementary school teacher. *Dino Dan* goes all in, however, with the fantasy of a rainbow spectrum of dinosaur colouration, celebrating the gay plumage of its early birds.

I have been studying my grandson's progress with dinosaurs since his birth in 2013. The first full-length film he saw in a theatre (at around age 2 1/2) was *The Good Dinosaur* (Peter Sohn, 2015). He was just as absorbed in our family-size bag of popcorn as he was in the movie. About sixty minutes into the film, he leaned over to me and said 'Grandpa, no more dinosaurs', and his dad took him out to the lobby where he promptly threw up all the popcorn. But because he was just acquiring language himself, I think he loved the table-turning premise of the film, which portrays an alternate history in which, thanks to a timely failure of a meteorite to wipe them out, dinosaurs have evolved into talking, thinking creatures, and human beings have not yet acquired language. Rudolph Zallinger's great mural, *The Age of Reptiles*, adorns his room. But at age six, I think he has outgrown them. Last year he was into the solar system, and could name all the moons of Saturn. This year it has been world travel via *The Magic Treehouse*, and we will be taking him to Paris next spring, where I hope to introduce him to the real monsters of the Jardin des Plantes. He is making rapid progress as a builder of sandcastles.

VB & NC: Picking up on the theme of outgrowing or losing interest, you have described dinosaurs as on the way to becoming a dead metaphor, culturally moribund. Do you still hold with this last viewpoint, or are you less willing to close the book on dinosaurs now?

WJTM: I think they have been dead for a very long time, both literally and figuratively. But that just means we will never be able to close the book on them. All books are books of the dead, keeping alive the mortal voices of their authors and images of their subjects long after they are gone.

VB & NC: Many thanks for taking the time to talk to us. It's been a great pleasure.

W.J.T Mitchell is Gaylord Donnelley distinguished service professor of English and art history at the University of Chicago. He is the senior editor of *Critical Inquiry*.

Valérie Bienvenue is a doctoral candidate in the Department of History of Art and Film Studies at the Université de Montréal. Her thesis critically examines human–equine relations through the prism of modern art and visual culture. Prior to her academic career, she worked for ten years in equestrian circles, including teaching bareback riding and rehabilitating horses suffering from physical and psychological trauma. She is the author of several articles and book chapters.

Nicholas Chare is professor of art history in the Department of History of Art and Film Studies at the Université de Montréal. He is the author of *After Francis Bacon* (2012). In 2017, with Sébastien Lévesque and Silvestra Mariniello, he founded the baccalaureate (BACCAP) in visual cultures at the Université de Montréal.

Notes

1. (VB & NC) This conversation was conducted by email from 5 September 2019 to 15 January 2020. All notes are our own. Mitchell, *The Last Dinosaur Book*.
2. Ibid., 58.
3. Snow, *The Two Cultures and the Scientific Revolution*.
4. Mitchell, *The Last Dinosaur Book*, 225.
5. Ibid., 182.
6. Mitchell, 'Planetary Madness'.
7. Mitchell, *The Last Dinosaur Book*, 54.
8. Mitchell, *What Do Pictures Want?*, 167.
9. Ibid., 90.
10. Mitchell, *Image Science*, 36.
11. Ibid., 37.
12. Mitchell, *What Do Pictures Want?*, 315–316.
13. Mitchell, *The Last Dinosaur Book*, 214.
14. Mitchell, *What Do Pictures Want?*, 86.
15. We have adopted the pronoun 'them' here rather than 'it' as a means to resist objectifying the thylacine. Harris's paper was read to the Linnean Society on 21 April 1807. It was published in 1808. Harris, 'Description of Two New Species of Didelphis'.

Bibliography

Harris, G.P. 'Description of Two New Species of Didelphis from Van Diemen's Land'. *Transactions of the Linnean Society of London* 9 (1808): 174–78.
Mitchell, W.J.T. *Image Science*. Chicago: University of Chicago Press, 2015.
———. *The Last Dinosaur Book*. Chicago: University of Chicago Press, 1998.
———. 'Planetary Madness: Globalising the Ship of Fools', in Alexander Streitberger and Hilde van Gelder (eds), *'Disassembled' Images: Allan Sekula and Contemporary Art* (Leuven: Leuven University Press, 2019), 21–38.
———. *What Do Pictures Want?* Chicago: University of Chicago Press, 2005.
Snow, C.P. *The Two Cultures and the Scientific Revolution*. Cambridge: Cambridge University Press, 1959.

Chapter 2

Visualizing Extinction

Harriet Ritvo in Conversation

Harriet Ritvo

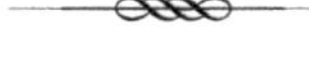

Valérie Bienvenue: You recently published a series of short essays under the title 'Extinction' for the Visualizing Climate and Loss project hosted by the Harvard Center for History and Economics.[1] The essays obviously form part of a broader project of 'visualizing', but can you tell me a little about what informed your own decisions in terms of envisioning extinction? You focus, for example, predominantly on mammals although you also write about the dodo. The mammoth became extinct in prehistory, but others much more recently. Some other mammals such as the bison, the red squirrel and the tiger are not yet extinct. Several of the extinct, such as the dodo, the quagga, and the thylacine have become iconic. What rationales underpin these specific choices, which seem consciously wide-ranging but, for example, avoid plants (although you have discussed flora in your work in the past)?

Harriet Ritvo: The Visualizing Climate and Loss website understands its subject very expansively, and its audience is imagined with similar generosity. It describes itself as 'a platform for thinking with history about change, loss, and daily life – and for thinking about what is to be done'. Since loss (as opposed to extinction) is subjective, in choosing examples I tried to represent a range of human interactions and human responses. All the species that I discussed have inspired significant human interest, whether synchronous or retrospective, as well as the desire, depending on particular circumstances, to protect, lament or reconstitute. The preponderance of mammals reflects the human tendency to respond most strongly to the animals that seem closest to us, not just taxonomically or phylogenetically, but also chronologically. For example, the *T. rex* (along with many other dinosaurs) has inspired a great deal of interest among both specialists and members of the general public since its discovery over a century ago. But *Jurassic Park* to the contrary notwithstanding, that fascination has seldom been expressed in the context of nostalgia or regret.

The only non-mammal included in 'Extinction' is the dodo, a bird that very few humans ever observed alive, as it survived for less than a century after Dutch sailors had noted it on the island of Mauritius, to which its range was restricted. Indeed, relatively few people even observed it dead, and the (partial) specimen preserved in the Oxford Museum of Natural History includes the only remaining bits of dodo soft tissue. (Dodo bones have recently become somewhat less scarce, as a trove of them was excavated from an ancient swamp in 2019, along with the bones of other extinct Mauritian animals.) The relatively high profile of the dodo has thus reflected its symbolic or representational presence, rather than remembered interaction or deeply felt loss; especially since the appearance of John Tenniel's image in *Alice's Adventures in Wonderland* (1865),[2] the dodo has tended to evoke laughter as well as loss.

It has always been understood that dodos disappeared as a result of their contact with humans, an experience that links them to almost all of the other animals discussed in 'Extinction'. The only possible exception is the mammoth, which disappeared much earlier than the others, although not before it had coexisted with our species for thousands of years. The cause of its extinction, along with that of many other large mammals, around the end of the last Pleistocene glaciation (although an isolated population survived on Wrangel Island until about 3,700 years ago), remains the subject of controversy, with human agency and climate changes as the likeliest – and not mutually exclusive – explanations. Like many controversies, this one reflects current politics as well as current science. The demise of the aurochs can be dated with unusual precision to 1627, because the only remaining individuals had previously received a degree of royal protection, although, as with many species currently on the brink of extinction, evidently not enough. In the nineteenth century there was some confusion about whether the extant European bison was in fact the aurochs (it was not); subsequently the aurochs has emerged as the symbol of a vanished Europe. The quagga of southern Africa and the American bison had both been numerous at the beginning of the nineteenth century, and both approached extinction after 1860 for similar reasons – human exploitation and human expansion into their habitats. The quagga disappeared without inspiring much regret; one indication is that although several of them were displayed in European zoos, and of course many died in their native range, very few skeletons or hides have been preserved in museums. The American bison, because of its iconic status, was the subject of one of the most successful rescue efforts, and became the national mammal of the United States in 2016 (the bald eagle, saved from extinction more recently when DDT was banned, remains the national bird of the United States). By the time Europeans began to settle Australia, the mainland thylacines had disappeared; the only surviving population inhabited the southern island of Tasmania. They met the fate that has often awaited predators who share (or who are imagined to share) the same hunting predilections as humans. They received official protection in 1936, a few months before the last-known individual died in the Hobart Zoo; Tasmanians have subsequently embraced the thylacine as their mascot. Tigers are not yet extinct, although human hunting and habitat encroachment have greatly reduced their wild populations from historical levels; more of them live in captivity outside South Asia than roam free within it. Even the few remaining tigers spark the desire to eliminate competition that led to the extinction of the thylacine, and to attempts, often officially sponsored, to eliminate wolves, wolverines and other carnivores. The British red squirrel is not particularly British, and it is not endangered or even threatened over most of its large Eurasian range. But in Britain it has been pushed to Scotland and the extreme north-west of England as a result of competition with the grey squirrel, introduced from North America, which also carries a disease that is lethal to the red squirrels.

VB: Linked with the question of the animals that you discuss in 'Extinction', in your book *The Dawn of Green*[3] you note in the context of a discussion of Cumbrian fauna that 'large mammals tend to be the most compelling representative of nature, wild or otherwise'. The same could also be said of large mammals in relation to the representation of extinction. Your absorbing and persuasive discussion of the British red squirrel (a relatively small rodent) shows, however, that tendencies in terms of what is seen as compelling are not inevitabilities.

HR: Humans tend to feel the strongest connection to the other animals that they resemble most closely, which is why we are more likely to mourn (and indeed to notice) the diminution or disappearance of large to medium-sized mammals than those of other creatures. But as has been the case with the dodo, metaphor or metonymy can lend less impressive creatures a figurative cachet disproportionate to their size. Thus, the red squirrel's prominent place in British affections owes much to Beatrix Potter's *The Tale of Squirrel Nutkin*,[4] which chronicles the close escape (with his life but not his tail) of the cheeky title character. He has become so widely appreciated in his native country, as well as in some other places, that when, in 2006 (long before Brexit) the UK Heritage Lottery Fund awarded £626,000 to protect what supporters affectionately termed 'the real Squirrel Nutkin', the grant drew praise from across the entire political spectrum – not just the mainstream parliamentary parties, but also the (self-described) 'patriotic nationalist' British National Party (BNP).

VB: Coming back to the British red squirrel, you note that the extremist BNP supports efforts to protect the squirrel as part of 'Britain's iconic wildlife'. In this instance, a locally endangered species becomes bound up with issues of national identity and nationalism. The politics of national identity (such as the national bird of the United States that you mention, the bald eagle) and regional identity (such as the extinct Honshu wolf), as they are played out through specific animal species, invites reflection on how animal conservation and animal imagery can be co-opted to serve nationalist ideologies. You gesture towards this phenomenon in your discussion of the American bison which, as you note, has been signed into law as the 'national mammal' of the United States. But can you say a little more about the history of nationalism as it intersects with animal histories?

HR: It turns out to be difficult to disentangle natural history from politics. The overlap is manifest in book titles like *The Breeding Birds of Quebec*[5] and *Guide to the Mammals of Pennsylvania*,[6] where political boundaries implicitly substitute for the limits of natural ranges, and in the characterization of animals like the red squirrel as extinct in a specific nation or province, when the same species is thriving elsewhere, often quite nearby. It has deep historical roots. For example, the British naturalist Thomas Pennant lamented in a preface of 1784, soon after the conclusion of the American Revolution: '[T]his Work was designed as a sketch of the Zoology of North America. I, though, I had a right to the attempt, at a time I had the honor of calling myself a fellow-subject with that respectable part of our former great empire; but when the fatal and humiliating hour arrived, which deprived Britain of power, strength and glory ... I could no longer support my clame [*sic*] of entitling myself its humble zoologist'. (George III could not have expressed himself with deeper feeling; Pennant published his survey under the title of *Arctic Zoology*,[7] an acknowledgement of Britain's remaining North American possession.)

Animals have often served as national symbols, whether designated by a legislature or sanctified by tradition. Animals selected for this purpose, like those chosen as mascots for sports teams, are usually large or aggressive or both; the kangaroo and emu who support the shield on the Australian coat of arms are unusually pacific (if relatively large). Like the British lion, such symbols do not need to be native to the nation that they represent – indeed, like the British unicorn, they do not even need to be real – but native animals turn out to be more resonant with nationalist ideologies. The most striking twentieth-century example of this resonance was the desire of some German zoologists and politicians to resurrect the vanished aurochs (the ancestor of all domesticated cattle), which they understood as the representative of a wilder, stronger, nobler, and purer Europe. In the 1920s and 1930s, the brothers, Heinz and Lutz Heck, both zoo directors, produced factitious aurochs by 'breeding back' from various European cattle breeds that possessed characteristics that they identified as primitive. The symbolism surrounding this effort was readily incorporated into Nazi ideology, and Hermann Goering dreamed of introducing aurochs into the forests of eastern Poland, to supplement their extant native deer and elk. Descendants of these Heck cattle still survive, mostly standing in for their Pleistocene forebears in attempts to reconstitute ancient landscapes, such as Oostvaardersplassen in the Netherlands. Nevertheless, some of their original aura apparently remains; when an English farmer imported a small herd of them in 2009, they were attacked in the media as Nazi cows. A more recent attempt to recreate the aurochs – one that is continental rather than national – has been undertaken by the Taurus Foundation, which has successfully introduced its breed of Tauros cattle in several European national parks.

VB: I'm really glad you brought up the Oostvaardersplassen. As you know, there's been considerable controversy about this effort at 'rewilding'. In 2018, the Staatsbosbeheer (the forestry commission) shot hundreds of animals, including Heck cattle, that were dying of starvation after a harsh winter. The carcasses were left in situ to simulate natural processes. The sight of dead and emaciated deer and cattle disturbed visitors and led to protests. Is the modern vision of nature too tame to embrace the harsh realities that accompany 'rewilding' initiatives such as the Oostvaardersplassen? Or is 'rewilding' itself a misnomer? Is 'wilderness' readily geographically and historically transplantable? I'd be interested to hear your views.

HR: Wilderness, whether reconstituted or not, turns out to be elusive. As William Cronon persuasively argued several decades ago in his essay, 'The Trouble with Wilderness; or, Getting Back to the Wrong Nature',[8] neither the idea of wilderness nor its material embodiments exist outside the sphere of human culture and human influence. Most rewilding efforts aim to restore a set of pre-existing – usually pre-agricultural – conditions. But it is always difficult to decide exactly which conditions should be the target, as pre-agricultural environments were not static (the somewhat mis-leading implication of terms like 'climax forest'), but altered in response to changes in climate and in the distribution of plant and animal species. Such attempts to recreate previous ecosystems, like attempts to preserve those in national parks and similar spaces, also have to confront the con-stant impact of a variety of anthropogenic influences, including, but not limited to, the ubiquitous presence of humans. The public response to the dead deer and cattle at Oostvaardersplassen, and the reluctance of the re-serve managers to let them die 'natural' if painful deaths from exposure and starvation, suggest that at least some people have become reluctant to con-front the most troubling aspects of the experience of wildlife. The absence of large predators from the reserve implicitly makes a similar point, as well as suggesting an explanation for the overpopulation of ungulates that led to this crisis. But the frequent scenes of struggle and death in nature doc-umentaries make a rather different point about contemporary sensibilities.

VB: The devastating effects of overhunting, the hunter as a recent agent of extinction, is a theme in your article 'Animal Planet',[9] and one you also explore in *The Animal Estate*.[10] Hunting for sport was common among colonialists. The hunting trophy (the head or skin) confirmed humans in their position of superiority over other creatures. Nowadays, hunting as a pastime is less common (although still popular, for example, here in Que-bec) and sometimes openly criticized. Are there equivalent contemporary practices that now work to secure humans as the 'top animals'? In 'Animal Planet',[11] for instance, you mention photographers as 'hunters transformed to suit modern sensibilities'. Are contemporary efforts to depict and doc-ument animals sometimes motivated by a need for mastery as much as a drive to know and understand? Can the two motivations be separated?

HR: Hunting remains popular in many places, and many hunters continue to be advocates of wildlife protection, as they have been since the late nineteenth-century origins of the wildlife conservation movement, al-though their ultimate agenda differs somewhat from that of many en-vironmentalists. This trajectory can be seen in the history of the NGO charity Flora and Fauna International, which defines its current focus as 'protecting biodiversity . . . which underpins healthy ecosystems and is crit-ical for the life-support systems that humans and all other species rely on'.[12] It was founded in the early twentieth century by a group that in-cluded big game hunters as well as naturalists (then as now, not mutually exclusive categories) as the Society for the Preservation of the Wild Fauna of the Empire. It has borne a series of names in the intervening century, mirroring shifts in the politics of wildlife conservation. In 1978, when the society published a brief history to commemorate its seventy-fifth anniversary, its name was the Fauna Preservation Society, and the book's title was *The Penitent Butchers*.[13]

The association between traditional hunting and its spectatorial analogue is also evoked by the phrase 'with gun and camera', which has often appeared in the titles of narratives of big game hunting. When travellers confine themselves to the camera, the trophies that they bring home are different, although, depending on the skill and reflexes of the photographer, they may nevertheless suggest the appeal of violence, albeit violence inflicted without human intervention, and therefore as a result of competition or predation that would presumably have happened whether or not humans had observed and documented it. Like the readers of these narratives, contemporary viewers also seem to relish animal combat and death. In consequence, wildlife documentaries routinely feature scenes (occasionally, if surreptitiously, staged) of struggle and predation, although (non-documentary) movies that include such scenes normally include the disclaimer, conferred by American Humane (formerly the American Humane Association), which has a contract with the film industry, that 'No Animals Were Harmed in the Making of This Film'.

VB: 'Extinction' is generously illustrated and includes numerous reproductions of artworks, artefacts and photographs to represent or visualize the animals you discuss. In the case of the dodo, you note that recent scientific investigations have shown that the plump animal represented in many artworks and models does not match the bird's real appearance, which was likely far thinner. This caused the Oxford Museum of Natural History to revise its display. In *The Platypus and the Mermaid*,[14] you also reproduce an engraving of two marsupials (the thylacine and the Tasmanian devil), which were drawn and named to resemble other carnivores, a dog and a bear respectively. These too are not anatomically correct – yet, like the fattened dodo, the 'errors' in their portrayal seem useful historically as an index to European attitudes and ideas regarding these animals. As a historian, how do you approach visual representations of animals from the past, extinct or otherwise? What is their status as historical evidence? Should their perceived inexactitudes be dismissed as 'bad science' or now considered alongside contemporary visions of the animals for the insights they can provide about feelings and viewpoints in the past?

HR: Albrecht Dürer's famous image of an armour-plated rhinoceros, based on a written description of a living one that had been displayed in Lisbon (so that at least the description was based on actual observation) exemplifies the challenges that European naturalists encountered in reconstructing the likeness of a living animal from a combination of organic remains and the accounts of travellers. Whether preserved in the form of dried bones and skins, or (at least for smaller creatures) suspended in alcohol solutions, shape, colour and size were likely to alter in transit. It was often not possible to tell whether the specimen was male or female, juvenile or adult. And interpretation of remains often reflected speculations or assumptions about which more familiar animals the exotic specimen was related to. Thus, the American opossums that were transported to Europe, some living and some preserved, beginning in the late fifteenth century, were variously compared to foxes, bats and apes, depending on which body part was considered to be diagnostic. The pouch was noticed, but it was not recognized as significant until centuries later, when naturalists encountered many more pouches among the Australian fauna. In addition, it was not always clear that specimens were genuine or that descriptions were credible. For example, mermaids composed of the front end of a monkey and the back end of a salmon surfaced repeatedly well into the nineteenth century. Such constructed creatures inspired the scepticism that caused George Shaw, who published the first scientific description of the platypus, to attack the specimen that arrived at the British Museum in 1799 with scissors, on the grounds that the bill must have been sewn on.

Occasionally, as on the voyages of Captain Cook, the collectors and observers were expert naturalists, and so the specimens, descriptions and images they sent back were especially accurate and reliable; however, more often this was not the case. In the mid-nineteenth century, therefore, the British Association for the Advancement of Science published guidebooks designed to encourage the colonial officials, military officers, and other amateurs who might find themselves in exotic locations, to record their observations and collect their specimens in the ways that would be most useful to metropolitan naturalists.

The difficulty that experts had in interpreting the results of amateur observation and collection was not confined to other species. Thus, beginning in 1874, and with editions continuing to appear into the twentieth century, the British Association for the Advancement of Science published the conveniently pocket-sized *Notes and Queries on Anthropology*,[15] the object of which was to 'promote accurate anthropological observation on the part of travellers, and to enable those who are not anthropologists themselves to supply the information that is wanted for the scientific study of anthropology at home'.

However inaccurate or misguided they may seem from the perspective of current zoology, such inexpertly produced images offer significant evidence about the availability of evidence in earlier periods, as well as about the underlying assumptions and understandings that conditioned the interpretation of that evidence. Since science is an evolving and collaborative enterprise, expert consensus routinely changes in response to increasing information and evolving theory. Superseded interpretations should be evaluated by the standards of their own time, rather than by those of later periods. (If people cling to such interpretations long after they have been definitively superseded, then the judgement of 'bad science' might be appropriate.)

VB: You have written extensively about animals in history, and your essays here can be seen as a continuation of that work. Is there, however, a difference in how you approach writing about extinct animals? Are there specific historiographic challenges that come with writing about animals that, like the past itself, are now only accessible by way of representation?

HR: I think that your question contains its own answer. The sources of information about extinct species are different (and less abundant) than the sources of information about extant ones; similar distinctions exist between sources of information about rare species and common ones, and between sources of information about wild species and domesticated ones. Indeed, the fact that most domesticated species have received their own species-level taxonomic designation, even though they continue to interbreed successfully with their wild ancestors (if they are not extinct) is an indication of the relative intensity of our connection with them. In general, for species whose relationship with humans has been infrequent or non-existent, sources tend to be scientific – whether archaeological, palaeontological, archaeozoological or zoological – rather than conventionally historical or literary or artistic. This means that humanists sometimes need to interpret them through the lens of an intermediate contemporary discipline, and sometimes, adding an extra layer of complexity, through an earlier version of such a discipline.

VB: It sounds like historians who study animals, extinct or otherwise, frequently need to adopt a qualitatively different approach to their research compared to historians who limit themselves to human history, because the sources of evidence are different. 'Animal history' is becoming an accepted subfield in history. How do you understand that term? Is it helpful to distinguish history about animals from other kinds of history? Do you view your own work as 'animal history'?

HR: It depends what you mean by animal history. As historians have increasingly recognized the importance of integrating non-humans into their accounts, other animals have been mainstreamed into works that would be categorized as social history or cultural history. But there are also many historians who focus primarily on other animals. Because of the nature of the historical record, both groups primarily depend on sources produced by people. And even non-written sources, such as those provided by taphonomy, rarely offer direct insight into the experience of the original possessors of the bones. Of course, the same sources can yield a variety of insights, depending on the perspective from which they are interrogated. A further challenge to animal historians is how to integrate the insights from zoology, ethology and animal behaviour studies into their work. History is not the only discipline that has increasingly acknowledged the importance of other species. Most of the disciplines in the social sciences and humanities now include animal-related essays in their journals and animal-related panels in their conferences. In addition, there is a very lively interdisciplinary field called animal studies, which overlaps to some extent but not completely with scholarship in the standard disciplines. It is closest in spirit to cultural studies and philosophy.

VB: Coming back to extinction, the idea of de-extinction is much debated. As you note, the availability of frozen woolly mammoth DNA has generated the desire to create artificial worlds such as the Pleistocene Park in Siberia – a project with echoes of Steven Spielberg's fictional *Jurassic Park*. There have also been efforts to clone a thylacine; and the Quagga Project, while having perhaps less 'spectacular' aspirations to the Pleistocene Park, can also be seen as symptomatic of a wish to make good the loss of a species. How do you think these efforts to undo past actions and events should be understood? Are these projects about refusing loss, advancing science, providing entertainment, or something else altogether?

HR: As you suggest, the motivations behind de-extinction are extremely varied, from the sentimental and frivolous, to the environmental and scientific. In *How to Clone a Mammoth: The Science of De-Extinction*,[16] biologist Beth Shapiro expansively defines her perspective as that of an 'enthusiastic realist'. She says: 'I believe that de-extinction is in many cases scientifically and ethically unjustified. However, I also believe that de-extinction technology has great potential to become an important tool for conserving species and habitats that are threatened in the present day'. The quagga is a promising target for de-extinction on several grounds. It is so similar to the still numerous Burchell's zebra that they have recently been reclassified as belonging to the same species. Thus, when Lutz Heck visited its former range in southern Africa after the Second World War, he suggested that the same back-breeding techniques that he had used in his quest to reconstitute the aurochs might also revive the quagga. Several decades later his suggestion was implemented by the Quagga Project, and they have managed to produce zebras with noticeably reduced striping.

Of course, opinions vary about exactly what has been achieved and how good it is, let alone about whether it should have been attempted in the first place. For example, the Grant Museum of Zoology in London, which owns one of only seven extant quagga skeletons, calls the project 'extremely controversial' as 'the result wouldn't actually be a quagga genetically, it would just be a plains zebra artificially selected to look like one'.[17] It then suggests an alternative means of resurrection: 'to extract DNA from bone marrow and remaining taxidermy specimens and use it for cloning by injecting this DNA into a zebra egg'[18] (you can see why the Quagga Project may have chosen the low road). But quagga advocates are not dismayed by such punctiliousness and scepticism. In a sense they are beside the point – that is, they don't address the fundamental impulse or perhaps the fundamental longing that motivates many de-extinction endeavours. As an American commentator for the Nature Conservancy put it, after admitting all the scientific problems, herds of resurrected semi-quaggas offered 'inspiration' and 'hope'; seeing them 'did not seem terribly different from seeing bison on a private ranch, or black-footed ferrets that had been introduced after captive breeding'.[19]

VB: In your Introduction to 'Extinction'[20] you draw attention to aspects of extinction that do not create consensus. You suggest that even if the dominant narrative of the disappearance of the dinosaurs has mainstream acceptance, certain institutions such as the Creation Museum in Kentucky offer an alternative timeframe for it. In an era when 'alternative facts' are given significant media attention, and objective standards for truth are frequently ignored (e.g. climate change science), the stakes involved in 'visualizing climate and loss' seem particularly high. What strategies do you think scientists and historians need to adopt to resist attacks upon their knowledge and understanding of environmental concerns, including the interrelated issues of extinction and global warming? Should we be striving for greater dialogue across the humanities and natural sciences now that intellectuals across all disciplines seem subject to vitriol and mistrust?

HR: The current climate of mistrust of expertise has unfortunately been conspicuous in responses to the coronavirus pandemic, as well as, previously, in resistance to vaccinations, and rejection of the evidence for climate change and evolution. It is very troublesome to see facts discussed and queried as if they are opinions – although, of course, not all opinions are equally debatable or equally well grounded. As a faculty member at MIT, I have taught a range of undergraduate classes in which most of students were science or engineering majors. One issue that inevitably arises, no matter whether the topic of the class is British history or the relationships between humans and other animals, is how to evaluate interpretations (or, to put it a different way, why one should not dismiss anything that is not clearly a fact). Dialogue across the humanities and natural sciences always seems to be a good thing, but I am not sure that it is the remedy for the very widespread and profound rejection of intellectual authority that features constantly in the news. Although there are exceptions in both groups, most humanists and scientists respect the expertise of other disciplines. In a way, the problem may lie with the recently prominent denotation of 'elite', the connotation of which invites resentment whatever it denotes, to refer to people with greater access to information, rather than to people with more money or elevated status. The challenge, then, would be to convert information from a perceived threat to a perceived benefit.

VB: Thank you, Harriet. It's been a great pleasure for us to talk together.

Harriet Ritvo is the Arthur J. Conner professor of history emeritus at the Massachusetts Institute of Technology. A past president of the American Society for Environmental History, she is editor of the 'Animals, History, Culture' book series (Johns Hopkins University Press), and co-editor of the 'Flows, Migrations, and Exchanges' book series (University of North Carolina Press). Her research has focused on the history of natural history, the history of human relations with other animals, and environmental history, especially in Britain and the British Empire. Her current work engages issues of wildness and domestication. Her books include *The Animal Estate: The English and Other Creatures in the Victorian Age* (1987), *The Platypus and the Mermaid, and Other Figments of the Classifying Imagination* (1997), *The Dawn of Green: Manchester, Thirlmere, and Modern Environmentalism* (2009), and *Noble Cows and Hybrid Zebras: Essays on Animals and History* (2010).

Valérie Bienvenue is a doctoral candidate in the Department of History of Art and Film Studies at the Université de Montréal. Her thesis critically examines human–equine relations through the prism of modern art and visual culture. Prior to her academic career, she worked for ten years in equestrian circles, including teaching bareback riding and rehabilitating horses suffering from physical and psychological trauma. She is the author of several articles and book chapters.

Notes

1. (VB) This conversation was conducted by email from 28 November 2019 to 31 August 2020. All notes are my own. Ritvo, 'Extinction'.
2. Carroll, *Alice's Adventures In Wonderland*.
3. Ritvo, *The Dawn of Green*.
4. Potter, *The Tale of Squirrel Nutkin*.
5. Gauthier and Aubry, *The Breeding Birds of Québec*.
6. Merritt and Matinko, *Guide to the Mammals of Pennsylvania*.
7. Pennant, *Arctic Zoology*.
8. Cronon, 'The Trouble with Wilderness'.
9. Ritvo, 'Animal Planet'.
10. Ritvo, *The Animal Estate*.
11. Ritvo, 'Animal Planet'.
12. 'About Us'.
13. Fitter, *The Penitent Butchers*.
14. Ritvo, *The Platypus*.
15. Committee of the Royal Anthropological Institute of Great Britain and Ireland, *Notes and Queries on Anthropology*.
16. Shapiro, *How to Clone a Mammoth*.
17. 'Quagga Skeleton'.
18. Ibid.
19. Miller, 'Quagga'.
20. Ritvo, 'Extinction'.

Bibliography

'About Us'. *Fauna and Flora International*. Retrieved 19 January 2021 from https://www .fauna-flora.org/about.

Carroll, Lewis. *Alice's Adventures In Wonderland*. With Forty-Two Illustrations by John Tenniel. London: Macmillan and Co, 1865.

Committee of the Royal Anthropological Institute of Great Britain and Ireland. *Notes and Queries on Anthropology*. London: Routledge and Kegan Paul, 1967 [1951] reprint.

Cronon, William. 'The Trouble with Wilderness: or, Getting Back to the Wrong Nature'. *Environmental History* 1(1) (1996): 7–28.

Fitter, Richard. *The Penitent Butchers*. Fauna Preservation Society, 1978.

Gauthier, Jean, and Yves Aubry. *The Breeding Birds of Québec: Atlas of the Breeding Birds of Southern Québec*. Montreal: Association québécoise des groupes d'ornithologoques, 1996.

Merritt, Joseph Francis, and Ruth Anne Matinko. *Guide to the Mammals of Pennsylvania*. Pittsburgh, PA: University of Pittsburgh Press, 1987.

Miller, Matthew L. 'Quagga: Can an Extinct Animal Be Bred Back into Existence?' *Cool Green Science*, 13 October 2014. Retrieved 19 January 2021 from https://blog.nature .org/science/2014/10/13/quagga-can-an-extinct-animal-be-bred-back-into-existence.

Pennant, Thomas. *Arctic Zoology*. London: R. Faulder, 1792.

Potter, Beatrix. *The Tale of Squirrel Nutkin* [based on the Original and Authorized Edition]. London: F. Warne, 2017.

'Quagga Skeleton'. University College London. Retrieved 19 January 2021 from https://www.ucl.ac.uk/culture/grant-museum-zoology/quagga-skeleton.

Ritvo, Harriet. *The Animal Estate: The English and Other Creatures in the Victorian Age*. Cambridge, MA: Harvard University Press, 1987.

———. 'Animal Planet'. *Environmental History* 9(2) (2004): 204–20.

———. *The Dawn of Green: Manchester, Thirlmere, and Modern Environmentalism*. Chicago: University of Chicago Press, 2009.

———. 'Extinction'. Visualizing Climate and Loss. Retrieved 19 January 2021 from https://histecon.fas.harvard.edu/climate-loss/extinction/extinction.html.

———. *The Platypus and the Mermaid and Other Figments of the Classifying Imagination*. Cambridge, MA: Harvard University Press, 1997.

Shapiro, Beth Alison. *How to Clone a Mammoth: The Science of De-Extinction*. Princeton, NJ: Princeton University Press, 2015.

Chapter 3

'Putting Nature Back Together Again'

Stuart Pimm in Conversation

Stuart Pimm

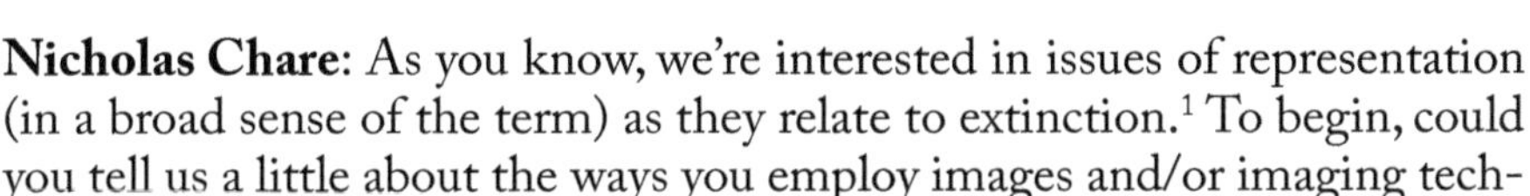

Nicholas Chare: As you know, we're interested in issues of representation (in a broad sense of the term) as they relate to extinction.[1] To begin, could you tell us a little about the ways you employ images and/or imaging techniques in your own work?

Stuart Pimm: There are two sides to this; on one I use a huge number of remote sensing maps and visualizations, on the other I also use pictures of gloriously charismatic animals. So where do you want to go with that?

NC: Do you find that you ever combine both? Do you think that you use them for different reasons?

SP: Yes, absolutely. One of the things that I try to instil in the classes I teach is that these days we face the challenge of those who violate science, violate facts, violate common sense, and do so for a catchy soundbite. We have to be cognizant of all this, and we have to recognize that we need to be compelling communicators of what we do. And then the challenge becomes how to intersect good communication with good science. You want to make sure the science is good, and it's credible, and it's justified. But it also has to be something that people can quickly understand. I do think that's a challenge, and I think good imagery can be very important in that context. If there's one thing that's certain, it is that a difficult graph or chart isn't going to convince anybody. But interestingly, you know, a good map can. My group spends a lot of time worrying in general about how we make our imagery, particularly our maps, so that we can clearly show people what we're doing. Most people can understand maps, but far fewer are good at looking at charts and tables and such things.

NC: So charts and tables are for a particular readership then, for a scientific audience?

SP: When you phoned, I was in the middle of working on paperwork where the key elements are tables, and there is a need to get them right. But even there, we scientists are incredibly busy people, and if you pick up a paper and you can't understand it, then you're likely to put it down again. If you pick up a paper and you understand the graphics, the tables, the charts, the maps, the figures, it's much more likely to have an impact. My group does rather well at that, because we go to a lot of trouble to make sure that our graphical items are easy to understand.

Valérie Bienvenue: Just briefly, could you give us a sense of what makes for a good chart or table? What might form a bad use of graphics or visual mapping?

SP: Near enough, a good chart or table should make sense even if it's labelled in another language! A scatter plot showing that Y increases as X increases makes sense in almost any language, albeit with simple labels of what Y and X are. We produce maps with red showing where there are more species, blue fewer. Everyone gets those – including people who find charts hard work! Bad charts need pages of explanation of text. Brilliant charts explain several ideas simply and intuitively – and I collect those when I come across them, to use as models for my own work.

NC: And when would the other kind of imagery come into play? The 'charismatic animals', the striking pictures of animals or birds that you include in some of your articles?

SP: One of the things that I worry about a lot – I'm working on a paper on giant pandas now – is that it's easy to include a drop-dead gorgeous picture of a giant panda that you want to go up to and cuddle. In reality, it would probably scratch the hell out of you if you did! So it's good to have that kind of imagery, but you want to use it in a sensible way. If you want to work on giant pandas that's fine, but how do you communicate biodiversity and species going extinct without constantly showing pictures of giant pandas and other things. How do we get the balance right, to get the species portrayed, but not rely on it as a crutch?

NC: How do we get away from that difficulty? The fact that certain animals have become iconic, like the panda, and are taken to be striking or viewed as cuddly, whether that's the reality or not. . . Does the existence of what might be called 'species aesthetics' – the way people fixate on particular animals because they view them as more beautiful or cute or visually interesting than others – create a problem?

SP: Yes, and there's both good and bad sides to that. Working with my Chinese colleagues, we know that all the efforts put into protecting giant pandas protect something like 70 per cent of all China's endemic birds, and mammals and amphibians at the same time. So that's good, the panda is a very real umbrella species. . . But yes, the danger is that because some species become so familiar, we tend to think that they are not in trouble.

I have a colleague, Brian Hare, who works on chimpanzees. Chimpanzees are often portrayed as almost human and are given human attributes. Brian shows that this actually diminishes people's expectation or understanding that they are really critically endangered. So the more you make a species familiar, the more you show quite adorable movies of fifteen baby giant pandas playing with each other in the captive breeding facility outside of Chengdu, the less likely you are to realize that pandas are extremely rare because of massive habitat destruction. So there is definitely a problem in how you handle that. It can cut both ways. It can engage people and it can give them a false sense of security.

NC: I hadn't actually thought of it that way myself, the fact that an animal becoming iconic causes a predicament in terms of people seeing it all the time and therefore believing that there isn't a problem in terms of its endangerment. Isn't there also the dimension that there are certain kinds of species, like small mammals, like rodents, that people have a negative perception of and prefer not to see or think about?

SP: Yes, that's an issue too. For many years I worked on an endangered bird in the Everglades in Florida called the Cape Sable seaside sparrow.[2] The fact it's called a sparrow doesn't help, because people think of sparrows as little obscure brown birds. In fact, it *is* a little obscure brown bird. On the other hand, the reality that it's declining dramatically is the best indication that we have that we are mismanaging the Everglades on a very large scale. It could be worse, I could work on some sort of species of rat, which would have even worse connotations.

NC: At least a rat registers with people, while plants, for example, seem very much overlooked. I guess orchids have a certain appeal because there's a perception that they're beautiful. Is it possible to conceive of a plant becoming iconic in conservation terms?

SP: I think that the kind of species that always comes up when the politicians are trying to trash the Endangered Species Act is called the Furbish's lousewort.[3] You know with a name like lousewort, you've got an uphill battle! Yes, it's difficult to engage people. My conservation group, a nonprofit that I run in my spare time, Saving Nature, is working in the western Andes in an area that's got exceptional plant diversity, and it doesn't help that there are more species of orchids than you can shake a stick at. Nor that one of the really interesting genera of orchids there is called *Dracula*.[4] And *Dracula* orchids can be spectacularly beautiful – purple and black and white. They are really funny looking, beautiful orchids, and you sigh in relief because you've got a plant that is as charismatic as could be, and which has a great name.

NC: So in a sense, now that issues around naming are recognized, memorable names can become important in a conservation context. Wasn't there a recently discovered primate nicknamed Skywalker?

SP: I've seen it! It's an absolutely wonderful gibbon.[5] The population that I saw is now down to two adults and a youngster. I think there might be a couple more adults, but there's no more of them. Five individuals in this very isolated population in Yunnan. And I was watching and photographing them in early May. And yes, it looks like Luke Skywalker in the latest movie, with a sort of hood over his head and scowling.

NC: But is that choice of name simply because some of the scientists involved like *Star Wars*, or is there a desire to exploit the publicity that such a name will generate?

SP: I know those guys very well. I do not think there is a desire to exploit. If you saw a picture of this thing and the latest *Star Wars* movie, you'd make the connection easily. It's got this sort of brooding, hooded look about it. And incidentally, that's not its scientific name. It's just a name that people have put on it because it resonates. And when I saw them, and I looked for them several times – this year we looked for them several times, finally found them on the last day – yes, that's a really good common name.

NC: It has helped its profile, in the sense that people have heard of the animal through the 'media buzz' generated by that name.[6] The issue of naming is something that you think about quite frequently in your work although more in the context of the many things that haven't yet been named and some of the issues arising from that. Could you talk a little about naming in that context?

SP: Where we work, particularly in the western Andes, we are discovering new species all the time. And I have absolutely no shame in trying to sell the naming rights of those species so that we can raise money to protect their habitats. My elder daughter is a plant taxonomist, who describes new species of orchids. And this is a subject that we do not discuss, because she thinks her dad is totally wrong on this and that it cheapens the profession. And I fight back and say that if you look back over the history of naming, all sorts of bird species were named after Lord Rothschild and Lord Derby and other wealthy donors. And she says: 'Well, Dad, that's not now'. And so we agree to not talk about it. My feeling is that if it saves the species from extinction, it's all worth it.

NC: Staying with naming, in your work on plants with Peter Raven you've considered the difficulties that come with invisibility, with undiscovered plants not having a name and being absent from the taxonomic list.[7] Could you talk a little about how that poses problems for conservation?

SP: Yes, there are two problems. One of them is how many more species are there that remain to be discovered. And when it comes to plants, it turns out that Peter and I agree. We don't always agree! Peter's approach to find out how many remain to be discovered is to ask all his plant taxonomist friends, and my approach is to build mathematical models of the rates of species descriptions. In this case, we have a happy convergence. We both agree that about 15 per cent more species of plants remain undiscovered. That's a relief to me because it means I know that my elder daughter will have a career in naming new species of orchids and that she's not going to run out of them anytime soon. But more seriously, the question then is, if you know how many are out there, where are they? Because as a practical conservation biologist, I want to know where those missing species are, because I want to add them to the list of species so we that can prioritize the right places for conservation.

NC: Is it difficult in terms of securing funding when you're talking about something that you don't even know is there for certain? We've already discussed issues raised by endangered animals becoming iconic, but what about in the case of undiscovered species when there is nothing to represent, at least not yet?

SP: The interesting thing, it could work against you; it could well be we predict that there are lots of missing species in places we don't expect. But it's working out. The places where we believe the missing species to be are already the places we think are important. And what that means is that the places we already think are important are even more important than we think they are. Of my top three priorities where I think missing species of plants occur, the first would be the northern Andes of South America, the second would be the Atlantic coastal forests in Brazil, and the third would be the mountains in the eastern Himalayas and south-western China, many of which are already safeguarded by the fact that we're protecting pandas. We're probably not going to describe many new species of plants in Connecticut, and so I think basically what we know about the patterns of missing species helps us reinforce the conservation priorities that we already have.

NC: But is it difficult to build momentum for conservation when you're talking about hypothetical species?

SP: I'm not sure. I do think there's a dimension to this that's exciting. In one of the Saving Nature projects that we have in the western Andes, a colleague sent me an email a couple of days ago that said, 'Look we have six new species of frogs. We collected them and we had no idea what they were. We took a toe snip, ran the DNA and there's nothing like these species'. That we have new species such as these, some of which are quite exciting, I think adds to what we're doing. We're working in that area, we're buying up land, we're restoring land, and the land contains species that we have not yet named scientifically. That's very exciting. And I think our donors understand and appreciate that too.

NC: So narratives regarding the continuing possibility of discovery, of the excitement of something being out there that we do not yet know, can be helpful?

SP: Absolutely, and I still live in hope that one day, on some remote mountain top, I'll see a bird and I'll say 'I have no idea what that is'. And nobody else will have an idea what it is either. There's still part of me that fantasizes about being a naturalist back in the 1880s.

NC: Although I guess the naturalists back in the 1880s probably did a lot of harm as well as good?

SP: That may be true. I would have been equipped with my shotgun, and whatever it was I would have shot it...! But you know, there's still that frustrated adventurer within us. I think that most of the people that I know, there's a part of us that says we'd love to go into a new part of the world and find something really exciting.

NC: Staying with the topic of undiscovered species, for some of these species that you go in search of, you arrive too late. In one of your articles, 'Species, extinct before we know them', which you co-authored with Alexander Lees, you discuss unknown species of birds in Brazil's coastal rainforests, and you refer to some of the kinds of evidence we have for the existence of birds for which there are no physical specimens.[8] Obviously in that context, representations, be they textual or visual, hold great importance. Can you tell us more about that research?

SP: Yes, one of the things that has worried me greatly, because I like to put numbers on things, is how many species have gone extinct before we knew what they were. And I've done that for places like Pacific islands. If you look across the Pacific, there is a scattering of very remote islands that have either a species of rail, a species of pigeon, or a species of parrot – and sometimes all three. And so you can say, where did rails get to, where did parrots get to, where did pigeons get to? The answer is almost everywhere. And then you ask how many islands still have parrots, or pigeons, or rails? The answer is not many. And you can say that we, we being Polynesians, probably wiped out a couple of thousand species of birds spread across the Pacific. In a place like Brazil, we know that in the last couple of hundred years the deforestation has wiped out 95 per cent of the coastal rain forest and that some species have gone extinct. But we don't always know what the species that went extinct are, that's the challenge. And that paper that I did with Lees noted that there have been one or two things that people have discovered by going back to the museums and looking at specimens and realizing that those specimens weren't the species that people thought they were, but were something else. They were given a scientific name posthumously. And I would love to have a method to estimate how many species like that there are, but I have no idea how to do that. It's clear, however, that there could have been a substantial number of species that got wiped out in coastal Brazil and in Madagascar, the Philippines, and a variety of other places, where humans came in and their actions did massive environmental harm, and we've never found the species for which depictions exist. For some of the Caribbean islands, there were oil paintings done in the 1600s depicting species, and we have no idea what they are – but they are probably birds that are now extinct.[9]

NC: So am I right in thinking that in Brazil, as well as revisiting museum holdings, researchers found eyewitness accounts and drawings that enabled them to posthumously identify birds that had gone extinct?

SP: In this particular case, it's a kind of Ovenbird.[10] They knew there was this rare species of Ovenbird up in the north-east, and when they went to the museum specimens and realized that the specimens that had been put into the drawer, labelled whatever this thing was, actually consisted of two clearly quite separate species, and they had just overlooked it.

NC: So one of those misidentified birds in the drawer then became the taxon for the species, is that right?

SP: Exactly so.

NC: With other birds where there's no physical specimen, like those birds in the Caribbean, they'll never become a species because there's no taxon, is that the case? A painting of the bird wouldn't do?

SP: That's an interesting question, because you might wonder if it's just a bad painting, and that could indeed be the case. With some of the Caribbean specimens it's pretty obvious you are looking at a new species. There's a macaw, it might have been the Cuban macaw or the Jamaican macaw, just a bloody big parrot, and you look at the oil painting, and it's clearly a large macaw, it's absolutely unmistakable.[11] It was obviously found and shot and stuffed on that particular island. So in those cases, I think we can be pretty sure that that's what it is.

The first map to have the word 'America' on it is from 1507, and it's by the cartographer Martin Waldseemüller. He produced a map of the New World and labelled South America 'America'. It's the first time the word America appears. And on the map in the Americas, there's a drawing of a bird, and it's clearly a macaw; it's got a face like a macaw, and a long tail. It's labelled in Latin, red parrot [*rubei psittaci*].[12] What fascinates me is that here's this guy producing a state-of-the-art map, defining America, and he has time to put a parrot on his artwork. I mean it's clear that people were fascinated by biodiversity when they first encountered it. You know, some sailor arrives from Spain or Portugal on the Brazilian coast, and suddenly this bloody big macaw flies overhead. It's not surprising there's a drawing of it. They're loud and they're noisy and they're big, and people from Europe had not seen anything remotely like them.

NC: Has the Caribbean macaw you mentioned got a Latin binomial?

SP: I think one of those does, but you can imagine that people get hot under the collar debating whether or not that's the right thing to do.

NC: Has it happened that a species has been identified based solely on anecdotal evidence, or must there be a physical specimen for it to be named?

SP: I am 99 per cent sure that there are species that have been recognized from anecdotal evidence.[13]

NC: So in certain circumstances materials such as drawings and paintings of extinct species can assume considerable value when it comes to scientific classification? I knew there was a strong emphasis on the specimen, on the taxon, but clearly in the absence of a physical specimen, representation can serve an important function as a means of identifying a species.

SP: Yes, I'm just trying to think how easy it would be for me to look that up, but I do believe it shouldn't be too hard to find the list of extinct birds, and then work out which ones actually do not have material, do not have skins, do not have skeletons to go with them.

VB: And I imagine that with newly discovered critically endangered species there might be arguments made against procuring a physical specimen to use for classification purposes, unless or until one of the species dies naturally. Perhaps in those circumstances images such as photographs would offer a valuable alternative for classification purposes... I know that for some of your work in the Pacific islands, there are fossilized remains that you can turn to in order to identify extinct species.

SP: Yes, this is how we know where rails and pigeons and parrots were. A lot of our knowledge comes from that kind of material. And there are some islands where people heard rails. Rails, birds such as corncrakes, tend to make funny sorts of noises at night. So people would arrive at an island and they'd hear this thing calling in the night, and from it they knew it was the Something Island rail, and they would never see one. So in addition to the fossil record, there's one or two sort of putative species claimed in that way – birds heard rather than seen.

NC: Ear-witness accounts of species, that's really interesting. Going back to the Ovenbird, you mentioned that specimens held in a museum collection enabled the ultimate identification of two distinct species. In the paper 'Can we defy nature's end?', which you were lead author for, you discuss museums and herbaria in the context of knowing enough about biodiversity and how to protect it.[14] Clearly museums are important knowledge resources and form a crucial interface between stakeholders and the wider public, but do you think institutions such as natural history museums can do more to raise awareness of environmental science and conservation issues? Or should they be doing things differently?

SP: There's clearly a downside. There are certainly some museum collections that have an excessive list of rare specimens, and I do think that's a serious problem. In some parts of the world there are many people who don't want collectors coming anywhere near the places where endangered species live. So there is still a real and present danger from overzealous collecting. On the other hand, museum collections are the basis on which we make our taxonomic catalogues, so we can't do without them. What we can do without is an excessive zeal when it comes to collecting.

NC: I hadn't thought of museums in those terms before. I didn't know that there was still a problem with overzealous collecting. That's something that I had associated more with the Victorian naturalists, who would go out and shoot various animals and bring them back for the collections. That said, I was in a museum in the United States recently where they pulled out a draw filled with numbats, with twenty or so examples of the same species.[15] It does pose the question, how many examples do you need of a given species?

SP: The argument is that we need to understand the geographical variation of the species. That's OK, but why do you need twenty specimens from one place? And you know, I am not compelled by the argument that the need for that science trumps the need to make sure you don't completely eliminate that species. I mean it's a problem for birds, but it's really a problem for amphibians and reptiles, that might have very tiny populations. In those circumstances, you might be able to go out in the space of an hour or two and collect a significant fraction of the world's population. There are some legitimate concerns there that lead to very energetic debates between conservation biologists like me on the one hand, and museum scientists on the other.

VB: Yes, perhaps part of the problem is the kind of knowledge of animals that is privileged, with morphology emphasized over behaviour? The truth of a species is perceived to lie in its bones and, nowadays, its DNA, and not in how it actually lives and acts in a given habitat. Does part of this debate therefore revolve around the kind of knowledge that you can glean from a specimen, which is very different from the kind of knowledge you can glean from studying a species in situ, in the wild?

SP: Exactly so, and I would argue that there's a huge amount of information that you cannot get from the species if it's dead.

NC: So, a problem is the enduring drive to collect, the need to become comprehensive. There's an ongoing competition to have the best collection.

SP: Yes.

NC: But there are also immensely positive things to be said about museums. For example, here at the Montréal Science Centre we recently had an exhibition that had toured from Australia, from the Australian Museum, about spiders, 'Spiders: From Fear to Fascination', which aimed to overcome aversion to arachnids. Can't museums play an important role in that kind of way?

SP: Yes, without museum collections we wouldn't have the basis for what we do. Most of the conservation I do is based on birds, because we know birds best. And I worry a great deal about how the actions that we take might lack appropriate representation because we're concentrating on birds. But we have learned an enormous amount about how to do conservation, and where to do conservation, because we know birds so well. Part of that is due to the fact there are millions of birdwatchers. But part of that is also that by 1900 we had 90 per cent of the world's birds described. The period from 1815 to 1900 took us from 10 per cent of the world's bird species being known to 90 per cent. Lord Rothschild, Lord Derby, those wonderful Victorian collectors, incredibly eccentric, sent out their chaps to every corner of the planet to shoot, stuff, and bring specimens back. And because they did that, and because they established those collections, we've got a good taxonomic catalogue and we can do the kinds of conservation that I do. So, I totally understand the importance of having collections, for birds, but especially for insects, which are really hard to identify. In this day and age, you can identify a bird from a photograph, but you can't begin to understand, say butterfly diversity, unless you've collected the thing and stuck a pin through it.

NC: The taxonomic catalogue, the museum as a repository for scientists, is one dimension of a museum's role, certainly incredibly important, but surely of equal importance is the public-facing side, how that collection is used and displayed.

SP: I think that museums, zoos too, have an important role in engaging the public, to show them how spectacular biodiversity is. I'm hugely fortunate that I can walk through the Brazilian rain forest, I can go to New Guinea, I can see all these wonderful species, but for a great many people they're only ever going to see them as museum specimens or as live animals in the zoo.

VB: I was wondering, Stuart, if you could tell us a little bit about the word 'extinction', what it means to you? Is the way it is sometimes framed in scientific discourse one that obscures human responsibility for species disappearances?

SP: My story is very simple. After I got my PhD, I was very conscious of the fact that I would never want to work in a place like Hawai'i, because I knew that the Hawaiian Islands were so badly beaten up ecologically. I thought that I needed to go and study the ecology in places where the ecology is pristine: the Amazon, the desert, places like that. And in one of those sorts of events that make you believe in magic fairies that fly around and tap you on your shoulder, I found myself out in Hawai'i, and it totally changed my life. I went out for an eight-month period in 1978, expecting to do fieldwork six days out of every eight. I was confident that anything that was in Hawai'i, any bird that was in Hawai'i, I would see. I was a very good fieldworker, and I thought I would learn the species there very quickly, and their calls. But, I didn't: there were some species that I did not see, however hard I tried. That absolutely grabbed my attention. I realized that the species were going extinct, that some of the species that people thought might survive were almost certainly extinct. I was working on a species that had dramatically declined. And those experiences made me what I am today, which is a conservation biologist.

A few years later, Michael Soulé – I didn't then know who he was then – phoned up, introduced himself, and said, 'I'd like you to come to a meeting on conservation biology', and I said 'What's that?' And he said 'Well, whatever "that" is, you're doing it'.[16] And so I was there at four o'clock on a Thursday afternoon when the Society of Conservation Biology was voted into existence. So for me, the experiences of working out in Hawai'i and seeing some species that are now extinct, not seeing other species that I wanted to see, species that were either already extinct or that are extinct by now, absolutely changed my whole career.

NC: Picking up on Valérie's question, there's a moment in your article 'The Dodo Went Extinct (And Other Ecological Myths)' where you state that the dodo did not go extinct, rather 'humanity bludgeoned it into oblivion', as if you thought that the term extinction was too cold and detached, and that it didn't really capture the violence behind what happened in Mauritius.[17]

SP: Exactly so, and I got that definition from, shame of shame, the Oxford English Dictionary, of which I have three copies: one at work, one at home, and one online. So here we have the great repository of the English language, and it says that the dodo 'became extinct'. It's as if it was the dodo's own bloody fault that it went extinct. Well, that's not the case. We drove it to extinction. This isn't something about which we should in any way be passive.

NC: No, your choice of words certainly leaves no ambiguity as to how the dodo ceased to exist in contrast to 'went extinct' which, in the context, seems an inadequate euphemism that deflects attention from the violent historical reality.[18] Linked with the need to be alert to language, is there a danger that sometimes in scientific discourse extinction is framed in a way that risks obscuring that human responsibility behind species disappearances? It seems to us that that's what you're working against in the article by choosing to foreground the dodo's fate in that way.

SP: Absolutely. One of the experiences that I encounter when I'm testifying in some congressional committee in the US is people saying, 'Well, species have always gone extinct!' That's true but we're driving species to extinction a thousand times faster than they would go extinct naturally. You always find people out there who say: 'Well, *c'est dommage, tant pis*, extinction is part of life. Let's just get on with it'. Well, no.

NC: Unsurprisingly, given you've told us that you've got three copies of the Oxford English Dictionary, your choice of language in your writings seem very careful and considered. You're well aware of the power that words have. In some of your writings, you mention 'wimp species'.[19] How did that kind of negative description of species come about?

SP: There was a time when Sir David Attenborough was rather dismissive of bird extinctions, because he thought they were mostly birds on islands. The notion is that birds on islands are like dodos, they're sort of wimpy, they're not well protected. I mean, look at the great auk. All you had to do was row up to an island, the islands off New England and Canada, and bludgeon the things. That gave rise to the idea 'what a bunch of stupid species'. And that attitude is out there, that these extinct species, as it were, were wimpy, that they had it coming to them, they were really never going to survive the modern era. And I think that's a view that we have to reject very forcefully.

NC: I guess that reinforces how even scientific language regarding extinction can't escape the presence of stereotypes. They're there, and you have to work against them.

SP: The answer is yes. We need to be very careful and we do not want to stereotype species. We don't want to stereotype dodos as being stupid, we don't want to stereotype chimpanzees as being just clever little subhumans, and they really are friends and pets, and they're OK. That's not true.

NC: And these do seem to be issues of representation in the broad sense. Scientists have got to negotiate this issue, the different ways of representing species.

SP: I think you're absolutely right. I think we have to be very careful about how we look at this.

VB: Speaking of looking carefully, could you tell us a little bit about your passion for birdwatching, and how, if at all, it informs your approaches in your research?

SP: Well, I grew up in a home in the north of England where my parents loved to hike and walk and explore nature whenever they could. I became a fanatic birdwatcher when I was twelve years old. That's sixty years ago – I haven't changed much. And I do think that birdwatching and other kinds of natural history really does give you an extraordinary insight into what's going on in the environment. I would not have become a conservation biologist if I had not gone out to Hawai'i as a birdwatcher and realized what a desperate state the Hawaiian birds were in.

NC: Are there qualities that birdwatching as a leisure activity, a pastime, necessitates that then play into your approach to conservation, to your research?

SP: Absolutely. I think that what it does is it's the one hobby where you can go anywhere and have an almost immediate assessment of the environment around you. If I'm out in the field with Stephanie, my elder daughter, who is a plant taxonomist, she is a very good tropical biologist but it's hard work for her when she finds a plant. She will look at the plant and have a very good idea of what family it is, and perhaps what genus it is, and then she'll whip out her guide to flora, and maybe after a bit of work she'll figure out what it is. Whereas for my walk this morning – three kilometres for my cup of coffee at the nearby Bean Traders and three kilometres back – I can rattle off the list of species that I saw, the list of species that I heard. The fact that there were some wood thrushes singing tells me it's actually quite a big patch of forest I'm walking past, because wood thrushes don't do very well in small patches. I heard barred owls calling the other day. That's a good sign, because to have barred owls you have to have prey around. On and on and on. And I can do that essentially anywhere in the world. I can go out with my ears and my binoculars, and I can come up with a list of species that can tell me a lot about that environment. And I think that's sort of a unique feature of birdwatchers that's hard to replicate within the other taxa.

VB: It seems to me that your passion is giving you sensory superpowers! It's great that you can listen and look, and then extrapolate so much information from what's around you.

SP: I hadn't sort of put it in the superpower context!

VB: Super vision … I think most people don't notice this richness of nature that is often present in our everyday surroundings.

SP: As I walk out to get my cup of coffee in the morning, I see people with their headsets on, and I think, 'You dummies, you're missing everything!' And for me, I want to be hearing it. Is that a species I haven't heard before? What's happened to my great crested flycatcher? You know, there are all sorts of things going on that you can be alert about. As a birdwatcher, I can have experiences of this kind anywhere in the world.

NC: I think you're telling us what a scientist can bring to the experience of birdwatching, in that you can hear the wood thrush, and upon hearing it, based on your knowledge of its habitat, you can estimate the density of wood thrushes in a particular area. That sounds like a scientist at work as they watch and listen to birds. Is there anything that goes the other direction? Is there a way that the birdwatching informs the scientist? I don't watch birds myself, so I don't know what kind of qualities are needed for it. In my ornithological imaginings, I envisage qualities like patience, or perseverance, or having a particular eye for detail to be able to identify species as being important?

SP: I'm not sure I have any of those qualities. I'm not sure I'm particularly patient, and I'm not sure I have got a great eye for detail, but what I definitely do have, formed over sixty years of being a birdwatcher, is a passion for birds. And a fascination for them. All the people that I think of as being my peers share this fascination … Paul Ehrlich is an enthusiastic birdwatcher. For much of his life he was an enthusiastic butterfly watcher. Peter Raven is enthusiastic about plants. Tom Lovejoy is enthusiastic about birds. Jared Diamond is enthusiastic about birds. Pat Wright is enthusiastic about lemurs. I've been in the field with all of those people, and being in the field with them is fantastic because you're constantly seeing nature through their eyes. I learn a huge amount of stuff with my elder daughter. It's a very different experience looking at plants. You go out there and you're seeing stuff continuously, and I think that absolutely shapes the science that I do. It's those experiences in the field that lead to the scientific questions. I became a conservation biologist asking the obvious question: why do some species go extinct and others not? Those first conservation questions that I asked came directly from field observations, or the lack of them.

VB: There's clearly continuity across birdwatching and your field research in terms of that first-hand encounter with nature, and the questioning and the quest for answers it can provoke. Staying with birdwatching, you talked earlier about your formative experiences in Hawai'i, about expecting to see some species, striving to see them, but then never doing so. Some birds you failed to see you now believe to be extinct. With that in mind, are you left feeling melancholy at the current state of affairs regarding the conservation of biodiversity?

SP: People tend to phone me up to talk because they know that when Al Gore says 'Species are going extinct a thousand times faster than they should', that he got that from me. What drives what I do is an enormous sense of optimism. I believe that we can prevent species from going extinct, that we can solve environmental problems. Sometimes journalists say, 'How on earth do you get up in the morning when you're the purveyor of bad environmental news?' I get up in the morning because I want to effect solutions. And so I think the deep sense of loss I have when a species goes extinct is very much weighed against the deep sense of accomplishment I get from the kinds of things we do at my non-profit, Saving Nature – things like rebuilding habitat corridors between isolated forest fragments to put nature back together again. And that image of restoration and of bringing species back, that's the powerful image that drives what I do every day.

Stuart Pimm is the Doris Duke chair of conservation ecology at Duke University. He is a world leader in the study of present-day extinctions, and what can be done to prevent them. Pimm received his BSc from Oxford University (1971) and his PhD from New Mexico State University (1974). He is the author of 350 scientific papers and four books. He directs Saving Nature, a 501(c)(3) non-profit, which uses funds for carbon emission offsets to fund local conservation groups to restore degraded lands in areas of exceptional tropical biodiversity. His international honours include the International Cosmos Prize (2019), the Tyler Prize for Environmental Achievement (2010), and the Dr A.H. Heineken Prize for Environmental Sciences from the Royal Netherlands Academy of Arts and Sciences (2006).

Valérie Bienvenue is a doctoral candidate in the Department of History of Art and Film Studies at the Université de Montréal. Her thesis critically examines human–equine relations through the prism of modern art and visual culture. Prior to her academic career, she worked for ten years in equestrian circles, including teaching bareback riding and rehabilitating horses suffering from physical and psychological trauma. She is the author of several articles and book chapters.

Nicholas Chare is professor of art history in the Department of History of Art and Film Studies at the Université de Montréal. He is the author of *After Francis Bacon* (2012). In 2017, with Sébastien Lévesque and Silvestra Mariniello, he founded the baccalaureate (BACCAP) in visual cultures at the Université de Montréal.

Notes

1. This interview was conducted by telephone on 22 September 2019 and then by email. We are very grateful to David Sume for transcribing the initial phone conversation.
2. See, for example, Boulton et al., 'Endangered Cape Sable Seaside Sparrow'; Curnutt et al., 'Population Dynamics'; Pimm et al., *Sparrow in the Grass*.
3. The Endangered Species Act of 1973 (ESA) 'recognized that some fish, wildlife, and plants were endangered by economic growth and development' and thus 'provided for conserving endangered and threatened species and the ecosystems they depend upon'. See Yellowstone National Park, 'Wolves for Yellowstone?', 2. For a discussion of Furbish's lousewort in the context of the ESA, see Macior, 'The Furbish Lousewort'. For a general discussion of the plant and its rarity, see Fiedler and Ahouse, 'Hierarchies of Cause', 36–38.
4. For a copiously illustrated discussion of Dracula orchids in the western Andes, see Orejuela-Gartner, 'Orchids of the Cloud Forests'.

5. The gibbon (*Hoolock tianxing*) is described in Fei-Fan et al., 'Description of a New Species'. Fei-Fan et al. suggest the name 'Skywalker' – of which the name 'Tianxing' is a pinyin translation – derives from the 'unique locomotory mode of gibbons' (ibid.: 9). Mark Hamill, however, tweeted his pride at having the gibbon named after his Luke Skywalker character, and media reports suggested the name was inspired by the scientists being fans of *Star Wars*.

6. The high public profile the gibbon attracted upon its discovery is discussed in Fan and Bartlett, 'Overlooked Small Apes'.

7. See Pimm and Raven, 'The Fate of the World's Plants'. See also Pimm and Joppa, 'How Many Plant Species'. Pimm has recently revisited the issue of taxonomic comprehensiveness in 'What We Need to Know to Prevent a Mass Extinction of Plant Species'.

8. Lees and Pimm, 'Species, Extinct Before We Know Them?'

9. A noted example of such artworks, albeit one where an identification of the bird has been offered, is the Guadeloupe macaw (*Ara guadeloupensis*), which is portrayed in an engraving by Sébastien Le Clerc for Du Tetre, *Histoire générale des Antilles habitées*, in a plate that appears between pages 246 and 247 (the plate is missing from some digitized copies of the manuscript that appear online). There are also numerous textual references to this macaw. See Clark, 'Note on the Guadeloupe Macaw', 377; Wiley and Kirwan, 'The Extinct Macaws of the West Indies'.

10. Two specimens of the bird, the cryptic treehunter (*Chiclocolaptes mazarbarnetti*), had been wrongly identified as the Alagoas foliage-gleaner by the National Museum of Brazil. See Lee and Pimm, 'Species, Extinct Before We Know Them?', R-179.

11. A 1765 painting by L.J. Robins has variously been claimed to depict the Cuban macaw and the hypothetical extinct species the Jamaican red macaw. Samuel Turvey discusses another painting from 1765 of a red macaw, this time by John Lindsay, that is now in the collections of Bristol City Museum and Art Gallery. Turvey speculates that the macaw in the painting was likely an ornamental bird traded to the island. Turvey, 'A New Historical Record of Macaws'.

12. Beneath the parrot, Waldseemüller has written '*rubei itaci*', with the first few letters of *psitacci* elided. Charles Short and Charlton T. Lewis provide Pliny's *Natural History* as an early example of the use of the word *psitaccus*. See Short and Lewis, *A Latin Dictionary*, 1483. We are indebted to Kristine Tanton for sharing her insights regarding Waldseemüller's map. The macaw also occurs as a subject in modern art. Georgina Moura Andrade du Albuquerque, for instance, includes the bird in her undated oil on canvas work *Moças e Arara* [Young women and macaw]. We are grateful to Camila de Oliveira Savoi for drawing our attention to this painting. Manoel Santiago also includes a macaw in his oil painting *Tatuagem* [Tattoo] (1929), where he portrays a bare-breasted Indigenous woman lying languorous in a hammock, the bird perched on her upraised left hand. Here the macaw is seemingly employed as a primitivist stereotype, used to symbolize the woman's proximity to nature. *Tatuagem* is analysed at length in Neto, 'Manoel Santiago vai a Paris'.

13. The Highland mangabey (*Rungwecebus kipunji*), for example, was named in 2005 on the basis of a photograph, although genetic material was subsequently sourced in 2006 from a specimen killed by a farmer. See Davenport et al., 'A New Genus of African Monkey'. A similar situation occurred with the macaque *Macaca munzala*. See Sinha et al., '*Macaca munzala*'. In the context of birds, the holotype for *Liocichla bugunorum* was a bird that was photographed and from which some plumage was obtained before it was released; see Atheyra, 'A new species of *Liocichla*'. The flatworm *Leptoplana mediterranea* was described from drawings first published in 1884 and acknowledged as a species in 2012; see Krell and Marshall, 'New Species Described from Photographs'.

In botany, type specimens may occasionally be illustrations because of the difficulty of preserving some species of plant. In this context, some of the engravings by Pierre-Joseph Redouté of plants from the *Liliaceae* family serve as holotypes; see Daston, 'Type Specimens and Scientific Memory', 160.

14. Pimm et al., 'Can We Defy Nature's End?' For a fresh examination of biodiversity management, see Pimm, 'What Is Biodiversity Conservation?'
15. For a discussion of numbats and their rarity, see Friend and Thomas, 'Conservation of the Numbat'.
16. Michael Soulé was a conservation biologist and a co-founder of the Society for Conservation Biology. Soulé outlines his vision for conservation biology in 'What Is Conservation Biology?'
17. Pimm, 'The Dodo Went Extinct', 196.
18. For a discussion of how dictionaries embody values through 'the apparently neutral and non-partisan process of defining and arranging information', see Rifkin, 'Ingres and the Academic Dictionary', 265.
19. See, for example, Pimm et al., 'Bird Extinctions in the Central Pacific'. When first used (31), 'wimps' is in scare quotes to signal it is a problematic term. In the context of a discussion of recently extinct and endangered birds on Pacific islands, Michael Jeffries refers to 'wimp species' (which he places in scare quotes) as species 'vulnerable to human pressures' (Jeffries, *Biodiversity and Conservation*, 115).

Bibliography

Atheyra, Ramana. 'A New Species of *Liocichla* (Aves: Timaliidae) from Eaglenest Wildlife Sanctuary, Arunachel Pradesh, India'. *Indian Birds* 2(4) (2006): 82–94.

Boulton, Rebecca L., et al. 'Endangered Cape Sable Seaside Sparrow Survival'. *The Journal of Wildlife Management* 73(4) (2009): 530–37.

Clark, Austin. 'Note on the Guadeloupe Macaw (*Ara guadeloupensis*)'. *The Auk* 51(33) (1934): 377.

Curnutt, John L., et al. 'Population Dynamics of the Endangered Cape Sable Seaside Sparrow'. *Animal Conservation* 1(1) (1998): 11–21.

Daston, Lorraine. 'Type Specimens and Scientific Memory'. *Critical Inquiry* 31(1) (2004), 153–82.

Davenport, Tim R.B., et al. 'A New Genus of African Monkey, *Rungwecebus*: Morphology, Ecology, and Molecular Phylogenetics'. *Science* 312(5778) (2006): 1378–81.

Du Tetre, Jean-Baptiste. *Histoire générale des Antilles habitées par les François*. Paris: Thomas Iolly, 1667.

Fan, Pengfei, and Thad Q. Bartlett. 'Overlooked Small Apes Need More Attention!' *American Journal of Primatology* 79(6) (2017): 1–4.

Fei-Fan, Peng, et al. 'Description of a New Species of *Hoolock* Gibbon (Primates: Hylobatidae) Based on Integrative Taxonomy'. *American Journal of Primatology* 79(5) (2017): 1–15.

Fiedler, Peggy, and Jeremy Ahouse. 'Hierarchies of Cause: Towards an Understanding of Rarity in Vascular Plant Species', in Peggy Fiedler and Subodh Jain (eds), *Conservation Biology: The Theory and Practice of Nature Conservation, Preservation and Management* (Boston: Springer, 1992), 23–47.

Friend, J. Anthony, and Neil D. Thomas. 'Conservation of the Numbat (*Myrmecobius fasciatus*)', in Menna Jones, Chris Dickman and Mike Archer (eds), *Predators with Pouches:*

The Biology of Carnivorous Marsupials (Collingwood, VIC: CSIRO Publishing, 2003), 452–63.

Jeffries, Michael. *Biodiversity and Conservation*. London: Routledge, 1997.

Krell, Frank-Thorsten, and Stephen A. Marshall. 'New Species Described from Photographs: Yes? No? Sometimes? A Fierce Debate and a New Declaration of the ICZN'. *Insect Systematics and Diversity* 1(1) (2017): 3–19.

Lees, Alexander C., and Stuart L. Pimm. 'Species, Extinct Before We Know Them?' *Current Biology* 25(5) (2015): R177–R180.

Macior, Lazarus. 'The Furbish Lousewort: Weed, Weapon, or Wonder?' *The American Biology Teacher* 43(6) (1981): 323–26.

Neto, João Augusto Silva. 'Manoel Santiago vai a Paris'. *Faces do História* 5(2) (2018): 64–84.

Orejuela-Gartner, Jorge. 'Orchids of the Cloud Forests of Southwestern Colombia, and Opportunities for their Conservation'. *European Journal of Environmental Sciences* 2(1) (2012): 19–32.

Pimm, Stuart L. 'The Dodo Went Extinct (and Other Ecological Myths)'. *Annals of the Missouri Botanical Garden* 89(2) (2002): 190–98.

———. 'What Is Biodiversity Conservation?' *Ambio* 50 (2021): 976–980.

———. 'What We Need to Know to Prevent a Mass Extinction of Plant Species'. *Plants, People, Planet* 3(1) (2020): 7–15.

Pimm, Stuart L., and Lucas N. Joppa. 'How Many Plant Species Are There, Where Are They, and at What Rate Are They Going Extinct?' *Annals of the Missouri Botanical Garden* 100(3) (2015): 170–77.

Pimm, Stuart L., and Peter Raven. 'The Fate of the World's Plants'. *Trends in Ecology and Evolution* 32(5) (2017): 317–20.

Pimm, Stuart L., et al. 'Bird Extinctions in the Central Pacific'. *Philosophical Transactions: Biological Sciences* 344(1307) (1994): 27–33.

Pimm, Stuart L., et al. 'Can We Defy Nature's End?' *Science* 293(5538) (2001): 2207–8.

Pimm. Stuart L., et al. *Sparrow in the Grass: A Report on the First Ten Years of Research on the Cape Sable Seaside Sparrow (Ammodramus maritimus mirabilis)*. Homestead: Everglades National Park Service, 2002.

Rifkin, Adrian. 'Ingres and the Academic Dictionary: An Essay on Ideology and Stupefaction in the Social Formation of the "Artist"', in Steve Edwards (ed.), *Communards and Other Cultural Histories: Essays by Adrian Rifkin* (Chicago: Haymarket, 2018), 250–71.

Short, Charles, and Charlton T. Lewis. *A Latin Dictionary*. London: Oxford University Press, 1879.

Sinha, Anindya, et al. '*Macaca munzala*: A New Species from Western Arunachal Pradesh, Northeastern India'. *International Journal of Primatology* 26(4) (2005): 977–89.

Soulé, Michael E. 'What Is Conservation Biology?' *BioScience* 35(11) (1985): 727–34.

Turvey, Samuel. 'A New Historical Record of Macaws on Jamaica'. *Archives of Natural History* 37(2) (2010): 348–51.

Wiley, James, and Guy Kirwan. 'The Extinct Macaws of the West Indies, with Special Reference to the Cuban Macaw *Ara tricolor*'. *Bulletin of the British Ornithologist's Club* 133(2) (2013): 125–56.

Yellowstone National Park. 'Wolves for Yellowstone?: A Report to the United States Congress, Volume II Research and Analysis'. National Park Service, 1990.

Part II

Indigenous Peoples and Extinction

Chapter 4

The Beothuk, the Great Auk and the Newfoundland Wolf

Animal and Human Genocide in Canada's Easternmost Province

Nicholas Chare

———⊗⊗⊗———

Introduction: Interconnections

In 1937, the zoologists Glover M. Allen and Thomas Barbour published an article, 'The Newfoundland Wolf', that used the cranial measurements of wolf skulls held at the Museum of Comparative Zoology in Cambridge, Massachusetts, to argue that the wolves of Newfoundland formed a distinct subspecies. The skulls entered the museum collection in June 1865, provided by the trapper J.M. Nelson. He furnished the institution with two complete skeletons and two additional skulls. Their Newfoundland provenance is noted in the museum's acquisitions catalogue, but, as Allen and Barbour observe, '[n]o additional particulars are given'.[1] It is unknown where it was in Newfoundland that Nelson killed the wolves, or precisely when.

Much of Allen and Barbour's article is taken up with issues of nomenclature. The trinomial they propose for the island subspecies is *Canis lupus beothicus*, chosen to honour the 'the now extinct aborigines of Newfoundland, the Beothuks [*sic*]'.[2] Allen and Barbour also considered the Newfoundland Wolf extinct, remarking: 'At the present time the Newfoundland Wolf is probably quite gone'.[3] Additionally, they reflect on the paucity of information available about the animal: 'There is little recorded concerning the Newfoundland Wolf'.[4] Much of the historical information they are able to provide simply records killings of wolves, including accounts of slayings of individual wolves in 1894 and 1911. Both these reports of wolf killings have been identified as potentially referring to a wolf that William Whiteway shot.

In either 1894 or 1911, Whiteway killed a wolf on Gaff Topsail, a tor located north-east of Hind's Lake in Newfoundland's interior.[5] That the region including the topsails was a home to wolves is reinforced by the presence of a body of water called Wolf Pond to the south of Gaff Topsail.[6] The wolf Whiteway shot was subsequently skinned, with the skin fashioned into a rug that was probably used as a wall hanging. In 1952, Whiteway's rug was sold by his brother, Herbert, to the naturalist Leslie Tuck, who was acting on behalf of the Newfoundland Natural History Society.[7] In 1958, the skin was loaned to the National Museum in Ottawa. There, at the request of the society, it was transformed into a mount. The wolf was returned to Newfoundland in 1980. The mount is now exhibited alongside the wolf's skull at The Rooms museum in St John's, the provincial capital of Newfoundland and Labrador (Illustration 4.1).

The area where Whiteway killed the wolf includes varied terrain; the topsails are an open, barren and rocky environment, whereas the vicinity of Hind's Lake is forested. Hind's Lake, which seems to take its name from the numerous caribou in the area, forms one of several lakes in close

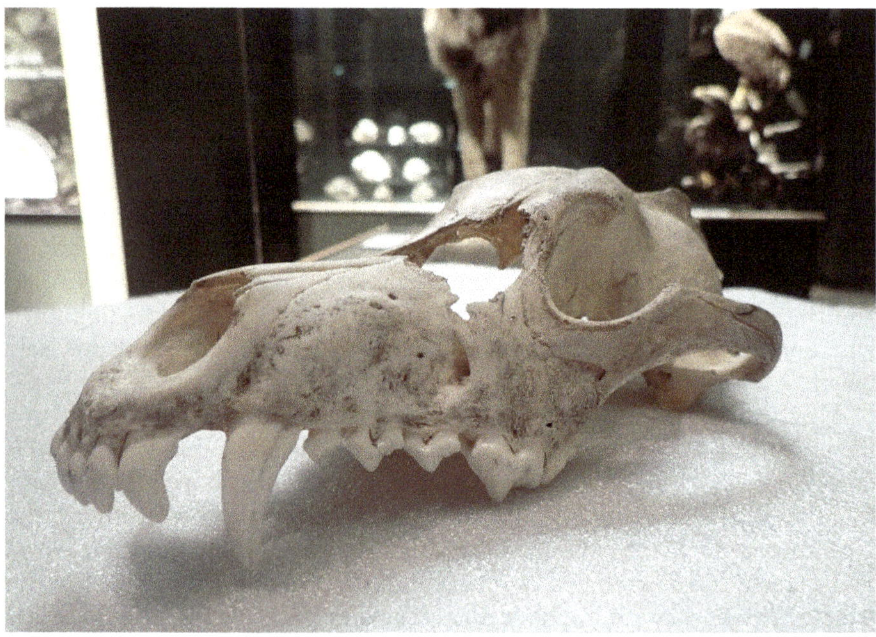

Illustration 4.1 Newfoundland Wolf (*Canis lupus beothucus*), skull from the 'Whiteway' wolf skin. Newfoundland, Gaff Topsails, 1894. NFM MA-8. The Rooms Corporation of Newfoundland and Labrador, Canada. © Nicholas Chare.

proximity. These include Buchans Lake and Red Indian Lake. Buchans Lake and a river that links it to Red Indian Lake are both named after Lieutenant David Buchan, a naval officer who made several expeditions to Newfoundland's interior in the early nineteenth century.[8] Red Indian Lake derives its toponym from a Beothuk community that used to winter on its shores. The Beothuk, a First Nations people, smeared their bodies (and their possessions) with red ochre, a practice that led some European settler colonists to refer to them as Red Indians.[9] The lake undoubtedly had a Beothuk name, but what that was is unknown. A small stream that runs into the lake is named Shanawdithit Brook.[10] In his *Geography of Newfoundland*, James Howley explains that the brook takes its name from 'Shawnadithit [*sic*], [a] Red Indian Woman [who] lived several years in St John's, and became quite civilized'.[11]

Shanawdithit, who died in 1829, was claimed by some to be the last of the Beothuk. She was captured by furriers at Badger Bay on the east coast in April 1823. Red Indian Lake was part of her community's territory. Nowadays, Shanawdithit is well known for the drawings she produced while held captive. Many were made with the encouragement of William Eppes (also Epps) Cormack, an entrepreneur and explorer who took considerable interest in the Beothuk and intended to publish a history about them. Several drawings feature the lake, and one portrays Badger Bay and its environs. The majority of Shanawdithit's extant works are now housed at The Rooms.[12] Reproductions of details from some of these are shown on panels in the main exhibition space, close to the alcove where the Newfoundland wolf is displayed. The upper section of a drawing titled by the museum as 'The Taking of Demasduit' but by Cormack as 'The Taking of Mary March' (Mary March was the European name given to Shanawdithit's aunt, Demasduit), appears as part of an information panel titled 'Last of the Beothuk'.

Reproductions of some of the drawings are also accessible in drawers that form part of a display dedicated to Shanawdithit. One of these represents David Buchan carrying the coffin containing Demasduit back to Red Indian Lake. In his description of this work in *The Beothucks or Red Indians*, James Howley observes that a 'very interesting new feature on this sketch is a black dotted line, on the same side reaching a long way up the lake to a cove which would seem to represent the mouth of Shanawdithit Brook'.[13] Here Shanawdithit is credited with noting a geographical feature which certainly did not possess her name while she was alive. If the brook possessed a Beothuk name, she would have known it and it was that which she drew. Through identifying the name of the feature as hers, Howley nominally takes her land from her.

Howley's prejudicial descriptions can be read as being on a continuum with Cormack's practice of overwriting Shanawdithit's drawings with partial explanations. Fiona Polack, who provides a pioneering close analysis of the drawings, reads these glosses as sensitive additions on Cormack's part, carefully positioned so as not to impinge on Shanawdithit's pictures.[14] She views him as often operating in complicity with the Beothuk artist, affirming her worldview. In 1824, Cormack published a report of his 1822 expedition to Newfoundland's interior of the island in the *Edinburgh Philosophical Journal*. A foldout map in the journal includes a rudimentary rendering of 'Red Indian's [*sic*] Lake'. Writing of the 'Red Indians', Cormack notes that they 'are not numerous' and that he 'discovered no traces of them'.[15] In the same issue of the journal, the Revered John Fleming published his 'Gleanings of Natural History, during a Voyage along the Coast of Scotland in 1821', in which he recalls observing a great auk (*Alca impennis*) that had been captured off St Kilda, an archipelago on the western edge of Scotland.[16] The great auk was one of the seabirds that sustained the island economy, their eggs and flesh being an important source of food until the nineteenth century, by which time the breeding colony had disappeared. Solitary auks were still occasionally encountered in St Kilda in the early nineteenth century. The last recorded great auk in the area was caught circa 1840 on the island of Stac-an-Armin and killed shortly afterwards for fear the bird was bewitched.[17]

On the other side of the Atlantic, the Beothuk were also historically known to consume eggs of the great auk. They gathered these from Funk Island, a small isle about 60 km from the Newfoundland mainland, and other locations.[18] As Ingvar Svanberg notes, the auk was 'harvested by … the Beothuks [*sic*] without any effect on the colony'.[19] Arrowheads and a paddle were found on the island at a spot now named Indian gulch.[20] The eggs were probably used in soups and sausages.[21] Joseph Banks mentions a pudding made of 'Eggs & Dears hair' that was baked in the sun.[22] One of the drawings that Shanawdithit made while staying with Cormack, known as *Sketch VII*, depicts different cooking and storage utensils and foodstuffs (Illustration 4.2). Some of the inked explanations for various objects were initially penned by Cormack with the drawing upside down and were subsequently crossed out. Cormack was clearly unable to orient himself with ease in the Beothuk world. A sack made of sealskin and a seal stomach stuffed with intestines are portrayed criss-crossed by sanguine (the pigment probably derives from red ochre). Cormack has traced over the sack's pencil outlines with ink, simultaneously accentuating and obscuring Shanawdithit's drawing.[23] There are also vessels made of birch rind used in the cooking and drying of eggs.

By the time Shanawdithit made her sketch, the great auk colony on Funk Island had been destroyed by rapacious Europeans seeking the bird's feathers and the Beothuk no longer occupied eastern coastal areas. Their

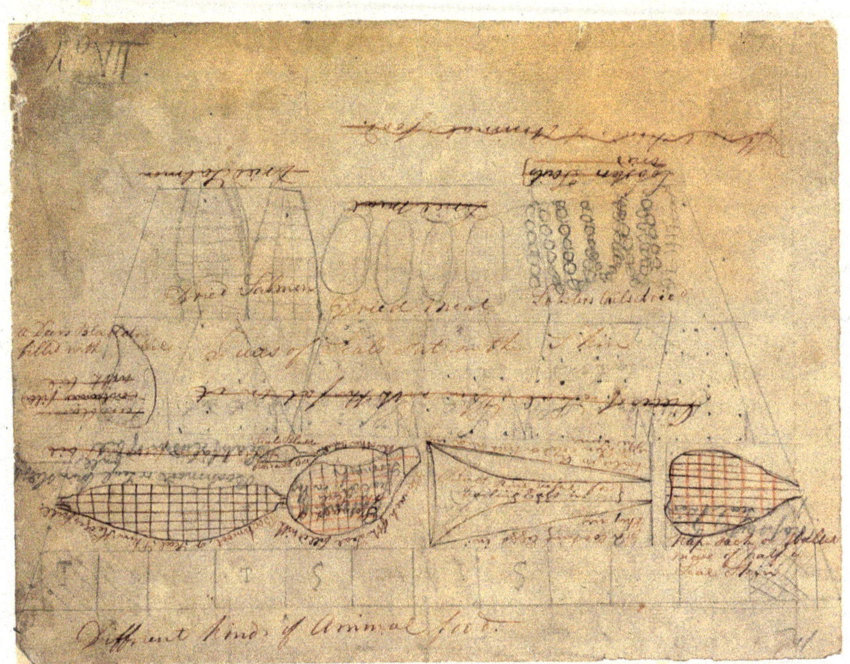

Illustration 4.2 Shanawdithit, *Sketch VII*: 'Different Kinds of Animal Food', 1828. VIIIA-561. The Rooms Corporation of Newfoundland and Labrador, Canada.

last known effort to visit Funk Island was on 30 July 1792, roughly a decade before Shanawdithit's birth. That attempt was abandoned when the Beothuk were shot at by John McDonald, who was heading to the island to gather eggs for himself.[24] William Montevecchi and Leslie Tuck note that Shanawdithit furnished Beothuk names for many of the bird species of Newfoundland but not for the great auk; cut off from the sea and access to breeding colonies, the auk was, perhaps, no longer part of the Beothuk world by the time Shanawdithit was abducted. Montevecchi and Tuck suggest that '[t]he Great Auks, like the Beothuks, were abused by some early settlers, and the birds and natives became extinct at about the same time'.[25] The Rooms displays a great auk skeleton a short distance from the Beothuk exhibits, victims of what I will suggest are animal and human genocide therefore appearing almost side by side. The auk skeleton is assembled from bones collected from Funk Island, potentially from birds bludgeoned to death for their feathers. Its label makes no reference to the recorded atrocities perpetrated at Funk. Similarly, the label for the Newfoundland wolf avoids mentioning the bounty scheme that was instituted against the animal.[26]

The great auk and the Newfoundland wolf are now extinct and although people with Beothuk lineage are alive today the genocidal violence perpetrated against the people clearly had a devastating effect on their culture. In this chapter, I will examine these three instances of genocide in their singularity and consider the role played by negative representations in enabling such violence.

Genocide against the Beothuk in Newfoundland

The Beothuk were continually subject to random acts of violence like the 1792 attack. Robert Jameson's footnote to Cormack's account of his voyage to the interior recalls that Lord Bathurst had told him of the Beothuk that 'there was reason to believe that our people had put them to death without sufficient provocation'.[27] Place names that were once in use in Newfoundland, such as Bloody Bay and Bloody Point, registered massacres of the Beothuk by Europeans.[28] The massacre at Bloody Reach, a stretch of land forming part of the inner reaches of Bonavista Bay, took place in about 1800 and reportedly involved 'three or four hundred Beothuks [being] driven onto a long point of land near their favourite sealing site and being shot down like deer'.[29] The comparison of the killing with that of non-human animals is not isolated. Arthur Grenke, for instance, suggests the colonists who usurped Beothuk lands, like those who violently annexed Yana territory in mid-nineteenth-century California, 'slaughtered Natives as if they were animals'.[30] Like the Yana, the Beothuk were gradually forced from their traditional fishing grounds and thus denied access to the salmon that was an important part of their diet. Writing in 1770, George Cartwright links his fear of the Beothuk becoming 'totally extinct in a few years' with their lack of access to salmon and the pillaging by colonists of bird eggs from surrounding islands.[31] Mass killings and individual murders, coupled with diminishing territory and access to essential resources, contributed to the Beothuk's decline and the disappearance of their culture.

Adam Jones lists the Beothuk among his examples of 'Genocides of Indigenous Peoples', suggesting that they were 'hounded to complete extinction'.[32] The extinction of an entire people is unnecessary for there to be genocide.[33] As the Shoah, which is often taken as paradigmatic, demonstrates, the intent to destroy coupled with the substantial destruction of a given group is sufficient for genocide to occur. Genocide is often defined as state-sponsored violence designed to exterminate a group based on their ethnicity or religion. As Jones notes, however, in settler colonial contexts 'non-state actors may play a dominant role'.[34] He recognizes that genocidal violence towards Indigenous peoples is frequently characterized by 'a large number of relatively small massacres [that are] not necessarily centrally directed [and] generally separated from each other spatially and temporally'.[35] In this context, there are some parallels between, for example, the genocide perpetrated against Indigenous groups in lutruwita/Tasmania and that perpetrated in Newfoundland.

Both Newfoundland and lutruwita had regional governments which paid lip service to safeguarding the lives of their Indigenous populations yet failed to intervene in any meaningful way to prevent wholesale violence against them. In his thoughtful analysis of genocide in lutruwita, Tom Lawson brings to the fore contradictory messages transmitted by the island's regional government, which at once called for kindness or forbearance towards Indigenous people and sanctioned the use of force against them in order to drive them from seized land.[36]

Drawing on Colin Tatz's research into how some Australian genocides occurred unsupported (materially speaking) by government and state, Lawson signals how genocide in lutruwita was, at least initially, perpetrated by individual settlers. The situation in Newfoundland has similarities to 'private genocide' of the kind Tatz identifies. Tatz foregrounds the role of what he terms 'private settlers' in the perpetration of Australian genocides.[37] These actions were uncoordinated by the colonial government (although they may indirectly have benefitted the government).[38] For Tatz, colonial authorities were initially complicit as bystanders because they failed to intervene when massacres were taking place and then took a more active role in punitive expeditions. Lawson outlines just such a situation in lutruwita, where the regional government outwardly decried settler violence against Indigenous groups while benefiting from its effects (through extended territorial control) before a shift in rhetoric and the mobilization of troops provided the framework through which the 'exterminatory desires' of individual settlers could be channelled.[39]

In Newfoundland, the situation was different. In the eighteenth century, several proclamations were issued calling for an end to violence between settlers and Indigenous groups. Governor John Byron proclaimed on 8 July 1769 that those guilty of murdering 'native Indians' should be sent for trial in England for their capital crimes. On 15 July 1772, Governor Molyneaux Shuldham sent an order against the 'barbarous murders committed on the natives' to Robert Carter, a justice of the peace in Ferryland on the Avalon Peninsula.[40] Governor John Campbell issued a proclamation on 14 September 1785 condemning the murders of Beothuk in the north of Newfoundland and calling on civil and military officers to bring those responsible to justice. On 16 October 1786, the succeeding governor, Rear Admiral John Elliott (also Elliot), issued another proclamation that berated settlers for having behaved inhumanely to the 'Native Indians' (the

Beothuk) and for killing them. Echoing Byron nearly twenty years previously, Elliott stated that anyone caught committing such crimes would be sent to England for trial. Shortly after his arrival in St John's, Governor William Waldegrave issued another proclamation.

Waldegrave subsequently wrote a letter to the Duke of Portland on 25 October 1797 noting the proclamation and inviting the duke to judge 'whether some serious steps ought not to be undertaken, in order to save from destruction the sad remains of this unhappy persecuted race of people'.[41] He clearly knew proclamations did little to deter violence against the Beothuk.[42] Their ineffectiveness was also recognized by the magistrate John Bland. In a letter to John Rance dated 16 August 1797, which appears motivated by Waldegrave's proclamation, Bland mentioned government efforts to legislate against violence towards the Indigenous population in terms that reveal he felt it was no more than empty rhetoric:

> Proclamations unless accompanied with some strong measure it is to be apprehended will operate to no salutary purpose upon a class of men who regard the Indian as fair game, and who destroy him with no more remorse than they shoot a deer. A repugnance to touch the life of a fellow creature would hardly appear to be an original principle in our northern hunters. . . . It is to be feared . . . that without the interference and aid of Government, no plan that humanity or local knowledge can devise will be likely to succeed.[43]

From correspondence such as this, it is clear that there were government calls for restraint yet these were ineffectual and their inadequacy was well known. There were, however, no military campaigns mounted against the Beothuk, despite the low-level resistance they engaged in towards their colonial oppressors.[44] They were not actively pursued by the state but instead abandoned to their fate at the hands of individual settlers. Cormack blamed the regional administration for the fate of the Beothuk, writing to Bishop John Inglis that '[i]t is a melancholy reflection that our Local Government has been such that under it the extirpation of a whole Tribe of primitive [*sic*] fellow creatures has taken place'.[45]

In a letter dated 1 September 1797, John Bland wrote at length to the governor, William Waldegrave, concerning the Beothuk. Towards the end of the missive, he opines:

> It ought to be remembered that these savages have a natural right to this island, and every invasion of a natural right is a violation of the principles of justice. They have been progressively driven from South to North, and though their removal had been produced by a slow and silent operation [it has had] all the effect of violent compulsion. In proportion as their means of procuring subsistence became narrowed their population must

necessarily have decreased and before the lapse of another century, the English nation, like the Spanish may have affixed to its character the indelible reproach of having extirpated a whole race of people![46]

Here, Bland clearly recognizes that denial of resources was tantamount to murder. In piecemeal fashion, European settlers annexed land and undermined the Beothuk means of survival. Genocide therefore involved both low-level killings of individuals and small groups, and the slow yet steady theft of Beothuk land and its accompanying resources. Both these actions were recognized as contributing to the group's destruction. Forced to retreat to Newfoundland's interior, the Beothuk lost access to valuable coastal resources, including, as already mentioned, Funk Island.

One of Shanawdithit's drawings, *Sketch V*, gives graphic expression to the regular atrocities being perpetrated against the Beothuk (Illustration 4.3).[47] It depicts the murder of a Beothuk woman by a party of settler colonists, including John Peyton Senior. Cormack has written on the drawing in ink: 'Showing that the murder of them was going on in 1816c'. If

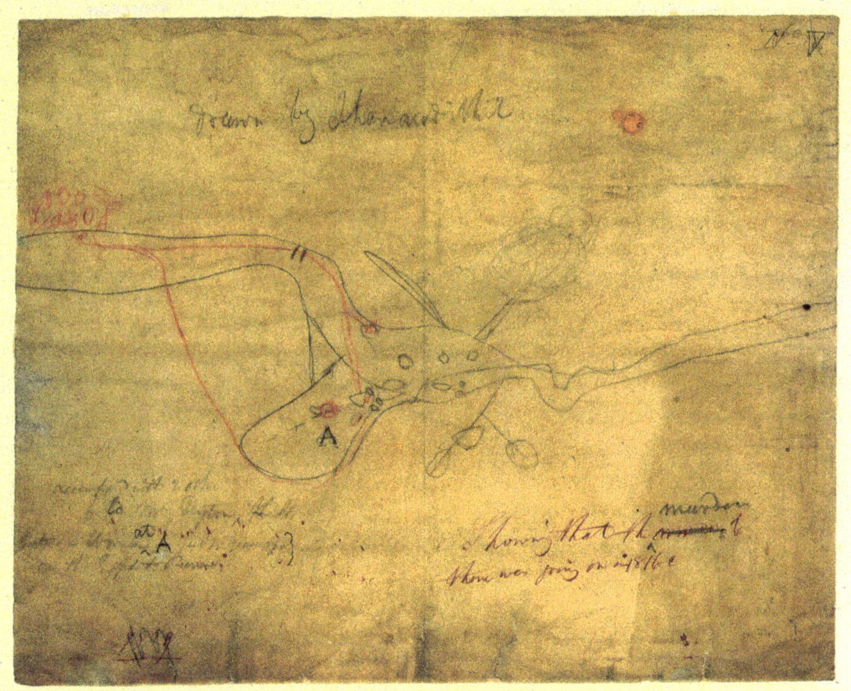

Illustration 4.3 Shanawdithit, *Sketch V*, 'Killing of a Beothuk Woman at the Exploits River. Showing That the Murder of Them Was Going on in 1816c', 1829. VIIIA-559. The Rooms Corporation of Newfoundland and Labrador, Canada.

Cormack's numbering of the series of drawings is taken to be accurate, this is the last of the overtly historical drawings produced by Shanawdithit. It is dated in ink '1829', but this has been crossed out in pencil. In it Shanawdithit attests to an event that Cormack would probably have known nothing about if she had not volunteered the account. This makes the work different to the first three historical drawings, which show what were well-known encounters between the Beothuk and Europeans (and were therefore feasibly made at Cormack's prompting), and to the fourth drawing, which is a kind of graphic population census. *Sketch V*, which carefully records physical features of this stretch of the Exploits River, identifies where the woman was murdered. A pencilled letter 'A' marks the spot.[48] In crime scene sketches, a genre of drawing that emerged later in the nineteenth century, letters would often be used to indicate the place were a murder victim was discovered.[49] Shanawdithit's drawing renders the murdered woman as a red circle scored by horizontal lines. As red was a spiritually significant colour for the Beothuk, the use of sanguine embodies continuing devotion to Beothuk beliefs while also potentially connoting bloodshed. Other red lines trace the lines of flight of fellow Beothuk, probably including Shanawdithit herself.

The drawing is necessarily schematized to provide clarity regarding the events that transpired, the movements and actions of those concerned. Topography is nevertheless carefully recorded. On one level, this offers Shanawdithit a means by which to return virtually to her homeland: drawing the riverbank involves a kind of revisiting of the place. On another, the precision also renders this compelling evidence of a crime, providing important locational specifics. It is clear Shanawdithit also named the murderer, with Cormack noting 'Old Mr Peyton' killed the woman.[50] As John Peyton Senior may still have been alive when the visual testimony was produced (he died sometime in 1829), the sketch can be viewed as a holding to account. More than this, however, the drawing provides a visible affirmation of survival. *Sketch V* is a survivor account of genocide. As her hand traces the woman's tragic death, Shanawdithit simultaneously signals the continuing existence of the Beothuk. Although identified as a 'doomed race', Shanawdithit's drawings as a whole provide a living history of the Beothuk and a reassertion of their culture.[51]

Animal Genocide in Newfoundland

John Maunder initially attributed the disappearance of the Newfoundland wolf to a bounty scheme instituted by the colonial government. He therefore saw a direct correlation between the targeted killing of specific animals for their membership of a group and their extinction. Maunder does not refer to the bounty scheme as animal genocide but others such as Barry Lopez have described the broader treatment of Turtle Island/North America's lupine population in terms that imply genocide. Lopez, for instance, titles a chapter of *Of Wolves and Men* that discusses wolf bounty schemes 'An American Pogrom'. In the chapter, he refers to the poisoning of wolves in the period 1875 to 1895 as having generated a holocaust.[52] The loaded terms 'holocaust' and 'pogrom' clearly invite comparison between the eradication of wolves and the destruction of Europe's Jews by the Nazis. Without saying as much, Lopez is strongly implying that what occurred to Turtle Island/North America's wolves (including, by implication, those of Newfoundland) was genocide. Other authors, such as Lynn Jacobs, have directly referred to the campaigns of extermination waged against Turtle Island/North America's wolves as genocide.[53] The persecution of the American bison (*Bison bison*) has been referred to in similar terms. Tasha Hubbard, whose work I will return to, makes the argument that 'the slaughter of the buffalo constitutes an act of genocide'.[54]

Maunder, however, revised his opinion about what caused the extinction of Newfoundland's wolves, blaming their disappearance predominantly on a severe decline in caribou numbers between 1915 and 1925.[55] He believes that the relatively small number of bounty claims discount it as a key factor in rendering the wolf extinct. He does not factor in the large number of wolves potentially intentionally killed yet with no bounty being claimed. If poison, for example, was employed (a common practice) then the corpses of dead wolves may never have been recovered. That there was historically regular use of poisons such as strychnine in Newfoundland is strongly suggested by repeated efforts in the 1880s to introduce legislation designed to control their sale.[56] This legislation was not motivated by any desire to protect wolves. Its aim was to safeguard the fur trade and animals of the chase. The effects of wanton use of substances such as strychnine was nonetheless clearly recognized. Dr Crowdy, a member of the legislative council, spoke in support of the bill. A summary of his speech records that

> apart from the danger of indiscriminate and unintelligent use of poisonous drugs, and apart also from the inducement of obtaining by such powerful weapons of offence the fur of wild animals, [Crowdy] was doubtful whether on economic grounds we have a right to encourage the killing of animals by such painful means. Strychnine is an exceedingly bitter and disagreeable substance, and animals often taking it in insufficient quantity do not die immediately but take to the water or their coverts, dying a most painful death, the result being that a large proportion of them are heartlessly destroyed and the fur is lost.[57]

It is probable that wolves were targeted in multiple ways, including shooting and poisoning. Even if the extinction of the wolf cannot be attributed with certainty to the campaign of annihilation waged against it, it still provides a clear example of animal genocide. As previously discussed, genocides do not have to cause the total destruction of a given group to be so named.

The great auk was not subject to systematic efforts aimed at the bird's destruction. Auks were instead 'overharvested', wantonly slaughtered. In his 1794 journal, Aaron Thomas provides a graphic account of this slaughter. He describes men plucking the best feathers from living birds and then releasing them: 'with [their] Skin half naked and torn off, to perish at [their] *leasure*'.[58] Thomas also states that auks were burned alive as fuel.[59] The mass killings of the auk that were carried out without knowingly endangering the species cannot be categorized as genocidal. It is clear, however, that a point was reached when it was recognized that such killings were compromising the bird's future existence. At this moment, which precedes Thomas's account, mass murder morphed into genocide. Writing in his journal in July 1785, George Cartwright noted that the auk colony at Funk Island was imperilled:

> [I]t has been customary of late years for several crews of men to live all the summer on that island, for the sole purpose of killing birds for the sake of their feathers; the destruction which they have made is incredible. If a stop is not soon put to that practice, the whole breed will be diminished to almost nothing, particularly the penguins; for this is now the only island they have left to breed upon; all others lying so near the shores of Newfoundland, they are continually robbed.[60]

In response to concerns such as those voiced by Cartwright, efforts were made to limit egg harvesting on the island. The secretary Aaron Graham issued a proclamation on behalf of Governor John Campbell on 28 October 1784 forbidding the destruction of birds on Funk Island solely for the procurement of their feathers.[61] Birds could only be killed with written permission from a magistrate. The next governor, John Elliott, continued Campbell's efforts. On 4 August 1786, he authorized the high sheriff, Nicholas Lechmere, to seize parcels of feathers and barrels of birds sourced from Funk. The next day he also authorized the justices of the peace of Harbour Grace and Trinity to use whatever means necessary to identify those responsible for killing birds on Funk for their feathers.[62] On 20 August, the governor communicated with the justices of the peace of St John's to advise them that a number of men had been caught, by Captain Pellew of HMS Winchelsea, destroying birds on Funk. The feathers of the birds they had slaughtered were seized and the men were being brought to St John's to face justice.

On 4 September 1792, Governor Richard King issued another broader proclamation condemning the killing of birds on the shores of Newfoundland or the islands adjacent to it (such as Funk) purely for their feathers. King recognized the value of seabirds as fishing bait and as food. He also suggested that 'they are useful in Fogs by warning vessels that they are near Land'.[63] Six men were found guilty of defying King's proclamation and were sentenced to a public whipping.[64] Their names and the details of their punishment were also posted across the districts of the island. Notwithstanding this instance of punishment, killing for feathers clearly continued. In 1810, Governor John Thomas Duckworth issued another proclamation, which noted the failure of the preceding ones to act as a deterrent against 'persons who are known to frequent the Penguin Islands for the express purpose of destroying them, and often merely on account of their feathers'.[65] Duckworth outlawed the killing of the penguins (great auks) and the taking of their eggs for any purpose – a sign of their great rarity.

Atrocities perpetrated against the great auks of Funk Island have been disavowed or downplayed in recent scholarship. Errol Fuller notes that the veracity of accounts such as that of Thomas are questioned nowadays.[66] Montevecchi and Tuck, for example, discount Thomas's testimony as rampant exaggeration.[67] Why they come to this conclusion is unclear, given that humans have historically behaved in similarly barbaric ways to each other.[68] The persecution of the Beothuk by John Peyton Senior, for instance, led John Bland to write that '[t]he stories told of this man would shock humanity to relate and for the sake of humanity it is to be wished are not true'.[69] Bland's fear that if the stories were reliable they would be a stain on humanity, explains his reluctance to credit them.

Suffering and Genocide

Genocide is far less frequently used as a term to describe the intent to exterminate groups of animals, be they species, subspecies or local populations. The best-known employment of the term 'animal genocide' is probably Jacques Derrida's in *L'animal que donc je suis* [The animal that therefore I am/The animal that I therefore follow]. The idea is also discussed in an earlier work involving Derrida, *De quoi demain. . . .* This text forms a series of conversations with Élisabeth Roudinesco, one of which is about violence against animals.[70] Derrida links slaughterhouses with genocide but advises Roudinesco that he is hesitant to use the term because of its inevitable associations with the Holocaust. He recalls being told indignantly by a member of an American Jewish university audience that he knew what genocide was and it was not, by implication, the killing of animals. Derrida therefore proffers the word only to take it back (*'retirons le mot'*).[71] Yet after withdrawing the word, he once more reaffirms its validity: 'But you see what I am talking about'.[72] As a Sephardic Jew, Derrida was painfully aware of the Shoah and its legacy. His refusal to abandon genocide as a suitable term to refer to aspects of animal cruelty therefore merits serious reflection.

By the time *L'animal que donc je suis* was published, Derrida was clearly no longer undecided about the term. He makes unambiguous reference to animal genocides, noting that 'the number of species on their way to disappearing because of Man takes the breath away'.[73] He then indicates what animal genocide means to him: '[T]he annihilation of species is certainly at work but by way of the installation and exploitation of a diabolical, almost interminable, artificial survival, they are eradicated through their survival and their overcrowding even, enduring conditions that people of the past would have judged as monstrous and outside all assumed norms of fair life'.[74] This conception of genocide, destruction through a lifeless life, is not one generally associated with the term. Derrida, however, asks his audience to imagine the Nazis artificially inseminating and overbreeding homosexuals, Jews, Roma and Sinti before then subjecting them to the same forced experimentation and modes of murder they actually endured.[75] In this understanding, animals suffer the same fate as occurred in the human genocide that was the Holocaust, only by a more circuitous route. There is terminological slippage here, however, as genocide as it is usually interpreted aims at the total destruction of a group rather than a perpetual preservation accompanied by regular mass killing. Mass killing might, in fact, be a better term for the situation Derrida describes.

For Derrida, the human and the animal cannot readily be divided, to do so is fatal.[76] In this sense, when he refers to 'animal genocide' he is not speaking of a category of genocide distinct from human genocide. To accept the animal in the human is to accept that genocide is always already implicated with the question of the animal. Coming from a different perspective, Hubbard also challenges the human–animal divide associated with Cartesian thought. Hubbard emphasizes that in an Indigenous paradigm the concept of people can include 'other-than-human animals'.[77] She notes that 'animals-as-people' are found throughout Indigenous epistemologies. Bison were viewed as people by some First Nations peoples. Hubbard argues that as 'humans do not hold exclusive title to personhood', they also do not hold exclusive title to genocide. Although she predominantly engages with the fate of the buffalo or bison, Hubbard also suggests that wolves provide another example of genocide.[78]

Neither Derrida nor Hubbard make reference to what could be termed the canonical literature on genocide. This is understandable, given that the standard literature is written from a largely Euro-Western and anthropocentric perspective and, although highly varied in outlook, views genocide as a solely human issue. Derrida and Hubbard nonetheless seem to share Raphael Lemkin's belief that genocide is characterized by group persecution. Lemkin suggests that '[g]enocide is merely a comprehensive term for the most violent manifestation of intergroup conflict'.[79] Both humans and animals are sometimes conceived as groups. In the life sciences, animal groupings are primarily based on morphology. Non-scientific grouping of animals often attends to form in a more restricted sense, focusing on external physical appearance, grouping individuals that look the same. Animals are also sometimes described as living in 'social groups'. Shared cultural characteristics have been attributed to some such social groupings in a move that potentially accords non-human animals something like ethnicity.[80] Although non-human animals may not themselves possess a concept equivalent to that of the group (a recognition that they are part of the assemblage of more than one related entity, or a recognition that they are perceiving such an assemblage), it is clear that humans regularly impose groupings on them; and animal genocide, the singling out of a specific group of animals for extermination, is bound up with such acts.

Despite both humans and non-human animals being grouped, there is a clear reticence to see these groupings as of the same order. This reticence may be linked to fears that extending genocide to animals somehow diminishes the horrors of human genocide. Such a perspective is anthropocentric but understandable. Yet even from within an anthropocentric perspective, there should be space for animal genocide to exist alongside human genocide. Human genocides are usually studied in ways that respect their singularity. As James Hatley observes in the context of a discussion of the violence of genocide, each genocide commands a 'unique responsibility' and 'makes a particular claim for our attention that is incomparable with all other claims'.[81] Hatley goes on to suggest that in confronting an act of genocide, 'our first duty is not to classify and compare but simply to respond'.[82] This response involves, first and foremost, attentiveness to another's suffering.[83] Acknowledging animal genocide is to recognize and respond to an event of suffering, to instances of pain, terror and death. For Derrida, the fact that animals suffer cannot be doubted. The resultant compassion that such suffering generates can be disavowed but never done away with.[84] This suffering does not surpass or displace human suffering. As Matthew Calarco notes during a discussion of Derrida's reflections regarding animal and human genocide, 'the very difficult task for thought . . . is to bear the burden of thinking through both kinds of suffering in their respective singularity *and* to notice relevant similarities and parallel logics at work where they exist'.[85]

Conclusion: Images and Enmity

There do seem to be parallel logics at work in relation to the animal and human genocides perpetrated in Newfoundland. These logics relate to how the Beothuk and the Newfoundland wolf were represented within settler colonial culture. Both were the target of violence because of behaviours attributed to them. The Beothuk were viewed as robbers with 'their petty thefts … regarded by their invaders as crimes of the darkest dye, quite sufficient to justify the unsparing use of the strong arm for their extermination'.[86] Moses Harvey's description, while retrospective, gives insight into how settler colonists constructed a negative image of the Beothuk as evil and iniquitous and then used this as a means to justify violence against them. He states that the Beothuk were regarded as 'vermin'.[87] This descriptor, suggesting the Beothuk were akin to a nuisance animal, positions them as similar in settler colonist eyes to the Newfoundland wolf. The great auk, by contrast, was not labelled as troublesome although the species did suffer from deleterious depictions, often being judged 'clumsy' and 'stupid'. This perceived lack of grace and intellect may have helped cultivate indifference to the bird's fate, yet it would not have led to their active persecution.

Negative constructions of the wolf are common. In *Newfoundland in 1842*, for example, Richard Bonnycastle refers to the destructiveness of the wolves, writing in a footnote that the wolf 'grows very large, is frequently traced near the capital, and does much injury to cattle; a price is put by the legislature on its head'.[88] Much of the destruction attributed to wolves may have actually been caused by wild dogs. Problems with wild dogs were long-standing on the island. On 22 September 1784, for instance, the colonial secretary Aaron Graham issued a proclamation on behalf of Governor John Campbell that permitted the inhabitants of the north shore of Conception Bay on the south-east coast to kill any dogs that attacked cattle or destroyed fish.[89] Wolves were also blamed for the abduction of a child. In the winter of 1838, a wolf thought to be a child-killer was slain in the vicinity of St John's and then displayed: 'It was kept for some time at a house, and a charge of sixpence made from every person who went to view it'.[90] As the boy disappeared without trace, attributing his fate to wolves is conjecture; yet it reveals much about how the animals were viewed. Wolves were also frequently given human attributes such as a desire for retribution. Thomas writes that 'the Wolf is a revengefull [*sic*] Animal'.[91] The psychological characteristics accorded to wolves in the popular imaginary were primarily negative: they were vengeful, vicious and wanton.[92]

In his discussion of genocide and mass killing, Erwin Staub notes that 'authorities have great power to define reality and shape the people's perception of the victims'.[93] Often, however, the shaping of perception is more insidious, difficult to trace back to a specific group or organization. In Newfoundland, negative perceptions of the Beothuk and of wolves emerged in and through everyday discourse. There were campaigns against the Beothuk and the wolves by particular individuals and groups of stakeholders, yet a more diffuse negativity fuelled by rumour also operated and manifested itself culturally. Staub additionally emphasizes the role of culture in constructing 'shared explanations and images of the world, shared values and goals, a shared symbolic environment'.[94] He suggests that sources such as 'art and literature' contribute to creating the cultural characteristics that make group violence possible.[95] In the case of the Beothuk and the Newfoundland wolf, however, it was primarily through popular verbal representation, word of mouth, that negative images were developed and disseminated.

These negative images worked to devalue the Beothuk and the Newfoundland wolf in the eyes of settler colonists. They contributed to the ingroup–outgroup differentiation that Staub identifies as potentially preparative for the perpetration of violence. For Staub, '[d]evaluation makes mistreatment likely'.[96] The use of derogatory terms can create antagonism and help to lay the psychological groundwork for genocide. The linking of wolves with cattle and sheep losses indicates a process of scapegoating. Hardships become more readily comprehensible when a clear cause is identified. As Staub explains, '[f]inding a scapegoat makes people believe their problems can be predicted and controlled; and it eliminates one's own responsibility, thereby diminishing guilt and enhancing self-esteem'.[97] Scapegoating as a process of devaluation of an outgroup also enhances the value of the ingroup. Staub only considers violence in a human context, but his reference to the figure of the scapegoat already implies animals have historically formed a convenient outgroup.[98]

Processes of devaluation reduce humans and animals to the status of objects rather than beings 'with feelings and suffering like our own'.[99] Verbal representations contributed to this objectification, an objectification that facilitated genocide. In the case of the Newfoundland wolf, genocide contributed to extinction. For this reason, animal genocide should be considered a historical extinction driver. The two-volume supplement to *Grizmek's Animal Life Encyclopedia* that is dedicated to extinction lists various anthropogenic extinction drivers.[100] Genocide, however, is not one of them. In the case of the Newfoundland wolf and other animals such as the Falkland Islands wolf (*Dusicyon australis*) and the thylacine (*Thylacinus cynocephalus*) it should be. The genocide perpetrated against the wolves of Newfoundland was aided by negative representations of the species. Their fate therefore also draws attention to the prospective role of representation in animal extinctions and, potentially, in their prevention.

Acknowledgements

I am grateful to the Social Sciences and Humanities Research Council of Canada for funding the research for this chapter. I am also indebted to Nathalie Djan-Chékar, Mark Ferguson and John Maunder for their help and assistance.

Nicholas Chare is professor of art history in the Department of History of Art and Film Studies at the Université de Montréal. He is the author of *After Francis Bacon* (2012). In 2017, with Sébastien Lévesque and Silvestra Mariniello, he founded the baccalaureate (BACCAP) in visual cultures at the Université de Montréal.

Notes

1. Allen and Barbour, 'Newfoundland Wolf', 229. The ledger entry manifests several revisions. The remains were initially simply identified in black ink as *Canis lupus*; this was subsequently corrected in what looks to be a different hand to *Canis occidentalis* and then, finally, in pencil in another hand, to *Canis lupus beothucus*. A similar process of renaming occurs in the entry for the skin of a wolf from Newfoundland donated to the museum by the Reverend Elwood Worcester on 20 October 1932. The name is listed as *Canis lupus beothucus* but the *lupus* is written across a patch of correction fluid and the *beothucus* is in a different, less slanting hand. Worcester recalled killing the wolf in about 1896. The geolocation for the origin of the remains (listed initially in ink in the ledger as just 'Newfoundland', then on 21 June 1985, in pencil, as from north of Grand Lake) is now considered by the curatorial assistant John Mewherter (5 August 2020) to be a wooded area north of Adie's Pond and the farming community of Cormack (named after William Eppes Cormack).
2. Allen and Barbour, 'Newfoundland Wolf', 230.
3. Ibid., 234.
4. Ibid., 233.
5. Both dates have been given for the killing of the wolf. John Maunder, an authority on the wolf, accepts the date of 1894 yet claims the taxidermist who prepared the rug was William Henry Ewing, even though Ewing's taxidermy business was not founded until 1896. A possible explanation is that Ewing accepted occasional taxidermy commissions while working as a cordwainer. For a discussion of Ewing's career as a taxidermist, see Anonymous, 'William H. Ewing'.
6. A newspaper report from 1906 mentions a pack of six wolves in the Topsails following migrating 'deer' (caribou) northward. The reporter ends by noting: 'The Government would do well to have these ferocious animals destroyed' ('A Pack of Wolves', *Evening Telegram*, 30 March 1906, 4). For an examination of caribou migration in the region of Buchans Plateau and the Topsails, see Mahoney and Schaefer, 'Long-Term Changes'.

7. For a comprehensive history of the wolf skin, see Maunder's 1982 article 'The New-foundland Wolf'. A revised version (1991) of this article is available on the website of The Rooms: https://www.therooms.ca/the-newfoundland-wolf-0.

8. Buchans Plateau (known as Area 62 of Newfoundland's Caribou Management Areas), which encompasses terrain including Buchans Lake, Cormacks Lake (named in honour of William Eppes Cormack) and Red Indian Lake, is also named after the lieutenant.

9. As John Cartwright explains in 'An Account of the Red Indians of Newfoundland': 'The epithet of red is given to these [illegible] of indians from their universal practice of colouring their garments, their canoes, bows, arrows, and every other utensil belonging to them with red oker' (Cartwright, 'An Account of the Red Indians of Newfoundland', unpaginated handwritten manuscript, Rare Books Collection, A.C. Hunter Library, St John's, Newfoundland 971.6 C24). I am grateful to John Griffin for sharing his insights regarding this manuscript. Part of Cartwright's account was published in Howley's *The Beothucks or Red Indians*, where the passage just cited is inaccurately transcribed. Writing about 150 years earlier, Richard Whitbourne also refers to the 'natives' having 'great store of red Oaker, which they use to colour their Bodies, Bowes and Arrowes, and Cannowes withal', unpaginated (but would be 72).

10. The brook was mentioned repeatedly by Kevin Parsons in his contribution to a debate on the repatriation of the remains of the Beothuk couple Demasduit and Nonosbaw-sut from Scotland to Newfoundland. For Parsons, the brook's name can potentially serve as a reminder of the Beothuk people, although he admits he was ignorant of its origins for many years. See Hansard Forty-Eighth General Assembly of Newfound-land and Labrador 158:12 (2016): 544–45.

11. Howley, *Geography of Newfoundland*, 23.

12. The precise number of drawings by Shanawdithit that survive is open to debate. For further discussion of the drawings and their attributions, see Chare, 'Shanawdithit's Drawings', 92.

13. Howley, *The Beothucks or Red Indians*, 243.

14. Polack, 'Reading Shanawdithit's Drawings', unpaginated.

15. Cormack, 'Account of a Journey Across the Island of Newfoundland', 161.

16. Fleming, 'Gleanings of a Natural History, during a Voyage along the Coast of Scotland.'

17. See Harvie-Brown and Buckley, *A Vertebrate Fauna of the Outer Hebrides*, 159.

18. For a summary of Beothuk great auk hunting and egg collecting, see Kristensen, 'Bird Hunting and Egg Collecting', 19.

19. Svanberg, 'The Great Auk', 311.

20. Birkhead, *Great Auk Islands*, 87.

21. In his 1620 *A Discourse and Discovery of New-Found-Land*, Richard Whitbourne mentions hearing of pots full of hardened egg yolks (of which species of bird is unknown) that were used by the Beothuk in broth. Pagination of Whitbourne's book stops at 69. The discussion of Beothuk cuisine occurs in a subsequent unpaginated conclusion on what would be page 72. John Cartwright refers to eggs in general as being used to make cakes (which were baked in the sun) and as an ingredient in 'a sort of pudding'. Cartwright, 'An Account of the Red Indians', unpaginated.

22. Banks states that eggs for the pudding were sourced from 'as far as fung [Funk] or Penguin Island'. Banks, *Joseph Banks in Newfoundland*, 132.

23. Both the ink and the ochre are emotive materials. Iron gall ink was introduced to Turtle Island/North America by Europeans and is a settler colonial technology, one employed here to seek to control visual significance. For Shanawdithit, red ochre was culturally expressive. During a discussion about the 'colour of art history', students in my 2021

course *Écrire l'histoire de l'art* suggested that writing about Shanawdithit in the settler medium of black ink raises ethical issues given her preference for using sanguine when representing family and kin and that using red ink would therefore be preferable.

24. Pulling, *Reports and Letters*, 120.
25. Montevecchi and Tuck, *Newfoundland Birds*, 42.
26. The label makes reference to 'hunting' as a contributing factor to the wolf's extinction but not to systematic efforts at eradication.
27. Cormack, 'Account of a Journey Across the Island of Newfoundland', 156.
28. Howley describes Bloody Point as being near Hant's Harbour, but his reference to their being four hundred Beothuk massacred at the location means it is probably another name for Bloody Reach, which is quite close by.
29. Grenke, *God, Greed and Genocide*, 173.
30. Ibid., 170.
31. Cartwright, *Journal of Transactions and Events*, Volume I, 7.
32. Jones, *Genocide*, 188.
33. There are people with Beothuk ancestry alive in Newfoundland (and elsewhere) today. Fresh research in genetics and oral history has demonstrated that destruction was not total.
34. Jones, *Genocide*, 29. Recently, Sidney Harring has examined the central role of 'civilians' in the genocide against the Beothuk. See Harring, '"Shooting a Black Duck"', 63–66.
35. Ibid., 33.
36. Lawson, *The Last Man*, 48.
37. Tatz, 'Genocide in Australia', 23.
38. Lawson's observation that clearly demarcating where state-sponsored actions end and individual actions begin is often fraught with difficulty, is important in this context. See Lawson, *The Last Man*, 19.
39. Ibid., 49.
40. 'Orders and Proclamations', Colonial Secretary's Letter Book Volume 5 (1771–1774), GN2/1/A, Memorial University of Newfoundland, 19–21.
41. D'Alberti Papers, Vol. 07, 1797 (Correspondence, incoming and outgoing, between the Colonial Office and the Governor's Office in Newfoundland), Memorial University of Newfoundland, 337.
42. In a letter dated 5 September 1797, the magistrate D'Ewes Coke had observed to Waldegrave that all proclamations (not solely those related to the Beothuk) lacked legal strength.
43. D'Alberti Papers, Vol. 06, 1797 (see Note 40 above), 178–79.
44. The Beothuk were regularly accused of theft and destruction of settler property.
45. Letter from William Cormack to John Inglis, 26 October 1828. 06.09.005 COLL-262. William Epps Cormack Papers in the Howley Family Papers, Archives and Special Collections, Memorial University, St John's, Newfoundland. In a passage that he subsequently excised, Cormack then goes on to discuss the specific attitude of the government towards the recently founded Boeothick Institution and the underlying humaneness of most government representatives.
46. D'Alberti Papers, Vol. 07, 1797 (see Note 40 above), 31.
47. For a detailed analysis of this drawing, see Chare, 'In Her Hands', 292–94.
48. All text in the drawings is usually attributed to Cormack but I believe that, occasionally, isolated individual letters may have been added by Shanawdithit. Cormack, or others before him, may have taught her the alphabet and the beginnings of how to write.

49. See Daniel, 'Découverte du crime et besoins de l'enquête'.
50. The presence of two others is added by Cormack as an after thought, suggesting the key information was Peyton's culpability.
51. For an in-depth analysis of the discourse of 'doomed races' as it related to Turtle Island/North America, see Chapter 3 of Brantlinger's *Dark Vanishings*. For a discussion of Sketch VI, for instance, as the assertion of a living Beothuk culture, see Chare, 'In Her Hands', 295.
52. Lopez, *Of Wolves and Men*, 180.
53. Jacobs, *Waste of the West*, 270. Looking beyond a purely Turtle Island/North American context, Carla Frecerro also notes that '[f]or a long time humans have intended genocide for wolves'. Frecerro, 'A Race of Wolves', 116.
54. Hubbard, 'Buffalo Genocide', 293. I am grateful to Valérie Bienvenue for bringing this text to my attention.
55. Maunder, 'The Newfoundland Wolf', Revised version (1991), unpaginated. See Note 7.
56. In this period, there was an unsuccessful effort by George Skelton to propose legislation to regulate the sale of poison.
57. 'Second Reading of Bill to Control and Regulate the Sale and Use of Poisons', *Evening Telegram*, 22 April 1886, 3–4.
58. Emphasis in the original. Thomas, *The Newfoundland Journals*, 27.
59. Ibid. The artist Walton Ford gives powerful visual expression to the animal genocide perpetrated against the great auk in his 1998 painting *Funk Island*, which portrays numerous auks surging into an inferno. The painting is analysed at length in Merola, 'Assembling the Archive'.
60. Cartwright, *Journal of Transactions and Events*, Volume III, 55.
61. For a discussion of punishments meted out for the theft of eggs from Funk Island, see Post, 'Newfoundland, Reeveland', 179–82.
62. 'Orders and Proclamations', Colonial Secretary's Letter Book Volume 11 (1785–1789), GN2/1/A, Memorial University of Newfoundland, 16–18.
63. D'Alberti Papers, Vol. 04, 1789–1792 (see Note 40 above), 203.
64. The men were Joseph Barber, Daniel Coffee, Richard Fitzgerald, Michael Lines, Edward Shea and John Shea. See D'Alberti Papers, Vol. 04, 1789–1792 (see Note 40 above), 249–50.
65. D'Alberti Papers, Vol. 20, 1810 (see Note 40 above), 248.
66. Fuller, *The Great Auk*, 66–68.
67. Montevecchi and Tuck, *Newfoundland Birds*, 147.
68. See, for instance, Chapter 6 of Chare and Williams, *The Auschwitz Sonderkommando*.
69. D'Alberti Papers, Vol. 07, 1797 (see Note 40 above), 27.
70. I am grateful to Tom Tyler for providing me with a copy of this text.
71. Derrida and Roudinesco, *De quoi demain...*, 122.
72. Ibid.
73. Derrida, *L'animal que donc je suis*, 46.
74. Ibid., 47.
75. Ibid.
76. See Turner's discussion of Derrida's thinking regarding the 'animal question' in her 'Introduction' to *The Animal Question in Deconstruction*, 2.
77. Hubbard, 'Buffalo Genocide', 294.
78. Ibid., 302.
79. Lemkin, *Lemkin on Genocide*, 35.
80. See, for example, Whitehead's observations regarding sperm whale cultures in *Sperm Whales*, 286–315.

81. Hatley, *Suffering Witness*, 2.
82. Ibid.
83. Hatley, *Suffering Witness*, 3.
84. Derrida, *L'animal que donc je suis*, 50.
85. Calarco, 'Thinking through Animals', 10.
86. Harvey, 'Memoirs of an Extinct Race', 504.
87. Ibid., 505.
88. Bonnycastle, *Newfoundland in 1842: Volume 1*, 224–25.
89. 'Orders and Proclamations', Colonial Secretary's Letter Book Volume 10 (1783–1785), GN2/1/A, Memorial University of Newfoundland, 78–79.
90. Tocque, *Wandering Thoughts, or Solitary Hours*, 112.
91. Thomas, *The Newfoundland Journals*, 131.
92. Philip Tocque, however, praises wolves for their affection towards each other. See *Wandering Thoughts, or Solitary Hours*, 115.
93. Staub, *The Roots of Evil*, 19.
94. Ibid., 51.
95. Ibid., 52.
96. Ibid., 60.
97. Ibid., 48.
98. For a brief discussion of the historical role of goats as scapegoats, see DeMello, *Animals and Society*, 312–13.
99. Staub, *The Roots of Evil*, 62.
100. MacLeod, *Grizmek's Animal Life Encyclopedia*.

Bibliography

Allen, Glover M., and Thomas Barbour. 'The Newfoundland Wolf'. *Journal of Mammalogy* 18(2) (1937): 229–34.

Anonymous. 'William H. Ewing: Shoemaker, Taxidermist and Furrier'. *Newfoundland Government Bulletin* 4(3) (1971): 15.

Banks, Joseph. *Joseph Banks in Newfoundland and Labrador, 1766: His Diary, Manuscripts and Collections*. Edited by A.M. Lysaght. Berkeley: University of California Press, 1971.

Birkhead, Tim R. *Great Auk Islands: A Field Biologist in the Arctic*. Berkhamsted: Poyser, 1993.

Bonnycastle, Richard Henry. *Newfoundland in 1842*, 2 vols. London: Henry Colburn, 1842.

Brantlinger, Patrick. *Dark Vanishings: Discourse of the Extinction of Primitive Races, 1800–1930*. Ithaca, NY: Cornell University Press, 2003.

Calarco, Matthew. 'Thinking through Animals: Reflections on the Ethical and Political Stakes of the Question of the Animal in Derrida'. *Oxford Literary Review* 29 (2007): 1–15.

Cartwright, George. *A Journal of Transactions and Events during a Residence of Nearly Sixteen Years on the Coast of Labrador*, Volume I. Newark: Allin and Ridge, 1792.

———. *A Journal of Transactions and Events During a Residence of Nearly Sixteen Years on the Coast of Labrador*, Volume III. Newark: Allin and Ridge, 1792.

Chare, Nicholas. 'In Her Hands: Affect, Encounter and Gestures of Wit(h)nessing in Shanawdithit's Drawings'. *parallax* 26(3) (2020): 286–302.

———. 'Shanawdithit's Drawings', in Nicholas Chare and Mitchell B. Frank (eds), *History and Art History: Looking Past Disciplines* (New York: Routledge, 2021), 87–101.

Chare, Nicholas, and Dominic Williams. *The Auschwitz Sonderkommando: Testimonies, Histories, Representations*. Cham: Palgrave Macmillan, 2019.

Cormack, William Epps. 'Account of a Journey across the Island of Newfoundland'. *The Edinburgh Philosophical Journal* 10 (1824): 156–62.

Daniel, Marina. 'Découverte du crime et besoins de l'enquête: Le dessin judiciaire en Seine-Inférieure au XIXe siècle'. *Sociétés & Représentations* 2(18) (2004): 109–22.

DeMello, Margo. *Animals and Society: An Introduction to Human–Animal Studies*. New York: Columbia University Press, 2012.

Derrida, Jacques. *L'animal que donc je suis*. Paris: Galilée, 2006.

Derrida, Jacques, and Élisabeth Roudinesco. *De quoi demain. . . : Dialogue*. Paris: Éditions Galilée, 2001.

Fleming, John. 'Gleanings of a Natural History, during a Voyage along the Coast of Scotland'. *The Edinburgh Philosophical Journal* 10 (1824): 95–101.

Frecerro, Carla. 'A Race of Wolves'. *Yale French Studies* 127 (2015): 110–23.

Fuller, Errol. *The Great Auk*. New York: Harry N. Abrams, 1999.

Grenke, Arthur. *God, Greed and Genocide: The Holocaust through the Centuries*. Washington, DC: New Academia Publishing, 2005.

Hansard Forty-Eighth General Assembly of Newfoundland and Labrador 158(12) (2016).

Harring, Sidney L. '"Shooting a Black Duck": Genocidal Settler Violence against Indigenous Peoples and the Creation of Canada', in Mohamed Adhikari (ed.), *Civilian-Driven Violence and the Genocide of Indigenous Peoples in Settler Societies* (London: Routledge, 2019), 61–85.

Harvey, Moses. 'Memoirs of an Extinct Race; or, the Red Indians of Newfoundland'. *The Maritime Monthly: A Magazine of Literature, Science and Art* 5(6) (1875): 498–514.

Harvie-Brown, J.A., and T.E. Buckley. *A Vertebrate Fauna of the Outer Hebrides*. Edinburgh: David Douglas, 1888.

Hatley, James. *Suffering Witness: The Quandary of Responsibility after the Irreparable*. New York: SUNY Press, 2000.

Howley, James P. *The Beothucks or Red Indians*. Cambridge: Cambridge University Press, 1915.

———. *Geography of Newfoundland*. London: Edward Stanford, 1876.

Hubbard, Tasha. 'Buffalo Genocide in Nineteenth-Century North America: "Kill, Skin, and Sell"', in Andrew Woolford, Jeff Benvenuto and Alexander Laban Hinton (eds), *Colonial Genocide in Indigenous North America* (Durham, NC: Duke University Press, 2014), 292–305.

Jacobs, Lynn B. *Waste of the West: Public Lands Ranching*. Berkeley: University of California Press, 2008.

Jones, Adam. *Genocide: A Complete Introduction*. 3rd edition. London: Routledge, 2017.

Kristensen, Todd J. 'Bird Hunting and Egg Collecting among the Beothuk, Mi'lmaq, and Innu'. *The Osprey* 39(1) (2008): 18–23.

Lawson, Tom. *The Last Man: A British Genocide in Tasmania*. London: I.B. Tauris, 2014.

Lemkin, Raphael. *Lemkin on Genocide*. Lanham, MD: Lexington, 2012.

Lopez, Barry. *Of Wolves and Men*. New York: Scribner, 1978.

MacLeod, Norman (ed.). *Grizmek's Animal Life Encyclopedia; Extinction*, 2 vols. Farmington, MI: Gale, 2013.

Mahoney, Stephen P., and James A. Schaefer. 'Long-Term Changes in Demography and Migration of Newfoundland Caribou'. *Journal of Mammalogy* 83(4) (2002): 957–63.

Maunder, John. 'The Newfoundland Wolf'. *The Osprey* 13(2) (1982): 36–46.

Merola, Nicole M. 'Assembling the Archive: Close(ly) Reading Great Auk Extinction with Walton Ford', in Helena Feder (ed.), *Close Reading the Anthropocene* (London: Routledge, 2021), 45–59.

Montevecchi, William A., and Leslie M. Tuck. *Newfoundland Birds: Exploitation, Study, Conservation.* Cambridge, MA: Nuttall Ornithological Club, 1987.

Polack, Fiona. 'Reading Shanawdithit's Drawings'. *Journal of Colonialism and Colonial History* 14(3) (2013): unpaginated.

Post, Andy. 'Newfoundland, Reeveland: Chief Justice Reeves as a Conservative Reformer, 1791–1793'. Unpublished MA thesis, Memorial University of Newfoundland, 2019.

Pulling, George Christopher. *Reports and Letters by George Christopher Pulling Relating to the Beothuk Indians of Newfoundland.* St John's, NL: Breakwater, 1989.

Staub, Ervin. *The Roots of Evil: The Origins of Genocide and Other Group Violence.* Cambridge: Cambridge University Press, 1989.

Svanberg, Ingvar. 'The Great Auk', in Andrew J. Hund (ed.), *Antarctica and the Arctic Circle: A Geographic Encyclopedia of the Earth's Polar Regions, Volume 1* (Santa Barbara, CA: ABC-CLIO, 2014), 311–12.

Tatz, Colin. 'Genocide in Australia: By Accident or Design?' *Indigenous Human Rights and History: Occasional Papers* 1(1) (2011): 1–100.

Thomas, Aaron. *The Newfoundland Journals of Aaron Thomas, 1794.* Ontario: Longmans, 1968.

Tocque, Philip. *Wandering Thoughts, or Solitary Hours.* London: Thomas Richardson and Son, 1846.

Turner, Lynn. 'Introduction', in Lynn Turner (ed.), *The Animal Question in Deconstruction* (Edinburgh: Edinburgh University Press, 2013), 1–8.

Whitbourne, Richard. *A Discourse and Discovery of New-Found-Land.* London: Felix Kyngston, 1620.

Whitehead, Hal. *Sperm Whales: Social Evolution in the Ocean.* Chicago: University of Chicago Press, 2003.

Chapter 5

Cultural Memory of Recent Extinctions

A Chinese Perspective

Samuel T. Turvey

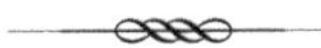

Insights about Extinction from Indigenous Knowledge

Determining the amount of biodiversity loss that has already taken place as a result of past human pressures is integral to conservation science, management and policy. Data on historical species extinctions and population declines can identify unsustainable human–environmental interactions and taxonomic, ecological and geographic patterns of vulnerability and resilience, and thus make predictive hypotheses about the likely impact of future anthropogenic activities to guide conservation priority-setting.[1] Understanding the extent to which landscapes have been disrupted from historical baseline conditions is also necessary to inform appropriate management, restoration and rewilding strategies.[2]

However, although humans have impacted biodiversity for millennia, long-term environmental datasets that can capture the duration and outcome of past anthropogenic activities are rarely used in conservation, with 'long-term' in ecology typically meaning decadal or multi-decadal.[3] This data constraint has led to a widespread 'extinction filter' effect, whereby information on past extinctions is excluded from analysis of human-caused biodiversity loss, thus limiting and biasing our understanding of human impacts through time.[4] Even recent environmental baselines, based on systematically collected ecological data sampled directly by trained scientists, remain unavailable to assess changes in the status of many threatened taxa and conservation 'hotspots'.[5] Assessing the information-content of alternative data sources, and their ability to provide new insights into past human-mediated biodiversity loss, is thus of substantial importance for evidence-based conservation.

Most traditional (Indigenous and/or rural) communities around the world possess a rich body of knowledge about local biodiversity.[6] This Indigenous knowledge is increasingly recognized as having invaluable applications for conservation, including monitoring of target species, environmental conditions and threats,[7] identification of sustainable environmental resource management strategies,[8] and provision of social frameworks to support biodiversity preservation based on local value systems.[9] Indigenous knowledge can be subdivided into two broad categories: local ecological knowledge (LEK), representing experiential knowledge derived from an individual's lived interactions with their environment; and traditional ecological knowledge (TEK) or 'cultural memory', the cumulative body of knowledge, beliefs, values and traditions about the natural world that is passed down between generations.[10] Both of these knowledge domains have the potential to provide novel conservation-relevant information about past biodiversity loss across different timescales.

LEK can provide information about species that have been encountered during an individual's lifetime but might now be regionally extinct. Quantitative analysis of last-sighting datasets can assess whether species are likely to be extinct, and model the timing of extinction events.[11] In many cases, LEK has provided the most recent records for now-extinct species that were otherwise the subject of limited historical monitoring or survey effort (e.g. mammals from central Australian deserts).[12] Indeed, in some instances LEK might provide the only possible evidence for recent survival of species that are otherwise recorded only from older fossil or archaeological archives. For example, Woods et al.[13] collected several reports from Hispaniola, dating from the 1970s or earlier, of a large arboreal animal locally called the comadreja, which did not match the description of any living native mammal but might represent one of the large extinct rodents (possibly *Plagiodontia velozi*) known from the island's recent fossil record: 'One man … even demanded why we had no drawings or photographs of the comadreja to show him. It was many years, he said, since he had seen the animal, and he would have liked to have seen it again'.[14]

Although in some cases respondents are aware that now-extinct species have declined and disappeared from their local environments,[15] LEK about such species is not necessarily associated with awareness that they are extinct, but instead simply that individual respondents have not encountered them for some time. Conversely, TEK has the potential to record information about past environmental conditions and changes, often framed explicitly in terms of differences from present conditions. On occasion, such information might be extremely ancient, particularly in isolated non-literate societies with a strong attachment to their landscape; as an example, Aboriginal stories about coastal inundation from around Australia seemingly refer to postglacial sea-level rise that occurred over seven thousand years ago.[16] Whereas conflict with Christian theology meant that the possibility of extinction only became accepted in Western thought around two hundred years ago,[17] the TEK of some non-Western Indigenous cultures includes ideas about past species losses expressed in terms similar to the scientific concept of extinction. For example, the Yukaghir people of the Russian Far East thought that mammoth remains preserved in permafrost represented animals called 'xolhut' that had existed long ago, but disappeared because they ate all the trees in northern Siberia and turned the landscape into tundra.[18] Other Indigenous cultures also recognize the possibility that species may disappear locally.[19]

Strange creatures in folktales of numerous Indigenous cultures have been interpreted as possible cultural memories of species that became extinct centuries or millennia ago. For example, an ogre in Madagascar folklore with a human face and an animal's body, and that cannot move across smooth rock outcrops, is suggested to be a representation of a sloth lemur (*Palaeopropithecus ingens*), which has been extinct for several centuries;[20] and tales about giant beavers in north-eastern Algonquian legends have been proposed as folk memories of a real species of giant beaver (*Castoroides ohioensis*), which became extinct around ten thousand years ago.[21] However, these potential identifications are highly speculative, with storytelling an obvious alternative hypothesis to explain the origin of such accounts. Overall, the potential for TEK to reconstruct past biodiversity baselines and provide information about extinct species has been surprisingly understudied.

Conversely, the existence of several sociocultural phenomena caution against the likelihood of widespread cultural retention of TEK about extinct species. Historical records reveal that local species concepts can change rapidly following extinction events, taking on fantastical and non-natural characteristics, or with the names of vanished species becoming transferred to other taxa. For example, reports of dodos (*Raphus cucullatus*) from Mauritius during the second half of the seventeenth century actually referred to the island's other flightless bird, the (also now extinct) Mauritius red hen (*Aphanapteryx bonasia*), as 'true' dodos were probably already extinct by this time.[22] This is just one instance of a wider issue: TEK does not provide a fixed record of the past, but instead constitutes an inherently dynamic body of information that is continuously updated and changed with successive generations.[23] Furthermore, TEK in Indigenous communities around the world is now experiencing erosion and loss of content associated with globalization, exposure to Western cultural and economic norms, and concomitant disruption of traditional social-ecological systems.[24] These major sociocultural changes are often associated with changing patterns and levels of intergenerational communication; this process can lead to a phenomenon known as 'shifting baseline syndrome', whereby lack of knowledge transfer between generations leads to younger people having less awareness of biodiversity patterns from the recent past, and incorrectly interpreting more degraded environmental conditions as the norm.[25] Therefore, even if extinct species were regarded as sufficiently 'important' at a cultural level to become incorporated within TEK when they still existed, this in itself provides no guarantee that any awareness about them will remain once they are gone.

It is thus essential to critically evaluate the ability of TEK to reconstruct historical biodiversity baselines and reveal patterns and dynamics of ecosystem change over time – an important role that this body of knowledge can potentially contribute to the modern conservation toolkit. Such evaluation should include assessment of whether extinct species are remembered by local communities; if so, how long they are remembered for; whether there is variation in knowledge between different species and systems, and what determines such variation; and which types of information are most likely to persist. Such research must be conducted within systems for which independent data exist on the identity and timing of regional extinctions, as a benchmark against which to compare the information content of TEK. It should also be conducted in a region with high biodiversity and high anthropogenic threats, in order to determine the direct practical benefit that TEK can provide towards strengthening the conservation evidence-base by supplying historical baselines on the vulnerability or resilience of biodiversity through time.

China: A Study System for Human-Caused Extinction

China is a huge (~9.6 million km^2) 'megadiverse' country containing 14 per cent of the world's vertebrate species, and covering a wide range of ecosystems.[26] China also has a long history of human occupation, and has suffered biodiversity loss associated with increasing human overpopulation, resource overexploitation and habitat modification throughout recent millennia and the historical era.[27] These anthropogenic impacts have escalated over the past century. Although national efforts are underway to protect regional biodiversity, natural resources and ecosystem services,[28] China's biota has experienced extensive recent population declines and extirpations, and even extinctions of endemic species.[29] Overall, Chinese terrestrial vertebrate populations decreased by 50 per cent between 1970 and 2010.[30]

Throughout history, Chinese thought has been primarily interested in the workings of human society, and has usually considered nature only in terms of its utility for philosophical or political analogies and moral insights. However, this perspective is grounded in a holistic, interconnected worldview whereby human actions can influence the workings of nature, often with unexpected and undesirable consequences. Both Confucian and Daoist thinkers promoted moral and moderate usage of natural resources to avoid their depletion, thus demonstrating an understanding of concepts associated with biodiversity loss.[31] There is little evidence for specific awareness or understanding of extinction in historical Chinese texts, although a passage written by the third-century BCE Legalist philosopher Han Fei has been interpreted as indicating recognition of local species disappearances: 'People rarely see living elephants, but if they obtain the bones of a dead elephant, they can imagine a living elephant based on their form. Because of this, everything people use to form an idea or mental image is called "elephant"'[32] [the Chinese word 象 *xiang* means both 'elephant' and 'appearance/image']. Regional species losses and former occurrences of now-vanished animals are also reported in several Imperial Era gazetteers or *difangzhi*, which served as handover documents for civil servants and often included considerable information on local biodiversity and environmental resources.[33]

However, environmental attitudes and knowledge of biodiversity loss recorded in historical Chinese texts written by scholars and officials may have little relationship with the TEK of rural communities across China that interact directly with nature. Indeed, in contrast to the relatively homogeneous body of formal Han Chinese thought and narratives of nature, China includes a diverse range of ethnic groups, each with distinct cultures and traditions, bodies of Indigenous knowledge, and interactions with their environments.[34] This extensive sociocultural variation makes it difficult to infer likely patterns of TEK about extinct species within communities in different Chinese landscapes that have experienced historical and recent biodiversity loss, or whether general patterns of cultural memory about extinction might exist across China.

Here, I summarize the findings of a series of interview surveys conducted in rural communities across different parts of China by myself and colleagues, which provide new insights into regional cultural memory of several recently extinct aquatic and terrestrial vertebrates. These studies establish a new understanding of patterns and levels of biodiversity-related TEK loss and retention within different social-ecological systems across China, and also provide a wider comparative baseline on the usefulness of

TEK to reconstruct past environments and the timing and dynamics of species losses.

Case Study 1: Yangtze Freshwater Megafauna

The Yangtze River, Asia's longest river system, supports high levels of aquatic biodiversity but also provides food, other resources and ecosystem services for 10 per cent of the world's human population, and has experienced severe degradation associated with overpopulation and industrialization.[35] The endemic Yangtze River dolphin or baiji (*Lipotes vexillifer*; Illustration 5.1A) declined rapidly during the 1980s and 1990s due to incidental by-catch in fishing gear, vessel strikes and wider-scale habitat degradation, and although a few individuals persisted into the twenty-first century, a range-wide survey in late 2006 failed to detect any surviving dolphins.[36] The river's fisheries have also experienced severe declines, with several fish species now regionally or globally extinct. These casualties include the Yangtze paddlefish (*Psephurus gladius*), probably the largest freshwater fish in the world (mature individuals were estimated to reach up to seven metres), which was impacted by overfishing and construction of dams that prevented these fish from reaching their upstream spawning grounds. Paddlefish catches declined rapidly from the 1980s, with few records after the mid-1990s, and the species is estimated to have become extinct by 2005.[37] Both species were culturally or economically important within Yangtze fishing communities. Baiji held an almost sacred status;[38] they were the subject of many local stories and legends,[39] and killing them

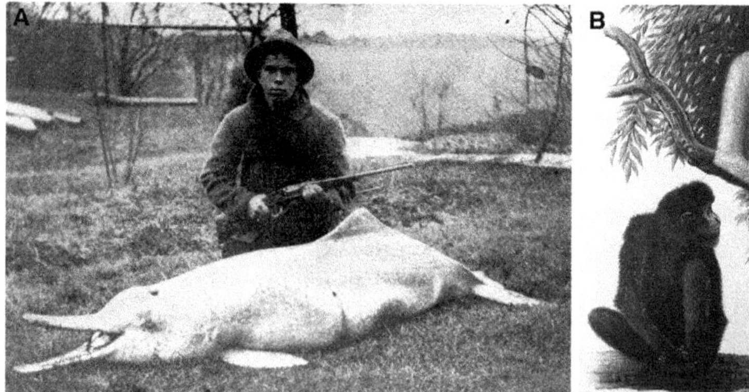
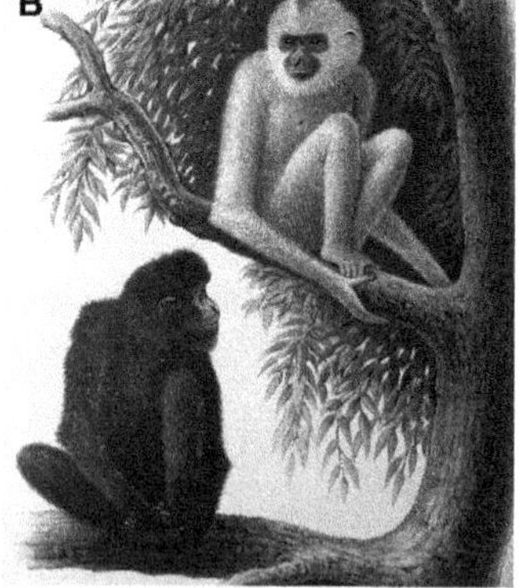

Illustration 5.1 Recent species extinctions and declines in China. (A) Yangtze River dolphin or baiji (*Lipotes vexillifer*) (from Hoy, 'The "White-Flag" Dolphin'). (B) Hainan gibbon (*Nomascus hainanus*) (from Pocock, 'Observations Upon a Female Specimen').

was considered to result in bad fortune.[40] Paddlefish were conversely targeted for food, and were formerly widely available in riverside markets, with ~25 tonnes harvested annually during the 1970s.[41]

A large-scale interview survey was conducted in February–November 2008 (Illustration 5.2A), collecting data from 599 fishers (age range: 22–90) in twenty-seven fishing communities along a 1,700 km transect of the middle-lower Yangtze channel and around the appended Dongting and Poyang lake systems, primarily to ascertain whether any baiji might still survive but also collecting wider local knowledge about the Yangtze ecosystem. Respondents across all age classes were strongly aware of the Yangtze's overall resource depletion and environmental degradation, but showed statistically significant age-related differences in experience and awareness of particular species. Younger fishers were, unsurprisingly, less likely than older fishers to have seen or caught baiji or paddlefish; however, they were also less likely to have even heard of either species, despite being prompted during interviews with photographic cue cards, appropriate local names and verbal descriptions.[42]

When measured in terms of the calendar year when they started fishing, the number of new fishers who had not heard of baiji or paddlefish increased markedly from the 1980s onwards, corresponding with the timing of major population declines in both species. In total, 10.3 per cent of all interviewed fishers had not heard of baiji, and 33.8 per cent had not heard of paddlefish, but these percentages increased to 15.6 and 56.8 respectively in respondents who had started fishing from the 1980s onwards, and to 19.2 and 67.0 in those who had started fishing from the 1990s onwards. The difference in awareness levels between these species is likely to reflect

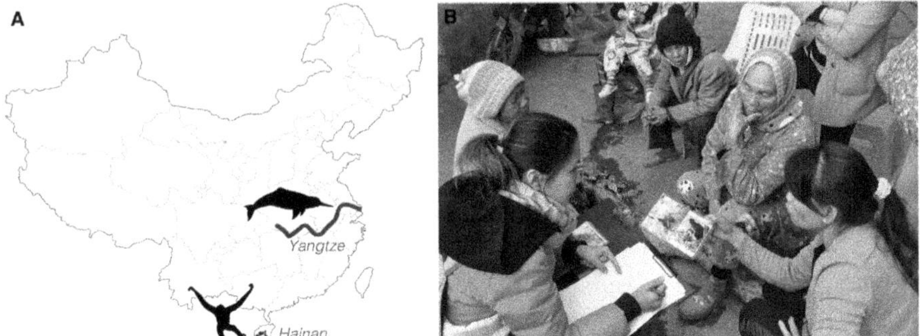

Illustration 5.2 (A) Map of China highlighting the middle-lower Yangtze River and the montane forests of Hainan, where interviews were conducted with rural respondents to collect data on freshwater megafauna and Hainanese mammals, respectively. (B) Conducting interviews in a rural community near Bawangling National Nature Reserve, Hainan. Photograph courtesy of Helen Nash.

the extensive regional media publicity that promoted past baiji conservation efforts (e.g. television programmes, local government and fisheries publicity, newspapers, posters), rather than differences in TEK within fishing communities. Of the subset of fishers who had never seen these species but were still aware of them, 60.8 per cent had heard about baiji from regional publicity, versus only 47.6 per cent from local community members; in contrast, 80.4 per cent had heard about paddlefish from local community members, versus only 19.6 per cent from regional publicity.[43]

Case Study 2: Gibbons and Other Mammals on Hainan

Hainan, China's southernmost province, is a 33,920 km² subtropical/tropical island in the South China Sea. Different environmental archives (recent fossil record, historical *difangzhi* gazetteer record, specimens collected by nineteenth-century naturalists) document progressive anthropogenic depletion of its biodiversity, and the former occurrence of numerous mammal species that are now regionally extinct. Species such as elephants, tapir, wild buffalo and tigers apparently disappeared from Hainan due to human impacts in recent prehistory, whereas Père David's deer (*Elaphurus davidianus*) persisted until at least the nineteenth century, and two large carnivores (wolf, *Canis lupus*; dhole, *Cuon alpinus*) persisted into the mid-twentieth century.[44] This biodiversity loss is ongoing, and the endemic Hainan gibbon (*Nomascus hainanus*; Illustration 5.1B) is now one of the world's rarest mammals, with a global population of only ~35 individuals. Gibbons were esteemed in local cultures across China, including on Hainan,[45] and were formerly distributed across much of Hainan's forested interior, but underwent a precipitous range collapse during the 1960s and 1970s caused by habitat loss and hunting. They only survive today as a remnant population within Bawangling National Nature Reserve.[46]

A large-scale interview survey was conducted in January–April 2015 (Illustration 5.2A and 5.2B) to collect data on sightings and knowledge of gibbons and other mammals, from 709 respondents (age range: 20–94) in seventy villages situated around Bawangling and six other nature reserves that contained gibbon populations during the 1970s or later.[47] Most respondents belonged to Indigenous Li or Miao ethnic minorities, and communities across the study area used animal and plant resources collected inside reserves and were thus familiar with local forest biodiversity. For respondents living close to the last surviving gibbon population at Bawangling, 73.8 per cent recognized photographs of gibbons and 65.4 per cent were familiar with the standard Chinese word for gibbon (*changbiyuan*), whereas these percentages were 30.6–70.3 and 23.0–80.2 respectively for respondents living near the six reserves where gibbons had disappeared during living memory. Some but not all of these levels of local awareness differed statistically from the levels observed at Bawangling.[48]

All respondents were asked an open-ended question about their TEK of gibbons ('Have you heard any stories about gibbons or anything else about them, such as uses?'), with 99 respondents (14 per cent) providing information. Older respondents were statistically more likely to report gibbon-specific TEK; the mean age of respondents reporting TEK was 57.6, while that of other respondents was 49.0, and only three respondents younger than 30 reported TEK. Gibbon-specific TEK included information on their ecology, behaviour and other natural history; practical/utilitarian knowledge (e.g. hunting, traditional medicine); cultural values/attitudes; and folktales, including six different folktale categories. Interestingly, there were statistically significant differences in TEK content between landscapes with different histories of gibbon survival or extinction. Respondents from Bawangling and the two landscapes where gibbons apparently died out most recently (Jiaxi, Yinggeling) were more likely to report gibbon-related folktales, whereas respondents from landscapes where gibbons had been absent for several decades were instead more likely to report knowledge about practical information or interactions with gibbons (e.g. natural history or hunting information)[49].

Respondents were also asked whether they knew about any animals that had existed in the past but no longer occurred locally, or knew any old stories about animals that had only existed in the past. Only 20 out of 709 respondents (2.8 per cent) named specific animals, including pangolin, bear, parrot, snake/python, turtle, gibbon, tiger, wild pig and muntjac. Of these species, only tigers (named by a single respondent) are regionally extinct across the study area. These are culturally significant zodiac animals across China,[50] but they disappeared thousands of years ago on Hainan, which was described as 'without horses or tigers' as early as 80 CE.[51] Most of the other responses were probably 'cued' from previous interview questions about specific named animals (pangolin, bear, gibbon, wild pig).[52]

Why Is There Little Evidence for Cultural Memory of Extinct Species?

These large-scale systematic investigations of the TEK content of natural resource users in China, conducted across different social-ecological systems and distinct local cultures, reveal a common pattern of shifting baseline syndrome and surprisingly little cultural memory of past extinctions, even for species that until recently were culturally important (e.g.

baiji, gibbons) or economically important (e.g. paddlefish). Rapid loss of Indigenous knowledge about recently extinct animals is observed amongst younger community members within both Yangtze fishing communities and Hainanese forest-users immediately following regional species disappearances, even while older community members who remember encountering these species are still alive. Indeed, much of the knowledge that younger community members possess about such species is demonstrated to have been obtained from external sources rather than intergenerational knowledge transfer within their communities. These surveys also reveal that Hainan's Indigenous knowledge record is an extremely poor source of information on older extinctions that took place before the lifetime of living respondents, with seemingly no local awareness of formerly occurring species that are likely to have been culturally significant (e.g. large carnivores) and that disappeared as recently as the mid-twentieth century. Different domains of TEK relating to recently extinct species appear to be lost at different rates, with folklore about gibbons vanishing first in Hainan; this pattern might reflect the fact that older members of Indigenous communities are typically the 'cultural repositories' of stories, traditions and worldviews, whereas knowledge about practical interactions with biodiversity (e.g. hunting methods) might be shared more widely with younger community members.[53]

These findings indicate that, at least for rural communities surveyed across China, TEK is unfortunately of little direct use for reconstructing biodiversity baselines and human-caused turnover, even for the very recent past. Indeed, Turvey et al.[54] noted that once species such as baiji and paddlefish 'ceased to be encountered on a fairly regular basis, they immediately started to become forgotten by local communities. The adage "out of sight, out of mind" may appropriately apply to community awareness of rare, cryptic, and recently extinct species'. Conversely, the flipside of this finding is that the existence of cultural knowledge and traditions about otherwise-cryptic species might indicate that such species still persist in local landscapes, thus providing an indirect tool for guiding further conservation research. However, reconstructing past ecosystem states and faunal change across recent centuries will likely require the use of alternative environmental archives (e.g. the fossil or archaeological records), at least in China.

Why has local knowledge about vanished species been lost so quickly in rural communities across China? And is this pattern likely to constitute a phenomenon specific to this region, or is it a more general finding about the quality of Indigenous knowledge at a global scale? Are individual species somehow not 'important' enough to be remembered for long in cultural memory? These are challenging questions to answer, and appear to have rarely even been raised by researchers.

Consideration of China's recent sociocultural history suggests that this pattern of TEK erosion might be a regional-specific consequence of the huge social upheavals experienced across the country over the past century, when traditional cultural systems and community social-ecological dynamics were widely disrupted by demands for political conformity and national policies such as collectivization.[55] For example, Indigenous communities on Hainan have been progressively undermined and acculturated by the Chinese state, with their cultural distinctiveness increasingly weakened and assimilated with Han lifestyles, values and beliefs; this loss of traditional social structures and identities escalated during the second half of the twentieth century.[56] Most rural communities across China are therefore unlikely to fulfil the sociocultural criteria suggested to be necessary to support long-term retention of knowledge about past environmental conditions, such as cultural and political isolation.[57]

Other Indigenous cultures might therefore be predicted to possess more TEK about regionally extinct species than is now observed within China. However, the escalating erosion of TEK in rural and traditional communities around the world means that any such as-yet-unknown Indigenous record of past faunas and environmental baselines is also now in the process of being lost. Indeed, many cultures from which past anthropologists and ethnographers recorded local ideas about extinction and folktales about ancient environmental change are themselves on the verge of extinction, or already gone. For example, Siberian Yukaghir languages are now classified as moribund, and cultural disruption has led to the disappearance of most of Australia's Aboriginal languages.[58] It would be instructive to see whether such examples of historically recorded TEK still persist in these communities today.

Few social studies have been conducted into changing community perceptions in the immediate aftermath of extinction events. We would inevitably expect to see some loss of cultural knowledge over time about vanished species, differential recall about different types of vanished species, and variation across communities in who retains knowledge about past biodiversity. Wider-scale comparative studies across multiple Indigenous communities and sociocultural contexts are still needed to understand the patterns and determinants of such variation. However, the Chinese case studies investigated here provide the sobering insight that, under cultural contexts that are increasingly widespread around the world, there might be minimal awareness of recent biodiversity loss within those communities that interact directly with local biodiversity, and who might therefore be unaware of the potentially irreversible impacts of human actions on the natural environment. Combatting the continued erosion of Indigenous knowledge about past environmental baselines should therefore be recognized as an important goal within the wider global mission to conserve biocultural diversity.

Acknowledgements

I thank the fishing communities of the Yangtze River and the Indigenous communities of Hainan for sharing their knowledge, and my many collaborators in China for assistance and support with fieldwork. I also thank Helen Nash for permission to use the photograph in Illustration 5.2.

Samuel T. Turvey is a professor at the Institute of Zoology, Zoological Society of London, where he has worked since 2004. He has previously been based at the University of Canterbury (New Zealand) and the University of Oxford. His research uses 'non-standard' ecological archives, including Indigenous knowledge and historical baselines, to reconstruct human-caused biodiversity loss over time and to inform current-day conservation. He also conducts applied conservation research on highly threatened species to identify effective management and recovery strategies for tiny vulnerable animal populations. He has worked in China for over twenty years and is involved with long-term conservation efforts for Chinese species such as the Hainan gibbon, the Chinese giant salamander and the Yangtze finless porpoise.

Notes

1. Dietl and Flessa, 'Conservation Paleobiology'; Rick and Lockwood, 'Integrating Paleobiology, Archeology, and History'.
2. Hall, *Restoration and History*.
3. Bonebrake et al., 'Population Decline Assessment'.
4. Balmford, 'Extinction Filters and Current Resilience'.
5. Boakes et al., 'Distorted Views of Biodiversity'.
6. Berkes, Colding and Folke, 'Rediscovery of Traditional Ecological Knowledge'; Berkes, *Sacred Ecology*.
7. Anadón, Gimenez and Ballestar, 'Linking Local Ecological Knowledge'; Turvey et al., 'Interview-Based Sighting Histories'.
8. Gómez-Baggethun, Corbera and Reyes-García, 'Traditional Ecological Knowledge'; Mistry and Berardi, 'Bridging Indigenous and Scientific Knowledge'.
9. Colding and Folke, 'Social Taboos'; Shen et al., 'Viable Contribution'.
10. Berkes, Colding and Folke, 'Rediscovery of Traditional Ecological Knowledge'.

11. Boakes et al., 'Distorted Views of Biodiversity'; Boakes et al., 'Inferring Species Extinction'.
12. Burbidge et al., 'Aboriginal Knowledge'.
13. Woods, Ottenwalder and Oliver, 'Lost Mammals of the Greater Antilles'.
14. Ibid., 40.
15. Ziembicki, Woinarski and Mackey, 'Evaluating the Status of Species'; Turvey et al., 'Interview-Based Sighting Histories'.
16. Nunn and Reid, 'Aboriginal Memories of Inundation'.
17. Rudwick, *The Meaning of Fossils*.
18. Jochelson, 'Some Notes on the Traditions'.
19. Berkes, *Sacred Ecology*.
20. Godfrey and Jungers, 'The Extinct Sloth Lemurs'.
21. Beck, 'The Giant Beaver'.
22. Cheke, 'Establishing Extinction Dates'.
23. Berkes, Colding and Folke, 'Rediscovery of Traditional Ecological Knowledge'.
24. Reyes-García et al., 'Evidence of Traditional Knowledge Loss'; Tang and McGavin, 'A Classification of Threats'.
25. Papworth et al., 'Evidence for Shifting Baseline Syndrome'.
26. Xie et al., *Living Planet Report*.
27. Shapiro, *Mao's War against Nature*; Shapiro, *China's Environmental Challenges*; Marks, *China*.
28. Xu et al., 'Strengthening Protected Areas'.
29. Turvey et al., 'First Human-Caused Extinction'; Turvey, Crees and Di Fonzo, 'Historical Data as a Baseline'.
30. Xie et al., *Living Planet Report*.
31. Sterckx, *Chinese Thought*.
32. Lander and Brunson, 'Wild Mammals of Ancient North China', 232.
33. Wen, *The Distributions and Changes*; Turvey, Crees and Di Fonzo, 'Historical Data as a Baseline'.
34. Hathaway, 'Global Environmentalism'; Zhang et al., 'Influence of Traditional Ecological Knowledge'.
35. Wang, Liu and Wang, 'The Yangtze River Floodplain'.
36. Turvey et al., 'First Human-Caused Extinction'.
37. Zhang et al., 'Extinction'.
38. Pilleri, 'The Chinese River Dolphin'.
39. Zhou and Zhang, *Baiji*.
40. Pope, *China's Animal Frontier*.
41. Zhang et al., 'Extinction'.
42. Turvey et al., 'Rapidly Shifting Baselines in Yangtze'.
43. Ibid.
44. Turvey et al., 'Complementarity, Completeness and Quality'.
45. Van Gulik, *The Gibbon in China*; Turvey, Bryant and McClune, 'Differential Loss'.
46. Zhou et al., 'Hainan Black-Crested Gibbon'; Turvey et al., 'How Many Remnant Gibbon Populations'.
47. Turvey et al., 'How Many Remnant Gibbon Populations'.
48. Ibid.
49. Turvey, Bryant and McClune, 'Differential Loss'.
50. Coggins, *The Tiger and the Pangolin*.
51. Marks, *China*.
52. Turvey et al., 'Complementarity, Completeness and Quality'.

53. Turvey, Bryant and McClune, 'Differential Loss'.
54. Turvey et al., 'Rapidly Shifting Baselines in Yangtze', 785.
55. Shapiro, *Mao's War against Nature*; Coggins, *The Tiger and the Pangolin*; Hathaway, 'Global Environmentalism'.
56. Mu, 'Music Loss among Ethnic Minorities'; Netting, 'The Deer Turned Her Head'.
57. Nunn and Reid, 'Aboriginal Memories of Inundation'.
58. Nettle and Romaine, *Vanishing Voices*; Eberhard, Simons and Fennig, *Ethnologue*.

Bibliography

Anadón, José Daniel, Andrés Gimenez and Rubén Ballestar. 'Linking Local Ecological Knowledge and Habitat Modelling to Predict Absolute Species Abundance on Large Scales'. *Biodiversity and Conservation* 19(5) (2009): 1443–54.

Balmford, Andrew. 'Extinction Filters and Current Resilience: The Significance of Past Selection Pressures for Conservation Biology'. *Trends in Ecology and Evolution* 11(5) (1996): 193–96.

Beck, Jane C. 'The Giant Beaver: A Prehistoric Memory?' *Ethnohistory* 19(2) (1972): 109–22.

Berkes, Fikret. *Sacred Ecology*. Fourth edition. New York: Routledge, 2018.

Berkes, Fikret, Johan Colding and Carl Folke. 'Rediscovery of Traditional Ecological Knowledge as Adaptive Management'. *Ecological Applications* 10(5) (2000): 1251–62.

Boakes, Elizabeth H., Tracy M. Rout and Ben Collen. 'Inferring Species Extinction: The Use of Sighting Records'. *Methods in Ecology and Evolution* 6(6) (2015): 678–87.

Boakes, Elizabeth H., et al. 'Distorted Views of Biodiversity: Spatial and Temporal Bias in Species Occurrence Data'. *PLoS Biology* 8(6) (2010): e1000385.

Bonebrake, Timothy C., et al. 'Population Decline Assessment, Historical Baselines, and Conservation'. *Conservation Letters* 3(6) (2010): 371–78.

Burbidge, Andrew A., et al. 'Aboriginal Knowledge of the Mammals of the Central Deserts of Australia'. *Australian Wildlife Research* 15(1) (1988): 9–39.

Cheke, Anthony S. 'Establishing Extinction Dates: The Curious Case of the Dodo *Raphus cucullatus* and the Red Hen *Aphanapteryx bonasia*'. *Ibis* 148(1) (2006): 155–58.

Coggins, Christopher. *The Tiger and the Pangolin: Nature, Culture, and Conservation in China*. Honolulu: University of Hawai'i Press, 2003.

Colding, Johan, and Carl Folke. 'Social Taboos: "Invisible" Systems of Local Resource Management and Biological Conservation'. *Ecological Applications* 11(2) (2001): 584–600.

Dietl, Gregory P., and Karl W. Flessa. 'Conservation Paleobiology: Putting the Dead to Work'. *Trends in Ecology and Evolution* 26(1) (2011): 30–37.

Eberhard, David M., Gary F. Simons and Charles D. Fennig (eds). *Ethnologue: Languages of the World*. Twenty-third edition. Dallas: SIL International, 2020.

Godfrey, Laurie R., and William L. Jungers. 'The Extinct Sloth Lemurs of Madagascar'. *Evolutionary Anthropology* 12(6) (2003): 252–63.

Gómez-Baggethun, Erik, Esteve Corbera and Victoria Reyes-García. 'Traditional Ecological Knowledge and Global Environmental Change: Research Findings and Policy Implications'. *Ecology and Society* 18(4) (2013): 72.

Hall, Marcus (ed.). *Restoration and History: The Search for a Usable Environmental Past*. New York: Routledge, 2010.

Hathaway, Michael. 'Global Environmentalism and the Emergence of Identity: The Politics of Cultural and Biological Diversity in China', in Genese M. Sodikoff (ed.), *The*

Anthropology of Extinction: Essays on Culture and Species Death (Bloomington: Indiana University Press, 2011), 103–23.

Hoy, Charles M. 'The "White-Flag" Dolphin of the Tung Ting Lake'. *The China Journal of Science & Arts* 1 (1923): 154–57.

Jochelson, Waldemar. 'Some Notes on the Traditions of the Natives of Northeastern Siberia about the Mammoth'. *The American Naturalist* 43(505) (1909): 48–50.

Lander, Brian, and Katherine Brunson. 'Wild Mammals of Ancient North China'. *Journal of Chinese History* 2(2) (2018): 291–312.

Marks, Robert B. *China: An Environmental History*. Second edition. Lanham, MD: Rowman & Littlefield, 2017.

Mistry, Jayalaxshmi, and Andrea Berardi. 'Bridging Indigenous and Scientific Knowledge'. *Science* 352(6291) (2016): 1274–75.

Mu, Yang. 'Music Loss among Ethnic Minorities in China: A Comparison of the Liu and Hui Peoples'. *Asian Music* 27(1) (1995): 103–31.

Netting, Nancy S. 'The Deer Turned Her Head: Ethnic Options for the Hainan Li'. *Bulletin of Concerned Asian Scholars* 29(2) (1997): 3–17.

Nettle, Daniel, and Suzanne Romaine. *Vanishing Voices: The Extinction of the World's Languages*. New York: Oxford University Press, 2000.

Nunn, Patrick D., and Nicholas J. Reid. 'Aboriginal Memories of Inundation of the Australian Coast Dating from More Than 7000 Years Ago'. *Australian Geographer* 47(1) (2015): 11–47.

Papworth, Sarah K., et al. 'Evidence for Shifting Baseline Syndrome in Conservation'. *Conservation Letters* 2(2) (2009): 93–100.

Pilleri, Giorgio. 'The Chinese River Dolphin (*Lipotes vexillifer*) in Poetry, Literature and Legend'. *Investigations on Cetacea* 10 (1979): 335–49.

Pocock, Reginald Innes. 'Observations Upon a Female Specimen of the Hainan Gibbon (*Hylobates hainanus*), Now Living in the Society's Gardens'. *Proceedings of the Zoological Society of London* 75(3) (1905): 169–80.

Pope, Clifford H. *China's Animal Frontier*. New York: Viking Press, 1940.

Reyes-García, Victoria, et al. 'Evidence of Traditional Knowledge Loss among a Contemporary Indigenous Society'. *Evolution and Human Behavior* 34(4) (2013): 249–57.

Rick, Torben C., and Rowan Lockwood. 'Integrating Paleobiology, Archeology, and History to Inform Biological Conservation'. *Conservation Biology* 27(1) (2013): 45–54.

Rudwick, Martin J.S. *The Meaning of Fossils: Episodes in the History of Palaeontology*. Chicago: University of Chicago Press, 1972.

Shapiro, Judith. *China's Environmental Challenges*. Second edition. Cambridge: Polity Press, 2016.

———. *Mao's War against Nature: Politics and the Environment in Revolutionary China*. Cambridge: Cambridge University Press, 2001.

Shen, Xiaoli, et al. 'Viable Contribution of Tibetan Sacred Mountains in Southwestern China to Forest Conservation'. *Conservation Biology* 29(6) (2015): 1518–26.

Sterckx, Roel. *Chinese Thought: From Confucius to Cook Ding*. London: Pelican, 2019.

Tang, Ruifei, and Michael C. McGavin. 'A Classification of Threats to Traditional Ecological Knowledge and Conservation Responses'. *Conservation & Society* 14(1) (2016): 57–70.

Turvey, Samuel T., Jessica V. Bryant and Katherine A. McClune. 'Differential Loss of Components of Traditional Ecological Knowledge Following a Primate Extinction Event'. *Royal Society Open Science* 5(6) (2018): 172352.

Turvey, Samuel T., Jennifer J. Crees and Martina M.I. Di Fonzo. 'Historical Data as a Baseline for Conservation: Reconstructing Long-Term Faunal Extinction Dynamics in Late Imperial-Modern China'. *Proceedings of the Royal Society B* 282(1813) 20151299.

Turvey, Samuel T., et al. 'Complementarity, Completeness and Quality of Long-Term Faunal Archives in an Asian Biodiversity Hotspot'. *Philosophical Transactions of the Royal Society B* 374(1788) 20190217.

———. 'How Many Remnant Gibbon Populations are Left on Hainan? Testing the Use of Local Ecological Knowledge to Detect Cryptic Threatened Primates'. *American Journal of Primatology* 79(2) (2017): e22593.

———. 'First Human-Caused Extinction of a Cetacean Species?' *Biology Letters* 3(5) (2007): 537–40.

———. 'Interview-Based Sighting Histories Can Inform Regional Conservation Prioritization for Highly Threatened Cryptic Species'. *Journal of Applied Ecology* 52(2) (2015): 422–33.

———. 'Rapidly Shifting Baselines in Yangtze Fishing Communities and Local Memory of Extinct Species'. *Conservation Biology* 24(3) (2010): 778–87.

van Gulik, Robert Hans. *The Gibbon in China: An Essay in Chinese Animal Lore*. Leiden: E.J. Brill, 1967.

Wang, Hongzhu, Xueqin Liu and Haijun Wang. 'The Yangtze River Floodplain: Threats and Rehabilitation'. *American Fisheries Society Symposium* 84 (2016): 263–91.

Wen, Rongsheng. *The Distributions and Changes of Rare Wild Animals in China*. Jinan: Shandong Science and Technology Press, 2009.

Woods, Charles A., José A. Ottenwalder and William L.R. Oliver. 'Lost Mammals of the Greater Antilles: The Summarised Findings of a Ten-Weeks Field Survey in the Dominican Republic, Haiti and Puerto Rico'. *Dodo* 22 (1985): 23–42.

Xie, Gaodi, et al. *Living Planet Report: China 2015. Development, Species and Ecological Civilization*. Beijing: WWF, 2015.

Xu, Weihua, et al. 'Strengthening Protected Areas for Biodiversity and Ecosystem Services in China'. *Proceedings of the National Academy of Sciences of the USA* 114(7) (2017): 1601–6.

Zhang, Hui, et al. 'Extinction of One of the World's Largest Freshwater Fishes: Lessons for Conserving the Endangered Yangtze Fauna'. *Science of the Total Environment* 710 (2020): 136242.

Zhang, Lu, et al. 'Influence of Traditional Ecological Knowledge on Conservation of the Skywalker Hoolock Gibbon (*Hoolock tianxing*) Outside Nature Reserves'. *Biological Conservation* 241 (2020): 108267.

Zhou, Jiang, et al. 'Hainan Black-Crested Gibbon is Headed for Extinction'. *International Journal of Primatology* 26(2) (2005): 453–65.

Zhou, Kaiya, and Xingduan Zhang. *Baiji: The Yangtze River Dolphin and Other Endangered Animals of China*. Washington, DC: Stone Wall Press, 1991.

Ziembicki, M.R., J.C.Z. Woinarski and B. Mackey. 'Evaluating the Status of Species Using Indigenous Knowledge: Novel Evidence for Major Native Mammal Declines in Northern Australia'. *Biological Conservation* 157 (2013): 78–92.

Chapter 6

Grief, Extinction and *Bilhaa* (Abalone)

hagwil hayetsk (Charles R. Menzies)

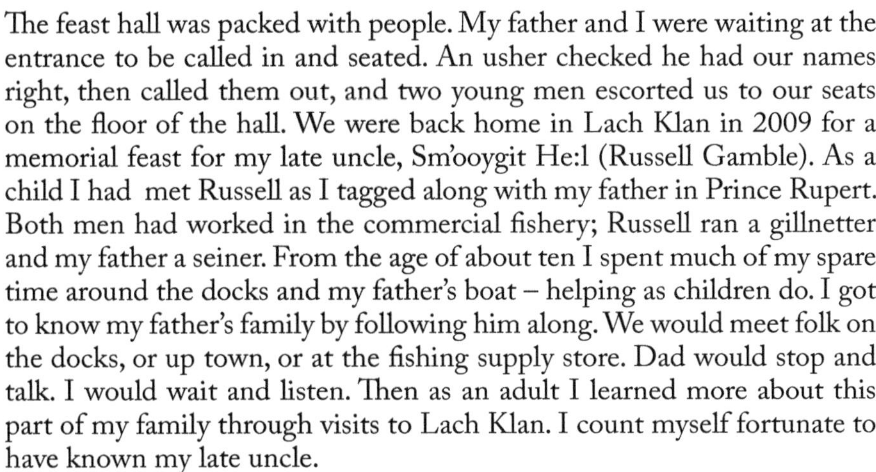

The feast hall was packed with people. My father and I were waiting at the entrance to be called in and seated. An usher checked he had our names right, then called them out, and two young men escorted us to our seats on the floor of the hall. We were back home in Lach Klan in 2009 for a memorial feast for my late uncle, Sm'ooygit He:l (Russell Gamble). As a child I had met Russell as I tagged along with my father in Prince Rupert. Both men had worked in the commercial fishery; Russell ran a gillnetter and my father a seiner. From the age of about ten I spent much of my spare time around the docks and my father's boat – helping as children do. I got to know my father's family by following him along. We would meet folk on the docks, or up town, or at the fishing supply store. Dad would stop and talk. I would wait and listen. Then as an adult I learned more about this part of my family through visits to Lach Klan. I count myself fortunate to have known my late uncle.

The trip my father and I took began on a plane from Vancouver to Prince Rupert, and then a boat to Lach Klan (Kitkatla Village). Like many other Indigenous people today (close to two-thirds) we do not live in our home village. Yet, Laxyuup Gitxaała (Gitxaała's national territory) is home. As we stepped off the boat at the village dock people greeted us and welcomed us home. We come home for many things; but coming home for funerals is a common reason. There is much important cultural work conducted at the memorial feast and the headstone moving feast that follows (albeit sometimes after an interval of many years). In addition to honouring the memory of those who have passed, hereditary names[1] are put on people at these feasts.

These are moments of grief, but also hope. Grief for the one we have lost, but recognition that their memory persists, and that their name will carry forward once it is placed upon the next person. Hope rests in the persistence of the hereditary name; its continuance is an enactment of our deep history and our continuing future.

Grief is palatable. It takes hold of oneself; it lingers, fades, and then returns. We know grief when we lose a loved one. Gone now more than a decade, I am still reminded of my uncle's presence and think of him. His memory lingers. It is not as though our loved ones will return, but the traces of their lives remain as reminders of their role in our lives and our Indigenous history. Grief honours our loss and helps us to keep faith with the memory of those who have left. But it is not just people that we grieve. The colonial expropriation of Gitxaała land and water – without treaty or agreement – has widened the scope of grief we feel. Places that are no longer accessible. Forests that have been logged to the ground. Unkempt landscapes left in the wake of development. Cherished foods like bilhaa (abalone, a single-shelled mollusc prized for its meat found in the inter-tidal zone) are scarce or non-existent due to colonial legal restrictions and non-Indigenous overexploitation.

The sense of loss and desire for bilhaa is clear in the words of the late Sm'oogyit Jeffrey Spencer, a Gitxaała elder and hereditary leader, in a No-vember 2001 interview: 'Seaweed and bilhaa and … ooh, I want to talk about bilhaa – chew it in my mouth. I never taste that for a long time. [laughter] Pretty hard to get. Not allowed to get it. Not allowed to get it. I just don't know why. I just don't know why'. Spencer's sense of grief and desire animate many of us as we survey the world we now find ourselves in. We hold memories of our ancestors within our life experiences as we chart ways to a shared future with those who came to colonize our world. The extirpation of bilhaa is a constant reminder of the wreckage wrought by colonialism. Yet, we know there is no going back to a moment before the settlers arrived. Our future necessarily involves reconciling to their pres-ence, and requires us to find a common future.

Gitxaała is an Indigenous nation whose history and national territory is located along the north coast of British Columbia, Canada. Laxyuup Gitxaała is comprised of the lineage territories of house groups that make up modern-day Gitxaała. Lach Klan, the primary contemporary village of Gitxaała, is located on Dolphin Island, about fifty kilometres to the south-west of Prince Rupert (Illustration 6.1). Laxyuup Gitxaała is an expanse of land and ocean that stretches from Ts'ibassa's oolichan processing territory along the shores of the Nass River and then south to the house territories

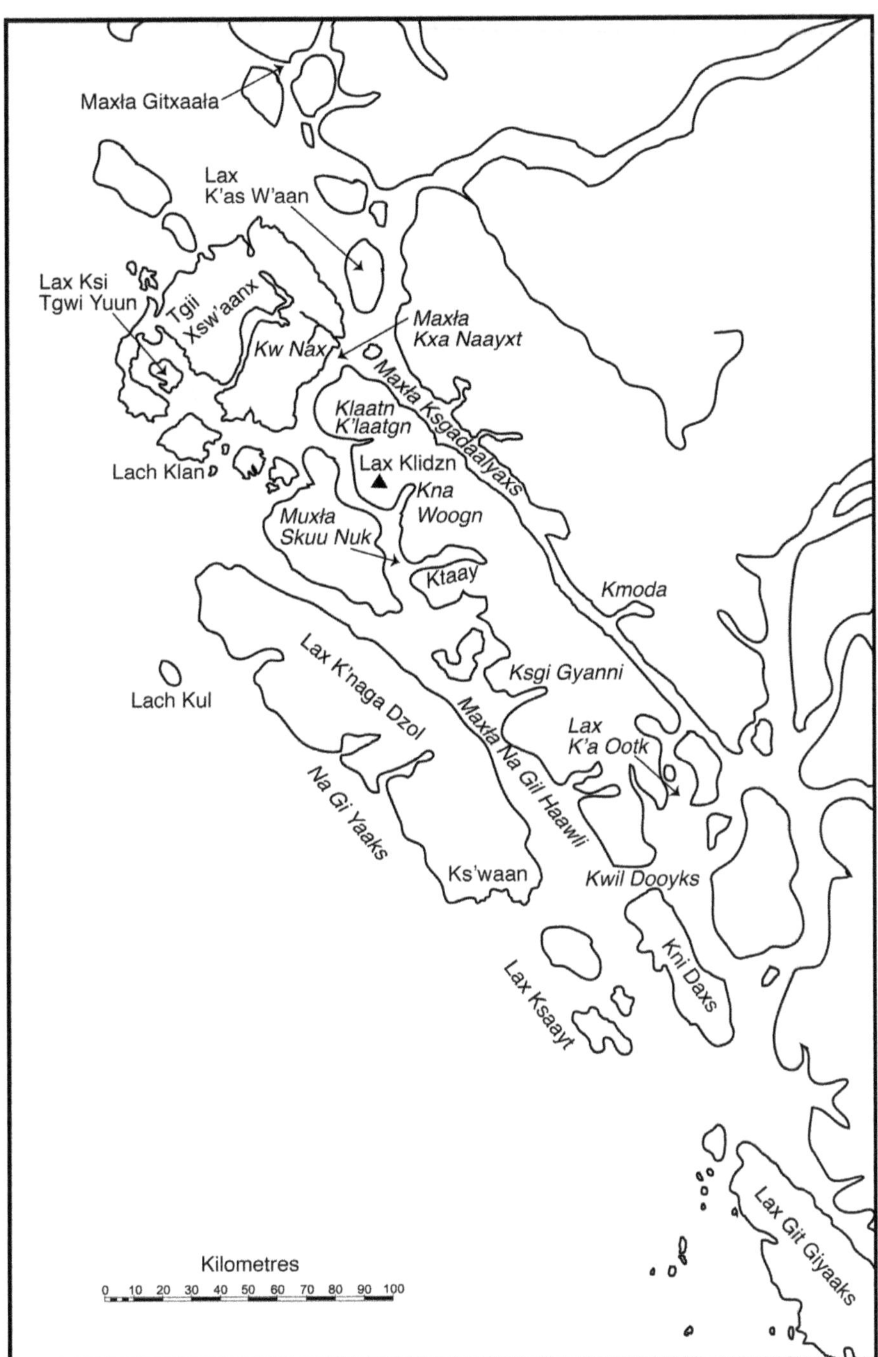

Illustration 6.1 Gitxaała place names. Created by Kenneth Campbell based on information provided by hagwil hayetsk (Charles R. Menzies).

of Txa-gyet, walp Gitnagunaks at ʾToʾtsip (Moore Island). The central core
of the laxyuup extends about 240 kilometres north to south from around
Prince Rupert into what non-Gitxaałans call the Great Bear Rainforest,
and includes the islands of K'tsm Kaawada (Porcher), Pitt, Lax K'naga Dzol
(Banks); Lax Ksaayɫ (the Estevan Group); Lax Git Giyaaks (Aristazabal Island); ʾToʾtsip (Moore Island); and portions of Kni Daxs (Campania Island)
(Illustration 6.1). The laxyuup also includes much of the mainland to the
east of Pitt Island around the laxyuup of He:l at Komodah (Lowe Inlet).

The legacy of colonialism has disrupted the customary respect accorded
to laxyuup Gitxaała. This has had implications for the biological health of
creatures such as bilhaa,[2] and for our community's well-being. Colonial
forces, arriving first as merchant adventurers in the late 1700s and then as
invasive colonists tirelessly working to displace and expropriate Indigenous
peoples, have imposed their own laws, and replaced and erased pre-existing
place names. Yet we persist and prosper.

In this chapter, a Gitxaała perspective is taken as the underlying and
guiding principle with regard to understanding the extirpation of bilhaa.
This is a story about bilhaa and grief, but also bilhaa in our future. Just as
hereditary names persist (sometimes changed) and are renewed through
being placed upon younger people, so too can we see a renewed place for
bilhaa in Gitxaała's future. This story concerns Gitxaała (and by extension the wider Indigenous world) and laxyuup Gitxaała. In what follows
I describe the place of bilhaa in Gitxaała's history; then the history of
twentieth-century commercial bilhaa fishery is outlined; and finally, I argue for a return to Indigenous place-based relations and values, which, unlike capitalist values, is the only way we can put a brake on the impending
mass extinctions in our global climate crisis.

The Place of Bilhaa in Gitxaała's History

Visiting with my father in late December 2014, I asked him when he first
remembered eating bilhaa. 'I was two', he said without a blink.[3] 'We always had them. Someone would get them. Ed [Smoygyet Tsibassa, Edward Gamble, my father's maternal grandfather] would bring them to us.
Family, friends, they got them. We would get them. When I ran my own
boat, I knew where to go. We always had them.'

Bilhaa have always been there (Illustration 6.2). We have always had
them, have always known how to find them, have always eaten them,
shared them and traded them. We have always known this. It is only in the
context of colonization that our memories have been challenged. Experts
on culture, history, language and laws think that they can correct us with

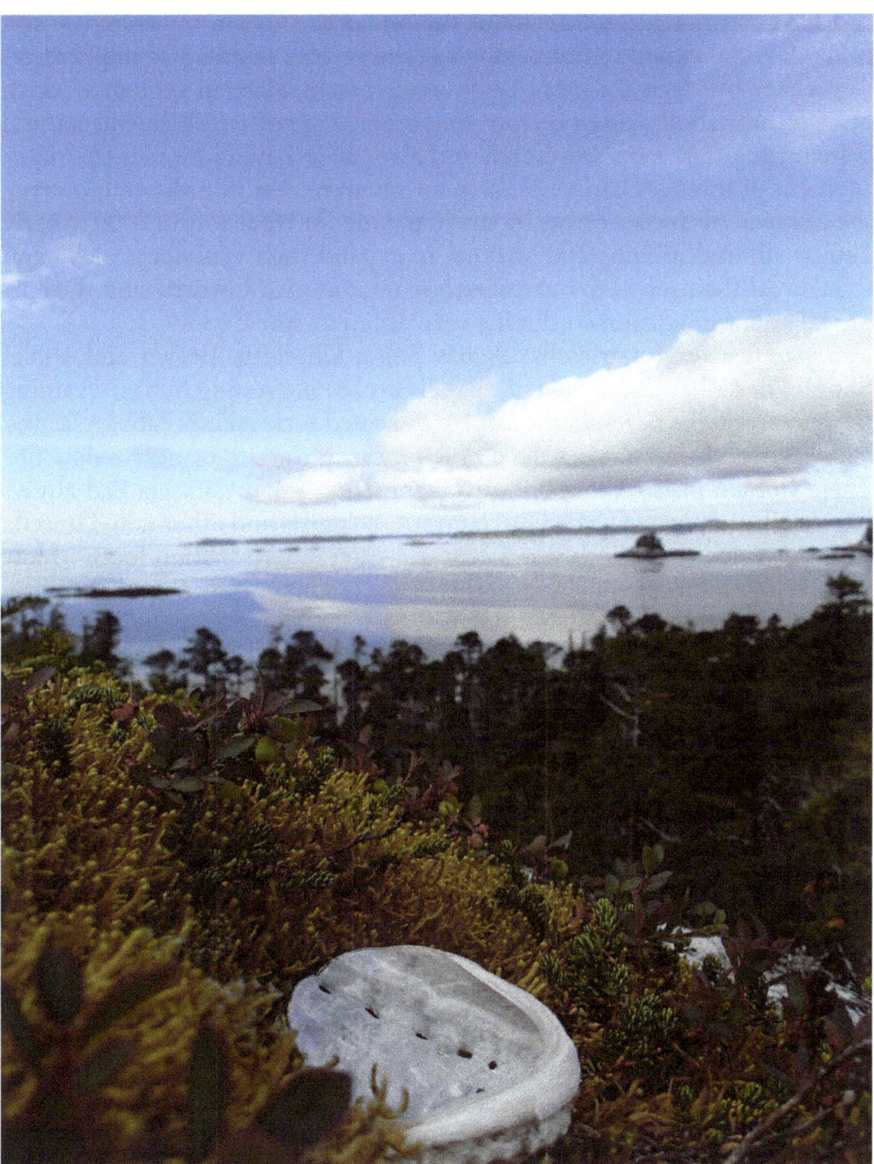

Illustration 6.2 Bilhaa onshore, overlooking Laxyuup Gitxaała. © hagwil hayetsk (Charles R. Menzies).

their external systems of disbelief, and compel us to deny what we know. This is one of the problems we have faced – the doubt and denial of the colonial experts who proclaim that for most of our history we did not and could not harvest, exchange or consume bilhaa.[4]

I have written previously about the bilhaa scepticism of ecologists and other colonial experts.[5] For ecologists, the pivotal moment is the extirpation of sea otters in the early 1800s, which resulted in an ecological crash of kelp forests following on from the removal of sea otters. In this natural system the otters kept sea urchin and abalone at bay, and under the management of otters rich, diverse, kelp forests were able to prosper. However, the removal of the sea otters by the maritime fur trade (late 1700s to early 1800s) allowed urchins and abalone to expand their ecological niche and thus reveal themselves to our ancestors. Yet, our own histories and the empirical archaeological record tell a very different story.

Sigidmnaa'nax (matriarchs) Agnes Shaw, Charlotte Brown and Violet Skog all describe in some detail the old ways of harvesting bilhaa: steaming the harvest on the beach in the sand with heated rocks, skunk cabbage leaves, and water, and then drying the cleaned meat in the sun or near a slow fire. These women insisted that bilhaa was something that Gitxaała had always harvested. They were puzzled by claims of ecologists and other non-Gitxaała who suggested that bilhaa was a post-sea-otter-extermination food.[6] "How could our grandmothers' grandmothers have taught us how to prepare bilhaa if it was something we only just learned to do", one matriarch asked.

Following the advice and suggestions of community knowledge holders, like the Sigidmnaa'nax named above, I had the opportunity to coordinate a community-based field archaeology project (2010–15), which, among other things, demonstrated conclusively the deep empirical evidence of the ancient and ongoing practice of bilhaa harvesting in laxyuup Gitxaała.[7] Our field archaeology programme combined university technical expertise with community knowledge holders in a collaborative coastal archaeology project. We travelled throughout laxyuup Gitxaała on board community-based fishing boats to a long list of old Gitxaała villages and contemporary fishing stations. At each of these locations we used the techniques of archaeology to examine the ancient soils. Later, at the University of British Columbia's archaeology labs, we sorted and identified the materials collected.

One village in particular overturned the settler stories of denial and scepticism. At the ancient village of Ks'waan we found the material evidence that archaeologists and ecologists had, until now, denied was possible: actual identifiable bilhaa shells in cultural soils in significant quantities. Putting the lie to the kelp forest myth (that it was only after the extermination of sea otter that Gitxaała discovered bilhaa), we revealed a faunal assemblage that supported the existence of the Gitxaała practice of bilhaa harvesting for at least two millennia. The faunal assemblage included marine mammals like seal, sea otter and sea lion; invertebrates such as urchin, chiton, and a host of snails; and deepwater fish such as halibut, cod and greenling. These are all species that reflect the ocean facing environment of deep-sea fishing and the coastal kelp forests that Gitxaała have stewarded for millennia.

Bilhaa, while conspicuously absent in the conversations of Northwest Coast archaeologists, turns up everywhere in Gitxaała conversations. Most telling is its presence in the cautionary tale about external research.[8] Here bilhaa figure as an object of a study conducted by community outsiders; government biologists who want to save bilhaa from overfishing yet whose research and management practices contributed to the extirpation of local bilhaa stocks in the first place.[9] But bilhaa is also part of our system of crests, our oral history, and, of course, a cherished food. Back home when we talk to each other about bilhaa, it is about experiences harvesting it, how it tastes, how the old people preserved it and traded it. Never is it spoken of as a question or presented as a doubt. For us, we know bilhaa has always been there. As my father said, 'we always had them'.

The presence, power, and importance of bilhaa in Gitxaała is recorded in our adawx (oral histories). Bilhaa are used on ceremonial regalia as historical references to adawx, and to denote power and prestige. The cultural importance of bilhaa plays a role in shaping resource-harvesting practices. In combination with the principle of syt güülum goot (being of one heart),[10] the high value placed on bilhaa as a symbol of prestige and rank acts to impose a cultural limitation on harvesting levels. This is so in two ways: firstly, the use of bilhaa as decoration and adornment is restricted to a minority of high-ranking community hereditary names and associated crests; secondly, the cultural importance of bilhaa as a signifier of rank obligates harvesters to treat bilhaa with respect such that unrestrained harvesting was and is a violation of social norms, and is subject to community sanction.

Throughout Gitxaała adawx are accounts of how bilhaa and bilhaa-adorned objects became important cultural markets. For example, 'Explanation of the Abalone Bow' is an adawx recorded by Franz Boas that describes how the Bilhaa Bow became a chief's crest.[11] In the narrative G-it-na-gun-a'ks, bilhaa also feature as an inlay on 'a good-sized box', which is one of several gifts exchanged between a naxnox, Na-gun-a'ks, and the people of Dzagam-sa'gisk.[12] Drawing upon his work up to that point,[13] Boas also notes that 'ear-ornaments of abalone shell' are mentioned in several important adawx. Viola Garfield, an early twentieth-century ethnographer who studied with Boas,[14] also notes: 'At any ceremonial, large wool ornaments with abalone-shell pendants were worn in the ears of the women who sing in the chief's choir, so that the status of each was clearly indicated to the tribes at large'. Anthropologist Marjorie Halpin documents how crests that were restricted to high-ranked individuals often had names that included shiny/shining/bright, and the individual's associated regalia would often use bilhaa shells to indicate their high status.

In a description of a mid-nineteenth-century feast, Halpin explains: 'We would have noted that the men who made the speeches wore the more elaborate headdresses, richly decorated with shining abalone'.[15] Bilhaa, in addition to being an important food, were also tightly incorporated within our cultural complex as a critical material manifestation of our cultural history and ancient spiritual practices.

All of these aspects of bilhaa for Gitxaała have been undermined in a reckless manner by non-Indigenous resource extraction and overexploitation. Our cultural practices and values continue, but our access to bilhaa has been undermined. The following section describes how the intersection of the global capitalist market for exotic seafood combines with inept resource management to extirpate bilhaa in laxyuup Gitxaała.

The Twentieth-Century Commercial Bilhaa Fishery

Bilhaa have not fared well under the commercial pressures of the global economy.[16] Traditional harvest practices and historical limitations in technology served to restrain human harvests to extreme low tides.[17] The advances of a capitalist market economy and twentieth-century developments in underwater diving have almost universally destroyed the global bilhaa stocks. This is no less true in the traditional territories of the Gitxaała where bilhaa had been harvested sustainably for millennia. Then, in the space of three decades, bilhaa were fished to the edge of extinction by a non-Indigenous commercial fishery.

The development of the non-Indigenous commercial dive fishery in British Columbia is a classic example of competitive greed combining with ineffectual resource management to decimate a resource.[18] Prior to 1972, bilhaa harvesting was unregulated.[19] Most harvesting in this period was either recreational or Indigenous. The pre-1972 annual harvest rates were estimated to be less than 20 tons.[20] However, the increased prices paid to fishermen for bilhaa, coupled with uncontrolled increases in fishing capacity, led to a rapid take-off of the non-Indigenous bilhaa fishery in the 1970s.[21] Annual rates of harvest quickly shot up to 481 tons in 1977 and then were held to 47 tons from 1985 to the closure of the fishery in 1990.[22] A significant portion – about half – of the non-Indigenous commercial fishing effort was concentrated on British Columbia's north coast in Gitxaała traditional territory.[23] During the period in which non-Indigenous commercial catch data was recorded (1977–90), the catch per unit effort declined by 46 per cent.[24] Canada's Department of Fisheries and Oceans (DFO) calculated that by 1984 the bilhaa stocks had been depleted by more than 75 per cent.[25]

The closure of the bilhaa fishery has had a significant impact upon Gtixaała people. It resulted in the loss of a critical food resource, the loss of a critical trade item, and an increase in surveillance of Indigenous harvesters. As with many other traditionally managed and harvested foods and materials, colonial legal restrictions and associated regulatory surveillance criminalized the harvest, exchange and consumption of bilhaa. This puts community members into harm's way if they persist in exercising their customary rights to pick bilhaa. These restrictions on Gitxaała followed in the wake of the disastrous non-Indigenous commercial bilhaa fishery, which had taken no account of Gitxaała's own long traditions and practices of respectful relations with beings like bilhaa.

Returning to Tradition under Gitxaała Authority

Capitalist forms of production are biased in favour of accumulation over sustainability.[26] So-called renewable resources, especially those considered common property, fair very poorly under regimes of the capitalist marketplace. While markets have existed within non-capitalist societies (such as the wealth-generating economy of my Gitxaała ancestors), capitalism differs from previous systems in the manner by which it is able to abstract value from the particular to the universal, and as such create insatiable demands for accumulation that act independently of the object of its exploitation. For renewable resources like bilhaa, this is particularly dangerous for their well-being.

Place-based management systems interfere with the universalizing tendencies of a capitalist marketplace. By instituting social constraints that limit harvesting practices (who, where, when) and investing them with cultural values (authority, prestige, obligation), Indigenous place-based systems of resource management have historically been able to sustain large volumes of harvest for lengthy durations.[27] For example, in examining Gitxaała's salmon harvesting practices,[28] it became apparent that Gitxaała have actively intervened in salmon ecology through modification of coastal watersheds to enhance spawning recovery. Furthermore, Gitxaała harvesting practices were oriented towards sustaining large harvests over long periods. Complementing Gitxaała knowledge-holders' descriptions of long-standing sustainable practices our adjoining archaeological research documented large stable harvests of salmon at multiple locations within laxyuup Gitxaała over the course of at least two millennia.[29] It is important to point out that Gitxaała had the technological capacity to extirpate salmon runs within the national territory. However, harvesting practices were kept within sustainable levels by social conventions that tied harvesting groups to particular locations, and did not allow for movement beyond a person's fishing station.

Gitxaała's bilhaa harvesting protocols restricted harvesting to times of the lowest tides. This method uses natural environmental cycles to regulate human access.[30] While there were rudimentary traps used to augment harvest from time to time (such as setting out a seal skin, blubber side up, below the low tide mark), these traps provided only a modest harvesting advantage, especially when compared against dive fisheries, which face practically no environmental impediments to their practice.

All Gitxaała harvesting protocols operate within a needs-based harvesting model.[31] The idea of community need is a long-standing guiding principle of Gitxaała harvesting. This is not simply a product of late modernity but can be documented through careful consideration of Gitxaała adawx in which excessive harvest is typically punished by the withdrawal of, say, salmon or bilhaa availability.[32] To understand this, one must appreciate that we see ourselves existing in social relations with the animals we harvest, and our relationship is in some sense analogous to kin relations. Our ability to harvest in a sustainable way is contingent upon our animal relatives offering themselves to us for our use, and this entangles us in relations of obligation to treat them with respect. When we demonstrate disrespect by overharvesting, they withdraw their availability to us. Part of rebuilding a respectful harvesting relationship will involve returning the practice of harvesting bilhaa to Gitxaała's own jurisdiction. Clearly there is little to lose in reverting to a Gitxaała-controlled bilhaa fishery as the non-Indigenous regulators have not been able to rehabilitate bilhaa – and, one fears, they are unlikely to be able to do so as long as they continue to operate under principles of open-access, free-market harvesting models.

Gitxaała's own fisheries department has for many years run monitoring programmes to observe, count and evaluate bilhaa stocks in laxyuup Gitxaała. What they have observed suggests that a Gitxaała-owned fishery is feasible and sustainable operating under Gitxaała protocols. Such a fishery could be tied to long-standing traditional methods and harvesting rules. It remains to be seen whether or not federal regulatory agencies have the maturity to relinquish their colonial controls, and reconcile themselves to the authority and jurisdiction and time-test protocols of Gitxaała's own harvesters.

Concluding Thoughts

Bilhaa face difficulties globally. They are a slow-growing marine mollusc, much prized for shell and meat, and this has made them vulnerable to abuse. Bilhaa hold an important sociocultural place in Gitxaała history, practice and foodways. They figure predominantly in oral histories. They adorn our regalia. They are tasty, and we delight in talk about cooking, eating and sharing them. All of this has been tarnished by our experience with European colonialism. Lands have been taken, practices have been demeaned and denigrated, our foods have been regulated and extirpated. We stand with grief as we survey the wreckage of capitalist colonialism in our laxyuup. Yet we can look to our deep history in laxyuup Gitxaała and take hope. There have been moments in the past where the way was lost. In those moments the relations between humans and other social beings were disrupted. But by remembering who we are and what our obligations are, as one among many beings, those ancient crises were resolved. We can return to respectful relations with bilhaa. We can do this by taking back our place as one among many, and return to honouring our obligations to treat bilhaa as relatives.

hagwil hayetsk (**Charles R. Menzies**) is a member of Gitxaała Nation and a professor of anthropology in the Department of Anthropology at the University of British Columbia in Vancouver, BC. His primary research interests are the production of anthropological films, natural resource management (primarily fisheries related), political economy, contemporary First Nations' issues, maritime anthropology and the archaeology of north coast BC.

Notes

1. Hereditary names have a deep history in Gtixaała and the wider Tsimshian worlds (Menzies, *People of the Saltwater*; Roth, *Becoming Tsimshian*). Names are like people in their own right. When a name is placed upon a person it is as though they are that name – today, historically, and in the future. There is a recognition of the historical specificity of each individual who carries the name, but they are like chapters in the life history of the name itself.
2. Menzies, 'Dm sibilhaa'nm da laxyuubm Gitxaała'; Menzies, 'Revisiting'.
3. Portions of this section on the place of bilhaa in Gitxaała's history are adapted from Menzies, 'Dm sibilhaa'nm da laxyuubm Gitxaała'; Menzies, 'Revisiting'.
4. Menzies, 'Dm sibilhaa'nm da laxyuubm Gitxaała'; Menzies, 'Revisiting'.
5. Menzies, 'Revisiting'.
6. Cf. Cannon and Burchell, 'Clam Growth-Stage Profiles'.
7. Menzies, 'Revisiting'.
8. Menzies, 'Putting Words into Action'; Menzies, 'Dm sibilhaa'nm da laxyuubm Gitxaała'.

9. Menzies, 'Dm sibilhaa'nm da laxyuubm Gitxaała.'
10. Menzies and Butler, 'Returning to Selective Fishing'.
11. Boas, 'Tsimshian Mythology', 284, 835.
12. Boas, 'Tsimshian Mythology'.
13. Ibid., 398.
14. Garfield, *Tsimshian Clan and Society*, 194.
15. Halpin, 'The Structure of Tsimshian Totemism',16. See also Halpin, 'The Tsimshian Crest System'.
16. The section on the history of bilhaa in the global economy is adapted from Menzies, 'Dm sibilhaa'nm da laxyuubm Gitxaała'.
17. Menzies, 'Dm sibilhaa'nm da laxyuubm Gitxaała'.
18. Gitxaała people have been involved in the development of the cash economy in British Columbia. See Menzies and Butler, 'Working in the Woods'; Menzies and Butler, 'The Indigenous Foundation'. However, Gitxaała were not involved in the commercial dive fishery for bilhaa.
19. Adkins, 'The British Columbia Fishery'.
20. Campbell, 'Review of Northern Abalone'.
21. Ibid.
22. Adkins, 'The British Columbia Fishery,' Campbell, 'Review of Northern Abalone'.
23. Adkins, 'The British Columbia Fishery'.
24. Campbell, 'Review of Northern Abalone'.
25. Ibid.
26. Rogers, *The Oceans are Emptying*.
27. Trosper, *Resilience, Reciprocity and Ecological Economics*.
28. Menzies and Butler, 'Returning to Selective Fishing'; Menzies, 'The Disturbed Environment'.
29. Menzies, 'Laxyuup Gitxaała'.
30. Menzies, 'Dm sibilhaa'nm da laxyuubm Gitxaała'.
31. Menzies and Butler, 'Returning to Selective Fishing'.
32. Menzies, *People of the Saltwater*.

Bibliography

Adkins, Bruce. 'The British Columbia Fishery for Northern Abalone, Haliotis kamtschatkana: Management from Inception to Closure and Beyond', in Alan Campbell (ed.), *Workshop on Rebuilding Abalone Stocks in British Columbia* (Nanaimo, Canada: Fisheries and Oceans Canada, Pacific Biological Station, 2000), 51.

Boas, Franz. 'Tsimshian Mythology', in F.W. Hodge and Franz Boas (eds), *Thirty-First Annual Report of the Bureau of American Ethnology to the Secretary of the Smithsonian Institution, 1909–1910* (Washington, DC: Government Printing Office, 1916), 29–1037.

Campbell, Alan. 'Review of Northern Abalone, Haliotis Kamtschatkana, Stock Status in British Columbia', in Alan Campbell (ed.), *Workshop on Rebuilding Abalone Stocks in British Columbia* (Nanaimo, Canada: Fisheries and Oceans Canada, Pacific Biological Station, 2000), 41–50.

Cannon, A., and M. Burchell. 'Clam Growth-Stage Profiles as a Measure of Harvest Intensity and Resource Management on the Central Coast of British Columbia'. *Journal of Archaeological Science* 36(4) (2009): 1050–60.

Garfield, Viola Edmundson. *Tsimshian Clan and Society*. Seattle: University of Washington, 1939.

Halpin, Marjorie. 'The Structure of Tsimshian Totemism', in Jay Miller and Carol M. Eastman (eds), *The Tsimshian and Their Neighbors of the North Pacific Coast* (Seattle: University of Washington Press, 1984), 16–35.

———. 'The Tsimshian Crest System: A Study Based on Museum Specimens and the Marius Barbeau and William Beynon Fieldnotes'. PhD dissertation, University of British Columbia, 1973.

Menzies, Charles R. 'The Disturbed Environment: The Indigenous Cultivation of Salmon', in Benedict J. Colombi and James F. Brooks (eds), *Keystone Nations: Indigenous Peoples and Salmon across the North Pacific* (Santa Fe, NM: School for Advanced Research, 2012), 161–82.

———. 'Dm sibilhaa'nm da laxyuubm Gitxaała: Picking Abalone in Gitxaała Territory'. *Human Organization* 69(3) (2010): 213–20.

———. 'Laxyuup Gitxaała: Gitxaała Territory through an Archaeological and Anthropological Lens'. Permit No. 2010-213. Final report, BC Archaeology Branch, Victoria, British Columbia, 2015.

———. *People of the Saltwater: An Ethnography of the Git lax m'oon*. Lincoln: University of Nebraska Press, 2016.

———. 'Putting Words into Action: Negotiating Collaborative Research in Gitxaała'. *Canadian Journal of Native Education* 27(3) (2004): 15–32.

———. 'Revisiting "Dm Silbilhaa'nm da Lasyuubm Gitxaała: Picking Abalone in Gitxaała Territory": Vindication, Appropriation, and Archaeology'. *BC Studies* 187 (2015): 129–53.

Menzies, Charles R., and Caroline F. Butler. 'The Indigenous Foundation of the Resource Economy of BC's North Coast'. *Labour/Le Travail* 61 (2008): 131–49.

———. 'Returning to Selective Fishing through Indigenous Fisheries Knowledge: The Example of K'moda, Gitxaala Territory'. *American Indian Quarterly* 33(3) (2007): 441–64.

———. 'Working in the Woods: Tsimshian Resource Workers and the Forest Industry of BC'. *American Indian Quarterly* 25(3) (2001): 409–30.

Rogers, Raymond. *The Oceans are Emptying: Fish Wars and Sustainability*. Montreal: Black Rose Books, 1995.

Roth, Christopher F. *Becoming Tsimshian: The Social Life of Names*. Seattle: University of Washington Press, 2008.

Trosper, R.L. *Resilience, Reciprocity and Ecological Economics: Northwest Coast Sustainability*. New York: Routledge Studies in Ecological Economics, 2009.

Part III

Representing Avian and Insect Extinctions

Chapter 7

Sparrows with Teeth and Claws?

Reconstructing the Cretaceous Enantiornithes
(Aves: Ornithothoraces)

Jingmai O'Connor

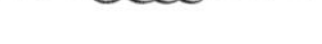

Introduction

Palaeontology – the study of ancient life – is a unique scientific discipline in that a strong dose of imagination and creativity is present, expressing itself through the creation of *in vivo* reconstructions of extinct creatures. Such reconstructions are important for engaging public interest in palaeontology. Although a trained scientist or devoted amateur may study a skeleton and be able to visualize what it would have looked like *in vivo* (alive), this requires a comprehensive knowledge of skeletal anatomy that can only be acquired through extensive training. Reconstructions are especially important visual tools if the fossil taxon is known from only a small portion of the skeleton. Fossils are notoriously rare and incomplete, and species have been named from a single bone or even less. As such, reconstructions are constantly transforming with discoveries of new material and data.

Many palaeontologists were set on their career path at an early age, stimulated by a childhood interest in prehistoric animals brought on by vivid reconstructions seen in books or on television. A number were inspired by the Jurassic Park movies, in which not only the visual appearance of Mesozoic reptiles was reconstructed, but also their movements and behaviours. Perhaps unsurprisingly, many palaeontologists dabble in art, a necessity for creating scientific figures, sometimes bringing their discoveries to life by creating their own reconstructions. Similarly, palaeoartists,

artists who have dedicated themselves to the reconstruction of ancient life, are typically avid amateur palaeontologists whose work is often detail oriented, strongly rooted in fossil evidence and the latest discoveries. Artists often work closely with palaeontologists to ensure their work is of the upmost accuracy. Others disregard the work of scientists and offer their own interpretations – these pseudoscientists are often widely recognized on the internet, but are detrimental to the general public's understanding of extinct life.

In recent decades, both exceptional new discoveries and the development of new methodologies have revolutionized the field of palaeontology, and provided rich new details that contribute to the increasingly life-like aspect of recent palaeo reconstructions. New methods include both the use of advanced technology, such as synchrotron-based computed tomographic (CT) scanning (three-dimensional X-ray images, for example),[1] and the novel application of methods typically utilized by other disciplines, such as Raman spectroscopy (which reveals chemical signatures) and the histochemical staining (a technique used by doctors and biologists to differentiate the structural elements of tissues by their colour or the intensity of staining by chemical dyes) of fossils.[2] The exceptional discovery of feathers (or protofeathers in some cases) preserved on various groups of dinosaur fossils uncovered primarily in Late Jurassic – Early Cretaceous deposits in north-eastern China[3] have revealed – to the dismay of many – that even *T. rex* might have been covered in 'dino fuzz' (protofeathers) during at least part of its lifetime,[4] and that oviraptorosaurs had small 'proto-wings' on their forelimbs,[5] while many deinonychosaurs had 'proto-wings' on their forelimbs and hindlimbs.[6] Raman spectroscopy has revealed the colour of dinosaur eggs,[7] and CT scans have provided unprecedented anatomical details, especially concerning internal structures like the semi-circular canals[8] and enigmatic structures like the avian predentary (a bone at the tip of the lower jaw in some Cretaceous birds that is absent in neornithines, the clade that includes all modern birds).[9]

These same deposits in north-eastern China that revealed that the ancestors to *T. rex* were covered in dino-fuzz have produced an enormous diversity of early bird fossils.[10] More than half of all known species of Mesozoic birds come from three formations and a period of the Early Cretaceous spanning approximately 10 million years (from 130 to 120 Ma).[11] These deposits reveal a diverse biota living in a system of ancient lakes and forests, punctuated by volcanic activity. This celebrated Jehol Biota has revealed more about the diversity and biology of Mesozoic birds than all other known avian-bearing strata combined.[12] The Jehol deposits represent Lagerstätten, meaning they are characterized by exceptional preservation, exemplified by the preservation of soft tissues such as feathers.[13]

Bird fossils are especially rare for two reasons: birds are small, their body size being constrained by flight; and their bones are hollow, a feature that is aerodynamically advantageous because of reduced body mass. As such, the Mesozoic fossil record of birds in most places is poor and fragmentary. Lacustrine (lake) deposits in particular represent taphonomic windows for the preservation of delicate fossils like birds. In contrast to most other Cretaceous avian bearing strata, fossils from the volcano-lacustrine Jehol Biota are typically nearly complete and fully articulated.[14] These spectacular fossils often preserve typically rare traces such as stomach contents and soft tissues most commonly in the form of feathers, but also including rare traces of organs such as lungs and ovaries. Embryos and juveniles have also been recovered.

One group of birds – the Enantiornithes[15] – dominated the Jehol Biota as well as nearly every known Cretaceous avifauna, with the exception of those from marine deposits.[16] Enantiornithines were the dominant clade of continental birds in the Cretaceous and are considered the first major avian radiation.[17] They account for approximately half of all the diversity of Mesozoic birds currently recognized. The Enantiornithes are a fairly derived Cretaceous clade. Their sister-taxon, the Ornithuromorpha, is the clade that modern birds (Neornithes) nest within. Together, these two clades are called the Ornithothoraces. Non-ornithothoracine birds (birds more primitive than the enantiornithines and ornithuromorphs such as *Jeholornis* and *Sapeornis*) are only definitively known in the Upper Jurassic Solnhofen Limestones in southern Germany, which produce *Archaeopteryx*, and the Lower Cretaceous deposits that record the Jehol Biota.[18]

Most data pertaining to the Enantiornithes is from the Early Cretaceous, where hundreds (if not thousands) of exceptional specimens from the Jehol Biota have revealed an enormous wealth of information. As a result, our understanding of this clade is strongly skewed to a single region over a relatively short period. Additional Early Cretaceous Lagerstätten in north-western China (the Xiagou Formation at Changma)[19] and Spain (Las Hoyas)[20] have also contributed important data, although the specimens from these deposits are far fewer and less complete.

By contrast, most Late Cretaceous enantiornithines are recognized from very incomplete specimens. Although a few partial skeletons have been found (e.g. *Mirarce, Neuquenornis, Elsornis, Yuornis*),[21] a majority of Late Cretaceous species are known from a few associated bones or even less than a single complete element.[22] However, these specimens are almost always preserved in three dimensions, revealing minute details such as muscle scars, whereas specimens from Early Cretaceous Lagerstätten are almost always crushed flat and do not preserve such details. As such, reconstructions of Early Cretaceous taxa are based on actual specimen data, whereas Late Cretaceous reconstructions must heavily borrow from skeletal and soft tissue details preserved in older fossils that can be phylogenetically inferred to have been present (as scientists infer the Late Cretaceous *T. rex* was likely feathered at some point in its life, based on basal tyrannosauroids from the Early Cretaceous).

Recently, ~100 Ma (early Late Cretaceous) amber from Myanmar has proved to be an unlikely new source of information regarding enantiornithines. A handful of birds have been recovered.[23] These are typically fragmentary, consisting of an isolated wing or a leg, although one nearly complete hatchling is known. These specimens are exceptionally well preserved in three-dimensions, and in most cases the skeletal remains are associated with soft tissues. Numerous isolated feathers have also been recovered, which are most likely referable to Enantiornithes – the only group of birds recovered with certainty from these deposits so far.

The Basic Bauplan

Avian lineages basal to enantiornithines possessed distinct characteristics that made them strikingly different from neornithines. Among these differences, *Archaeopteryx* and *Jeholornis* possessed elongate reptilian tails formed by over twenty free caudal vertebrae;[24] confuciusornithiforms (a group of Early Cretaceous birds from the Jehol Biota) had hands with three digits with large claws, and unusual long and narrow remiges (the flight feathers of the wing);[25] and *Archaeopteryx* and *Sapeornis* lacked an ossified sternum, suggesting that the breast muscle (the m. *pectoralis*) would have been small.[26] In Ornithothoraces the basic avian skeletal morphology was in place: the tail is abbreviated and ends in a distally fused element, the pygostyle, which is the bone that would have supported the tail feathers (rectrices) and associated musculature; the coracoid (a bone that connects the cranial edge to the sternum to the shoulder joint that supports the forelimb) is elongate and separate from the scapula; the furcula (wishbone) demarcates a narrow interclavicular angle; and the sternum is keeled with caudal trabeculae (bony processes that extend out from the caudolateral margins of the bone).[27] As such, enantiornithines would have strongly resembled living birds, with a few important differences, the most obvious being the presence of teeth and small manual (hand) claws in most taxa. Compared to ornithuromorphs, enantiornithine skeletal structure is notably more primitive. While both have a narrow furcula, that of Cretaceous ornithuromorphs was U-shaped and resembles that of living birds, whereas that of enantiornithines was Y-shaped, a morphology unique to this clade.[28] Similarly, the pygostyle of ornithuromorphs resembles that of living birds being small and ploughshare shaped, whereas that of enantiornithines was proportionately longer and more robust than that of neornithines.[29]

Compared to neornithines, cranial disparity was very limited in the Enantiornithes – all taxa were mesorostrine (the rostrum, the portion of the skull that in living birds forms the beak, accounts for 50–70 per cent the total skull length) whereas neornithines include brevirostral (30–50 per cent skull length) and longirostral (70–90 per cent) forms. Within the mesorostral range, the longipterygids (a diverse group of enantiornithines from the Jehol Biota), with their rostrum accounting for 60–65 per cent of the total skull length, represent a distinct departure from other enantiornithines, in which the rostrum is close to 50 per cent skull length.[30]

Most enantiornithines retain teeth, although at least some Late Cretaceous species, like *Gobipteryx* and *Yuornis*, were edentulous (teeth absent) like all living birds, and presumably had a rostrum covered in a keratinous sheath – the rhamphotheca.[31] Most commonly, teeth were fairly low in number (6–8 in each dentary), and simple and conical. However, pengornithids (a basal family of enantiornithines from the Jehol Biota) had higher tooth counts (~11 dentary teeth) and low-crowned teeth. The longipterygids, with their proportionately elongate rostra, had dentition restricted to the tip of their rostrum. Their tooth morphology varied from small peg-like in *Longirostravis* and *Rapaxavis*, to large and strongly recurved in *Longipteryx*. Bohaiornithids (the most diverse family of enantiornithines in the Jehol Biota) had proportionately robust teeth compared to other enantiornithines, which may suggest a more durophagous diet. Some enantiornithines (e.g. *Monoenantiornis*, *Sulcavis*) reveal enamel wrinkles that may have served to strengthen the teeth.[32]

Cranial morphology further indicates that cranial kinesis (relative movement between the upper jaw and braincase) was likely absent, indicating that certain feeding behaviours such as probing were absent in these birds.[33] The skull was robust compared to that of living birds, retaining a postorbital bone and free squamosal bone.[34] In neornithines, the premaxilla (the bone forming the tip of the rostrum) is expanded, while the maxilla (the bone forming the sides of the rostrum) is reduced. This results in the external nares (nostrils) being retracted (the exception being the Kiwi bird). The external nares in enantiornithines would have been more rostrally located, as the premaxilla was unexpanded in all known taxa except *Gobipteryx*.

In the postcranial skeleton, the morphology of the cervical vertebrae suggests the neck of enantiornithines would have lacked the mobility present in extant birds, because the articular surfaces were only incipiently heterocoelic. Heterocoeleous vertebrae, which have saddle-shaped articular surfaces, provide greater flexibility and are characteristic of the neck in living birds. They also lack the diversity of neck proportions observed in extant birds. The enantiornithine neck consisted of only 10–12 vertebrae, compared to 11–25 in neornithines. The sternum was much smaller in enantiornithines, and the keel (the ventrally projecting process of the sternum that provides expanded surface area for the attachment of the large flight muscles) was low and caudally limited in Early Cretaceous taxa, such that the m. *pectoralis* would have been smaller and less powerful in enantiornithines compared to most neornithines. The enantiornithine synsacrum (formed by fusion of the sacral vertebrae) was shorter than that of ornithuromorphs, formed by fewer vertebrae (usually 7 or 8, whereas in neornithines the number can exceed 19), and the pelvis was unfused – both to the synsacrum and, in most cases, at the level of the acetabulum (the hip socket for the femur). The ilium (one of the three bones that forms each side of the pelvis) was proportionately smaller than that of neornithines, and the pubes were not fully retroverted.[35] Although some taxa had proportionately long wings (e.g. *Longipteryx*), no enantiornithines with elongate hindlimbs are definitively known.[36] In most taxa the ratio of the forelimb to hindlimb is close to one.

Most enantiornithines were small birds similar to extant passerines (the large group of extant perching birds that includes all songbirds). Compared to species in the Early Cretaceous, Late Cretaceous taxa occupy a greater size range, ranging from very small forms the size of hummingbirds and up to the largest-known enantiornithines, roughly the size of Turkey vultures.[37] A vast majority of known enantiornithines preserve morphologies that suggest they were arboreal, including an elongate, reversed and distally located hallux (the first pedal digit, which is 'reversed' so that it opposes the rest of the digits in all perching birds) and large, curved pedal claws.[38] All known Early Cretaceous species are interpreted as primarily arboreal, whereas in the Late Cretaceous there appears to be a greater ecological diversity – although because specimens are much more incomplete, habitat is more difficult to ascertain. The Late Cretaceous *Elsornis* was probably flightless based on the shape and proportions of its humerus.[39]

The skeletal morphology of enantiornithines indicates they were clearly capable of powered flight, although skeletal diversity is more limited compared to neornithines indicating they had not achieved the diversity of flight styles observed in modern birds. Early Cretaceous species appear to be mostly intermittent fliers using bounding or flap-gliding flight,[40] although the Bohaiornithidae may have utilized brief continuous flapping.[41] Large Late Cretaceous taxa almost certainly utilized different flight styles, as intermittent flight is restricted to small taxa, although aerodynamic capabilities in Late Cretaceous taxa have yet to be explored, probably because they are mostly so incomplete. The appearance of advanced skeletal morphologies in the Late Cretaceous, such as an expanded sternal keel and increased pneumaticity (air-filled spaces within bones), indicate improved flight performance, which probably facilitated the increase in body size.

Plumage and Other Soft Tissues

Enantiornithine plumage is fairly well known due primarily to the numerous specimens from Early Cretaceous Lagerstätten preserving a halo of feathers and other soft tissues, resembling fresh roadkill.[42] This extends mostly to general wing shape, the extent of hindlimb feathering, and tail morphology (Illustration 7.1). Body feathers in other regions are poorly known due to overlap. Melanosomes are preserved allowing coloration to be at least partially determined through destructive sampling.

No Late Cretaceous specimen has well-preserved feathers, with the exception of Burmese amber specimens. These three-dimensional amber mummies have revealed that interpretations regarding the structure of the unusual so-called rachis-dominated tail feathers (the rachis is the central shaft of a feather) based on two-dimensional lithic specimens from Early Cretaceous Lagerstätten were completely wrong,[43] and exposed the presence of unusual scales with filamentous projections on the feet,[44] which owing to their extremely delicate morphology and lack of coloration, are unlikely to preserve in other depositional environments. The purpose of these scale-filaments is unknown, although a tactile function has been proposed.[45] These specimens also provide the best information regarding large-scale plumage patterns in the wing.[46]

Early Cretaceous enantiornithines had fairly short broad wings, hindlimb feathers that typically extended to the ankle, and most commonly a complete absence of rectrices (the larger feathers in the tail that are commonly used for steering), with only short contour feathers (feathers that cover the body) around the pygostyle.[47] Body feathers had a wispy appearance and are often described as rachis-less, although it is more likely that a short, thin rachis was present, but obscured by overlap between feathers. Body feathers on the dorsal surface of the body are generally longer than those on the ventral surface. Hindlimb feathers typically decrease in length distal in the limb, ending at the ankle, although fairly long crural feathers (feathers that cover the tibial portion of the leg) are documented in one specimen,[48] and short feathers extending down the tarsometatarsus (the bone of the foot that bears the toes) are present in another.[49] In one Burmese amber specimen, short feathers extend all the way down the toes,[50] as in some extant owls.

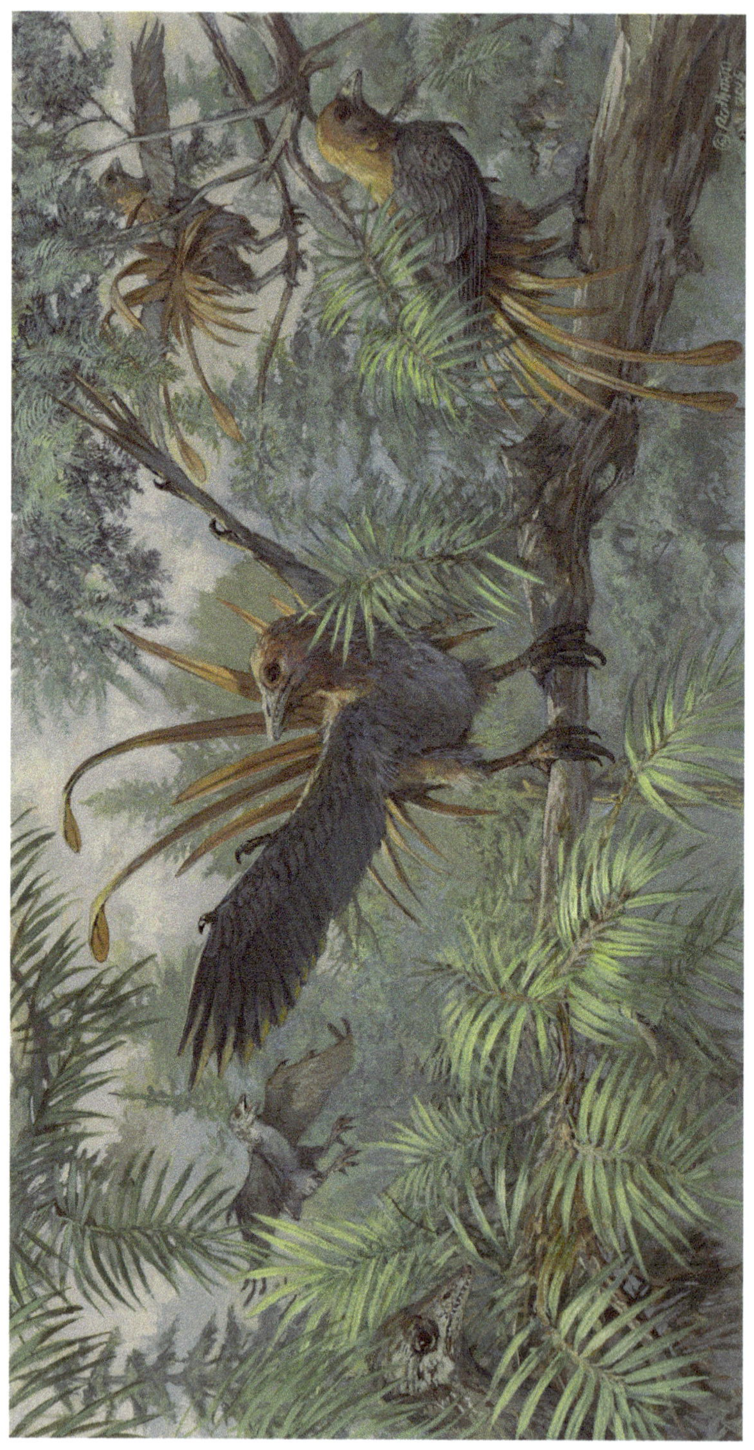

Illustration 7.1 Three male *Feitianius paradisi* from the Xiagou Formation displaying for two females in a forest dominated by *Torreya* sp. Acrylic painting, reconstructing *Feitianius* from life, by Michael Rothman. Image © M. Rothman.

The final appearance of some living birds is strongly linked to the orientation of the feathers (e.g. owls). It is possible that such structural morphologies cannot be detected in the fossil record. However, the limited morphological variation observed among enantiornithines compared to that in neornithines suggests such specialized feather morphologies were probably absent in this clade. Due to compression, the body feathers in specimens from lithic Lagerstätten are usually oriented at right angles to the bony surface, and it is nearly impossible to determine their original *in vivo* orientation.[51] As such, the feathers along the dorsal margin of the head project from the skull like a mohawk. In *Protopteryx*, these feathers were described as a cranial crest, interpreted as an *in vivo* feature.[52] However, experiments on extant birds strongly suggest that this is a taphonomic artefact.[53] Similarly, the hindlimb feathers projecting from the tibiotarsus (a compound leg bone formed by fusion of the tibia to the astragalus and calcaneum, the two proximal tarsal bones) have been described as small hindwings, although it is more likely that *in vivo* these feathers hung down, giving the legs a shaggy appearance.[54]

Elongate tail feathers, when present, most commonly consist of a pair of elongate rachis-dominated feathers (RDFs).[55] These unusual feathers consist of an extremely thin C-shaped rachis that, when preserved, flattened in lithic specimens appears proportionately wide, earning these feathers the name 'rachis-dominated'.[56] The feathers are most commonly racket-plumes, feathers in which the pennaceous, vaned portion is distally restricted so that the feather visually resembles a badminton racket, although fully pennaceous RDFs are present in some pengornithids. These tail feathers typically exceed the total body length of the bird. Most often these feathers occur as a single pair, although *Paraprotopteryx* preserves two pairs. These tail feathers have been interpreted as sexually dimorphic ornaments present only in males, which has been supported by their absence in specimens that are identified as female based on the preservation of female reproductive tissues[57] (Illustration 7.2).

A potentially aerodynamic tail shape is only documented in *Chiappeavis*, which preserves a short fan-shaped array of rectrices.[58] UV light photos illuminating the soft tissue surrounding the skeleton indicate an absence of rectricial bulbs – soft tissue structures that include the muscles responsible for tail fanning in extant birds.[59] Without rectricial bulbs to open and close the tail fan, the aerodynamic benefit of this tail shape would be limited.[60] The holotype and only known specimen of *Feitianius paradisi* from the Changma avifauna reveals an elaborate tail morphology consisting of several different feather morphotypes[61] (Illustration 7.1). In living birds, tails with such complex morphologies are most commonly associated with sexual dimorphism and polygamy, indicating the holotype of *Feitianius* is most likely male.

The main aerodynamic surface of the wing is formed by the remiges. Lift in the wing is augmented by patagia, flaps of skin that extend off the forelimb. The propatagium, the flap that extends between the shoulder and wrist, and the postpatagium, the skin flap that extends across the proximal portions of the remiges, evolved outside enantiornithines, being documented in more primitive non-ornithothoracine birds like confuciusornithiforms.[62] In addition to these two patagia, enantiornithines had another skin flap shared with modern birds but not present in more basal lineages. This skin flap, the alular patagium, extends between the alular digit (the bird equivalent of a thumb) and major digit (the longest digit of the avian hand).[63] This feature complemented the alula (also called the 'bastard wing'), a feathered structure only found in ornithothoracines, which extends off the alular digit and is important during take-off and landing.[64]

In the past, it was thought impossible to understand certain aspects of an extinct animal's biology, such as colour. However, in the twenty-first century, it has become possible to at least partially understand colour in extinct feathered organisms through the presence of preserved melanosomes, melanin containing mono-organelles (organelles are subunits within cells).[65] This has resulted in a distinct decrease in the variety of colours employed by artists to reconstruct these animals, and resulted in a rather dreary palette of browns, reddish browns, white, black, and grey (illustrations 7.1 and 7.2). However, this is most likely due to the fact that melanosome-based coloration only accounts for a narrow range of colours, and thus colour reconstructions based entirely on these structures are not providing the complete picture.

Structural colours are responsible for the spectacular range of hues observed in peacock plumage. These colours are imparted by the microstructure of the keratin matrix of the feather, and have yet to be reported in enantiornithines.[66] In most Cretaceous feathers all that is preserved is the decay-resistant melanosomes, while the keratin matrix is heavily degraded or completely lost, preventing identification of structural colour. However, in exceptional conditions structural colour can preserve, as it has been reported in one Late Jurassic non-avian paravian[67] and an Eocene feather,[68] leaving the potential for structural colour to be identified in exceptionally well-preserved enantiornithines in the future. Bright colours are also sometimes produced by pigments, such as carotenoids, which have yet to be found preserved in fossil feathers.[69]

Melanosome-based coloration has only been studied in a few enantiornithines: one specimen of *Protopteryx*,[70] the pengornithid *Yuanchuavis*,[71] and an indeterminant bohaiornithid.[72] These studies reveal the presence of elongate eumelanosomes (elongate melanosomes responsible for black colour) indicating the plumage was at least partially black. Melanosome-based coloration can vary enormously within a single feather, limiting the utility of this method, which requires a sample of the preserved feather to be extracted so it can be viewed using scanning electron microscopy.[73] Although spots and spangles are rarely preserved in Jehol fossils, such patterns have yet to be documented in any enantiornithine. These spots, usually on the distal tip of the feather, correspond to regions of increased melanization and not necessarily differences in coloration.[74] The tips of feathers are often heavily melanized to reduce feather wear. A juvenile enantiornithine from the Early Cretaceous of Brazil preserves unusual spots on the rachis of its paired RDFs, which have been interpreted as indicative of *in vivo* coloration.[75] However, the rachis is not commonly melanized and no analyses have been conducted to support the interpretation that these spots represent a true feature.

As such, the greatest wealth of information regarding plumage colour and patterning comes from specimens in Burmese amber, which include among their number several partial and complete wings, revealing overall patterns of light and dark coloration.[76] In these specimens the feathers are preserved in various shades of brown, ornamented with pale spots and bands. The brown appearance may be at least partially a product of the yellow of the amber, and might also be due to chemical alterations unique to this preservational medium. However, the large-scale patterns of light and dark areas are considered to reflect *in vivo* morphologies based on the overall consistency of the spots and stripes, whereas if light spots were inferred to be due to degradation their morphology would be expected to be more sporadic. The patterns revealed by these amber specimens suggest crypsis (patterns evolved to avoid detection by other animals), but this may be partially due to the young ontogenetic stage inferred for many of these specimens.[77] Juvenile neornithines are typically more drably coloured compared to their adult counterparts.[78]

Behaviour

Numerous behavioural aspects can be inferred from the fossil record. Tail plumage in at least one enantiornithine suggests polygamy, with males competing to mate with numerous females.[79] This interpretation is supported by the structure of the pygostyle, which suggests the presence of musculature for raising and depressing the ornamental tail feathers.[80] This suggests that male enantiornithines may have engaged in some form of display behaviour like that observed in pheasants (Illustration 7.1). Similarly, the lack of rectricial bulbs indicates that enantiornithines would not have actively fanned and contracted their tails during flight.[81] The rigid skull morphology, which suggests cranial kinesis was absent,[82] also indicates feeding strategies were limited, because behaviours like mud-probing require a specific form of cranial kinesis called distal rhynchokinesis.[83] The presence of teeth in most taxa also indicates that seed or nut-cracking was also unlikely to have been possible.[84]

A few morphologies are only found in enantiornithines – without extant analogues it is difficult to interpret what these morphologies indicate about behaviour. Two Burmese amber enantiornithines preserve pedal morphologies that are not utilized by the ten thousand (or more) species of living birds. Pedal morphology is a good indicator of ecology and this highlights the unique aspects of the enantiornithine radiation – although many specializations utilized by modern birds did not evolve in enantiornithines, this group evolved to utilize unique morphospace (morphospace refers to representations of the possible form, shape or structure an organism can assume) and may have occupied ecological niches not utilized by birds today, or utilized these ecological niches through different behaviours. For example, one Burmese enantiornithine, *Elektorornis*, preserves a hyperelongated third pedal digit.[85] This has been interpreted as indicative of a probing feeding adaptation. Some living birds also feed by probing but through the use of tools or an elongate tongue. The Burmese enantiornithine *Fortipesavis* preserves soft tissue traces that indicate the presence of an unusually wide fourth pedal digit,[86] which may have increased stability while utilizing mobile purchases similar to the syndactyl foot in alcedinids (kingfishers).[87]

Soft tissue traces indicate that enantiornithines, like modern birds, laid a single egg at a time.[88] Preserved eggs and nests suggest at least some enantiornithines utilized breeding colonies that were situated near a source of water.[89] A humid nesting environment is supported by the microstructure of the eggshell cuticle preserved in *Avimaia*[90] (Illustration 7.2). Eggs were inserted half buried into soft, muddy or sandy substrates, indicating that enantiornithines did not engage in egg turning.[91] This in turn suggest the chalaza, a membrane that keeps the developing embryo in place, was absent. Egg colour, which can now be determined in exceptional specimens using Raman spectroscopy to identify traces of the original pigments, supports inferences that the nests were open.[92] The single Mongolian enantiornithine egg studied so far reveals a pattern of brown speckles. The presence of adult bones together with perinatal bones suggests some amount of parental care, although this may be limited to nest attendance. The discovery of fledged late-stage embryos and hatchlings,[93] highly ossified embryonic remains,[94] and the large number of recovered ju-

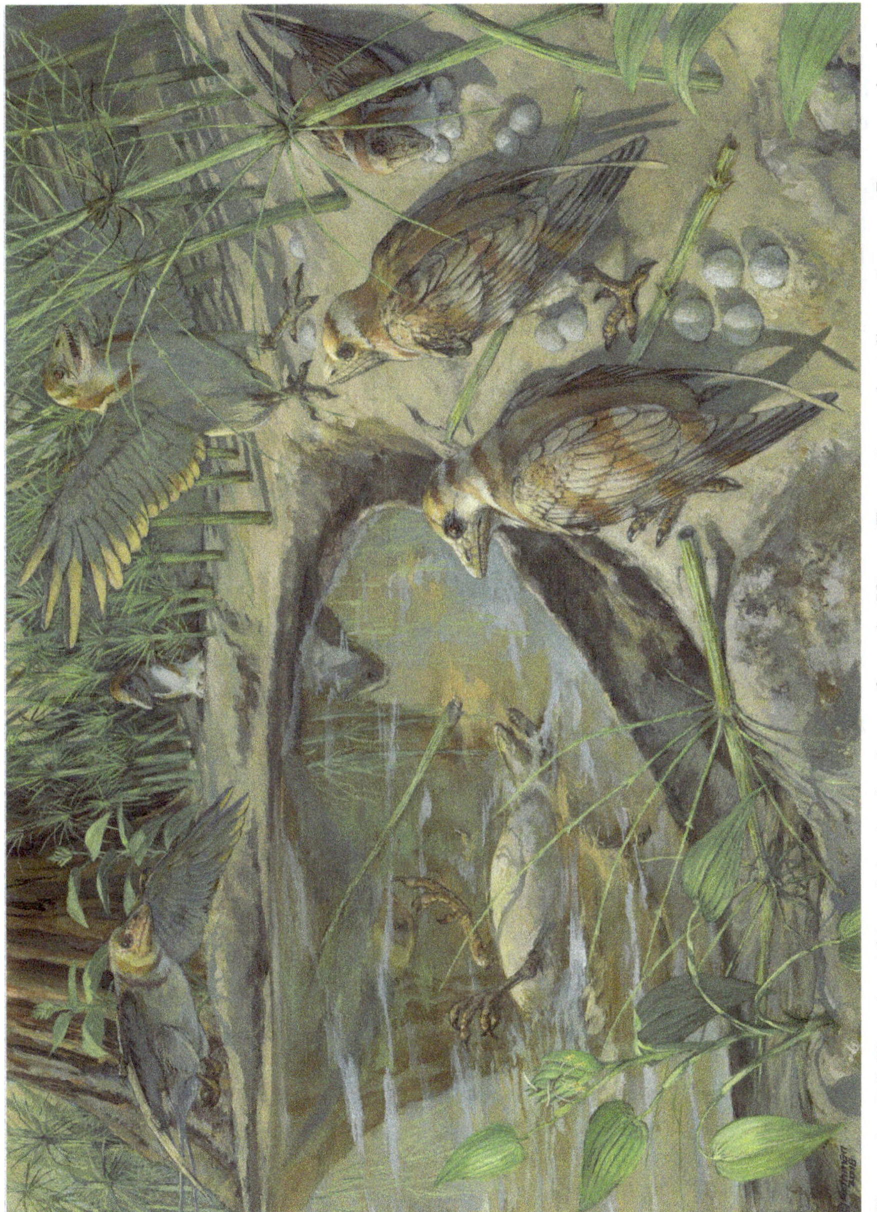

Illustration 7.2 Enantiornithine *Avimaia schweitzerae* from the Xiagou Formation, colonially nesting near Poaceae-related plants. Acrylic painting by Michael Rothman. Image © Jingmai O'Connor.

veniles,[95] all together indicate that enantiornithines were highly precocial, capable of flight from the moment they hatched and their feathers dried. This is supported by osteohistological studies of specimens from various ontogenetic stages ranging from late-stage embryo to adult, that indicate enantiornithines mostly grew slowly, taking many years to reach adult size, and achieving reproductive maturity before skeletal maturity.[96] As small arboreal birds, this level of precociality goes beyond even that observed in the so-called super-precocial extant megapodes from Australia, which – although capable of flight from the day they are free from their mound nests – rarely engage in such behaviour, being primarily ground-dwelling birds.[97]

Although traces of diet are abundantly preserved in the Jehol Biota, none that are unequivocal pertain to enantiornithines.[98] This suggests that most enantiornithines from this avifauna fed on soft food items unlikely to preserve. No specimen preserves gastroliths (small stones swallowed to aid digestion) forming a gastric mill further supporting inferences their diet was soft. However, this does somewhat conflict with the apparently durophagous (durophagous predators consume hard-shelled bearing organisms) tooth morphology observed in bohaiornithids. The only enantiornithine that preserves traces of diet is *Eoalulavis* from the Early Cretaceous of Spain, in which traces of aquatic invertebrates are present in the abdominal cavity.[99] This suggests aquatic feeding habits in at least some enantiornithines, which are also suggested by preserved colonial nesting sites, as such colonies are only utilized today by aquatic and semi-aquatic birds.

Habitat

Because enantiornithines are restricted to the Cretaceous, this limits the types of plants that would have coexisted with these birds. Although ferns and gymnosperms (plants with unenclosed seeds such as conifers) would have dominated in the Early Cretaceous, this period documents the sudden appearance of flowering plants (angiosperms) as well as the appearance of grasses.[100] Most enantiornithines are arboreal, meaning they would mostly have inhabited forested environments.[101] The palaeoenvironment of different geologic units can be reconstructed from the type of rock they consist of (which indicates the depositional environment), fossils they contain, and geochemical analyses that hint at temperature and seasonality. The Jehol has produced a rich flora including early flowering plants, and a diverse fauna of insects, mammals, pterosaurs, dinosaurs and other animals that together can be used to create a rich and detailed backdrop for reconstructions of enantiornithines from this biota in a region of lakes and temperate forest punctuated with volcanic activity.[102] The stomach contents of *Eoalulavis* from Spain suggest this taxon foraged near water, and the Las Hoyas locality is interpreted as an ancient wetland based on fossils and lithology.[103] In addition to feeding on other organisms, enantiornithines would have been preyed upon by others. One *Microraptor* (the so-called tetrapteryx dinosaur) preserves the remains of an enantiornithine in the stomach, indicating that this volant dromaeosaurid (a group of feathered dinosaurs closely related to birds) represented at least one of the many predators faced by Jehol enantiornithines.[104] Enantiornithine remains have also been found in the stomach of an ichthyosaur, although this occurrence is considered to have resulted from scavenging.[105] Notably, this association suggests that some Early Cretaceous enantiornithines lived in near marine habitats. The 100 My forests that produced Burmese amber were also near marine, as evidenced from preserved ammonites.

Conclusions

The enantiornithines have a short scientific history, discovered only forty years ago, and our understanding of this clade has grown enormously over the last three decades. Currently there is ample evidence to accurately reconstruct enantiornithines from their mode of flight, to the overall shape of their plumage, to their life cycle. A vast majority of available data is from the Early Cretaceous Jehol avifauna, with relatively little being known about enantiornithines from other localities. Diet and colour remain poorly known – diet from the lack of preserved evidence and colour from limited investigation. Only three melanosome-based studies have been published so far, and recent discoveries of Burmese amber enantiornithines have only been studied superficially at this time. Due to the high rate of discovery that continues in north-eastern China and to the application of new techniques, our scientific understanding of enantiornithines will continue to grow rapidly, providing rich new biological details with which to reconstruct these birds that dominated the Cretaceous aviary.

Jingmai O'Connor is an American paleontologist whose research focuses on the dinosaur–bird transition and the Mesozoic evolution of birds and other flying dinosaurs. Her research includes studies of exceptional soft tissues such as lung and ovary traces preserved in specimens from the 130–120 Ma Jehol Biota. She has authored over one hundred and thirty papers, including studies published in *Nature*, *Science*, *Current Biology*, and *PNAS*. She was the 2019 recipient of the Palaeontological Society's Charles Schuchert Award for outstanding palaeontologist under the age of 40. She is currently serving as the associate curator of fossil reptiles at the Field Museum of Natural History in Chicago.

Notes

1. Bailleul et al., 'Origin of the Avian Predentary'; Chapelle, Fernandez and Choiniere, 'Conserved In-Ovo Cranial Ossification'.
2. Wiemann, Yang and Norell, 'Dinosaur Egg Colour'; Bailleul et al., 'Confirmation of Ovarian Follicles'.
3. Norell and Xu, 'Feathered Dinosaurs'.
4. Xu et al., 'A Gigantic Feathered Dinosaur'.
5. Ji et al., 'Two Feathered Dinosaurs'.
6. Xu et al., 'Four-Winged Dinosaurs'; Hu et al., 'A Pre-Archaeopteryx Troodontid Theropod'.
7. Wiemann, Yang and Norell, 'Dinosaur Egg Colour'.
8. Georgi, Sipla and Forster, 'Turning Semicircular Canal Function'.
9. Bailleul et al., 'Origin of the Avian Predentary'.
10. Zhou, 'The Jehol Biota'.
11. Pan et al., 'The Jehol Biota'; Yang et al., 'The Appearance and Duration of the Jehol Biota'.
12. Chiappe and Meng, *Birds of Stone*.
13. Chang et al., *The Jehol Fossils*.
14. Ibid.
15. Walker, 'New Subclass of Birds'.
16. Zhou and Zhang, 'Mesozoic Birds of China'.
17. O'Connor, Chiappe and Bell, 'Pre-Modern Birds'.
18. Ibid.
19. O'Connor et al., 'A New Early Cretaceous Enantiornithine'; Bailleul et al., 'An Early Cretaceous Enantiornithine'.
20. Sanz et al., 'The Birds from the Lower Cretaceous of Las Hoyas'.
21. Chiappe and Calvo, '*Neuquenornis volans*'; Chiappe et al., 'A New Enantiornithine Bird'; Atterholt, Hutchison and O'Connor, 'The Most Complete Enantiornithine'; Xu et al., 'A New, Remarkably Preserved, Enantiornithine Bird'.
22. Chiappe, 'Enantiornithine (Aves) tarsometatarsi'; Panteleev, 'Morphology of the Coracoid'.
23. Xing et al., 'A Mid-Cretaceous Enantiornithine (Aves)'; Xing et al., 'A Fully Feathered Enantiornithine Foot and Wing'; Xing et al., 'A Mid-Cretaceous Enantiornithine Foot and Tail Feather'; Xing, McKellar and O'Connor, 'An Unusually Large Bird Wing'.
24. Zhou and Zhang, '*Jeholornis* Compared to *Archaeopteryx*'.
25. Chiappe et al., 'Anatomy and Systematics'; Falk et al., 'Laser Fluorescence'.
26. Zheng et al., 'On the Absence of Sternal Elements'.
27. O'Connor, Chiappe and Bell, 'Pre-Modern Birds'.

28. Ibid.
29. Wang and O'Connor, 'Morphological Coevolution'.
30. O'Connor and Chiappe, 'A Revision of Enantiornithine'.
31. Chiappe, Norell and Clark, 'A New Skull of *Gobipteryx Minuta*'; Xu et al., 'A New, Remarkably Preserved, Enantiornithine Bird'.
32. O'Connor, 'A Systematic Review of Enantiornithes'.
33. O'Connor and Chiappe, 'A Revision of Enantiornithine'.
34. Hu et al., 'New Anatomical Information'.
35. O'Connor, 'A Systematic Review of Enantiornithes'.
36. Zhang et al., 'Early Diversification of Birds'; O'Connor, 'A Systematic Review of Enantiornithes'.
37. Atterholt et al., 'The Most Complete Enantiornithine'; Xing et al., 'A New Enantiornithine'.
38. O'Connor, 'A Revised Look at *Liaoningornis Longidigitrus* (Aves)'.
39. Chiappe et al., 'A New Enantiornithine Bird'.
40. Liu et al., 'Flight Aerodynamics'; Serrano et al., 'Flight Reconstruction'; Chiappe et al., 'Anatomy and Flight Performance'.
41. Chiappe et al., 'New *Bohaiornis*-like Bird'.
42. O'Connor, 'The Plumage of Basal Birds'.
43. Xing et al., 'Ornamental Feathers'.
44. Xing et al., 'A Mid-Cretaceous Enantiornithine'.
45. Xing et al., 'A New Enantiornithine with Unusual Pedal Proportions'.
46. Xing et al., 'A New Enantiornithine (Aves)'.
47. O'Connor, 'The Plumage of Basal Birds'.
48. Zhang and Zhou, 'Leg Feathers'.
49. Chiappe and Meng, *Birds of Stone*.
50. Xing et al., 'A Fully Feathered Enantiornithine Foot and Wing'.
51. O'Connor, 'The Plumage of Basal Birds'.
52. Chiappe et al., 'Anatomy and Flight Performance'.
53. O'Connor et al., 'New Information on the Plumage of *Protopteryx*'.
54. O'Connor and Chang, 'Hindlimb Feathers'.
55. O'Connor, 'The Plumage of Basal Birds'.
56. Xing et al., 'Ornamental Feathers'.
57. Bailleul et al., 'An Early Cretaceous Enantiornithine'.
58. O'Connor et al., 'An Enantiornithine with a Fan-Shaped Tail'.
59. Chiappe and Meng, *Birds of Stone*.
60. O'Connor et al., 'An Enantiornithine with a Fan-Shaped Tail'.
61. O'Connor et al., 'A New Early Cretaceous Enantiornithine'.
62. Zheng et al., 'Exceptional Preservation of Soft Tissue'.
63. Navalón et al., 'Soft-Tissue and Dermal Arrangement'.
64. Sanz et al., 'An Early Cretaceous Bird from Spain'.
65. Vinther, 'A Guide to the Field of Palaeo Color'.
66. Ibid.
67. Hu et al., 'A Bony-Crested Jurassic Dinosaur'.
68. Vinther et al., 'Structural Coloration'.
69. Thomas et al., 'Seeking Carotenoid Pigments'.
70. O'Connor et al., 'New Information on the Plumage of *Protopteryx*'.
71. Wang et al. 'An Early Cretaceous Enantiornithine Bird'.
72. Peteya et al., 'The Plumage and Colouration'.
73. Zhang et al., 'Fossilized Melanosomes'.

74. Zheng et al., 'Exceptional Preservation of Soft Tissue'.
75. De Souza Carvalho et al., 'A Mesozoic Bird'.
76. Xing et al., 'A New Enantiornithine (Aves)'.
77. Ibid.
78. Gill, *Ornithology*.
79. O'Connor et al., 'A New Early Cretaceous Enantiornithine'.
80. Wang and O'Connor, 'Morphological Coevolution'.
81. O'Connor et al., 'An Enantiornithine with a Fan-Shaped Tail'.
82. O'Connor and Chiappe, 'A Revision of Enantiornithine'.
83. O'Connor, 'The Trophic Habits of Early Birds'.
84. O'Connor and Zhou, 'The Evolution of the Modern Avian Digestive System'.
85. Xing et al., 'A New Enantiornithine with Unusual Pedal Proportions'.
86. Xing et al., 'A Mid-Cretaceous Enantiornithine Foot and Tail Feather'.
87. Clark and O'Connor, 'Exploring the Ecomorphology of Two Cretaceous Enantiornithines'.
88. Zheng et al., 'Preservation of Ovarian Follicles'.
89. Varricchio and Jackson, 'Reproduction in Mesozoic Birds'.
90. Bailleul et al., 'An Early Cretaceous Enantiornithine (Aves)'.
91. Varricchio and Jackson, 'Reproduction in Mesozoic Birds'.
92. Wiemann et al., 'Dinosaur Egg Colour'.
93. Zhou and Zhang, 'A Precocial Avian Embryo'; Xing et al., 'A Mid-Cretaceous Enantiornithine (Aves)'.
94. Elzanowski, 'Embryonic Bird Skeletons'.
95. Chiappe, Ji and Ji, 'Juvenile Birds'.
96. Chinsamy, Chiappe and Dodson, 'Mesozoic Avian Bone Microstructure'; Chinsamy and Elzanowski, 'Evolution of Growth Pattern in Birds'.
97. Jones and Göth, *Mound-Builders*; Xing et al., 'A Mid-Cretaceous Enantiornithine (Aves)'.
98. O'Connor and Zhou, 'The Evolution of the Modern Avian Digestive System'; O'Connor, 'The Trophic Habits of Early Birds'.
99. Sanz et al., 'An Early Cretaceous Bird from Spain'.
100. Chang et al., *The Jehol Fossils*; Wu, You and Li, 'Dinosaur-Associated Poaceae Epidermis'.
101. O'Connor et al., 'Pre-Modern Birds'.
102. Chang et al., *The Jehol Fossils*.
103. Sanz et al., 'An Early Cretaceous Bird from Spain'.
104. O'Connor, Zhou and Xu, 'Additional Specimen of *Microraptor*'.
105. Kear, Boles and Smith, 'Unusual Gut Contents'.

Bibliography

Atterholt, Jessie.A., J. Howard Hutchison and Jingmai K. O'Connor. 'The Most Complete Enantiornithine from North America and a Phylogenetic Analysis of the Avisauridae'. *PeerJ* 6:e5910 (2018): 1–45.

Bailleul, Alida M., et al. 'Origin of the Avian Predentary and Evidence of a Unique Form of Cranial Kinesis in Cretaceous Ornithuromorphs'. *Proceedings of the National Academy of Sciences USA* 116(49) (2019): 24696–706.

———. 'An Early Cretaceous Enantiornithine (Aves) Preserving an Unlaid Egg and Probable Medullary Bone'. *Nature Communications* 10(1275) (2019): 1–10.

———. 'Confirmation of Ovarian Follicles in an Enantiornithine (Aves) from the Jehol Biota Using Soft Tissue Analyses'. *Communications Biology* 3(399) (2020): 1–8.

Chang, Mee-Mann, et al. (eds). *The Jehol Fossils: The Emergence of Feathered Dinosaurs, Beaked Birds and Flowering Plants*. Shanghai: Shanghai Scientific & Technical Publishers, 2003.

Chapelle, Kimberley E.J., Vincent Fernandez and Jonah N. Choiniere. 'Conserved In-Ovo Cranial Ossification Sequences of Extant Saurians Allow Estimation of Embryonic Dinosaur Developmental Stages'. *Scientific Reports* 10(4224) (2020): 1–10.

Chiappe, Luis M. 'Enantiornithine (Aves) Tarsometatarsi from the Cretaceous Lecho Formation of Northwestern Argentina'. *American Museum Novitates* 3083 (1993): 1–27.

Chiappe, Luis M., and Jorge O. Calvo. '*Neuquenornis Volans*, a New Late Cretaceous Bird (Enantiornithes: Avisauridae) from Patagonia, Argentina'. *Journal of Vertebrate Paleontology* 14(2) (1994): 230–46.

Chiappe, Luis M., and Qingjin Meng. *Birds of Stone*. Baltimore, MD: JHU Press, 2016.

Chiappe, Luis M., Shu-An Ji and Qiang Ji. 'Juvenile Birds from the Early Cretaceous of China: Implications for Enantiornithine Ontogeny'. *American Museum Novitates* 3594 (2007): 1–46.

Chiappe, Luis M., Mark A. Norell and James M. Clark. 'A New Skull of *Gobipteryx Minuta* (Aves: Enantiornithes) from the Cretaceous of the Gobi Desert'. *American Museum Novitates* 3346 (2001): 1–15.

Chiappe, Luis M., et al. 'Anatomy and Systematics of the Confuciusornithidae (Theropoda: Aves) from the Late Mesozoic of Northeastern China'. *Bulletin of the American Museum of Natural History* 242 (1999): 1–89.

———. 'Anatomy and Flight Performance of the Early Enantiornithine Bird *Protopteryx Fengningensis*: Information from New Specimens of the Early Cretaceous Huajiying Formation of China'. *The Anatomical Record* 303(4) (2019): 716–31.

———. 'New *Bohaiornis*-like Bird from the Cretaceous of China: Enantiornithine Interrelationships and Flight Performance'. *PeerJ* 7:e7846 (2019): 1–50.

———. 'A New Enantiornithine Bird from the Late Cretaceous of the Gobi Desert'. *Journal of Systematic Palaeontology* 5(2) (2006): 193–208.

Chinsamy, Anusuya, and Andrzej Elzanowski. 'Evolution of Growth Pattern in Birds'. *Nature* 412 (2001): 402–3.

Chinsamy, Anusuya, Luis M. Chiappe and Peter Dodson. 'Mesozoic Avian Bone Microstructure: Physiological Implications'. *Paleobiology* 21(4) (1995): 561–74.

Clark, Alexander D., and Jingmai K. O'Connor, 'Exploring the Ecomorphology of Two Cretaceous Enantiornithines with Unique Pedal Morphology'. *Frontiers in Ecology and Evolution* 9 (2021). https://doi.org/10.3389/fevo.2021.654156.

de Souza Carvalho, Ismar, et al. 'A Mesozoic Bird from Gondwana Preserving Feathers'. *Nature Communications* 6(7141) (2015): 1–5.

Elzanowski, Andrzej. 'Embryonic Bird Skeletons from the Late Cretaceous of Mongolia'. *Palaeontologica Polonica* 42 (1981): 147–79.

Falk, Amanda R., et al. 'Laser Fluorescence Illuminates the Soft Tissue and Life Habits of the Early Cretaceous Bird *Confuciusornis*'. *PLoS ONE* 11(12) (2016): 1–15.

Georgi, Justin A., Justin S. Sipla and Catherine A. Forster. 'Turning Semicircular Canal Function on Its Head: Dinosaurs and a Novel Vestibular Analysis'. *PLoS ONE* 8(3) (2013): e58517.

Gill, Frank B. *Ornithology*, 3rd Edition. New York: W.H. Freeman and Company, 2007.

Hu, Dongyu, et al. 'A Bony-Crested Jurassic Dinosaur with Evidence of Iridescent Plumage Highlights Complexity in Early Paravian Evolution'. *Nature Communications* 9(1) (2018): 217.

———. 'A Pre-Archaeopteryx Troodontid Theropod from China with Long Feathers on the Metatarsus'. *Nature* 461(7264) (2009): 640–43.

Hu, Han, et al. 'New Anatomical Information of Bohaiornithid *Longusunguis* Confirms Plesiomorphic Diapsid Skull Retained in Enantiornithes'. *Journal of Systematic Palaeontology* 18(18) (2020): 1481–95.

Ji, Qiang, et al. 'Two Feathered Dinosaurs from Northeastern China'. *Nature* 393(6687) (1998): 753–61.

Jones, Darryl, and Ann Göth. *Mound-Builders*. Collingwood, Australia: CSIRO Publishing, 2008.

Kear, Benjamin P., Walter E. Boles and Elizabeth T. Smith. 'Unusual Gut Contents in a Cretaceous Ichthyosaur'. *Proceedings of the Royal Society of London B* 270, suppl. 2 (2003): S206–S208.

Liu, Di, et al. 'Flight Aerodynamics in Enantiornithines: Information from a New Chinese Early Cretaceous Bird'. *PLoS ONE* 12(10) (2017): 1–18.

Navalón, Guillermo, et al. 'Soft-Tissue and Dermal Arrangement in the Wing of an Early Cretaceous Bird: Implications for the Evolution of Avian Flight'. *Scientific Reports* 5 (2015): 14864.

Norell, Mark A., and Xing Xu. 'Feathered Dinosaurs'. *Annual Review of Earth and Planetary Science* 33 (2005): 277–99.

O'Connor, Jingmai K. 'The Plumage of Basal Birds', in Christian Foth and Oliver W.M. Rauhut (eds), *The Evolution of Feathers* (Cham: Springer, 2020), 147–72.

———. 'A Revised Look at *Liaoningornis Longidigitrus* (Aves)'. *Vertebrata Palasiatica* 5(1) (2012): 25–37.

———. 'A Systematic Review of Enantiornithes (Aves: Ornithothoraces)'. PhD, University of Southern California, 2009.

———. 'The Trophic Habits of Early Birds'. *Palaeogeography, Palaeoclimatology, Palaeoecology* 513 (2019): 178–95.

O'Connor, Jingmai K, and H. Chang. 'Hindlimb Feathers in Paravians: Primarily 'Wings' or Ornaments?' *Biology Bulletin* 42(7) (2015): 616–21.

O'Connor, Jingmai K., and Luis M. Chiappe. 'A Revision of Enantiornithine (Aves: Ornithothoraces) Skull Morphology'. *Journal of Systematic Palaeontology* 9(1) (2011): 135–57.

O'Connor, Jingmai K., and Zhonghe Zhou. 'The Evolution of the Modern Avian Digestive System: Insights from Paravian Fossils from the Yanliao and Jehol Biotas'. *Palaeontology* 63(1) (2019): 13–27.

O'Connor, Jingmai K., Luis M. Chiappe and Alyssa Bell. 'Pre-Modern Birds: Avian Divergences in the Mesozoic', in Gareth D. Dyke and Gary Kaiser (eds), *Living Dinosaurs: The Evolutionary History of Birds* (Hoboken, NJ: John Wiley & Sons, 2011), 39–114.

O'Connor, Jingmai K., Zhonghe Zhou and Xing Xu. 'Additional Specimen of *Microraptor* Provides Unique Evidence of Dinosaurs Preying on Birds'. *Proceedings of the National Academy of Sciences USA* 108(49) (2011): 19662–65.

O'Connor, Jingmai K., et al. 'An Enantiornithine with a Fan-Shaped Tail, and the Evolution of the Rectricial Complex in Early Birds'. *Current Biology* 26(1) (2016): 114–19.

———. 'A New Early Cretaceous Enantiornithine (Aves: Ornithothoraces) from Northwestern China with Elaborate Tail Ornamentation'. *Journal of Vertebrate Paleontology* 36(1) (2016): e1054035.

———. 'New Information on the Plumage of *Protopteryx* (Aves: Enantiornithes) from a New Specimen'. *Cretaceous Research* 116 (2020): 1–17.

Pan, Yanhong, et al. 'The Jehol Biota: Definition and Distribution of Exceptionally Preserved Relicts of a Continental Early Cretaceous Ecosystem'. *Cretaceous Research* 44 (2013): 30–38.

Panteleev, A.V. 'Morphology of the Coracoid of Late Cretaceous Enantiornithines (Aves: Enantiornithes) from Dzharakuduk (Uzbekistan)'. *Paleontological Journal* 52(2) (2018): 201–7.

Peteya, Jennifer A., et al. 'The Plumage and Colouration of an Enantiornithine Bird from the Early Cretaceous of China'. *Palaeontology* 60(1) (2017): 55–71.

Sanz, José L., et al. 'The Birds from the Lower Cretaceous of Las Hoyas (Province of Cuenca, Spain)', in Luis M. Chiappe and Lawrence M. Witmer (eds), *Mesozoic Birds: Above the Heads of Dinosaurs* (Berkeley: University of California Press, 2002), 209–29.

———. 'An Early Cretaceous Bird from Spain and Its Implications for the Evolution of Avian Flight'. *Nature* 382 (1996): 442–45.

Serrano, Francisco J., et al. 'Flight Reconstruction of Two European Enantiornithines (Aves, Pygostylia) and the Achievement of Bounding Flight in Early Cretaceous Birds'. *Palaeontology* 61(3) (2018): 359–68.

Thomas, Daniel B., et al. 'Seeking Carotenoid Pigments in Amber-Preserved Fossil Feathers'. *Scientific Reports* 4(5226) (2014): 1–6.

Varricchio, David J., and Frankie D. Jackson. 'Reproduction in Mesozoic Birds and Evolution of the Modern Avian Reproductive Mode'. *The Auk* 133(4) (2016): 654–84.

Vinther, Jakob. 'A Guide to the Field of Palaeo Colour'. *Bioessays* 37(6) (2015): 643–56.

Vinther, Jakob, et al. 'Structural Coloration in a Fossil Feather'. *Biology Letters* 6 (2010): 128–31.

Walker, C.A. 'New Subclass of Birds from the Cretaceous of South America'. *Nature* 292(5818) (1981): 51–53.

Wang, Min, et al. 'An Early Cretaceous Enantiornithine Bird with a Pintail'. *Current Biology* (2021), in press. https://doi.org/10.1016/j.cub.2021.08.044.

Wang, W., and Jingmai K. O'Connor. 'Morphological Coevolution of the Pygostyle and Tail Feathers in Early Cretaceous Birds'. *Vertebrata Palasiatica* 55(4) (2017): 289–314.

Wiemann, Jasmina, Tzu-Ruei Yang and Mark A. Norell. 'Dinosaur Egg Colour Had a Single Evolutionary Origin'. *Nature* 563(7732) (2018): 555–58.

Wu, Yan, Hai-Lu You and Xiao-Qiang Li. 'Dinosaur-Associated Poaceae Epidermis and Phytoliths from the Early Cretaceous of China'. *National Science Review* 5(5) (2018): 721–27.

Xing, Lida., Ryan C. McKellar and Jingmai K. O'Connor. 'An Unusually Large Bird Wing in Mid-Cretaceous Burmese Amber'. *Cretaceous Research* 110 (2020): 1–5.

Xing, Lida, et al. 'A Fully Feathered Enantiornithine Foot and Wing Fragment Preserved in Mid-Cretaceous Burmese Amber'. *Scientific Reports* 9(927) (2019): 1–9.

———. 'A Mid-Cretaceous Enantiornithine (Aves) Hatchling Preserved in Burmese Amber with Unusual Plumage'. *Gondwana Research* 49 (2017): 264–77.

———. 'A Mid-Cretaceous Enantiornithine Foot and Tail Feather Preserved in Burmese Amber'. *Scientific Reports* 9(15513) (2019): 1–8.

———. 'A New Enantiornithine (Aves) Preserved in Mid-Cretaceous Burmese Amber Contributes to Growing Diversity of Cretaceous Plumage Patterns'. *Frontiers in Earth Science* 8(264) (2020): 1–11.

———. 'A New Enantiornithine with Unusual Pedal Proportions Found in Amber'. *Current Biology* 29(14) (2019): 2396–401.

———. 'Ornamental Feathers in Cretaceous Burmese Amber: Resolving the Enigma of Rachis-Dominated Feather Structure'. *Journal of Palaeogeography* 7(13) (2018): 1–18.

Xu, Li et al., 'A New, Remarkably Preserved, Enantiornithine Bird from the Upper Cretaceous Qiupa Formation of Henan (Central China) and Convergent Evolution between Enantiornithines and Modern Birds'. *Geological Magazine* 158(11) (2021): 2087–94.

Xu, Xing, et al. 'Four-Winged Dinosaurs from China'. *Nature* 421 (2003): 335–40.

———. 'A Gigantic Feathered Dinosaur from the Lower Cretaceous of China'. *Nature* 484 (2012): 92–95.

Yang, Saihong, et al. 'The Appearance and Duration of the Jehol Biota: Constraint from SIMS U-Pb Zircon Dating for the Huajiying Formation in Northern China'. *Proceedings of the National Academy of Sciences USA* 117(25) (2020): 14299–305.

Yu, Tingting, Richard Kelly, Lin Mu, Andrew Ross, Jim Kennedy, Pierre Broly, Fangyuan Xia, Haichun Zhang, Bo Wang, and David Dilcher. 'An Ammonite Trapped in Burmese Amber'. *Proceedings of the National Academy of Sciences* 116, no. 23 (2019): 11345–11350.

Zhang, Fucheng, and Zhonghe Zhou. 'Leg Feathers in an Early Cretaceous Bird'. *Nature* 431 (2004): 925.

Zhang, Fucheng, et al. 'Early Diversification of Birds: Evidence from a New Opposite Bird'. *Chinese Science Bulletin* 46(11) (2001): 945–49.

———. 'Fossilized Melanosomes and the Colour of Cretaceous Dinosaurs and Birds'. *Nature* 463 (2010): 1075–78.

Zheng, Xiaoting, et al. 'Exceptional Preservation of Soft Tissue in a New Specimen of Eoconfuciusornis and Its Biological Implications'. *National Science Review* 4(3) (2017): 441–52.

———. 'On the Absence of Sternal Elements in Anchiornis (Paraves) and Sapeornis (Aves) and the Complex Early Evolution of the Avian Sternum'. *Proceedings of the National Academy of Sciences USA* 111(38) (2014): 13900–5.

———. 'Preservation of Ovarian Follicles Reveals Early Evolution of Avian Reproductive Behaviour'. *Nature* 495 (2013): 507–11.

Zhou, Zhonghe. 'The Jehol Biota, an Early Cretaceous Terrestrial Lagerstätte: New Discoveries and Implications'. *National Science Review* 1(4) (2014): 543–59.

Zhou, Zhonghe, and Fucheng Zhang. '*Jeholornis* Compared to *Archaeopteryx*, with a New Understanding of the Earliest Avian Evolution'. *Naturwissenschaften* 90(5) (2003): 220–25.

———. 'A Precocial Avian Embryo from the Lower Cretaceous of China'. *Science* 306(5696) (2004): 653.

———. 'Mesozoic Birds of China: A Synoptic Review'. *Vertebrata PalAsiatica* 44(1) (2006): 74–98.

Chapter 8

Rare Birds and Rare Books

The Species as Work of Art

Gordon M. Sayre

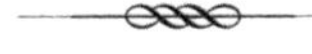

Of the nearly five hundred bird species that John James Audubon painted in the four hundred and thirty-five plates of his *Birds of America*, six have become extinct since that great work was published, and at least two, the great auk and the California condor, were not actual portraits of specimens he had shot in the field.[1] Audubon proudly claimed, 'I have never drawn from a stuffed specimen', and directed his engravers to add the phrase 'drawn from nature' alongside his signature on most of the *Birds of America* plates, so these two species undercut his statements of method.[2] Why did he nonetheless include the two birds in his series? The story of these two species can teach us a lot about early efforts towards bird conservation, and about the role of art and media in humans' conception of endangered species.

The great auk was the first widely publicized and fetishized endangered species in the anglophone world. Audubon himself associated the great auk with extinction. While hunting bison along the Missouri River in 1843 he wrote: 'this cannot last . . . before many years the Buffalo, like the Great Auk, will have disappeared; surely this should not be permitted'.[3] As a large flightless seabird nesting in dense and highly social colonies on small islands in the North Atlantic, unable to reach the steeper, more inaccessible cliffs and sea stacks where other birds found protection from predators, the great auk was (like the dodo two centuries earlier in the Indian Ocean) highly vulnerable to hungry sailors and fisherman. Like the passenger pigeon, a species better known to American birders and fans of Audubon, the great auk's dense flocks made its roosting sites juicy targets for hunters, and encouraged mass slaughters with little thought of the bird's rarity. Funk Island, near the north-east corner of Newfoundland, was one such site, home to the largest known colony of great auks in the Americas (it still hosts large colonies of guillemots, murres, and other species). Because it was an early landfall for ships crossing the stormy North Atlantic, and close to the cod fisheries that had been exploited since the 1400s, the birds were slaughtered there repeatedly.

The great auk's rarity, large size, strikingly patterned eggs, and picturesque habitat near northern Europe gave it such notoriety that Audubon must have felt obligated to include it in *Birds of America*, even if he could not obtain a specimen because the birds no longer lived on or near the continent. During a voyage to Newfoundland in 1833 to collect those North Atlantic seabirds he needed to complete the *Birds of America*, Audubon tried to find a great auk. He wrote in his *Ornithological Biography* of the species this pathetic story:

> The only authentic account of the occurrence of this bird on our coast that I possess, was obtained from Mr Henry Havell, brother to my engraver, who, when on his passage from New York to England, hooked a great auk on the banks of Newfoundland, in extremely boisterous weather. On being hauled on board, it was left at liberty on the deck. It walked very awkwardly, often tumbling over, bit every one within reach of its powerful bill, and refused food of all kinds.[4]

Audubon must have used stuffed great auk specimens preserved in England or Scotland as models for his painting (Illustration 8.1).

The bird still survived on the European side of the Atlantic for a short time after Audubon's work appeared. St Kilda, west of the Outer Hebrides, was one island group where great auks were often found, and where the ethnographic interest of the local human population had enhanced the lure for bird collectors and other readers of books like Martin Martin's *A Late Voyage to St. Kilda, the Remotest of All the Hebrides, or Western Isles of Scotland* (1698).[5] The inaccessibility of these islands, the short summer nesting season of the great auk, and the absence of photography in the early nineteenth century meant the birds were not individuated by those who 'collected' or killed them. These collectors did not seek to bond with the live birds, but they treasured the specimen commodities, such as eggs, which became valuable artefacts. In *The Great Auk*, bird artist Errol Fuller includes photos and provenance for each of seventy-six remaining eggs of the species, and each of eighty stuffed specimens (as well as several skeletons). Most of Fuller's entries have a photograph and description, and the distinctive patterns that helped to make the large eggs so collectable

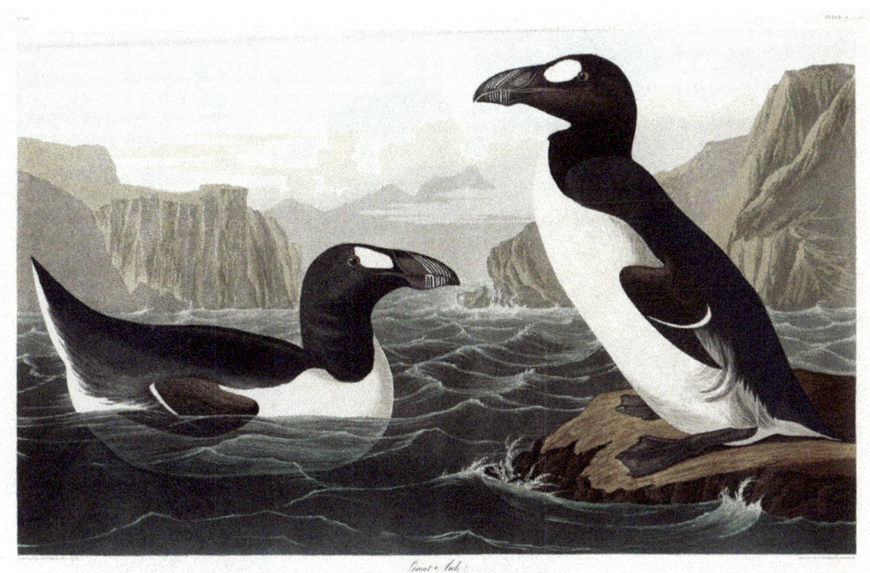

Illustration 8.1 John James Audubon, *Great auk*, 1836. Hand-coloured aquatint.

enabled him to cross-reference old photographs to the individual eggs on display today. Fuller's catalogue of species artefacts resembles research by art historians and bibliophiles on the paintings and publications by Audubon and by Mark Catesby, and, as we shall see, the work of conservation biologists on the California condor.

The California condor that Audubon painted was based upon descriptions and paintings by John Kirk Townsend, a friend from Philadelphia who had travelled with fur trade entrepreneur Nathaniel J. Wyeth to the lower Columbia River in 1834–35.[6] Townsend returned just in time for Audubon to paint the *Californian vulture* in 1838 and publish it among the last three sets of *The Birds of America* plates, sent to subscribers in 1839 (Illustration 8.2).

The California condor, the largest North American bird measured by wingspan, was fairly common along the Pacific Coast and inland valleys at that time, but its population then declined steadily for a century and half. In the twentieth century its remaining habitat in southern California was invaded by miners, ranchers and real estate developers, and it nearly went extinct in the 1980s. Today there are about five hundred birds alive.

This chapter examines these two bird species, and Audubon's artistic methods of representing them, as a means to better understand conflicts between the motives of collection and conservation, between economic possession and aesthetic pleasure, that afflict scientists, tourists and artists

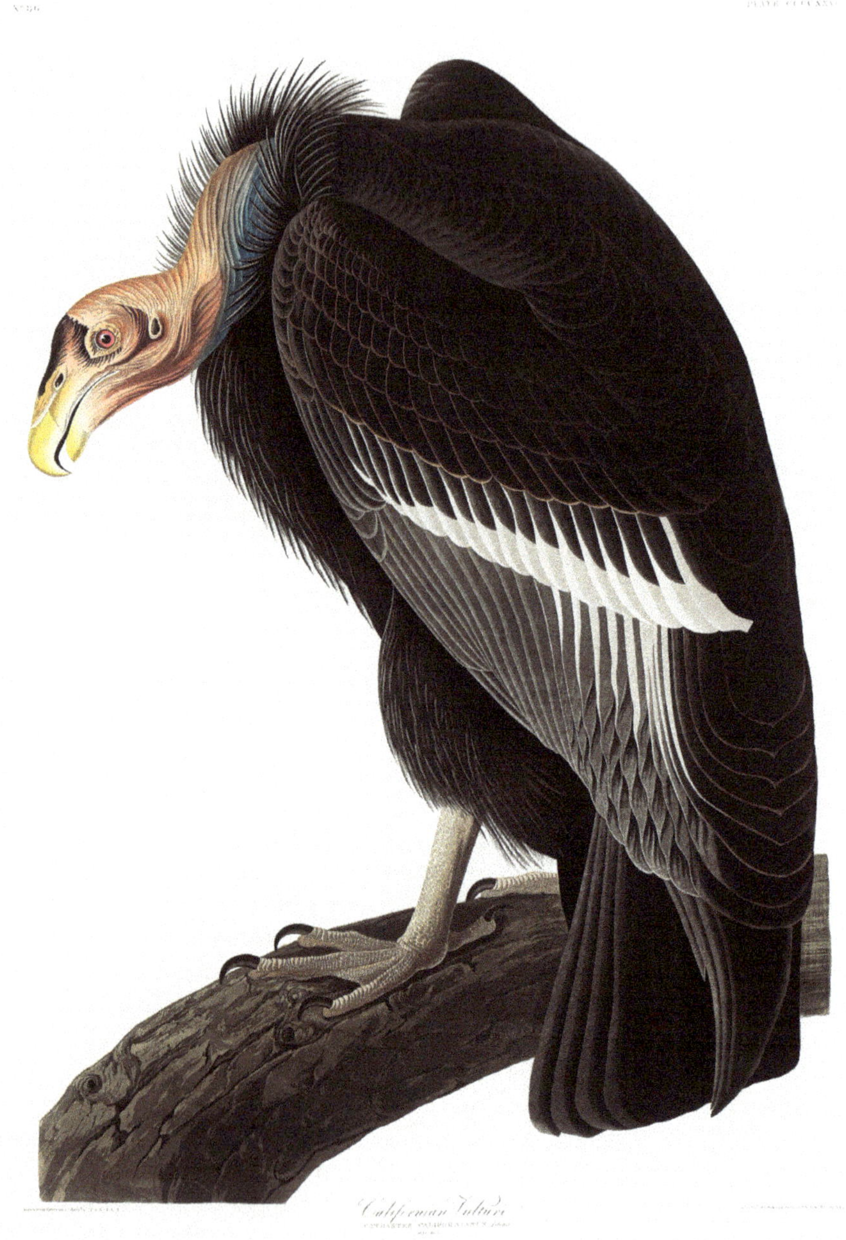

Illustration 8.2 John James Audubon, *Californian vulture*, 1838. Hand-coloured aquatint.

alike. I argue that endangered species are treasured and protected accord-ing to principles of value that also structure the market for works of visual art, and that the aura of singularity that differentiates an original painting from its printed reproductions evolved in the eighteenth and nineteenth centuries alongside the concept of rare animal species and their artefactual specimens.

The great auk illustrates the dilemma of conservation and collection, because its extinction was a direct consequence of hunting by or for collec-tors, who were well aware of the threat they posed to the species' survival:

> Paradoxically, amidst the rise of natural history and discussions of po-tential extinction, the great auk's rarity and threatened status accelerated its demise. The remaining birds and their eggs became a form of exotic animal capital, commodified and highly valued in the burgeoning mar-ket for specimens powered by museums and private collectors. Hunters scoured remote islands, braving dangerous seas and rocky cliffs in pursuit of birds and their large, striking eggs. The death of what were claimed to be the final great auks is well recorded. It took place on a small island [Edley] off Iceland in 1844, when a pair of birds were caught, clubbed and sent to a Belgian museum for stuffing and display.[7]

At the time Audubon was publishing and promoting his work there in the 1830s, bird fanciers and egg collectors in Britain knew the great auk was greatly endangered. One clue was that the prices paid for its eggs were among the highest of all in an oology market that received as much media attention then as auctions of masterpiece paintings and classic cars do today. The prestige of egg and specimen collecting was so great among the ruling classes that it was difficult to stop, even when the entire species was at risk. Audubon adopted the persona of a woodsman and hunter, as well as an ornithological expert. While in Britain promoting and selling subscriptions for his books, he dressed like John Filson's Daniel Boone or J.F. Cooper's Leatherstocking. The macho mystique of the American frontier hunter helped to burnish his image in the eyes of his customers, but Audubon also reminded his audience that their aesthetic pleasure in beholding his images of birds was a result of his exhausting, messy, smelly work seeking, shooting, stuffing, posing, and painting, from before dawn to last light. He routinely shot many dozens of birds to obtain specimens for his painting and writing, and reminded his readers/clients of their shared responsibility for this carnage: 'I was truly sorry to rob them of their eggs, although impelled to do so by the love of science, which offers a convenient excuse for even worse acts'.[8] Audubon would not have heeded anyone's command to stop hunting a bird, yet he still felt overhunting should not be permitted. His contradictory impulses were common; until the early twentieth century, very few scientists were employed at academic institutions, but many collectors of bird eggs and skins thought of themselves as scientists empowered to collect and study bird specimens.[9] Joseph Kastner in his history of American birdwatching cites an influential 1902 pamphlet by Reginald Robbins, 'Bird Killing as a Method in Ornithology', which described ornithologists who 'speak up for protecting the lives of birds, but "using turgid talmudic arguments" they exempt themselves from this duty'.[10] The story of the California condor also demonstrates how scientists dedicated to protecting rare bird species display an insatiable desire to keep chasing after them, shooting, measuring and examining specimens, even if their methods no longer require more specimens of birds, just as Audubon often did in pursuit of his art.

The wildlife conservation movement arose in America and Britain in the late nineteenth century when activists aimed to persuade people to satisfy their desires for animals not by hunting or collecting animal bodies or the commodities derived from them (such as birds' eggs), but instead by collecting images or representations of these animals. The Audubon Society's drive to protect birds began with an effort to end the collecting of eggs and the killing of birds for feathers to use in hats, dresses and other fashionable attire, much like Audubon offered his art as a substitute for collecting living birds. The campaign was quite effective in changing the sartorial habits of bird lovers, but less so in changing the behaviour of bird scientists. The strategy of early Audubon societies was to appeal to conservationists as ethical consumers, similar to the early abolitionist movement a century before. Still today the politics of wildlife recreation are often defined by an ethical boundary between appropriative consumption (hunting, fishing) and aesthetic observation (birdwatching, nature photography). The leave-no-trace ethic, 'Leave only footprints, take only pictures', enforced in wilderness areas and by outing clubs like Outward Bound and the Sierra Club is another example of the principle. It implies that both footprints and pictures are ephemeral impacts, and that a picture of a bird can become an observer's unique possession, evidence of a singular event, yet equivalent to an indefinite number of other pictures of specimens of the same species.[11]

The commodification of animal species can take many forms, however, and I wish to argue that the distinction between appropriative consumption and passive observation is not so simple or obvious as it seems. For one thing, both academic ornithologists and a large and active group of amateur birdwatchers advocate for protection of rare birds and conservation of their habitats, but also work to protect their own claims of expertise and sovereignty over images and descriptions. Bird lovers have come to rely upon mass-media images of birds, in popular guidebooks, magazines and smartphone apps, as well as academic journals. Collecting high-quality representations of birds is no longer limited to those who can afford to subscribe to Catesby or Audubon's series of engravings. But the sense of rarity and connoisseurship of rare birds flaunted by nineteenth-century collectors can now be exercised by supporting the conservation of rare species like the whooping crane and the California condor. Textual and visual artefacts of birds have become objects of authority and marketing that in a previous era were contested over specimens now held in archives and natural history museums. The oldest major ornithology journal in the United States, founded in 1884 by the American Ornithologists Union, is entitled *The Auk*, and the group's other journal (published since 1900 by what is now known as the American Ornithological Society) is entitled *The Condor*. These two extinct birds have totemic importance, and inspire fetishist behaviour whereby images and specimens of the birds are enshrined with a value of their own.

To produce or collect images of animals does not eliminate the forces of supply and demand threatening endangered species, and may in fact exacerbate them. The commodification and fetishization of animal specimen products, such as bird feathers and eggs, shark fins, elephant ivory, rhinoceros horns and pangolin scales, continues to motivate the hunting, poaching and smuggling of endangered and protected species, as it has for centuries. Efforts to ban international trade in such items, notably the Convention on International Trade in Endangered Species of Flora and Fauna (CITES), have failed to reduce what has become a speculative investor-fuelled mania for the commodities. Similarly, environmentalists are naive if they assume that scientific and touristic activity in pursuit of animal observation and animal images does not also damage the habitats and endanger the livelihoods of charismatic megafauna.

This paradox seems symptomatic of modern consumer society, but it began in the eighteenth century, before extinction was widely understood. Natural history books, such as the ones Catesby and Audubon produced, existed on a continuum between bodily specimens and textual representations.[12] They created the most beautiful and valuable bird books of eighteenth- and nineteenth-century America respectively, books central to the phenomenon wherein images of birds drive a luxury art market in the modern era, largely replacing the market for bird commodities such as eggs and feathers.[13] For Catesby and Audubon, ornithologist-artists on the supply side of the exchange, the goal was to create vivid, hand-coloured engravings for their books, and sell them to wealthy subscribers who perceived the exclusivity of ownership. Because each specimen of the book had a patron, it had an aura, to use Walter Benjamin's concept, that shifted its status from one element of a set of identical works, to an individual work of art. Their success in publishing reproductions also enhanced the value of their originals. The paintings made by each artist and furnished to engravers were later nearly all acquired by a single institution. King George III acquired the Catesby watercolours in 1768 from a London dealer, and today they are held in the Royal Library at Windsor Castle, bound as three volumes rather than two with an appendix, but otherwise arranged to mimic the printed book, with the same text.[14] The original paintings that Audubon provided to his engravers in Britain were sold in 1863 by Audubon's widow, Lucy Bakewell Audubon, to the New York Historical Society, which bills itself on its website as 'the world's largest repository of Auduboniana'.

Studying the bird art by Catesby and Audubon reveals how the economic calculus of scarcity and value, the relentless force of capitalist markets, has structured books, pictures and other aesthetic representations of birds just as much as it has bird commodities like the feathers used for fashion accessories. And it continues to structure conservationists' assessments of the value of endangered birds, and priorities for protecting them.

The great auk became extinct just prior to the development of photography, and around the same time as Audubon was publishing *The Birds of America*. But contrary to the strategy of the Audubon societies in the late nineteenth century, it seems likely that photography would have done little to protect the great auk, as it has also served to circulate images of collectors' fetishes, such as the eggs in Fuller's book, as well as images of the wild birds. The value of an endangered species and the fetish for its specimens both increase with the circulation of such photographs. Some might argue that the photographs replace the specimens, but this seems true only in the economic logic that bad money drives out good.

The market for art antiquities has always relied on the dirty, risky and illegal work of looters who break into burial tombs, whether in Africa or the Americas, and the smugglers and dealers who evade customs controls and launder the provenance of these objects. Audubon at least was honest about his methods, and did not try to hide this messy business from the wealthy collectors to whom he sold his work, which he marketed as 'alive and moving', as more compelling or rewarding than the empty eggs and static stuffed specimens that had been fetishes for collectors of the great auk and other bird species.

Around the same time, the creation of mass-media reproduction technologies was changing the fine art and illustrated book markets. It became feasible to print indefinite numbers of copies of an illustrated book or of a famous painting. Whereas the woodblocks used for illustrations when printing was invented in the fifteenth century wore out after between five hundred and a thousand impressions, the later engraved metal plates and lithographs were more resilient, and advances in inks also brought down the price of colour images. Mark Catesby and John James Audubon's careers spanned a century (roughly 1730s to 1830s) during which image printing technologies advanced quickly, culminating in photography. Using similar methods for both making and marketing their work, they created limited, or rare, sets of their work, and imbued them with the aura of a singular original.

Linda Dugan Partridge writes that Audubon's *Birds of America* was designed

> to attract wealthy patrons with the appearance of opulence. Such luxury dictated not only the physical appointments of the book but also the presentation of the bird as a consumable good. This could be accomplished through representation of sensuous textures, sinuous contours, or feather coloration. It fed upper-middle-class tastes for other exotic imports (including even live birds and bird skins). The bird on the page undoubtedly ranked as a possession to be displayed beside other objects of art.[15]

The aptly named historian describes aesthetic features that birds share with luxury commodities and that could be conveyed in two-dimensional art. This is not exclusive to upper-middle-class moderns. The colours of birds' feathers and the durable aesthetic shapes of bills and claws have been luxury products in America for as long as people have lived there. Aztec artists collected feathers and ground them up to make colour tints for mosaic art.[16] The California condor in particular played an important role in the mythology of Native American peoples, including the Chumash of the Santa Barbara area. Chumash representatives secured permission in 1987 to be 'present for the trapping of the last three birds so that appropriate ceremonies could be performed'.[17] At the opposite end of the commodity spectrum, now extinct birds were once the source of food staples. Audubon wrote in his passenger pigeon biography: 'You may find several Indian towns, of not above 17 houses, that have more than 100 Gallons of Pigeons Oil, or Fat; they using it with Pulse, or Bread, as we do Butter'.[18]

Catesby's *Natural History of Carolina, Florida, and the Bahama Islands* and Audubon's *Birds of America* continued the tradition of birds as luxury art work. These two books are among the most valuable in Anglo-American publishing history. Recent sales at auction of complete sets of *The Birds of America* have set records for a printed book. Each artist travelled for years recruiting subscriptions from affluent individuals and from some of the same people and institutions who collected bird and other natural history specimens. Audubon, as we have seen, emphasized how his representations of birds were dependent upon his shooting and handling many specimens or models for his paintings. These bird books were not at all like a modern pocket-sized Audubon or Sibley's field guide. They were marketed as specimens from a limited population, not as textual arte-facts of potentially limitless supply. Catesby and Audubon pursued similar processes for reproducing, marketing and disseminating their work. Both worked in the field in America collecting (that is, shooting or netting) birds, and painting them in a mix of watercolour and other media. Both devoted great care to the engraving of their images; Catesby taught himself en-graving and did this work over fourteen years, while Audubon travelled to Scotland and England to hire the best available engravers and publishers – William Lizars in Edinburgh and Robert Havell in London. Audubon and Catesby both asked subscribers to pay two guineas for each set of a multi-instalment work that would be forthcoming on an uncertain sched-ule. Of Catesby's book, David R. Brigham reports: '*Natural History* was one of the most expensive publications of the eighteenth century, costing twenty-two guineas for a complete set'.[19] Catesby collected subscriptions for 166 copies from 155 subscribers, although evidently more were printed, because at his death his widow had some copies to sell as her legacy.

Audubon was an even more skilled marketer. He presented his method of painting as a means for viewers of his paintings (or buyers of the books) to see bird species 'alive and moving'. At the outset of his publication proj-ect in 1827 he decided to represent every bird in actual size, a decision that forced him to publish on the largest available paper, double elephant folio, measuring 39.5 by 29.5 inches. Subscribers to *The Birds of America* had to hire a bookbinder to collect the eighty-seven sets of five plates each, and sew them into bound volumes with covers. The resulting books are so large and heavy it is difficult to lift and open them. These are not guidebooks that a reader could take into the field in pursuit of the flying birds; they are great books at the centre of prestigious archives' permanent collections, and like sculptures displayed atop pedestals in a museum, the books in-habit special display and storage cases. For the largest bird species, such as the Californian vulture, the flamingo, great herons and some egrets, Audu-bon's goal of life-size representation results in an apparent imprisonment as the bird depicted folds its wings, lowers its head, or contorts its neck down to its feet so as to fit within the space of the paper. These large birds present very differently from the smaller ones, who cavort in flocks of up to five or six, either various species combined in one plate, or one species in a flock of male and female, younger and older specimens, arranged in a composition with tree leaves or flowers. For the largest birds, including the condor and great auk, the borders of the folio sheets and the bindings of the books resemble the cages at a captive breeding programme. The en-gravings hold the precious birds in stasis, so as to preserve the existence of their species, in contrast to their creator's goal of representing them 'alive and moving'. Jennifer Roberts, in a study of Audubon's commitment to depicting all the birds in actual size, has argued that Audubon desired his bird species to carry the authenticity and the enduring value of gold specie: 'Audubon hoped that his bird pictures might function like gold coins rather than paper tokens – so that, as they were transmitted through space, their essence might hold steady as embodied and intrinsic, knitted into the substance of their support'.[20]

Audubon wrote in his journal on 21 March 1827 of his work as 'a book that in fifty years will be sold at immense prices because of its rarity'.[21] He was more prescient about the rarity of his book than about the rarity of many species in America that he studied. He predicted, for instance, that deer would become rare but the passenger pigeon would remain numerous. *The Birds of America* was described in 2008 as an 'endangered book' by John N. Hoover, the president of the Bibliographical Society of America, when he wrote a review of an updated edition of Waldemar Fries' 1973 book, *The Double Elephant Folio*. Hoover wrote, based on Fries' research: '[O]f an estimated 200–225 complete sets produced by Audubon, 119 full sets of 435 plates are known along with 18 incomplete sets. Forty other complete and incomplete sets have been broken up, and at least twelve of these have occurred since Fries published the first edition of his guide to Audubon'. A book dealer or even a library might make a large profit from a complete set of the work by taking it apart and selling the plates individually, but this would amount to reducing the 'living' extant book to a series of commodified specimens.

An animal or plant species, by Ernst Mayr's standard definition, is a set of organisms of the same kind, living in a contiguous habitat, where they reproduce to perpetuate their population. Each individual organism is one element of the set, interchangeable with any other, and to observe an individual is, for the typical birdwatcher or wildlife tourist, to observe the species. A coin or stamp collector generally treats units of currency in much the same way, and species and money have subversive similarities beyond the common etymology in *specie*. The number of elements in the set (of animals or of bills or coins) is indefinite, and is hard to census or count without using high-tech cameras, sensors and telecommunications to capture or track the animals. Nineteenth-century naturalists like Audubon rarely tried to estimate bird populations and did not see their 'collecting' as reducing the number. This logic today supports the quasi-competitive activity of birders who maintain lists of the species they have observed, often subdivided by region or season. A similar logic structures the publication of bird guides, including Audubon's nationalistic *Birds of America*. Since the development of modern species taxonomy by Linnaeus in the early eighteenth century, descriptive field guides have supported the national and regional identities of human readers by assigning natural organisms to their given place. As human populations caused the populations of charismatic megafauna species to decline, however, efforts of humans to observe, study and conserve the animals intensified, and these efforts often become concentrated in small fractions of habitat that are legally protected and/or have been developed with touristic or scientific infrastructure. Species populations are now definite and delimited – mountain gorillas in the Ruwenzori Mountains, Uganda; giant pandas in Sichuan Province, China; orcas along Vancouver Island, Canada – all of these species have become emblems for tourism in those regions, and support the industry there. Guides become skilled at leading clients to view individual animals whose location is known, and conservation biologists study individual animals over their lifespan, as Jane Goodall did with chimpanzees in Tanzania, beginning in 1960. These small reserves function as outdoor museums for eco-tourists. Birds, especially seabirds and migratory birds, maintain more extended and diffuse habitats, however, which preserves the aleatory, sporting flavour of birdwatching. Only recently, as we shall see, have individual birds been named and particularized by the naturalists who work to conserve them.

In the history of wildlife conservation and tourism, the great auk and California condor exhibited important developments across 150 years, from the 1830s to the 1980s. The way each species was collected, represented, displayed and valued illustrates how endangered birds have been mediated and commodified, and how media technologies have shaped human understanding of rare animals. The great auk demonstrates the problem of the fetish, of humans valuing commodity specimens, such as collectable eggs, more highly than the species population itself. Because the auks were seabirds that roosted on inaccessible islands, they could not easily be represented or perceived through visual images or through casual tourism. The California condor, a very large bird with a habitat close to a major metropolitan area, is more accessible, and its recent history illustrates the problem of intimate commodification, whereby humans confuse their bonds with individual birds for the value of the broader species population. Each of these problems found reflection in visual media: painting and engraving in the time of Audubon and Catesby and the great auk, photography and film in the time of the California condor.

Conservationists and sportsmen have both argued since the late nineteenth century that wild animals are a public good; and whereas specimen commodities were limited in number and thus available only to elite collectors, members of the wider public could all enjoy the excitement of seeing charismatic birds and mammals. Zoos have been one venue for such public spectacle, and optical technology was also decisive for the early conservation movement. The portable camera was one such invention, but before 1900 it was too large and its exposure times too slow for capturing birds. The development of improved field glasses was instead the major catalyst for the reaction against shooting birds for collection. Field glasses introduced around 1900 were an improvement upon opera glasses, such as those Florence Merriam Bailey wrote about in *Birds through an Opera Glass*, which helped to popularize birdwatching as an activity that, unlike hunting, was widely appealing for women.[22] This led to a gender divide among birdwatchers, as Joseph Grinnell complained that the 'opera glass student' cannot take the place of the collector, because 'the skin record is essential'.[23] Also, guidebooks changed in the early 1900s to emphasize the visible features of birds that could be discerned from a distance, on the fly, through field glasses, compared to the enumeration of wing feathers, and other structural features perceptible only upon examining a dead specimen. Photographs were visual artefacts that could be reproduced indefinitely, unlike specimen fetishes, or the limited editions of engraved, hand-coloured prints that Catesby and Audubon sold to their subscribers. Through photography, anyone could possess a beautiful image of a rare species, and thus all could share in the treasure of rare birds and their preservation. With the California condor, however, the public stewardship of endangered species was carried to a new extreme.

The condor nests in cliffside caves in steep, arid canyons that are nearly as remote as the islands once inhabited by the great auk. In the early twentieth century, oologists did collect the condor's large eggs, and conservationists both followed the knowledge of egg collectors and tried to foil their depredations. As a carrion eater like other species in the *Vulturidae* family of New World vultures, the California condor is aesthetically disagreeable. It has a mostly featherless head, and employs techniques such as urinating on its own legs in order to control the bacteria that thrive in its food sources. In spite of its grotesque appearance, however, the California condor has been a very charismatic species and a compelling story among conservationists in wealthy, fast-growing, media-centric southern California.[24] Since the 1980s, media technologies have followed the condor closely and it has been valued as a spectacle, for its size and rarity, for the dangers its claws and beak have posed for the conservationists who climb into its nests and capture it, and for its association with a region that was heavily populated and developed in the twentieth century. Advancements in visual observation and tracking technologies made it possible for conservationists to follow individual birds across time and space, and thereby to invest value in living individuals, rather than in the abstracted concept of the species, or commodified specimens of the species, such as an egg.

The first sustained field research on the California condor was done by Carl Koford over four years sandwiched around his service in the Second World War. He completed his PhD dissertation at UC-Berkeley under the direction of Joseph Grinnell, whose opinions were quoted above. In addition to observing the birds through field glasses, Koford was able to locate and access the nests of several mating pairs, and he made regular visits to these nests to observe the length of gestation, the size of eggs, the growth and feeding of fledglings, and the full reproductive cycle of the bird. Female condors lay one egg at a time, and only if that egg is lost or destroyed will she lay another – a behaviour that has been exploited for captive breeding purposes. Koford travelled throughout the remaining habitat of the species, in the Sierra Madre mountains along the southern California coast, and in the southern Sierra Nevada mountains inland. In a 1953 report based on his doctoral dissertation, he arrived at a population estimate of just sixty birds, and expressed alarm at the species' low reproductive rate. His research helped to establish the Sespe Condor Sanctuary in the Los Padres National Forest in 1947, and as Noel and Helen Snyder, prominent condor ornithologists of the subsequent generation, wrote: 'Koford's recommendations for condor conservation were to become the standard for several decades, and by the 1970s Koford himself had become a cult figure for wilderness enthusiasts, famed for his espousal of noninterventionist techniques in studying and conserving endangered species'.[25] The Snyders point out an irony: for his research Koford made near daily visits to nests, and measured the size and weight of eggs and chicks, yet in his publications and activism he insisted that humans should not disturb condor nests or habitat, and that captive breeding would not be an effective method of increasing the species population because the birds tried to avoid humans. Afficionados of the great auk had sought to acquire its eggs and skins, even as they knew it was close to extinction in the 1830s. Carl Koford, in his visits to remote condor nests, wished to protect the eggs and increase the species, but he also sought to prevent others from repeating his close contact with the living birds. Intimate observation of the birds had replaced specimen commodities as the gold standard of bird lore, the valuable treasure that only a select few, whether scientists or tourists, could experience.

In the 1960s and 1970s, the Sierra Club, the Wilderness Society, and Friends of the Earth promoted an ascetic version of American conservation that demanded large nature preserves where humans could visit but not live or build. They also created the 'leave no trace' ethic that elevated observation of wildlife and experience of nature as a spiritual privilege that nourished the soul but did not return tangible, material rewards. The Wilderness Act of 1964 was a great success of this movement, and arguably has had its greatest impact in California. Almost 15 per cent of the state's land area is protected as wilderness, yet it is home to nearly 40 million people as of 2020, with an economy that would rank as the fifth largest among the world's nations. The mountains surrounding greater Los Angeles hold a ring of wilderness areas including the Sespe and Dick Smith, both created to protect the California condor. Whereas the condor builds its nests in cliffside caves accessible only to 'condorvationists', it feeds on carcasses it finds up to a hundred miles away on ranches and rangelands. These food sources are abundant, but eating them can be risky. Lead poisoning from shot lodged in deer and other wildlife killed by hunters, and coyotes poisoned by ranchers, has been the leading cause of condor mortality. The birds' lives encompass the geographic contrasts of modern California – from wilderness to some of the richest, most consumerist and unequal cities in the world.

The California condor has also lived amidst the modern developments of photography and film. William L. Finley, along with Carl Koford's advisor Joseph Grinnell, made the first detailed study of a nesting pair of condors in the San Gabriel Mountains above Los Angeles in 1906, and took fine photographs of those birds as well as of others in Oregon, where he lived near Portland and kept a California condor named 'General' as a pet. Finley was among the first American conservationists to produce wildlife motion pictures for education and advocacy, and his influence led to the creation of the Tule Lake, Malheur and Klamath national wildlife refuges. Noel and Helen Snyder's studies of the California condor also used photography to make an important breakthrough in the biopolitics of endangered species conservation. In 1982 the Snyders were part of a team that attached radio telemetry devices to the wings of California condors, enabling researchers to plot their movements, within range of radio towers or aircraft. The radio transmitters bore numbers that aided identification in photographs, and around this time Eric Johnson and Noel Snyder demonstrated that individual birds could be identified from photographs by virtue of distinctive patterns in their larger wing feathers. This method enabled them to solicit photographs from many volunteers and then to census and name each bird in the entire species population of just twenty-one individuals.[26] No longer did this endangered species consist of a set of functionally identical organisms; now each condor had a name and a casting photo, and was ready to be treated like a star.

Catesby and Audubon had published their writing and art by subscription. Each copy was destined for a specific buyer, and the work was so successful that each copy is still tracked and treasured. Waldemar Fries' *The Double Elephant Folio* had included a bibliography of the location and provenance of every extant copy, much like Errol Fuller had traced and photographed every extant great auk egg; and finally, the Snyders' team had photographed every surviving California condor.[27]

In April 1983 the first condor chick to be born in captivity was helped out of its shell at the San Diego zoo. Zookeepers named him Sisquoc, after one of the condor sanctuaries near Santa Barbara where public access is strictly limited. The new bird was greeted with manic publicity. For more than two decades condor conservation efforts had been riven by a debate between conservation biologists using 'hands-on' techniques that included removing eggs from the nests and hatching them in captivity, as Sisquoc was, and 'hands-off' environmentalists who insisted the birds were safest in their native habitat, and that wilderness preserves and bans on hunting were the best way to restore the species. The public's excitement at the birth of Sisquoc was a coup for the hands-on faction. Then, in the winter of 1984/85, 40 per cent of the wild population of condors died, most by unknown causes, and public opinion again shifted in favour of captive breeding programmes. The biologists then set out to locate and eventually to capture every surviving member of the species, a goal achieved by 1987. The techniques for raising condor chicks in captivity, apart from their parents (who were tasked with laying additional or 'relay' eggs in order to quickly boost the total population), involved keepers wearing condor hand puppets to feed the chicks. Similar techniques have been used for the whooping crane and other large bird species, whose chicks otherwise would imprint upon human caregivers and become incapable of life in the wild. The San Diego and Los Angeles zoos have cooperated in this work, at a cost of more than a million dollars a year, although it has also inspired large donations to the zoos.

The surviving California condors have become works of art, admired by the public and curated by zoo staff and by wildlife biologists, much like the extant sets of Audubon and Catesby bird books are conserved in libraries and museums, where portions of the art are exhibited and loaned out for travelling exhibitions. The population of condors has reached nearly five hundred in 2020. Just under half are in captivity, while the remainder have been introduced to the species' former habitat in southern California, as well as near Big Sur, in southern Utah and northern Arizona, and in Baja California Norte, Mexico. The California condor has been a high-profile success for captive breeding programmes, and fuelled plans for more such efforts on behalf of other endangered species. Conservation biologists have proposed more long-term captive breeding to maintain species *ex situ*, notably in response to the crisis of amphibians, as 'a new ark'. However, Noel Snyder and several co-authors published a paper in *Conservation Biology* on the ethical and practical limitations to captive breeding and reintroductions. The ark might be like a museum, but artworks require limited care and no feeding compared to birds. Most animals, even insects, will become domesticated – or if that term is too imprecise, will evolve behaviours adapted to life in captivity that differentiate them from the wild animals they are meant to supplement when released into the wild. Snyder and co-authors were all involved in captive breeding programmes, and yet they asserted that

> in practice the connection between captive breeding and habitat preservation is sometimes tenuous. Captive breeding can become an end in itself and may undermine rather than enhance habitat preservation by reducing the urgency with which this goal is pursued. The existence of a captive population can give a false impression that a species is safe, so that destruction of habitat and wild populations can proceed.[28]

Whereas ornithologists a century ago or longer sought to 'collect' or kill specimens of wild animal populations for museums or private collections, for the past fifty years wildlife biologists have often collected living animals for propagation in captivity, with the ostensible goal of protecting a species severely endangered in its natural habitat. These efforts have been supported by conservation movements and legal protections that establish wildlife as a public good, theoretically available to all as spectators, either in zoos or in the wild. Zoos have attracted public support and subsidies similar to art museums, centres for cultural edification through the open display of great works of art. The masterpieces that attract visitors to museums are valued for their rarity, much as endangered species lure visitors to zoos. Endangered species have become like works of art, dependent upon humans both for their protection and for their value. But humans lose sight of the fact that animals (like humans) cannot be held in archival stasis, cannot be taken out of their ecological habitat without losing the animated qualities that make them what they are. Species are defined by the process of reproduction, and therefore perhaps are better represented by the reproduction and dissemination of works of art, such as Audubon's engravings, than by static, singular, originary works of artistic genius.

Gordon M. Sayre is professor of English and folklore at the University of Oregon, and a specialist in colonial history and literature of North America. He is the translator of *The Memoir of Lieutenant Dumont* by Jean-François Benjamin Dumont de Montigny, written in 1747 and first published in French in 2008, and in English in 2012. His work on natural history, species and media has also appeared in the journals *Environmental Humanities* and *The William and Mary Quarterly*.

Notes

1. The extinct birds are the passenger pigeon (plate LXII), Carolina parakeet (plate XXVI), ivory-bill woodpecker (plate LXV), Bachman's warbler (plate CLXXXV), esquimaux curlew (plate CCVIII), and great auk (plate CCXLI). Several other species are apocryphal: 'the Carbonated Warbler, the Blue Mountain Warbler, the Cuvier's Kinglet, and the Small-Headed Flycatcher are not familiar to the average bird student … these species, like the Townsend's Bunting here, are mystery birds that ornithologists to this day have not been able to identify'. Slatkin, *John James Audubon*, 228.
2. Audubon's claim to have never drawn from a stuffed specimen appears in Audubon, 'Method of Drawing Birds', 48–54, which functioned as a prospectus for his project as he was soliciting subscriptions and arranging for its printing and engraving in Edinburgh. It is reprinted in Irmscher, *John James Audubon*, 753–58. On the catchphrase 'drawn from nature', see Irmscher, *The Poetics of Natural History*, 206–17.
3. Audubon, 'Missouri River Journal', 131.
4. Audubon, *Ornithological Biography*, 316. Fuller in *The Great Auk* includes a few other eyewitness reports of great auk behaviour (50–57), but not this one. He writes that Funk Island, a small rock islet forty miles off the NE corner of Newfoundland, was the 'largest Garefowl colony of recent historical times' (368), but that the population there was likely wiped out before 1800. See also Kalshoven, 'Piecing Together the Extinct Great Auk'.
5. Martin, *A Late Voyage to St. Kilda*.
6. Townsend published an entertaining narrative of his encounter with the Californian vulture and other species in his 'Popular Monograph of the Acciptrine Birds', 265–70.
7. Lorimer, 'On Auks and Awkwardness', 200. The same holds true for the American bison. Museums and collectors frantically tried to obtain stuffed specimens in the 1880s and 1890s when there was widespread belief that the species would become extinct.
8. *Birds of America* quarto edition, 5, 282, quoted in Irmscher, *Poetics of Natural History*, 208.
9. As recently as 2015 this conflict between scientific ornithology and the protection of rare bird species recurred in the Solomon Islands, where Christopher Filardi, working for the American Museum of Natural History in New York, captured the first photograph of a male moustached kingfisher and then collected (a gentle euphemism for euthanized) the specimen for further study. In an interview with the Audubon Society, Filardi said that finding the bird was like encountering a 'magnificent … ghost', and that the experience evoked in him a 'surreal, childlike sense of a mythical beast come to life'. Although Filardi explained that collecting the male specimen was necessary for the species' conservation, outrage quickly followed the news of the bird's death. Avian advocates launched a Change.org petition, calling for the museum to 'Stop Killing in the Name of Science'. Wright, 'Reordering Nature', 16.
10. Kastner, *A World of Watchers*, 97.
11. 'Matthew Brower locates the preference for the image of the animal over the body of the specimen at the moment of the rise of animal photography. But … that preference predates photography's advent, and can be found in the writings and drawings of Audubon.' Wright, 'Reordering Nature', 57.
12. Some natural history books were actually published with specimens of the flora they described affixed to their pages. See Secord, 'Pressed into Service'.
13. Catesby has become overshadowed by Audubon in the history of American ornithology. His artistic and publication methods were remarkably similar, despite working

almost exactly one century earlier. The best biography and introduction to Catesby is Frick and Stearns, *Mark Catesby*.

14. Many of these are reproduced in the exhibition catalogue: McBurney and Windsor Castle Royal Library, *Mark Catesby's Natural History of America*.
15. Partridge, 'By the Book', 272–73.
16. See de Acosta, *Natural and Moral History*; Durán, *The History of the Indies*.
17. Snyder and Snyder, *The California Condor*, 305.
18. Audubon, 'The Passenger Pigeon', 265.
19. Brigham, 'Mark Catesby and the Patronage', 93.
20. Roberts, *Transporting Visions*, 190.
21. Audubon, 'European Journals', 222.
22. Bailey, *Birds through an Opera Glass*. For a brief portrait of Bailey, see Wolfe, 'Overlooked No More'.
23. Grinnell, quoted in Kastner, *A World of Watchers*, 109.
24. John Nielsen in his book about the bird writes of his memories of growing up in Piru, a small town in the Sierra Madre mountains near the concentration of California condor nesting sites. Settler colonial societies often shared a strong urge to create a perception of local bonds, values and ethics, and thus tried to form a bond with an animal like the condor, which was emblematic of that bio-region. Nielsen, *Condor*.
25. Snyder and Snyder, *The California Condor*, 62.
26. Ibid., 137. The species was ideal for this method of identification, because it is so large, it soars over mountainous areas while only rarely flapping its wings, and it regrows wing feathers slowly over several seasons. The Snyders claimed the method would be effective on a population as great as sixty.
27. The individuation of endangered animals for the purposes of conservation appeals and fundraising was also used with respect to humpback whales by the organization Whalewatch in 1990. Its Whale Adoption Project offered donors an 'adoption certificate' for one of eight named whales, identified by notches and distinctive shapes in their fins, and described by their affective traits. See Alaimo, 'Cyborg and Ecofeminist Interventions', 140–41.
28. Snyder et al., 'Limitations of Captive Breeding', 345.

Bibliography

Alaimo, Stacy. 'Cyborg and Ecofeminist Interventions: Challenges for an Environmental Feminism'. *Feminist Studies* 20(1) (1994): 133–52.
Audubon, John James. 'European Journals', in Maria R. Audubon (ed.), *Audubon and his Journals*, Vol. 1 (New York: Dover Publications, 1960), 222.
———. 'Method of Drawing Birds Employed by J.J. Audubon, Esq., FRSE'. *Edinburgh Journal of Science* 8 (1828): 48–54.
———. 'Missouri River Journal', in Maria R. Audubon (ed.), *Audubon and his Journals*, Vol. 2, (New York: Dover Publications, 1960), 131.
———. *Ornithological Biography*, Vol. 4. Edinburgh: Adam and Charles Black, 1831. Retrieved 27 August 2020 from http://audubon.library.pitt.edu/
———. 'The Passenger Pigeon', in Christopher Irmscher (ed.), *John James Audubon: Writings and Drawings* (New York: Literary Classics of the United States, 2000), 260–68.
Bailey, Florence Merriam. *Birds Through an Opera Glass*. New York: Chautauqua Press, 1889.

Brigham, David R. 'Mark Catesby and the Patronage of Natural History in the First Half of the Eighteenth Century', in Amy R.W. Myers and Margaret Beck Pritchard (eds), *Empire's Nature: Mark Catesby's New World Vision* (Williamsburg, VA: Omohundro Institute and University of North Carolina Press, 1999), 91–146.

De Acosta, José. *Natural and Moral History of the Indies*. Translated by Frances Lopez-Morillas. Durham, NC: Duke University Press, 2002.

Durán, Diego. *The History of the Indies of New Spain*. Translated by Doris Heyden. Norman: University of Oklahoma Press, 1958.

Frick, George Frederick, and Raymond Phineas Stearns. *Mark Catesby: The Colonial Audubon*. Urbana: University of Illinois Press, 1961.

Fuller, Errol. *The Great Auk: The Extinction of the Original Penguin*. Charlestown, MA: Bunker Hill, 2003.

Irmscher, Christoph. *The Poetics of Natural History: From John Bartram to William James*. New Brunswick, NJ: Rutgers University Press, 1999.

——— (ed.). *John James Audubon: Writings and Drawings*. New York: Literary Classics of the United States, 2000.

Kalshoven, Petra Tjitske. 'Piecing Together the Extinct Great Auk'. *Environmental Humanities* 10(1) (2018): 150–70.

Kastner, Joseph. *A World of Watchers*. New York: Knopf, 1986.

Lorimer, Jamie. 'On Auks and Awkwardness'. *Environmental Humanities* 4(1) (2014): 195–205.

Martin, Martin. *A Late Voyage to St. Kilda, the Remotest of All the Hebrides, or Western Isles of Scotland: With a History of the Island, Natural, Moral, and Topographical*. London: printed for D. Brown and T. Goodwin at the Black Swan and Bible without Temple-Bar, 1698.

McBurney, Henrietta, and Windsor Castle Royal Library (eds). *Mark Catesby's Natural History of America: The Watercolours from the Royal Library, Windsor Castle*. London: Merrell Holberton, 1997.

Nielsen, John. *Condor: To the Brink and Back – the Life and Times of One Giant Bird*. New York: HarperCollins, 2006.

Partridge, Linda Dugan. 'By the Book: Audubon and the Tradition of Ornithological Illustration'. *Huntingdon Library Quarterly* 59(2/3) (1996): 269–301.

Roberts, Jennifer L. *Transporting Visions: The Movement of Images in Early America*. Berkeley: University of California Press, 2014.

Secord, Anne. 'Pressed into Service: Specimens, Space, and Seeing in Botanical Practice', in David N. Livingstone and Charles W.J. Withers (eds), *Geographies of Nineteenth-Century Science* (Chicago: The University of Chicago press, 2011), 283–310.

Slatkin, Carole Anne. *John James Audubon: The Watercolors for the Birds of America*. New York: Blaugrund and Stebbins, 1993.

Snyder, Helen, and Noel Snyder. *The California Condor: A Saga of Natural History and Conservation*. San Diego: Academic Press, 2000.

Snyder, Noah F.R., et al. 'Limitations of Captive Breeding in Endangered Species Recovery'. *Conservation Biology* 10(1) (1996): 338–48.

Townsend, J.K. 'Popular Monograph of the Accipitrine Birds of N.A. – No. II'. *The Literary Record and Journal of the Linnaean Association of Pennsylvania College* 4(12) (1848): 265–72.

Wolfe, Jonathan. 'Overlooked No More: Florence Merriam Bailey, Who Defined Modern Bird-Watching'. *The New York Times*, 17 July 2019.

Wright, Paula Ann. 'Reordering Nature: Romantic Science, Natural History, and Mountaineering'. PhD dissertation, Dept. of English, University of Oregon, 2019.

Chapter 9

The Species Revivalist Sublime

Encountering the Kauaʻi ʻŌʻō Bird in
Jakob Kudsk Steensen's *Re-Animated*

Sarah Bezan

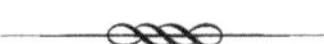

In his reflections on the use of virtual reality (VR) technologies in his artistic practice, Danish artist Jakob Kudsk Steensen describes the present ecological age as one infected with a 'collective anxiety that everything will vanish'.[1] As species, habitats and ecosystems have undergone radical changes that press at the limits of human perception and experience, the rise of VR and other eco-technological artworks have created the possibility of virtual spaces where, Kudsk Steensen says, '[extinct] species can live on as data'.[2] These technological developments have in turn ushered in a new set of parameters for ecological aesthetics and values. Heather Davis and Etienne Turpin suggest in their critical introduction to *Art in the Anthropocene* that the Anthropocene epoch is itself 'primarily a sensorial phenomenon',[3] as evidenced by the use of increasingly sophisticated visual and technological tools, including data visualization, satellite imagery, climate models, and the like.[4] The Anthropocene, they insist, has rapidly refashioned our sensorial and perceptive systems so that 'we daily experience what used to be a sublime moment'.[5]

While the sublime is for some eco-critics a seemingly outmoded concept freighted with the baggage of human exceptionalism and colonialism,[6] Davis and Turpin's observations suggest that the concept may nevertheless have ongoing relevance for interpreting the elevated affective and aesthetic experiences with nature that have been afforded by recent technological innovations. Philosophical anthropologist Jos de Mul, for instance, utilizes the sublime to articulate what he views as the '"second" or "next nature[s]"' of the '(bio)technological sublime'.[7] For de Mul, the (bio)technological sublime marks the shift from nature to technology that characterizes the nineteenth and twentieth centuries but also illuminates the emergence of 'converging technologies – information technology, biotechnology, nano-technology, and the neurosciences'.[8] Yet as we spiral further into the sixth mass extinction, these converging technologies place further demands upon the aesthetic and moral relationship between humans and natural environments. On a planet progressively emptied of its biodiversity, what does it mean to revive an extinct species through digital technologies?

As a response to this question, this chapter examines how Kudsk Steensen's VR installation *Re-Animated* (2018–19) stages a sublime virtual encounter with the Hawaiian 'ō'ō bird (a member of the Australo-Pacific honeyeaters family declared extinct in the late 1980s) from the island of Kaua'i. Extending de Mul's insights on the (bio)technological sublime, I analyse how extinction storytelling is made meaningful through VR artworks that engage with issues of species loss and revival. As I define it, the 'species revivalist sublime' is an affective experience that takes VR users beyond elegy and into a space of reflection upon the sublimity of (bio)technological potentials, in a virtual narrative in which users engage with the second natures of extinction. Through the user's encounter with virtual landscapes and digital objects, the species revivalist sublime engenders novel emotions, sensations and technologically (re)mediated memories with extinct species. It therefore finds its basis in an encounter with what is to come rather than merely a representation of what is past. The VR stage is not primarily a space of elegiac introspection: in *Re-Animated*, it is the (bio)technological site of the birth of 'a newly created species – one defined by its original vocal evocation, remixed by humans and partially composed of new bits of digital data'.[9] As I argue in this chapter, Kudsk Steensen's creative simulation of a sublime storyworld (or fictional space) of species revival exceeds normative elegiac responses to anthropogenic extinction. Taking users beyond these standard elegiac expressions, the species revivalist sublime explores how the second natures of extinction are not only seen and heard but also experienced through immersive digital technologies in the twenty-first century.

VR Encounters with Second Natures

A self-styled 'digital gardener' of virtual bio-architectural environments, Kudsk Steensen is an artist who strategically populates VR spaces with a wide range of real and imagined flora and fauna. Kudsk Steensen's virtual environments include Kensington Gardens and Hyde Park, French Polynesia's Bora Bora Island, the mountains of New Mexico, and the Peruvian rainforest (to name but a few). A number of these creative productions have their basis in photographic images taken by the artist, along with digital recordings and other historical or material archives sourced from museums. But what Kudsk Steensen creates is often beyond the realm of present-day experience. The global art project 'Catharsis' (2020), for example, 'collaps[es] various ecological timelines' to explore forest landscapes over a period of centuries, pristinely preserved and safely protected from the impact of human activities.[10] The act of experiencing the absence or presence of the human species in fragile environments is at the core of Kudsk Steensen's artistic philosophy, whether it be sensing and experiencing shifting ecosystems or responding to resounding species losses.

Re-Animated serves as a particularly poignant example of how technological environments can function as a site of encounter with extinct species. The VR installation comprises one cinematic film ('Re-Animated') and three 4K videos ('Arrival', 'Mating Call', and 'Bug Zapper'), which are screened in a studio space lined with wood chips to capture the sensation of the forest floor. Kudsk Steensen recreates the foliage and natural features of the island of Kaua'i through algorithms, but also supplements the VR environment with archival materials including photographs of preserved 'ō'ō birds held at the American Museum of Natural History and sound recordings of the last 'ō'ō from 1987 archived at the Macaulay Library in the Cornell Lab of Ornithology. Illustrating how extinct species live on in sound and image, as well as through VR experience, *Re-Animated* is an artwork that reflects upon the posthumous existence made possible by recent advancements in de-extinction science (a practice that promises to revive extinct species through the means of back-breeding, cloning, and synthetic biology). Although the 'ō'ō is not officially a candidate for de-extinction, curator Toke Lykkeberg writes that *Re-Animated* is a 'biotechnological laboratory' that 'meditates on our paradoxical techno-scientific trajectory'.[11] Kudsk Steensen can be described as a kind of de-extinction artist who revives key features of the 'ō'ō's anatomy, habitat, and even its final mating call.

Along with the distinctive sounds and images of the 'ō'ō, the installation features audio excerpts from Kudsk Steensen's interview with American ornithologist H. Douglas Pratt, a well-regarded bird illustrator and wildlife photographer with a record of research across the Hawaiian islands and the Pacific region. Pratt's own vivid memories of the extinct 'ō'ō bird, combined with the rare audio recordings he produced of its mating call, are interwoven into the VR user's experience. While Kudsk Steensen himself has no living memories of the 'ō'ō, Pratt's experience with the last dwindling populations of birds from the island of Kaua'i is instilled within his artistic renderings. It is mainly according to memory, rather than scientific data, that Pratt reanimates the ʻōʻō in his oil paintings. In his interview with Pratt, Kudsk Steensen asks how these paintings relate to Pratt's ornithological work. In response, Pratt explains:

> I work from memory, rather than from observation, in the sense that I know the form and feeling of experiencing the bird. But I also use photographs and specimens from the natural history museums I have connections to, when painting the birds. It is almost like a mummy lying there – there is no life. The memory and the photographs are what help you put life back into those feathers.[12]

Pratt's realistic reproduction of a 'lively' ʻōʻō is based on the form and feeling imparted through memory. Kudsk Steensen, on the other hand, recreates the form and feeling of the bird by listening to its song, remixing it, and 'convert[ing] it into something else that is mediated and transformed into something that does not follow the same biological rules of DNA and evolution', in a space where 'organic and virtual realities have started to rapidly interbreed'.[13]

As a clear departure from the naturalistic style of Pratt's ʻōʻō oil paintings, *Re-Animated* builds upon Kudsk Steensen's encounter with the recorded MP3 of the mating call and his photographs of dead ʻōʻō specimens. These sonic, visual and material remainders are the rudimentary elements of the species revivalist sublime. At no point in the user's experience of the virtual world of *Re-Animated* does the extinct bird ever appear to be realistically alive (i.e. in its 'first nature'); it is presented either as photographs of dead bird skins or as a monstrously oversized and unambiguously zombified bird. The effect of this is that users – who are also unlikely to have living memories of the ʻōʻō – are guided to reflect less on what has been lost and more on what has been remade. What this indicates is that while classic interpretations of elegy would begin from the point at which the dead speak in the form of a melancholic lyric or lament,[14] the species revivalist sublime transforms these expressions into a rumination of the voice itself, and the technologies of its transmission.

Although Kudsk Steensen moves his users away from a purely elegiac reflection on the extinct ʻōʻō, the user's novel encounters with the technologically mediated bird do not occur in a vacuum. To draw from Jody Berland's treatment of virtual menageries, the ʻōʻō bird appears as an 'animal emissary': a figure in dynamic relation to, and in historical context that encompasses the audience or human user of digital technologies.[15] As an animal emissary, the simulated ʻōʻō reveals how 'technological innovations (re)mediate our encounters with animals, just as animals (re)mediate our encounters with technology'.[16] Rather than produce an indulgent fantasy of species revival, Kudsk Steensen's voiceover narration furnishes the VR world with historical context that allows users to retrace the movements of European colonizers to the Pacific as they descend 12,000 feet to the forest floor of Kaua'i below. Combined with encounters with the simulated 'ō'ō bird, this historically grounded VR narrative demonstrates how anthropogenic changes have led to a range of technological mediations over the past several decades.

Expanding upon this range of technological mediations, *Re-Animated* exemplifies the manifold meanings of the virtual: as that which is, according to Marie-Laure Ryan, an immersive and interactive computer-generated environment; a narrative medium or art; and an expression of the 'real'.[17] As a computer-generated environment, *Re-Animated* provides the user with an opportunity to reflect on the technological elements of the VR space. These engagements with technology are made apparent in three ways: through (1) the sublime drone-like opening scenes of the cinematic film, leading to a mausoleum site; (2) the transformation of the 'ō'ō mating call into a floating, multidimensional sonic object; and (3) the user's encounter with zombified 'ō'ō birds and oversized bug zappers. In the following sections, I examine how these encounters go beyond traditional elegiac treatments of extinction by instead producing new experiences with the extinct and technologically revived 'ō'ō bird.

As a narrative, Kudsk Steensen's project illustrates how storyworlds of extinction in VR (as with all creative genres) are mediations to some degree; they are filtered through the narrative arcs and technological tools that shape human responses to extinction. Through encounters with the simulated objects, figures and spaces outlined above – some of which may be explored in slightly different ways due to each individual user's breath and voice picked up by the VR headset – the imagination of the user is activated in order to 'co-produce' the VR storyworld.[18] This co-produced storytelling process, which describes the melding of the narrative itself with the imagination of the audience, is one that is arguably shared across all narrative media (from novels to film), including VR.[19] In *Re-Animated*, this co-produced storytelling process revolves around the 'ō'ō animal emissary, which unites the user's reflections on the technological encounters with the narrative elements presented.

Lastly, the use of remediated digital data creatively simulates a virtual environment and a lost species that is, within the self-enclosed storyworld of the installation, entirely 'real'.[20] In other words, these virtual objects produce real feelings (and possibly even attitudinal changes and behaviours) in VR users. While *Re-Animated* may be perplexing or even unsettling to some who might view such projects as an alarming sign of a declining natural world that is only 'alive' in virtual reality environments, Kudsk

Steensen's VR project arguably provides the necessary space for exploring the emotional contours of species loss and revival. Following cognitive scientist David J. Chalmers's argument that virtual or digital objects are real and foster valuable experiences,[21] *Re-Animated* models encounters with extinct species that may impact upon the user's attitudes towards the second natures that have begun to reshape the cultural context of the sixth mass extinction crisis.

The Sublime Aesthetic of 'Arrival': Descending and Ascending Kaua'i

'Arrival', the first of three 4K videos of *Re-Animated*, firmly situates Kudsk Steensen's work within the aesthetic tradition of the sublime. As a reflection of the elevated or lofty experiences with natural environments that characterize this tradition,[22] the 10 minute 30 second video begins with the user's haunting descent upon the island of Kaua'i from an altitude of 12,000 feet, followed by a transcendental ascent towards a mausoleum structure at the peak of Alaka'i (a present-day wilderness preserve). During the initial descent, the island comes into view through the mist and fog, while the hypnotic and even-keeled narrator addresses the user in the second person: 'you descend, you fall through the air, through the ages'.[23] A white building no bigger than a Monopoly game piece emerges on the island's peak, still thousands of feet below. The narrator informs the user of their likeness to the explorers, the missionaries, the naturalists, and even 'the horses, the mosquitoes, the viruses' that altered the habitat of the island over the course of the eighteenth, nineteenth and twentieth centuries.[24] As the user traverses the empty landscape, the narrator introduces Reverend Dwight Baldwin, a nineteenth-century missionary doctor and patriarch of what would become Hawaii's first pineapple business and one of several sugarcane plantations. Reverend Baldwin is one of a number of nineteenth-century colonizers who witnessed the arrival of a 'tiny translucent bird' in 1826: the mosquito, a carrier of avian malaria that wiped out a large number of Hawaiian bird species, eventually pushing the 'ō'ō to higher and cooler elevations, towards the swampy plateau of Alaka'i. Descending through time and space from the eighteenth century to the present, the user experiences an acute sense of awe and wonder at the vastness of the temporal and spatial scale of the VR environment.

What begins with a *descent* to Kaua'i soon becomes an *ascent* led by Victorian naturalist Walter Rothschild, whose vast collections of preserved avian bird skins (including numbers of 'ō'ō birds) are housed at the American Museum of Natural History.[25] As an embodiment of natural scien-

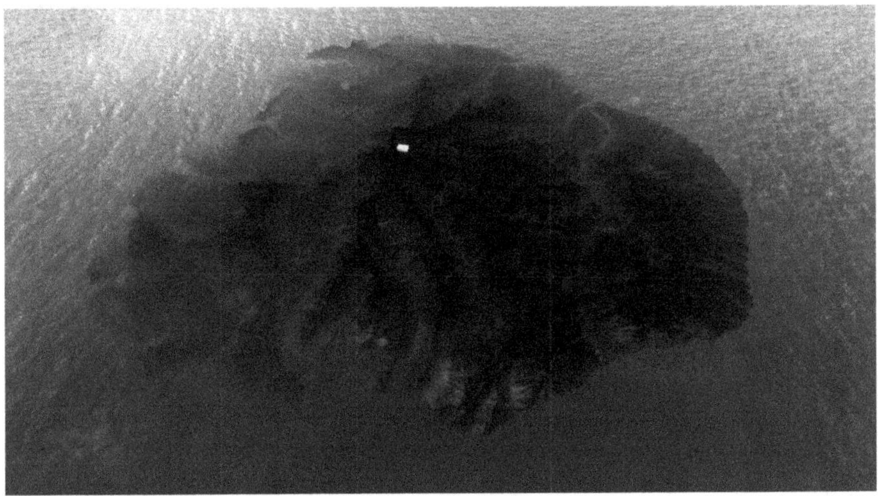

Illustration 9.1 Jakob Kudsk Steensen, 'Arrival', 2018–19. Screenshot of 10 min 30 sec video excerpt.

tific practices of preservation and collection, Rothschild is presented as an imaginary guide on a 'trail to immortality'. On this trail, the narrator explains, Rothschild finds an ʻōʻō, snaps its neck, and 'fuses' with the bird, which is then displayed within the mausoleum space. It is here that users are presented with their first sombre glimpse of the ʻōʻō, which appears not in 3D (as one would expect of a VR environment) but in 2D as a set of photographs on the mausoleum wall. As the film draws to a close, the walls of the mausoleum swarm with flies as the user is brought through the entrance of the structure. 'A bird preserved, a bird displayed, a bird forever', the narrator proclaims as the camera swiftly zooms towards one of two images depicting tagged ʻōʻō bird skins, their cotton-stuffed eyes eerily aglow.[26] In the film's final moments, the unceasingness of death and extinction permeates the mausoleum site, transforming the user's sublime response to the transcendental journey into a bitter contemplation of how the imperialistic practice of resource extraction and the pursuit of natural scientific knowledge have ultimately hastened the extinction of the ʻōʻō species.

By orienting users to the perspective of European figures like Baldwin and Rothschild, 'Arrival' demonstrates how the sublime aesthetic remains nostalgically entangled in the destructive fantasy of Western arrival into what were perceived to be boundless and untouched natural landscapes, replete with species as yet unknown to natural science. Kudsk Steensen's use of the sublime therefore allows for a broader examination of how Western aesthetic and natural scientific values have shaped understandings of species biodiversity since European contact. The relationship between these values has been well accounted for in the intellectual history of the sublime. As eco-critics have argued, the sublime is rooted in the hierarchical separation of humans and 'nature' (the basis of human exceptionalism); it is also, as environmental philosopher Emily Brady suggests, a tradition that historically represents 'an othering of nature, where nature is overpowered, conquered, and colonized'.[27] Yet *Re-Animated* exemplifies how the sublime might also be understood as a concept that is evolving according to ever-shifting environments. As Brady contends, the Anthropocene has given rise to more 'challenging aesthetic qualities in nature', with technology allowing 'us to approach places that remain wild to a great extent, in ways that still leave room for the sublime response'.[28] While the descent and ascent of the user through the island of Kaua'i may align with the aesthetic and philosophical traditions of the sublime, it also explicitly outlines a three-hundred-year colonial history that allows the artist to position a critique of the Anthropocene responses Brady describes. The VR environment of 'Arrival' thereby captures a new use of the sublime, interrogating how Western values inform the representation of ecological habitats affected by anthropogenic extinctions.

A further component of the user's transcendental ascension to the mausoleum space in 'Arrival' is how the sublime is utilized to facilitate, but also to expand upon, elegiac responses to extinction. The final moments of the film invoke an elegiac response, but do so in a way that meets Jesse Oak Taylor's standard for the genre of extinction elegy: as an artwork that 'both mediates the relationship between species and helps us conceptualize human existence in species terms'.[29] The film does not merely feature the loss of a charismatic species that already dominates cultural narratives – a pitfall Ursula Heise elucidates in her critical treatment of species elegy.[30] Instead, the film's species-level reflections on the impact of human colonizers and settlers over a three-hundred-year timeline induces users to observe evolutionary shifts over the course of several centuries. As I will discuss in the next section, Kudsk Steensen's other pieces ('Mating Call', 'Bug Zapper', and 'Re-Animated') utilize the species revivalist sublime to redirect expressions of mourning into novel encounters with the extinct 'ō'ō bird. But in 'Arrival' in particular, the sublime aesthetic initiates what is fundamentally a question of *belonging* across time and space; a sense in which the narrative scale of the sublime aesthetic allows individual users in the VR environment to consider the impact of humans on a timeline that exceeds the average human lifespan. This multiscale narrative frame occurs through what narratologist David Herman describes as 'allegorical projection', through which 'individual animals, or the traces left by their activity, can be used to figure forth species-level phenomena'.[31] Herman contends that these narratives 'provide resources for imagining and engaging with species-level processes, including those involving the loss or extinction of whole species'.[32] Kudsk Steensen's use of archival materials, including photographs of Rothschild's 'ō'ō bird skins, introduces a VR storyworld that extends these individual animal traces into a more robust and extensive narrative about species loss, in turn accounting for the ways in which the scientific and technological innovations of the human species figure into the story of the 'ō'ō bird extinction.

This multiscaled, species-level narrative frame does not, of course, provide an all-encompassing deep history of human-driven ecological changes on Kaua'i. As Daniel Lewis writes in *Belonging on an Island*, the history of human arrivants and subsequent extinctions across the Hawaiian islands is extensive, stretching over a period of more than a thousand years, with extinctions on the islands occurring naturally prior to human contact. According to Lewis, 'there are a variety of statistical and descriptive ways to slice the extinction pie', given the impact of pre-industrial Polynesian settlers around 1000 CE, which sparked a catastrophic wave of extinctions that continued on after the arrival of Captain Cook and other Westerners in 1778.[33] Considering the extent to which remote island habitats (where most species have come from somewhere else) can be said to host 'native' or endemic species, Lewis's argument is that there are a myriad of ways in which human and non-human arrivants can be said to 'belong' on the Hawaiian islands. The fact that 'Arrival' charts a limited temporal course through a three-hundred-year colonial timeline suggests that users must consider how structures of belonging emerge out of the Western aesthetic, natural scientific, and technological developments that have set the stage for the VR storyworld within which users find themselves. As I explain in the next section, it is through the VR environment that users discover alternative ways of belonging with, and relating to, extinct species in their second natures.

The Species Revivalist Sublime: Encountering the 'Ō'ō Bird in VR Environments

I have argued thus far that *Re-Animated* expands the affective register of extinction beyond more familiar expressions of grief and into the realm of awe and wonder at the sublimity of (bio)technological revival. Elegiac expression can therefore be understood as that which hinges upon representations of what is lost, while the species revivalist sublime can be defined as the arousal of novel emotions and sensations through virtual encounters with the objects and landscapes of the extinction storyworld. In further outlining my interpretation of the species revivalist sublime in this section, I discuss how the shifting emotional resonances associated with representations and encounters with extinction are explored in Kudsk Steensen's artwork through the juxtaposition of 2D and 3D forms. As we have seen with 'Arrival', the 2D photographs of dead bird skins on the mausoleum wall portray *pictorial* or *representational space*, while the 3D mausoleum site models an *embodied space* that accommodates immersive and experiential encounters with digital objects and landscapes.[34] Unlike 'Arrival', however, encounters with the 'ō'ō bird in the 4K videos 'Mating Call' and 'Bug Zapper', as well as in the cinematic film 'Re-Animated' enable the user to experience the second natures of extinct species. While the 2D photographic representations of dead 'ō'ō birds are characterized by their realism (thereby eliciting a sense of grief at the fullness of what has been lost), the 3D objects of the film and videos are fantastical and distorted, requiring the emotional responsivity and imaginative co-production of users as they contemplate the role of technology in mediating these encounters.

In 'Mating Call', for instance, users encounter 'ō'ō birdsong as a floating, multidimensional sonic object. Kudsk Steensen's depiction of the mating call of a 3D form restores the audible remainders of birdsong into the simulated environment of Kaua'i through digital tools. In doing so, the artist constructs a technological space within which to re-orient the user's relationship to the revived 'ō'ō. In 'Bug Zapper' and 'Re-Animated', on the other hand, the user's movements through the VR space are determined by oversized zombified birds and bug zappers, which counter the

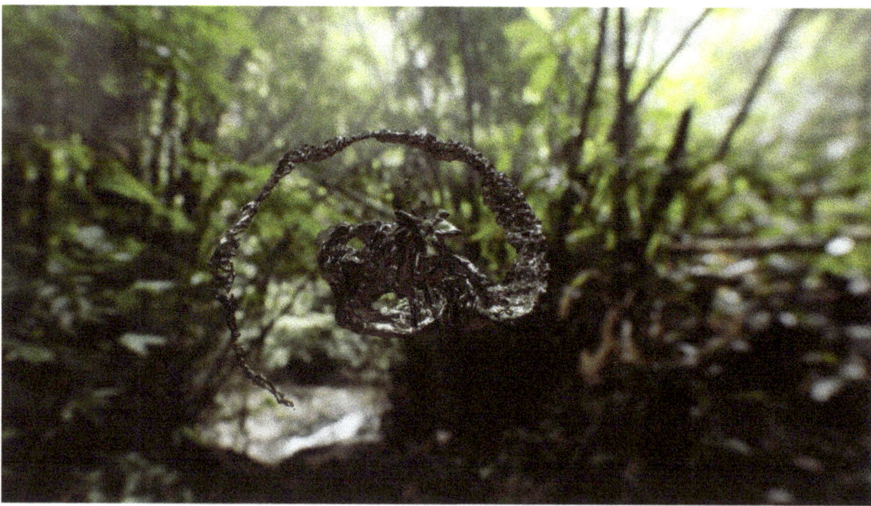

Illustration 9.2 Jakob Kudsk Steensen, 'Kaua'i 'ō'ō Mating Call', 2018–19. Screenshot of 3 min 25 sec video excerpt.

lofty, elevated perspective of 'Arrival' by giving the user an insect-sized ground view of the forest floor. Through these encounters, *Re-Animated* creatively reconfigures the sounds of lost species while also exploring the shifting temporal and spatial scales of extinction's second natures. These emotionally responsive, co-productive, and embodied VR encounters re-draw the parameters around which users of digital technologies relate to extinct species.

The 3 minute 25 second video 'Mating Call' transforms the mournful birdsong of the last 'ō'ō into a digital object that can be approached by the user in VR space. This digital object is based on Pratt's recording of the mating call, which has been archived in the Macaulay Library and has also been made widely available on YouTube, where it has to date garnered well over a million views and more than a thousand elegiac tributes in the comments section.[35] The MP3 audio recording of the last 'ō'ō continues to elicit a strong elegiac response because listeners recognize it as a one-sided communication; a message transmitted to a mate that no longer exists. Yet by foregrounding the 'ō'ō mating call as the voice of extinction, Pratt's acoustically engineered sound also dissociates it from the creaturely vocal-izations of its ecological habitat. In this way, the recording sets up a bina-ried relationship between the foreground and background of multispecies soundscapes. The problem with this is that the 'sonic individualization' of birdsong in nature recordings has, according to Jody Berland, 'eliminated the background noises of nature', along with layers of musical phrases and responses that might 'speak to us of the complexity of presence, interac-tion, noise and loss'.[36] Blending the foreground and background sounds of nature recordings through VR technologies can address these complexities while also accounting for the ways in which the biological and technolog-ical realms intersect across species lines.

In its depiction of 'ō'ō birdsong, 'Mating Call' digitally reconnects the 'ō'ō call with a sonorous VR space that amalgamates the foreground and background of multispecies soundscapes. This in turn allows the user to serve as an *interactive responder* to the mating call. Unlike the passive lis-tener of the MP3 recording, the VR user is confronted with a distorted and incomplete 3D sonic object that metonymically stands in for the body of a bird that is distinctly absent. In the absence of the body of the bird, however, the sonic object takes on a larger-than-life quality that showcases its embedded relations with the forest and its inhabitants, including the VR user who encounters it. By activating multiple senses in the user, from sight and hearing to a sense of movement in space (also known as pro-prioception), the user learns how, in Berland's terms, to 'situatedly listen'.[37]

This act of situated listening is achieved through the VR user's embod-ied and immersive encounter with the 3D mating call, which can only be heard, seen and spatially encountered in the forest environment that appears halfway through the video. The video begins in the mausoleum, where the user follows the camera as it lingers over the familiar 2D photographs of 'ōʻō bird skins. The buzz of flies and the symphonious chirps, clicks and tweaks of forest creatures fill the mausoleum space. After sixty seconds, the user encounters an entirely new object: a feathered orb that descends from the sky through a hole in the mausoleum ceiling. A percussive cracking similar to the snapping of twigs can be heard as the orb and the mausoleum walls melt away, revealing silvery sonographic ribbons of sound that undulate and swell in time with the mating call. Juxtaposed with the self-enclosed vacuity of Pratt's audio recording and the funereal realism of the 2D photographs on display in the mausoleum, the 3D sonic object unites 'ōʻō birdsong together with the creaturely operatics of the simulated forest before they slowly fade away from view.

While the user's embodied and immersive response to the mating call does not (and indeed could not) take the place of the mate of the 'ōʻō, this novel encounter with the bird activates multiple senses that reorient the user to the dynamic dimensions of a 3D sonic object that has been situated in a virtually created natural environment. In doing so, it taps into a more expansive range of feelings (from grief to awe) that characterize human responses to the bird's (bio)technological expressions. As the summation of gestures, expressions and songs that have been preserved and revived through digital tools, these surges of sound reveal the way in which birdsong functions as a sophisticated 'form of code' that is distinctly (bio)technological in and of itself.[38] It also elucidates how an interactive, embodied, and experiential encounter with the mating call of the 'ōʻō can offer a new perspective on the mediated relationship between human technologies and extinct species. In a VR space where the user not only grieves but creates new memories with the extinct bird, Kudsk Steensen makes it possible for the user to reflect upon the sublimity of the bird's (bio)technological afterlife.

The multidimensionality of VR has the effect of blending the foreground and background of multispecies landscapes, but it also opens up the static framing of sonic preservations in ways that audio recording alone cannot. David Michael's critique of nature recording as a documentary genre, for instance, reveals how nature recording often belies a 'unique historical obsession with the documentation and transportation of place' that ultimately 'imposes a frame on a boundless, limitless process. This frame freezes a portion of our environment to be appreciated as a work in itself: an object'.[39] As a result, Michael continues, nature recordings seem to suggest that an environment 'can be preserved and communicated as an objective reality, able to reproduce a subjective experience in the listener of 'being there'.[40] According to Michael, however, the static framing and objectification of nature in audio recording means that it is incapable of representing complex phenomenological reality.[41] While VR cannot solve this problem entirely, I would suggest that as an experiential medium, VR challenges purely aestheticized views of nature that institute a hierarchical relationship between 'nature' and the human observer. As I have shown, 'Mating Call' does not aim to merely objectify birdsong but rather to situate and embed it within a simulated natural environment. In this way, Kudsk Steensen's VR project speculatively explores a broader spectrum of emotions, sensations and experiences in users by considering the role of technology in mediating encounters with extinct species. While audio recording may maintain the veneer of all-encompassing objectivity, the value of Kudsk Steensen's work lies in its potential for generating alternative realities and affective relations with extinct species within the VR storyworld.

'Bug Zapper' and 'Re-Animated' explore these shifting frames of reality and affective relations by positioning the user into an insect-sized ground view of the forest floor. Further extending the multiscalar narrative established in 'Arrival', the highly speculative and imaginative video and film recreates the second natures of extinction through non-human perspectives and timescales. In 'Bug Zapper', for instance, the user is presented with a series of subtitles that narrate the user's movement through the VR space as the forest transitions from day to night. Floating across the forest landscape, the 5 minute 45 second video likens the user to a bug gazing at a screen, 'steering toward its own end, hypnotized by the buzz of a bright bug zapper'.[42] The bug zapper, which is suspended from an entangled mass of floating branches awash with purple light, is stationed inside of the mausoleum adjacent to the 2D photographs of dead 'ō'ō birds. In this VR space, Kudsk Steensen creates a retroactive fantasy in which the user – imagined as one of the malaria-carrying mosquitos that brought an end to the 'ō'ō species – is exterminated. In this way, the user enters into a kind of immortal technological realm where fantastically overscaled digital objects and otherworldly landscapes pulsate with the potential of (bio)technological revival.

The thrumming electric zing of bug zappers that appear in both 'Bug Zapper' and in the longer cinematic film 'Re-Animated' is reminiscent of the kind of gothic galvanism that is prominently featured in Mary Shelley's classic novel *Frankenstein* (1818). Appearing for the first time in the 14 minute and 16 second 'Re-Animated' film (and also, for the first time in the *Re-Animated* project, entirely away from the mausoleum), the

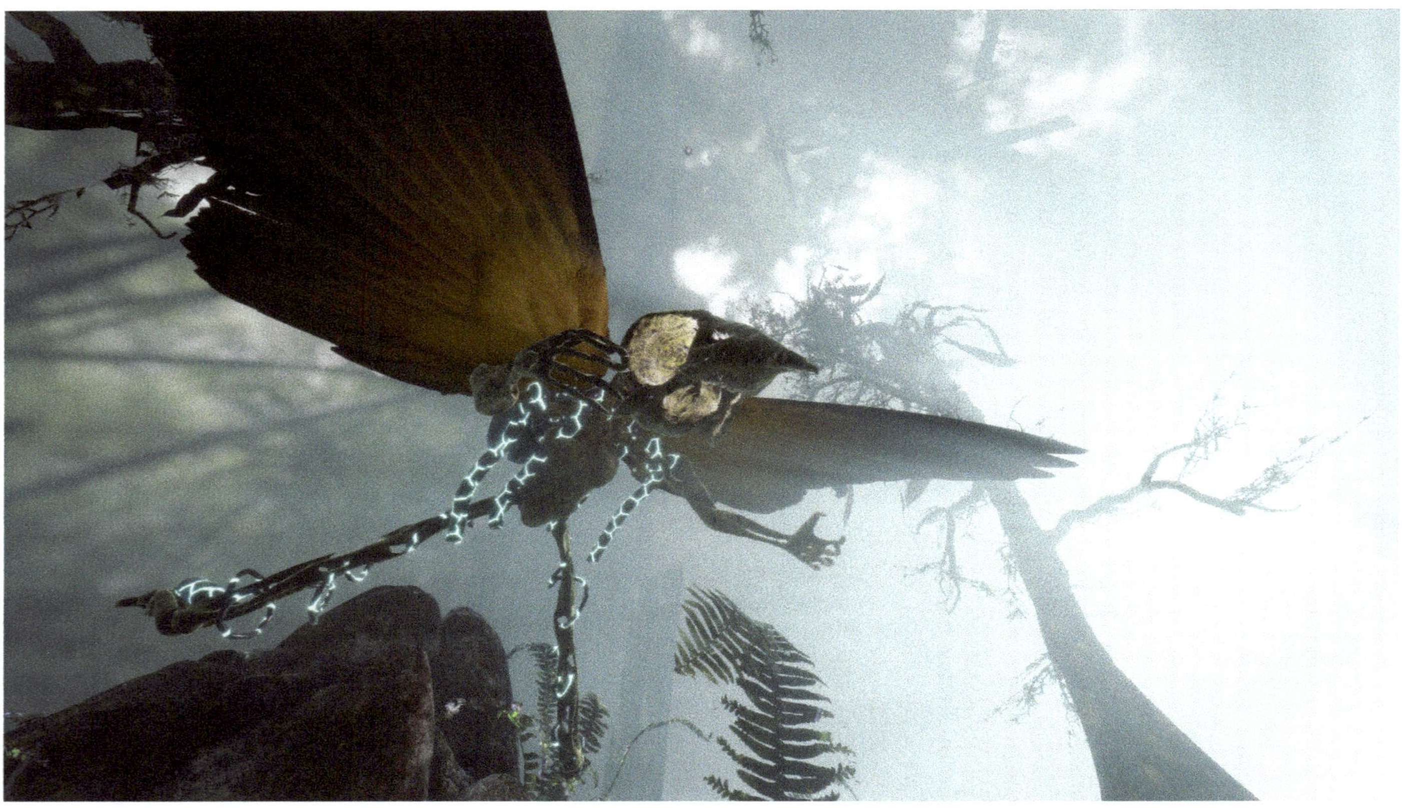

Illustration 9.3 Jakob Kudsk Steensen, 'Re-Animated', 2018–19. Screenshot of 14 min 16 sec video excerpt.

(bio)technological resurrection of the ʻōʻō bird is presented to the user as a disassembled and oversized zombie. Accompanied by ambient algorithmic sound and intermittent zaps of electricity, users journey through a watery cave (marked in neon green ink with elegiac messages from YouTube) where they encounter an oversized assembly of objects, including a bug zapper, a skeletonized ʻōʻō bird wing, large branches wired with electricity, a large white rock projected with flies, and a floating feathered orb that emits purple light. The camera pans the landscape from above and below, as if from the perspective of an insect. Suddenly, a billboard-sized 2D photograph of an ʻōʻō bird skin erupts from the ground like a tombstone, reminding the user of the organic remains that resist digital revival. Meanwhile, the feathered orb that initially appeared in 'Mating Call' shrinks and expands as it navigates the user through an underground tunnel towards a higher altitude. It is here, thousands of feet above sea level surrounded by a grey mist, that the film culminates in a sublime cloudscape suspended with the disparate components of the ʻōʻō bird: a pair of wings, a head, and humanoid-looking legs and arms. That the bird appears only in an assembly of parts, never entirely animated or lifelike, speaks to the magnificent scale of technological immortality that characterizes extinction's second natures. Yet it also illuminates how the species revivalist sublime resists revelling in pre-extinction fantasies of liveliness: in 'Re-Animated', the species revivalist sublime acknowledges what has been lost but also exposes the VR user to what has been remade, however impartially and incompletely, in the afterlives of digital datasets.

In juxtaposing 2D and 3D forms throughout his *Re-Animated* project, Kudsk Steensen does not allow VR users to become entirely enraptured by the sublimity of (bio)technological potentials. For Kudsk Steensen, the oversized and alluring bug zappers, ʻōʻō bird body parts, and floating feathered orb stand in for the process of extinction that is superlative by its very nature: human-created and yet greater than the human, and happening on a global scale and speed too rapid to comprehend with reason. Kudsk Steensen's positioning of users in multiple perspectives – from the colonizer of 'Arrival' who safely descends upon the island, to the mosquito that is drawn into the deadly structure of the electric bug zapper or confronted with an imposing photographic wall of a dead ʻōʻō bird – challenges and undermines feelings of mastery and control that might arise during their immersive encounters with the virtual environment. When users finally meet the extinct ʻōʻō in its second nature, towering above the forest floor in a cloudscape during the screening of the cinematic film, they are urged to reflect upon the unsettling feelings of awe and discomfort that arise from these shifts in perspective in which they are transformed from mourners to interactive responders and finally, to immortal witnesses.

Conclusion: The Future Aesthetics of Extinct Species

Tracked through embodied, immersive, and emotionally responsive experiences with the extinct 'ō'ō bird in the VR storyworlds of 'Arrival', 'Mating Call', 'Bug Zapper' and 'Re-Animated', the shifting range of affects that I have identified in Kudsk Steensen's eco-technological installation point to the emergence of alternative relations and structures of belonging with non-human animals in digitally reconstructed environments. As eco-artistic practice comes under closer analysis in the field of ecocriticism, these alternative relations bring greater awareness to the future aesthetics of extinct species. Emily Brady and Jonathan Prior suggest in their essay on environmental aesthetics and rewilding, for example, that the 'future aesthetics' of extinct and endangered species relies upon 'the role of imagination in the unfolding aesthetic narrative of rewilded places'.[43] But what about the future aesthetics of extinct species in virtual environments?

While extinction studies scholars like Ursula Heise, Jussi Parikka, Dolly Jørgensen and Thom van Dooren have focused on the role of grief and longing in extinction imaginaries, my aim in this chapter has been to plot out the emerging aesthetics of species revival in digital artistic practice. For an artist like Kudsk Steensen, the imagination is a key component of how revived species are featured as both 'physical and virtual forms' that 'mimic the ambience and material[ity] of digital worlds'.[44] However, Kudsk Steensen also rightly recognizes that the use of digital tools in the wake of the sixth mass extinction points to a collective anxiety about accelerating biodiversity declines.

Elizabeth Swanstrom assesses the significance of this collective anxiety in *Animal, Vegetable, Digital*. In her discussion of the intersection of digital technologies and 'natural signs', Swanstrom outlines the extent to which an ever-increasing number of species are becoming 'vanishing signs'. However, digital and new media art practices can, according to Swanstrom, have a positive effect by allowing audiences to 'access a dynamic and participatory version of nature' in which 'human and non-human agents [can] inflect and shape their shared environments, as well as each other'.[45] Elaborating on Swanstrom's observations, I propose that this participatory crossover between species lines and natural and virtual environments is what makes the species revivalist sublime meaningful. If it is the case that one of the defining features of the Anthropocene is that the realms of biology and technology have become altogether inseparable, then the species revivalist sublime can open up spaces for the production of novel emotions, sensations, and technologically (re)mediated memories with extinct species. To reconstruct a digital world with extinct species in this way would not be to neglect natural habitats populated with endangered ones, because both are entirely real and capable of generating authentic feelings, attitudinal changes, and behaviours. In outlining a new set of values for the future aesthetics of extinct species, Kudsk Steensen's eco-technological artwork shifts users' perspectives and generates a wider range of affective responses to species loss and revival. Kudsk Steensen's species revivalist sublime ultimately plots an emerging trajectory for extinction imaginaries that not only radically questions what is past but also speculatively imagines what is to come as the sixth mass extinction crisis unfolds.

Sarah Bezan is postdoctoral research associate in perceptions of biodiversity change at the University of York's Leverhulme Centre for Anthropocene Biodiversity in the UK. Her research focuses on the entangled social and ecological dimensions of species loss and revival in contemporary British, North American and Australian literature and visual culture. She is currently at work on two book projects: *Dead Darwin: Necro-Ecologies in Neo-Victorian Culture* (under advance contract with Manchester University Press), along with a second monograph (in progress) that examines species revivalist representations of the woolly mammoth, great auk, dodo, Steller's sea cow, thylacine, and Pinta Island tortoise.

Notes

1. Chen, 'You Can Slow Down Time in Virtual Reality'.
2. Vickers, 'Artist Profile'.
3. Davis and Turpin, 'Art and Death', 3.
4. Ibid., 3–4.
5. Ibid., 11.
6. Brady, *The Sublime in Modern Philosophy*, 195.
7. de Mul, 'The (Bio)Technological Sublime', 33.
8. Ibid., 37.
9. Kudsk Steensen, 'How a YouTube Video Brought an Extinct Bird back from the Dead'.
10. Kudsk Steensen, 'Catharsis'.
11. Lykkeberg, '*Re-Animated* Press Release'.
12. Kudsk Steensen, 'How a YouTube Video Brought an Extinct Bird back from the Dead'.
13. Ibid.
14. Oxford English Dictionary, 'elegy, n.'
15. Berland, *Virtual Menageries*, 9.
16. Ibid., 5.
17. Ryan, *Narrative as Virtual Reality*, 12.

18. Ryan, 'Story/Worlds/Media', 2.
19. Ibid., 1.
20. Chalmers, 'The Virtual and the Real', 309.
21. Ibid.
22. Oxford English Dictionary, 'sublime, n.'
23. Kudsk Steensen, 'Arrival', 1:13–18.
24. Ibid., 2:23–26.
25. 'Walter Rothschild's Unusual Birds'. *Natural History Museum*. Retrieved 28 December 2020 from https://www.nhm.ac.uk/our-science/collections/zoology-collections/bird-skin-collections/walter-rothschild-birds.html#:~:text=Before%20he%20died%20Rothschild%20sold,display%20in%20his%20public%20museum.
26. Kudsk Steensen, 'Arrival', 9:45–55.
27. Brady, *The Sublime in Modern Philosophy*, 194.
28. Ibid., 185, 187.
29. Oak Taylor, 'Tennyson's Elegy for the Anthropocene', 225.
30. Heise, *Imagining Extinction*, 50.
31. Herman, *Narratology beyond the Human*, 262.
32. Ibid.
33. Lewis, *Belonging on an Island*, 9–10.
34. Ryan, 'Introduction', 2.
35. Kaua'i 'ō'ō.
36. Berland, *Virtual Menageries*, 196.
37. Ibid.
38. De Bruyn, 'Anthropocene Audio', 161.
39. Michael, 'Toward a Dark Nature Recording', 206.
40. Ibid., 208.
41. Ibid.
42. Kudsk Steensen, 'Bug Zapper', 3:43–4:10.
43. Brady and Prior, 'Environmental Aesthetics and Rewilding', 33.
44. Vickers, 'Artist Profile'.
45. Swanstrom, *Animal, Vegetable, Digital*, 18.

Bibliography

Berland, Jody. *Virtual Menageries: Animals as Mediators in Network Cultures*. Cambridge: MIT Press, 2019.

Brady, Emily. *The Sublime in Modern Philosophy: Aesthetics, Ethics, and Nature*. Cambridge: Cambridge University Press, 2013.

Brady, Emily, and Jonathan Prior. 'Environmental Aesthetics and Rewilding'. *Environmental Values* 26 (2017): 31–51.

Chalmers, David J. 'The Virtual and the Real'. *Disputatio* IX(46) (Nov 2017): 309–52.

Chen, Jiayin. '"You Can Slow Down Time in Virtual Reality": Why This Artist Is Using VR to Recreate Lost Ecosystems in the Era of Climate Change'. *Artnet*, 8 January 2020. Retrieved 28 December 2020 from https://news.artnet.com/art-world/jakob-kudsk-steensen-vr-1748265.

Davis, Heather, and Etienne Turpin. 'Art and Death: Lives between the Fifth Assessment and the Sixth Extinction', in Heather Davis and Etienne Turpin (eds), *Art in the Anthro-*

pocene: Encounters among Aesthetics, Politics, Environments and Epistemologies (London: Open Humanities Press, 2015), 3–29.

De Bruyn, Ben. 'Anthopocene Audio: The Animal Soundtrack of the Contemporary Novel'. *Critique: Studies in Contemporary Fiction* 57(2) (2016): 151–65.

de Mul, Jos. 'The (Bio)Technological Sublime'. *Diogenes* 59(1–2) (2013): 32–40.

Heise, Ursula. *Imagining Extinction: The Cultural Meaning of Endangered Species*. Chicago: University of Chicago Press, 2016.

Herman, David. *Narratology beyond the Human: Storytelling and Animal Life*. Oxford: Oxford University Press, 2018.

Kaua'i 'ō'ō. YouTube video uploaded by Robert Davis, 11 March 2009. Retrieved 28 December 2020 from https://www.youtube.com/watch?v=nDRY0CmcYNU.

Kudsk Steensen, Jakob. 'Arrival'. *Jakob Kudsk Steensen Vimeo Channel*, 26 September 2018. Retrieved 15 January 2021 from https://vimeo.com/291992820.

———. 'Bug Zapper'. *Jakob Kudsk Steensen Vimeo Channel*, 26 September 2018. Retrieved 15 January 2021 from https://vimeo.com/291992820.

———. 'Catharsis'. *Jakob Kudsk Steensen*. Retrieved 28 December 2020 from http://www.jakobsteensen.com/#/new-page/.

———. 'How a YouTube Video Brought an Extinct Bird back from the Dead'. *Silica Magazine*, 25 May 2018. Retrieved 28 December 2020 from https://www.engadget.com/2018-05-25-kauai-oo-honeyeater-youtube-memorial-ghost-media.html.

———. 'Re-Animated Video & VR Installation, 2018–2019'. *Jakob Kudsk Steensen*. Retrieved 28 December 2020 from http://www.jakobsteensen.com/.

Lewis, Daniel. *Belonging on an Island: Birds, Extinction, and Evolution in Hawaii*. New Haven, CT: Yale University Press, 2018.

Lykkeberg, Toke. 'Re-Animated Press Release'. *Jakob Kudsk Steensen*. Retrieved 28 December 2020 from http://www.jakobsteensen.com/#/new-page-1/.

Michael, David. 'Toward a Dark Nature Recording'. *Organized Sound* 16(3) (2011): 206–10.

Oak Taylor, Jesse. 'Tennyson's Elegy for the Anthropocene: Genre, Form, and Species Being'. *Victorian Studies* 58(2) (2016): 224–33.

Ryan, Marie-Laure. *Narrative as Virtual Reality: Immersion and Interactivity in Literature and Electronic Media*. Baltimore, MD: Johns Hopkins University Press, 2003.

———. 'Introduction', in *Narrative as Virtual Reality 2: Revisiting Immersion and Interactivity in Literature and Electronic Media* (Baltimore, MD: Johns Hopkins University Press, 2015), 1–24.

———. 'Story/Worlds/Media: Turning the Instruments of Media-Conscious Narratology', in Marie-Laure Ryan and Jan-Noël Thon (eds), *Storyworlds across Media: Toward a Media-Conscious Narratology* (Lincoln: University of Nebraska Press, 2014), 25–49.

Swanstrom, Elizabeth. *Animal, Vegetable, Digital: Experiments in New Media Aesthetics and Environmental Poetics*. Tuscaloosa: University of Alabama Press, 2016.

Vickers, Ben. 'Artist Profile: Jakob Kudsk Steensen'. *Rhizome*, 14 December 2018. Retrieved 28 December 2020 from https://rhizome.org/editorial/2018/dec/14/artist-profile-jakob-kudsk-steensen/.

Chapter 10

Insects, Spiders, Snails and Empathy

Representing Invertebrate Extinctions in Natural History Museums

Pedro Cardoso

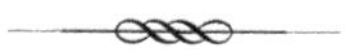

Museums as Testimonies to Species Extinctions

The origin of natural history museums can probably be traced back to the 'cabinets of curiosities, rooms filled with remarkable objects' that emerged in the sixteenth century. In such collections, exotic or strange animals and other living beings from around the world, including humans, were displayed to visitors. There was no scientific context offered, and fact and myth were often intertwined.[1] In the not too distant past, as a legacy of these curious times, two-headed cows formed centrepieces in what were otherwise modern natural history museums.

Joseph Grinnell (1877–1939), a North American zoologist and museum director, was probably one of the first scientists to start recording field observation data, and he also contributed to changing the perception of what a natural history museum should be.[2] According to Grinnell, a museum must serve as a repository of specimens and data documenting the composition of communities across space and time. Modern natural history museums are now repositories of countless specimens, often dating as far back as the eighteenth century, or much earlier if we include fossils, each with a given scientific context in the form of, at least, location and date of collection. These specimens provide evidence of past populations of species that are often already extinct from the sites where they were captured, or even worldwide. It is a fact that if we start digging into old and not so old collections, we will find the only known individuals of many species that are long gone and that cannot be recovered, the so-called Centinelan extinctions.[3] Museum collections therefore provide some of the most important testimonies that exist of past population declines and species extinctions,[4] and potentially also furnish the data and tools needed to prevent future extinctions.[5]

In this chapter, I delve into some of the main challenges involved in representing human-driven extinctions, specifically invertebrate extinctions, in museums. I also discuss some possible solutions to these challenges. Although, in the popular imagination, extinctions are usually associated with large mammals and birds (of which some of the better-known examples include mammoths and the dodo respectively), the majority of extinctions are occurring among invertebrates, involving insects, spiders, snails and their kin. Our major challenge is to cultivate sensitivity towards the lives of such species, which are frequently forgotten, seeking to foster empathy for them among museum visitors.

Museum Exhibitions as Venues for Extinction Showcase

The role of natural history and other kinds of museums does not begin and end in collection management and research.[6] Museums are also ideal venues to showcase nature to the general public, this often being the single function visitors are aware of. Most natural history museums dedicate at least some of their exhibition space to dinosaurs or other fauna that went extinct for non-anthropogenic reasons. As extinct species constitute 99 per cent of all species that ever existed, everyone is familiar with the concept that the disappearance of living forms is part of the natural rhythm of our planet. Museum exhibitions, however, also present broader opportunities to connect increasingly urban populations with contemporary phenomena, and foreground how current human actions influence the natural world. This influence often leads to species extinctions at levels many times higher than in any past era.

Building on in-house expertise on conservation biology, many museums around the world often host permanent or temporary exhibitions, the main theme of which is anthropogenic species extinctions, their causes and their consequences. Given the gravity of such a theme, it may have been deliberately avoided in the past. More recently, radical approaches to the subject have gained global attention. In 2019, for example, Bristol Museum & Art Gallery (UK) made worldwide headlines with their 'Extinction Voices' intervention. By cloaking all the threatened and extinct animals in black translucent veils, this intervention highlighted the disappearance of their kin due to habitat loss, pollution, poaching, and a myriad of other threats that the species may have faced. Often extinctions are portrayed alongside a related theme, such as climate change (Fig. 10.1). Given the major atten-

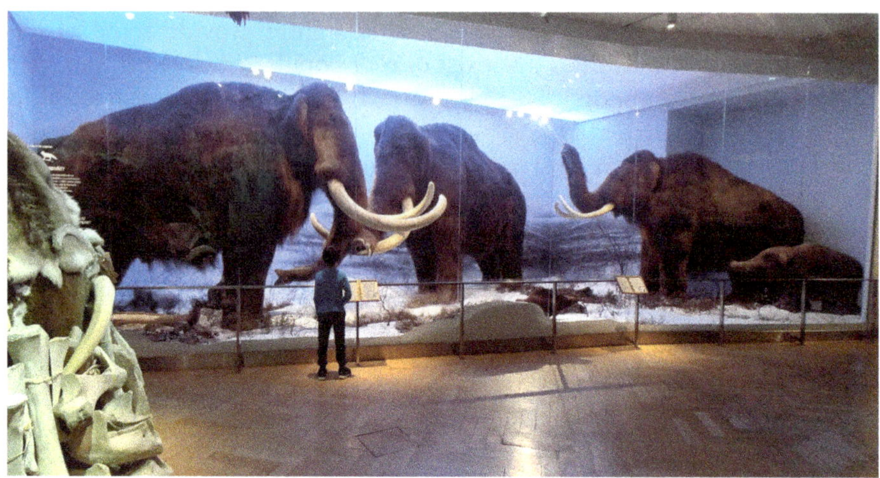

Illustration 10.1 Woolly mammoth exhibit, Finnish Museum of Natural History (Luomus), Helsinki, Finland. © Pedro Cardoso.

tion accorded to climate change by the public and in the media, an attention that is much greater than that given to species extinctions themselves, it serves as a way to connect us as visitors to the consequences of our own actions, and to the ecosystems of the millions of species that are affected by the current global warming crisis.

Invertebrate Extinctions around the World

When many of us think of extinctions, we imagine large mammals roaming across savannas and tropical forests. Most conservation biologists are as biased as the general public in this regard. But the truth is that these are just a diminutive part of the species in peril. Extinctions occur mainly among 'the little things that run the world'[7] – the invertebrates (animals without a backbone or bony skeleton), including insects, spiders, snails and their countless kin – and happen even on our doorsteps. Until recently, we were largely unaware that invertebrates could be imperilled to levels equivalent to, or even higher than, other taxa such as vertebrates, and that their loss would have consequences for our own well-being.

Given the lack of knowledge about, and monitoring of, invertebrates, relatively few extinctions are reported. Probably the best-known cases come from islands, where the evolution of unique and naturally rare species makes them particularly noticeable. Darwin himself was mainly inspired by island species to develop his evolution theory. The best documented declines on islands and elsewhere are reported for snails, as they leave shells after death, in what often constitutes important evidence of past populations. The entire genus *Carelia* was made up of twelve large species (with shells over 85 mm long) that were endemic to the island of Kaua'i (Hawai'i). The last specimens were seen almost seventy years ago, and now all species are extinct.[8] Several endemic beetles[9] and spiders[10] are probably extinct in the Azores, and doubtless many other species suffering similar fates remain undetected. On Madeira Island, the Madeiran large white butterfly (*Pieris brassicae wollastoni*) was last seen in the 1970s.[11] A complete list would need an entire book, or maybe even an encyclopedia, and would include examples from all species groups and geographical regions.

Fuelled by the recognition of declining numbers from specific regions,[12] concern over the fate of invertebrates has recently begun to gain traction in the non-scientific realm. In total, at least one million species are facing extinction in the coming decades, the majority of them invertebrates.[13] It is not only their vast numbers, but the dependency of ecosystems and of humanity on them, that makes the conservation and diversity of insects and other invertebrates critical for future generations.[14] A major challenge for now and in coming years is to draw attention to the beneficial contributions of nature to all people. Insects and other invertebrates are irreplaceable components in this, as is biodiversity in general.

Human activity is responsible for almost all current population declines and extinctions. The precise trends and drivers, and their respective importance, are mostly unquantified, but it is clear that six main factors contribute synergistically to decline or extinction: habitat loss, or its degradation or fragmentation; pollution, including harmful pesticides; the spread of invasive species; global climate change; direct overexploitation; and the coextinction of species dependent on other species.[15] If habitat loss has long been regarded as a major extinction driver,[16] pesticides and climate change have also recently been linked to major declines, namely of the better-studied pollinators. Invasive species are particularly relevant in island contexts.[17] On the other hand, coextinction might be the most important factor for many parasite taxa.[18] Two recent studies, which collected the opinions of more than five hundred experts on insects and spiders worldwide, confirmed that experts are not only worried, but that they also believe multiple common pressures are driving species extinctions worldwide.[19]

Humanity Depends on Invertebrates

With invertebrate extinctions, we lose much more than species. We lose abundance and biomass, diversity across space and time, which consequently causes homogenization, large parts of the tree of life, unique ecological functions and traits, and fundamental parts of extensive networks of biotic interactions. Such losses lead to the decline of key ecosystem services on which humanity depends. Insects and their kin contribute to provisioning services, supporting services, regulating services, and cultural services.[20] They change the structure, fertility, and spatial dynamics of soil, being a crucial element for maintaining biodiversity and food webs.[21] A large number of invertebrates provide medical or industrial products;[22] and in agroecosystems, invertebrates perform many different functions, such as pollination, nutrient and energy cycling, pest suppression, seed dispersal, and decomposition of organic matter, faeces and carrion.[23]

Despite their ubiquity, humanity's dependence on them, and the dire situation that many of them face, representations of invertebrate extinction intended for the general public are rare. Museums, science centres and similar venues have been laying the ground for such representations for some time, but not as systematically as for other groups. Probably more people are aware of what caused the end of the dinosaur era, even though it does not affect our lives, than of what is causing current invertebrate declines and extinctions. A quick online search for museum exhibitions specifically dedicated to extinctions will, in general, reveal multiple activities, but with few covering invertebrates in any meaningful way. This is not exclusive to museum exhibitions, but is common to conservation science in general, with these smaller animals being given only minimal attention and funding compared with other groups.[24]

Public Perceptions of Invertebrate Extinctions

Museums are now finally, albeit slowly, catching on to the importance of invertebrate extinctions, and beginning to represent some of the perils they face. The main difficulty that arises in such representations is probably one of creating empathy in a human audience for invertebrates. Being small, apparently insignificant, and often perceived as dangerous or as pests, insects, spiders, snails and others have a significant image problem that requires fixing. As a spider researcher, I regularly receive viral emails containing graphic pictures of a necrotic hand allegedly caused by a bite of the infamous brown recluse spider (*Loxosceles sp.*). Fortunately, these are fake, and, in fact, it is very rare for bites from this species to cause any kind of persistent harm.[25] And yet, such fake news definitely works against overturning the negative public perception surrounding spiders.

There are exceptions to the generally negative public perception of invertebrates though, with butterflies and bees viewed positively due to their perceived beauty and utility respectively.[26] Butterflies, with their often colourful wings, have always been regarded as symbols of beauty, and are valued for it. Their close cousins, dull brown moths, on the other hand, have always been feared, even though both groups are very similar. It is a tough break being sombre, coloured and an invertebrate.

Bees, as well as being colourful, have also accrued a positive perception due to their utility as pollinators. The bulk of our food depends on bees and other pollinators to enable reproduction and fruit production. There is growing awareness in the general public about this reality. Maybe in the future, dung beetles (which help to keep us in a dung-free world), earthworms (which prepare the soil for crops), spiders (which eat insect pests) and many other invertebrates will also be seen through a similarly positive lens.

Taking advantage of such optics, invertebrate extinctions are now often represented in ways that emphasize their utility. At Manchester Museum (UK), the exhibition 'After the bees' (2016–17) focused on bees as pollinators, and foregrounded how crucial they are to our own well-being. Another approach is to represent invertebrates as unique to a given neighbourhood, as was the case in the exhibition 'Azorean for millions of years', which depicted endemic Azorean insects for the public using large outdoor macrophotography. This use of extreme close-ups renders otherwise easily overlooked invertebrates highly visible and therefore difficult to ignore. Another way of nurturing a sense of connection between humans and invertebrates is by attributing common names to species, so that people can associate some characteristic of the species with a familiar concept.[27] The use of common names, along with other strategies to encourage empathy, form some of the best ways to raise invertebrate extinction awareness. People can name specific species of mammals and birds, but rarely can they name invertebrates. The use of carefully selected flagship species of invertebrates, the fates of which are foregrounded, could potentially help to draw attention to the bigger picture. Exhibitions with live specimens in recreated habitats, giant models that allow their intricate body structures to be viewed, and interaction with the scientists who study them and with their work, are some of the new ways being used to showcase invertebrates, their extinction, and why and how they should be saved.

Exhibitions involving the depiction of extinction are not the sole preserve of natural history museums. Art museums also frequently engage with extinction, often employing works of art in ways that leave more room for use of the imagination and for interpretation. When framed by a strong message, subjectivity and emotion can potentially play a central role in changing perceptions and attitudes. The Finnish Museum of Contemporary Art (Kiasma) recently (2019–20) showcased two exhibitions depicting global change and extinctions. 'Weather Report: Forecasting Future' was themed around the complex and varied relations that exist between the human and non-human in an age when climate change and mass extinction are threatening the future of life on Earth. 'Coexistence' dealt specifically with the question of the coexistence of humans, animals and nature. In a cooperation between Kiasma and Luomus, conservation biologists provided guided visits in which art pieces were interpreted through the prism of conservation science.[28] If it is difficult for the invertebrates themselves to engage the public, artists and scientists can give them a helping hand.

Moving Forward

The depiction of invertebrate extinctions in museums or similar venues is a work in progress. Even when they are known, invertebrates often endure an image problem; but usually unknown, most simply pass under the radar. Both reason and emotion must be mobilized if we are to improve the current situation. We can use an array of psychological tools[29] in an effort to overcome the 'public dilemma',[30] which is that invertebrates and their ecological services remain largely unknown to the general public.

Introducing museum visitors to those of us working with invertebrates gives a human dimension to a specific exhibition. There are many, including myself, who have first-hand experience of species extinctions and conservation. We may have lived in a tropical forest for many months identifying bugs and trying to learn a little bit more about them and their behaviour. I can share memories of expeditions to places including Brazil, Nicaragua, Tanzania and Ghana, where it is believed 80 per cent of invertebrates are still undescribed, waiting to be discovered; or memories of a recent project concerning one of the largest spiders in Europe, the Desertas Wolf Spider (Fig. 10.2), which only lives in a small valley of a small islet close to the island of Madeira, and under small rocks.[31] Thanks to an initiative led by Madeira Natural Park and Bristol Zoo (where you can visit it), this spider, which was on the brink of extinction ten years ago, is now recovering both in its native habitat and in a number of zoos across Europe. To save it we have had to swim (not voluntarily), climb cliffs and sleep in the remotest of places while doing scientific work to support its recovery, using our knowledge and expertise. Fieldwork does not always involve remote areas, however, and it is equally possible that a scientist may have been collecting specimens in the fields or forests near your own home, without even being noticed. During an exhibition at the Finnish Museum of Natural History titled 'See spider researchers in action', myself, my colleagues and some of our students had the opportunity to talk about activities such as these to anyone willing to listen and find out more about our work. Our talks

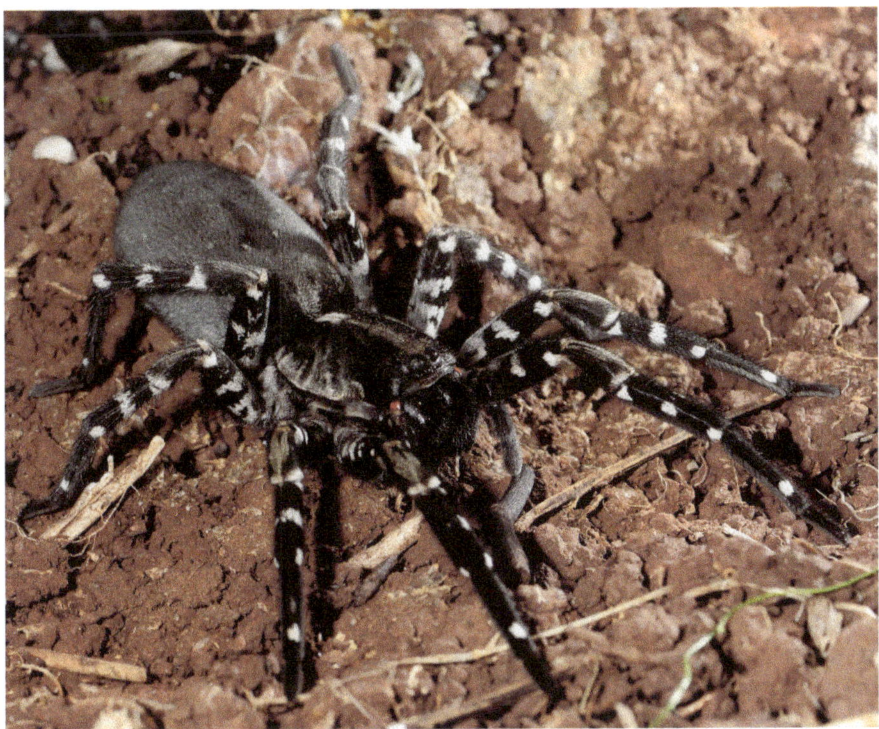

Illustration 10.2 The Desertas Wolf Spider (*Hogna ingens*), a critically endangered species with an ongoing recovery project. © Pedro Cardoso.

included looking under the microscope at many of the spiders we had collected and researched.

Sensory experiences that allow us to perceive reality at an appropriate scale bring small invertebrates into our world, even if we must use technological aids to supplement the limitations of our senses. New technologies, such as robotics and virtual or augmented reality (VR and AR respectively), are still underused in museums. Given their capacity to emphasize what cannot be seen by the naked eye, or to allow the stimulation of multiple senses simultaneously,[32] VR and AR can potentially show us invertebrates in new ways, giving them a more human dimension and, through this, fostering greater empathy for their fate. A system could be envisioned, for instance, where a visitor to the museum would be embedded in the world as it is seen through the lens of an insect, with gigantic grasses and even more gigantic humans walking around. A game might be developed in which the goal would be to survive the many perils an invertebrate could face, from destruction of habitat (which might be someone's backyard) to pesticides and unbearable heat. Maybe that visitor/player would have to help other insects to survive too, emphasizing the interconnection in an ecosystem. Through a judicious use of new tools and technologies, we can better grow empathy towards invertebrates, and thereby foster a collective will towards species preservation.

Pedro Cardoso is curator at the Finnish Museum of Natural History (Luomus), and adjunct professor in ecology at the University of Helsinki. As head of the Laboratory for Integrative Biodiversity Research (LIBRe, www.biodiversityresearch.org), he is currently mostly interested in understanding global drivers of extinction, and the distribution of species and communities across space and time. To help achieve this understanding, he is also developing new statistical and computational tools to quantify extinction risk and biodiversity at all levels: taxonomic, phylogenetic and functional.

Notes

1. Farrington, 'The Rise of Natural History Museums'.
2. Grinnell, 'Methods and Uses of a Research Museum'.
3. Wilson, 'The Little Things That Run the World'.
4. Shaffer, Fisher and Davidson, 'The Role of Natural History Collections'; Lister and Climate Change Research Group, 'Natural History Collections as Sources'.
5. Krishtalka and Humphrey, 'Can Natural History Museums Capture the Future?'
6. Miller et al., 'Evaluating the Conservation Mission'.
7. Wilson, 'The Little Things That Run the World'.
8. Solem, 'How Many Hawaiian Land Snail Species'.
9. Terzopoulou et al., 'Drivers of Extinction'.
10. Cardoso et al., 'Drivers of Diversity in Macaronesian Spiders'.
11. Gardiner, 'The Possible Cause of Extinction'.
12. Hallmann et al., 'More than 75 Percent'; Hallmann et al., 'Declining Abundance of Beetles'; Powney et al., 'Widespread Losses of Pollinating Insects'; Seibold et al., 'Arthropod Decline'; and many other studies.
13. Díaz et al., *Summary for Policymakers.*
14. Cardoso et al., 'Scientists' Warning to Humanity'; Samways et al., 'Solutions for Humanity'.
15. Cardoso et al., 'Scientists' Warning to Humanity'.
16. Foley et al., 'Global Consequences of Land Use'; Dirzo et al., 'Defaunation in the Anthropocene'; Habel, Samways and Schmitt, 'Mitigating the Precipitous Decline'.
17. Borges et al., 'Increase of Insular Exotic Arthropod Diversity'.
18. Dunn, 'Modern Insect Extinctions'; Dunn et al., 'The Sixth Mass Coextinction'.
19. Branco and Cardoso, 'An Expert-Based Assessment of Global Threats'; Milicic et al., 'Insect Threats and Conservation'.
20. Noriega et al., 'Research Trends in Ecosystem Services'.
21. Schowalter, Noriega and Tscharntke, 'Insect Effects on Ecosystem Services'.
22. Ratcliffe et al., 'Insect Natural Products and Processes'.

23. Schowalter, Noriega and Tscharntke, 'Insect Effects on Ecosystem Services'.
24. Cardoso et al., 'The Seven Impediments'; Mammola et al., 'Towards a Taxonomically Unbiased EU Biodiversity Strategy'.
25. Vetter, 'Spiders of the Genus *Loxosceles*'.
26. Sumner, Law and Cini, 'Why We Love Bees and Hate Wasps'.
27. Arroz et al., 'Bugs and Society I'.
28. Luomus is a contraction of *Luonnontieteellinen keskusmuseo* (Finnish Museum of Natural History). The museum is located in Helsinki.
29. Simaika and Samways, 'Insect Conservation Psychology'.
30. Cardoso et al., 'The Seven Impediments'.
31. Crespo et al., 'Assessing the Conservation Status'.
32. Rodrigues et al., 'An Initial Framework'.

Bibliography

Arroz, Ana Moura, et al. 'Bugs and Society I: Raising Awareness about Endemic Biodiversity', in Paula Castro et al. (eds), *Biodiversity and Education for Sustainable Development* (Dordrecht: Springer, 2016), 68–89.

Borges, Paulo A.V., et al. 'Increase of Insular Exotic Arthropod Diversity is a Fundamental Aspect of the Current Biodiversity Crisis'. *Insect Conservation and Diversity* 13(5) (2020): 508–18.

Branco, Vasco Veiga, and Pedro Cardoso. 'An Expert-Based Assessment of Global Threats and Conservation Measures for Spiders'. *Global Ecology and Conservation* 24 (2020): e01290.

Cardoso, Pedro, et al. 'Drivers of Diversity in Macaronesian Spiders and the Role of Species Extinctions'. *Journal of Biogeography* 37 (2010): 1034–46.

———. 'Scientists' Warning to Humanity on Insect Extinctions'. *Biological Conservation* 242 (2020): 108426.

———. 'The Seven Impediments in Invertebrate Conservation, and How to Overcome Them'. *Biological Conservation* 144 (2011): 2647–55.

Crespo, Luís Carlos, et al. 'Assessing the Conservation Status of the Strict Endemic Desertas Wolf Spider Hogna Ingens (Araneae, Lycosidae)'. *Journal for Nature Conservation* 22 (2014): 516–24.

Díaz, Sandra, et al. (eds). *Summary for Policymakers of the Global Assessment Report on Biodiversity and Ecosystem Services of the Intergovernmental Science-Policy Platform on Biodiversity and Ecosystem Services*. Bonn, Germany: IPBES Secretariat, 2019.

Dirzo, Rodolfo, et al. 'Defaunation in the Anthropocene'. *Science* 345(6195) (2014): 401–6.

Dunn, Robert R. 'Modern Insect Extinctions, the Neglected Majority'. *Conservation Biology* 19 (2005): 1030–36.

Dunn, Robert R., et al. 'The Sixth Mass Coextinction: Are Most Endangered Species Parasites and Mutualists?' *Proceedings of the Royal Society B* 276 (2009): 3037–45.

Farrington, Oliver Cummings. 'The Rise of Natural History Museums'. *Science* 42(10) (1915): 197–208.

Foley, Jonathan A., et al. 'Global Consequences of Land Use'. *Science* 309(5734) (2005): 570–74.

Gardiner, Brian O.C. 'The Possible Cause of Extinction of *Pieris Brassicae Wollastoni* Butler (Lepidoptera: Pieridae)'. *Entomologist's Gazette* 54 (2003): 267–68.

Grinnell, Joseph. 'The Methods and Uses of a Research Museum'. *The Popular Science Monthly* 77 (1910): 163–69.

Habel, Jan Christian, Michael J. Samways and Thomas Schmitt. 'Mitigating the Precipitous Decline of Terrestrial European Insects: Requirements for a New Strategy'. *Biodiversity and Conservation* 28 (2019): 1343–60.

Hallmann, Caspar A., et al. 'Declining Abundance of Beetles, Moths and Caddisflies in the Netherlands'. *Insect Conservation and Diversity* 13(2) (2020): 127–39.

———. 'More than 75 Percent Decline over 27 Years in Total Flying Insect Biomass in Protected Areas'. *PLoS One* 12 (2017): e0185809.

Krishtalka, Leonard, and Philip S. Humphrey. 'Can Natural History Museums Capture the Future?' *BioScience* 50(7) (2000): 611–17.

Lister, Adrian M., and Climate Change Research Group. 'Natural History Collections as Sources of Long-Term Datasets'. *Trends in Ecology and Evolution* 26(4) (2011): 153–54.

Mammola, Stefano, et al. 'Towards a Taxonomically Unbiased EU Biodiversity Strategy for 2030'. *Proceedings of the Royal Society B* 287(1940) (2020): 20202166.

Milicic, Marija, et al. 'Insect Threats and Conservation through the Lens of Global Experts'. *Conservation Letters* (2021): e12814.

Miller, Brian, et al. 'Evaluating the Conservation Mission of Zoos, Aquariums, Botanical Gardens, and Natural History Museums'. *Conservation Biology* 18(1) (2004): 86–93.

Noriega, Jorge Ari, et al. 'Research Trends in Ecosystem Services Provided by Insects'. *Basic and Applied Ecology* 26 (2018): 8–23.

Powney, Gary D., et al. 'Widespread Losses of Pollinating Insects in Britain'. *Nature Communications* 10 (2019): 1018.

Ratcliffe, Norman A., et al. 'Insect Natural Products and Processes: New Treatments for Human Disease'. *Insect Biochemistry and Molecular Biology* 41(10) (2011): 747–69.

Rodrigues, João M.F., et al. 'An Initial Framework to Develop a Mobile Five Human Senses Augmented Reality System for Museums', in João M.F. Rodrigues et al. (eds), *Handbook of Research on Technological Developments for Cultural Heritage and eTourism Applications* (Hershey, PA: IGI Global, 2018), 96–119.

Samways, Michael J., et al. 'Solutions for Humanity on How to Conserve Insects'. *Biological Conservation* 242 (2020): 108427.

Schowalter, T.D., J.A. Noriega and T. Tscharntke. 'Insect Effects on Ecosystem Services – Introduction'. *Basic and Applied Ecology* 26 (2018): 1–7.

Seibold, Sebastien, et al. 'Arthropod Decline in Grasslands and Forests Is Associated with Landscape-Level Drivers'. *Nature* 574 (2019): 671–74.

Shaffer, H. Bradley, Robert N. Fisher and Carlos Davidson. 'The Role of Natural History Collections in Documenting Species Declines'. *Trends in Ecology and Evolution* 13(1) (1998): 27–30.

Simaika, John P., and Michael J. Samways. 'Insect Conservation Psychology'. *Journal of Insect Conservation* 22 (2018): 635–42.

Solem, Alan. 'How Many Hawaiian Land Snail Species Are Left and What We Can Do for Them'. *Bishop Museum Occasional Papers* 30 (1990): 27–40.

Sumner, Seirian, Georgia Law and Alessandro Cini. 'Why We Love Bees and Hate Wasps'. *Ecological Entomology* 43(6) (2018): 836–45.

Terzopoulou, Sofia, et al. 'Drivers of Extinction: The Case of Azorean Beetles'. *Biology Letters* 11(6) (2015): 1–4.

Vetter, Richard S. 'Spiders of the Genus *Loxosceles*: A Review of Biological, Medical and Psychological Aspects Regarding Envenomations'. *Journal of Arachnology* 36(1) (2008): 150–63.

Wilson, E.O. 'The Little Things That Run the World (the Importance and Conservation of Invertebrates)'. *Conservation Biology* 1(4) (1987): 344–46.

Part IV

Representing Extinct Plants and Fungi

Chapter 11

Reconstructing Lycopsids
Lost to the Deep Past

Jeffrey P. Benca

―――⊗⊗⊗―――

Introduction: Plant Blindness and Its Influences
on Our View of Extinct Life Forms

Although plant communities define ecosystem structure in many modern landscapes in terms of biomass[1] and appearance, they receive less emphasis than animals in reconstructions of prehistoric landscapes. Plants associated temporally and palaeogeographically with prehistoric animals are often used as a background, serving to contextualize the focal animal subject, but they less often take on a central role in the ecosystems they help to comprise. Additionally, extinct plants have only rarely been depicted as focal subjects, independent of animals, or even their surrounding environment, and certainly not in meticulous detail – unlike, say, newly discovered dinosaurs. In many cases, long-extinct animals are reconstructed in great detail, while extinct plants are depicted in less detail, and vaguely resemble modern species[2] In summary: 'This form finds its most common expression as dinosaur art, characterized by a scrim of distant conifers, a pounded brown dirt foreground, and a centrepiece of fully realized dinosaurs in action poses. I call this form of dinosaur iconography "Monkey Puzzles and Parking Lots" for its regular reliance on stereotypical backgrounds that do not depict accurate vegetation'.[3] This is not an unexpected trend.

Plants are overlooked as organisms in their own right. The term 'plant blindness' was first used to describe a trend in the United States education system in which biology students perceive animals as having greater importance than plants, let alone other eukaryotes or prokaryotes.[4] This condition has been shaped, in part, by greater emphasis being placed by teachers on animals than plants. It has been argued further that plant blindness is just a facet of a much larger issue, which is that humans are conditioned to be everything-but-vertebrates-blind.[5] Given that 99 per cent of described terrestrial animal species are invertebrates,[6] overcoming plant blindness is a good start to creating a more empathetic connection to the diversity of life on Earth, which in turn could spur more inclusive efforts in conservation of biodiversity.[7] This literal and figurative anthropocentric view of plants as inanimate greenery is innate,[8] stemming from a snap-judgement decision-making process based on visibility bias in what is termed 'System 1'.[9] By contrast, those trained to study or appreciate plants beyond flashy flowers often engage in 'System 2',[10] a decision-making process that 'allocates attention to the effortful mental activities that command it, including complex computations'.[11] Plant blindness, moreover, has been found to be partially a physiological phenomenon, in that plants capture attention through the human visual system differently from animals.[12] In the case of 'seeing' plants, one has to grow accustomed to pausing to dedicate sufficient time and mental energy to observe their features, and therein to begin to appreciate their complexity, beauty and behavioural traits, as well as ultimately our full dependence on them as a foundation of many terrestrial food webs and living systems.[13]

Although extinct plants and the habitats they generated have and continue to be an integral part of palaeoartistic reconstructions, 'plant blindness', and overarching 'non-vertebrate blindness', have resulted in a vertebrate-centric visual culture. It is also natural to empathize more with organisms most similar to ourselves.[14] This is further compounded by the inherent difficulties of humans to comprehend 'deep time', a concept first described as a 'long Earth history' by Scottish geologist James Hutton,[15] and coined roughly two hundred years later by American author John McPhee.[16] Deep time (e.g. time spans of millions to billions of years), and the scale of geological and evolutionary processes over such long periods, are sublime to the human imagination. Accordingly, it is difficult to appreciate how profoundly organismal lineages, ecosystems and the appearance of landscapes have changed over such immense timespans. The art of depicting both the more familiar world and those alien worlds lost to the deep past is therefore crucial to contextualizing the history of life in modern ecosystems, and to learning how these systems themselves came to be.

While overcoming plant blindness in palaeoartistic reconstructions may seem a merely academic exercise, its consequences can profoundly impact our understanding of the history of life. The emphasis on which organisms are highlighted or emphasized in artistic depictions not only set a standard for which life forms should be considered important and interesting to the general public, but also to children who someday will comprise the next generation of palaeobiologists. If children were to grow up getting to see meticulously detailed depictions focusing on extraordinary and beautiful plants, invertebrates and other biota as often as they do charismatic mega-fauna, it is likely they would have a greater curiosity to better understand life forms that are very different from themselves.

Charismatic Plants in the Fossil Record

The underrepresented realm of artistically reconstructed ancient plant worlds is itself shaped by numerous forms of favouritism. For instance, there are entire industries today that capitalize on our obsession with growing and displaying showy, colourful, animal-pollinated flowers in gardens over non-flowering plants. However, few of the most iconic 'primeval' plant lineages bear flowers. This is because many of these plant lineages diverged prior to the occurrence of the first discernible flowering plants in the fossil record (the Early Cretaceous Period; ~130 million years ago).[17]

Plants popularized in prehistoric imagery tend to fall into two major categories in public perception. First, 'plants of the dinosaur days' or 'plants that dinosaurs ate' – a measure of a plant's identity and value based on its utility to vertebrates.[18] These consist predominantly of seed-bearing plant groups such as ginkgoaleans, cycadophytes (cycads and bennettites) and conifers. Second, are the 'earliest' or 'most primitive' plants category, which are invariably also lumped into the category of 'dinosaur food'. Most often, horsetails and ferns are given this recognition by the public, although nei-ther technically represents the earliest-diverging living vascular plant group.

Paradoxically, one of the most iconic, if not alien, groups of primeval plants is one the general public has seen in depictions but seldom heard of: lycopsids. This lineage of seed-free (spore-bearing) plants evolved a wide range of bizarre growth habits after diverging from all other vascular plants (those with lignified water- and nutrient-conducting tissues) over 415 million years ago. This division long precedes the divergence times for each of the remaining extant vascular plant lineages. Furthermore, lycopsids have survived all the major Phanerozoic mass extinctions since plants invaded land, and they persist to this day. They have always been evolu-tionary misfits, though are oddly ahead of their time compared to other vascular plants. Early in their evolutionary history, they became amongst the first vascular plant groups to evolve leaves, roots and a reproductive method that involves generating two different types of spores, giving rise to unisexual gametophytes (heterospory) that led to the development of structurally complex propagules in some lineages (e.g. lepidodendrids) that paralleled the evolution of seeds and pollen.[19] Moreover, lycopsids developed all these innovations independently of all other land plants, as the common ancestor to their group (lycophytes) and all other vascular plants (euphyllophytes) lacked roots, leaves and seeds.

In palaeoimagery, lycopsids have become a sort of botanical dinosaur, defying plant blindness, and somehow even eclipsing focal extinct vertebrates in reconstructions. In part, this is because most depicted extinct lycopsids occurred in the Silurian, the Devonian, and especially the Carboniferous Period (415–299 million years ago). In the late Silurian through to the Late Devonian, vertebrates were confined to aquatic realms, as they were all pre-tetrapod fishes. Therefore, terrestrial ecosystems were composed of microbial-, algal-, fungal-, lichen-, plant- and invertebrate-based food webs.[20] During the early stages of vascular plant diversification, lycopsids, along with their relatives and forerunners, were amongst the largest and most structurally complex multicellular organisms on land. From the Late Devonian through to the Carboniferous period – the Coal Age – arborescent (tree-forming) 'scale trees', or lepidodendrids, became iconic, towering statues, some species reaching over 50 metres (160 ft) tall, resembling gigantic telephone poles with open canopies of coral-like branches. These scale trees formed peculiarly sunlit, vast equatorial swamp forests,[21] whose remains comprise a considerable portion of the world's coal reserves. In Carboniferous swamp depictions, the giant lycopsids became a quintessential icon for the alien worlds dominated by giant arthropods before the age of dinosaurs. Given their historical and evolutionary significance, coupled with iconic visual status, how are these plants not more widely known by the general public?

One way to help to close gaps in the awareness of extinct plants may be for palaeoartists and/or scientists to make concerted efforts to visually reconstruct newly described fossilized plants or their structures in vivid detail and colour. Not only can compelling illustrations of the organism itself serve as a visual counterpart to technical descriptions, but colourful and photo-realistic illustrations capture the immediate attention of viewers (the snap-judgement visual bias of System 1),[22] and inspire more analytical viewing (System 2).[23] Such 'dinosaur-like' illustrations can, in theory, elevate plants to prehistoric animal-like recognition, and inspire renewed interest in extinct plants through public display.

Reconstruction: A Challenge at the Intersection of Scientific and Artistic Frontiers

Plants provide a conservative starting point for breathing life into ecosystems of the deep past. Although there are always exceptions to the rule, plants have largely behaved and functioned in a remarkably predictable manner over the past 500 million years. They tend to be photosynthetic autotrophs – meaning, plants manufacture their own foods by converting solar energy, water and carbon dioxide into chemical energy in the form of sugars. Furthermore, all body plans of green algae and land plants are generated by five cellular developmental processes,[24] yielding four body plans, of which one is multicellular, comprising all land plants. With just a small number of developmental processes added, a seemingly endless array of body plan variations have evolved in land plants, all stemming from contrasts in timing, location and planes of cell division.[25] Colour, while variable in modern plants, also tends to be conservative, with renditions of green being most typical of photosynthetic tissues. Additionally, variations of yellows, oranges, reds, pinks, bronzes and purples are produced by a variety of UV-absorbing pigments, such as flavonoids, carotenoids and anthocyanins. Such 'sunscreens' are either concentrated in emerging foliage, or residually expressed when chlorophyll is drained from leaves during senescence, like deciduous trees in autumn. Blue to grey hues in foliage (glaucousness) can also result in plants under intense sun exposure. Under these circumstances, plants can exude thick, protective layers of reflective epicuticular waxes (farina) secreted from their outermost cells (epidermis) and membrane (cuticle), or through producing reflective hairs or glands on their outer surfaces. Furthermore, macro- and micronutrient availability as well as the degree of solar exposure can impact which shade of green an individual plant manifests, and also lead to predictable responses in leaf colouration, size, shape, orientation and damage. These traits and responses have evolved independently across numerous plant lineages, and – insofar as they are caused by structural changes of leaf surfaces – are preserved to some degree in the fossil record, making coloured reconstructions less of a guessing game for extinct species.

Despite their developmental and functional stability over geologic time, fossilized plants, especially wholly extinct lineages, can present challenges for reconstruction. Sporophytes (the largest, most frequently preserved life stage in vascular plants) are composed of three general organ classes: shoots, roots, and lateral appendages (e.g. leaves, reproductive structures). Many plants routinely produce and shed their outermost tissue layers and lateral appendages over time, such as leaves, branches and bark, as well as reproductive propagules such as spores, pollen and seeds. As context, it is not normal for animals such as vertebrates to jettison their body parts across the landscape. These organs, in turn, must land in an appropriate depositional environment where they have a chance of being preserved as fossils. Such environments enable preservation of organic materials through suppressing the metabolism of microbial decomposers, such as cold temperatures, absence of oxygen, and extreme pH (highly basic or acidic environments). Plant parts can easily become damaged when travelling from the parent specimen to their site of preservation. Following burial, during the process of fossilization, what organic remains are still intact can be further altered by chemical and physical changes that occur as sediment becomes sedimentary rock over millions of years.

As a result of these processes, palaeobotanists are left with a fragmentary record requiring a skill set similar to a forensic investigator to piece together life forms of the past. They not only have the Herculean task of reconstructing organisms that no longer exist, but must often rely upon fragmentary, isolated organs to do so, and if possible, attempt to assemble extinct plants from these parts.[26] Furthermore, reconstructing extinct plants is heavily influenced by the 'pull of the recent' or 'pull of the present', in which views of modern plants shape how extinct species are reconstructed.[27] With plants that are long-extinct, unusual or fragmentary, using modern species as a blueprint for reconstruction can be either illuminating or limiting. One must carefully integrate detailed study of fossil morphology and anatomy with traits of nearest living equivalents to approach accurate reconstructions. Additionally, unlike vertebrates, it is also difficult to extrapolate the exact habit of an extinct plant beyond the organs that are preserved, due to the plasticity in plant body plan variations over the past 500 million years.

Whether the intact fossilized fragments of one organ match up with another, or belong to the same biological entity, requires detailed comparisons between new and revisited fossilized parts. Such comparisons depend upon observations of morphology (larger-scale, or macroscopic details), and, whenever possible, anatomy (cellular and microscopic details). When a fossilized plant is found to be attributed to a new species on the basis of morphology (morphospecies or morphotaxon), envisioning how the whole plant is three-dimensionally organized on the basis of two-dimensionally preserved compression fossils generates additional challenges to reconstruction.

Fortunately, the lycopsid branch of the tree of life has managed to survive multiple mass extinction events and has retained a strikingly conservative set of morphological traits.[28] Having undergone several major diversification events in the Late Paleozoic Era (~415–252 million years ago), only three lineages persist to this day: clubmosses and firmosses (Lycopodiaceae), spikemosses (Selaginellaceae), and quillworts (Isoëtaceae) (Illustration 11.1). Among the unifying characteristics of many extinct and extant members of this plant lineage are dichotomizing (bi-furcating) shoot and/or root systems, shoots bearing dense spirals of microphylls (small or simplified leaves containing only one vascular vein), and a single reniforme (kidney-shaped) sporangium [spore-bearing capsule] having a

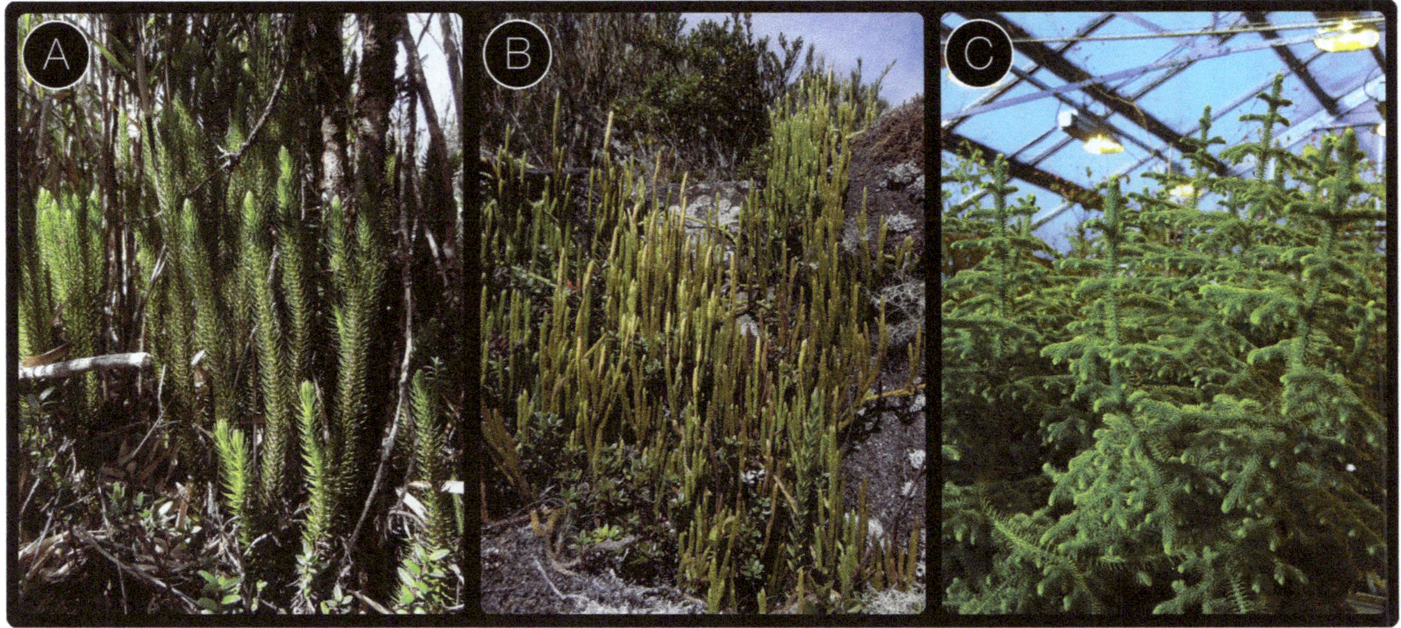

Illustration 11.1 Examples of living lycopsids (clubmosses and firmosses: Lycopodiaceae) from sunlit habitats similar to those used for *Leclercqia scolopendra* reconstruction. (A) *Phlegmariurus* (Herter) Holub sp., Cerro de la Muerte Massif National Park, Costa Rica; (B) *Lycopodium clavatum* var. *contiguum* (Klotzsch) Øllgaard, Cerro de la Muerte Massif National Park, Costa Rica; (C) *Palhinhaea* cf. *cernua* (Linnaeus) Vasconcellos & Franco (South African form), University of Washington Biology Greenhouse, Seattle, USA. Photos by author.

marginal clamshell-like opening on the adaxial (upper surface) per leaf. Despite maintaining conservative body plans through time, lycopsids can be challenging to portray faithfully. In particular, the multitudes of tightly inserted, spirally arranged needle- or awl-shaped leaves covering much of the shoot system can create substantial labour for artists, resulting in many reconstructions being generalized, resembling pipe cleaners or bottle brushes from a distance. In scientific descriptive illustrations, accurately placing the leaves is critical to demonstrating, as precisely as possible, how the plant may have appeared in life. Similar challenges can be presented when representing root morphology and habit in extinct lycopsids, as many modern species have roots densely clothed in fine wispy root hairs, and some extinct arborescent lineages had massive subterranean shoots (stigmaria) clothed in clouds of spirally arranged dichotomizing rootlets.[29] Through a case study reconstructing a shoot fragment of a Middle Devonian lycopsid, some challenges in extinct plant representation were overcome through an integrative study of morphological variation between extinct and extant lycopsids.

Reconstructing the Centipede Clubmoss

On Red Mountain, in Whatcom County of northern Washington State, an assemblage of black, two-dimensional, film-like compression fossils of early land plants were found on sheet-like slabs of early Middle Devonian (~375 million years)[30] sandstone.[31] Tiny, densely leaved lycopsid branch fragments were amongst the twig-like branches of numerous early relatives of ferns and seed plants. All these lycopsid fragments belonged to the genus, *Leclercqia* Banks, Bonamo and Grierson. However, amongst the fragments, there appeared to be two distinct morphotypes (morphologically different entities).

The leaves of *Leclercqia* are unusual for lycopsids. Rather than being simple and needle-like, like their closest living counterparts (clubmosses and firmosses, Lycopodiaceae; Illustration 11.1) those of this genus appear only simple at the base, but then divide into five to twelve segments away from the stem. In the type species, *L. complexa* Banks, Bonamo and Grierson, the leaf somewhat resembles the arched neck and long, curved bill of an ibis, adorned with pronghorn antlers in profile view – as is often their orientation of preservation in the fossil record. As such, the leaves attached to the stem look somewhat like miniature hunting trophies mounted to a wall. The result, in *L. complexa*, is a highly intricate, three-dimensionally pronged leaf with a central, downward-curved leaf segment and two pairs of lateral segments, splitting into prongs that project upwards and outwards.

At the time of collecting and describing the lycopsid fossils from Washington, only two species of *Leclercqia* had been described; *L. complexa* and *L. andrewsii* Gensel and Kasper. In *L. andrewsii*, the leaf base is flattened into one plane, curling upwards and dividing into five erect leaf divisions.[32] Because the leaves of all Washington specimens were oriented in the pronghorned-ibis-like, three-dimensional orientation, both morphotypes were most similar to *L. complexa*.[33]

Given some differences between fossils, it was not clear whether the more compacted branch fragments of *Leclercqia* with curved leaf bases, found in Washington, represented an oddly compressed *L. complexa*, a variety of that species, or an entirely new entity. In the scientific community, there are a range of considerations in determining which criteria are used to define a species of plant. Since fossils of old, extinct plants do not contain DNA, there is no way to compare their relation to each other or to modern species using cutting-edge molecular techniques. Morphology and anatomy therefore provide the only criteria upon which fossil plants can be assigned a species. Furthermore, some parts of plant bodies are assigned their own form genera names. For example, the branches, trunk, and root-bearing organs of the arborescent (tree-forming) lycopsids of the Carboniferous coal swamps all have different form genera names, even though they all come from the same tree specimen (and therefore a single species). These different form genera names exist because isolated organs of these plants were discovered at different times and only subsequently pieced together as belonging to the same organism. Additionally, it is well possible that, for instance, several distinct fossil species shared the same, morphologically indistinguishable type of root-bearing organs. As a result, the separate name for such root-bearing organs must be maintained alongside the species name for the whole plant.

Fortunately, there are no known form genera of *Leclercqia*, as these lycopsids are known exclusively from branch fragments. Traditionally, a new species of fossil plant could be assigned if it was qualitatively distinct enough in morphology from all other previously described species. For example, it might be argued that 'species X looks different from species Y and Z based on traits A, B and C. Therefore, species X can reasonably be described as a new entity, and it is hypothesized that it represents a separate species'. However, in the case of the two co-occurring *Leclercqia* morphotypes in the same sedimentary rocks of Washington State, there were clear similarities but also differences, and so a more detailed comparison was warranted. Rather than using a more conventional qualitative comparison, these *Leclercqia* morphotypes needed to be assessed by applying quantitative methods more often used in comparing variation between animals.

In order to more quantitatively determine whether the unknown morphotype represented a new species or variety, linear morphometric analyses were performed. This meant that a series of measurements were taken (in this case, the lengths, thicknesses and angles within a leaf) in leaves of as many intact specimens of the unknown morphotype as possible. These measurements were then compared to those taken on *L. complexa* species sampled from six continents. The same analyses were subsequently used to compare morphological variability between extant species and variants of modern clubmosses (Lycopodiaceae) grown in a greenhouse.[34]

It was found that the unknown *Leclercqia* morphotype from Washington was statistically significantly distinct from *L. complexa*.[35] Furthermore, these fossil lycopsids were as distinct from one another as the two modern clubmoss species compared are from each other. On the basis of these analyses, it was revealed that the unknown morphotype was very different from *L. complexa* in several key traits. The new morphotype was assigned to a new species: *Leclercqia scolopendra* Benca et Strömberg.[36] This species name means 'Centipede Clubmoss' based on the resemblance between its modular and tightly packed curved leaves with the legs of tropical centipedes in the genus *Scolopendra* (Scolopendridae).

Previous anatomical and morphological studies of *Leclercqia complexa* provided an excellent resource for reconstructing a structurally similar member of the genus in great detail by minimizing guesswork in interpreting fossils having less optimal preservation in Washington.[37] Typically, newly described fossilized plant species are reconstructed within the descriptive studies using detailed contour line drawings, rendered by hand. However, it became clear after several hand-drawn renditions based on the two-dimensional compression fossils of *L. scolopendra* that bringing life to this plant would present unique challenges. In short, something seemed to be lost in translation between the compression fossils and the initial illustrations, yielding reconstructions that were indistinguishable from *L. complexa*, which was inconsistent with the quantitative results.

By hand, it was difficult to anticipate exactly how the structurally intricate leaves of *L. scolopendra* would interplay amongst each other in three-dimensional space. Using the measurement data of a range of traits that quantified leaf shape, thickness, angle and orientation, it was possible to precisely render the leaves digitally in several different orientations using the vector software programme Adobe (San Jose, CA, USA) Illustrator CS6.[38]

Based on the spirally arranged scars left from detached leaves (resembling the inter-locking, ganoid scales of gar fishes) along the branch fragments of *L. scolopendra*, it was possible to render a stem fragment and know exactly where the leaves would be inserted (Illustration 11.2). After reconstructing anatomically precise leaves from multiple angles, it

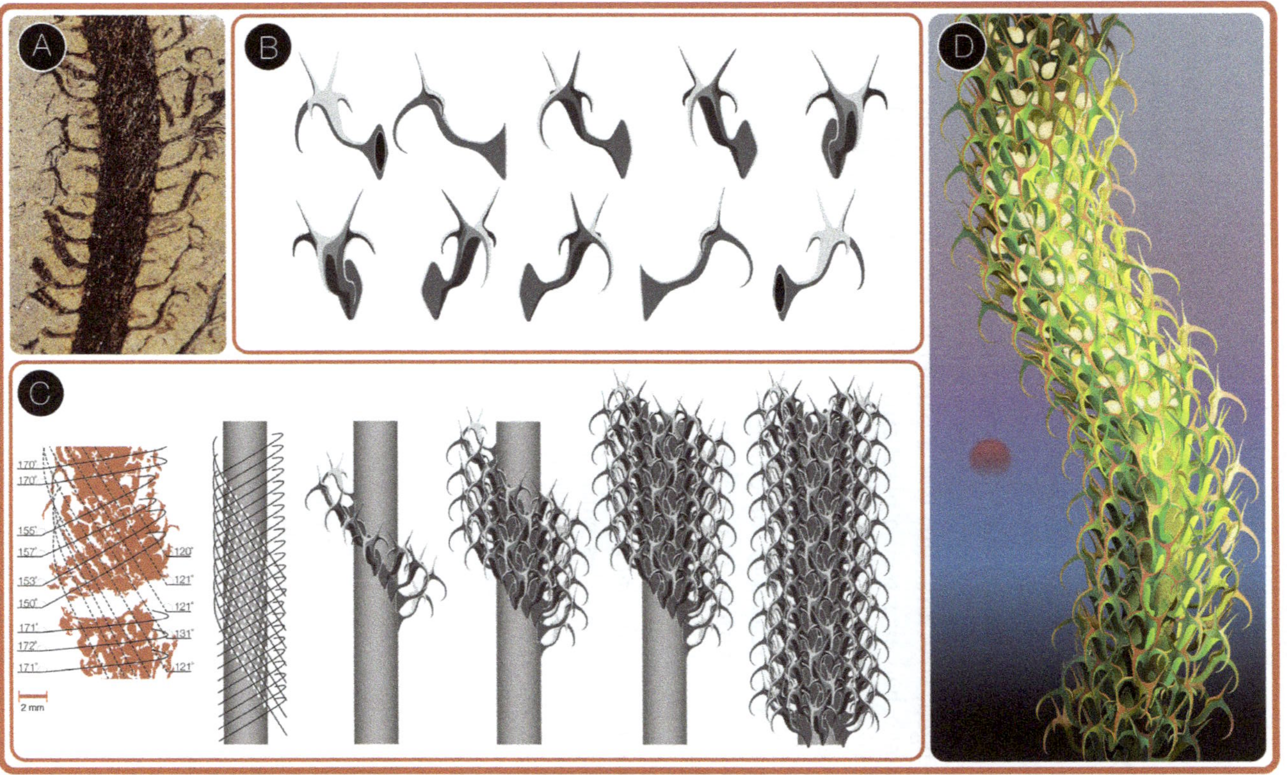

Illustration 11.2 Digital reconstruction process of *Leclercqia scolopendra* used to accompany the species description (Benca et al., 'Applying Morphometrics'). (A) Fossil fragment of *L. scolopendra*; (B) Digitally reconstructed leaves of *L. scolopendra* in different perspectives; (C) Process of reconstructing leaf arrangement (phyllotaxy) of *L. scolopendra* using fossilized leaf and sporangial insertion points; (D) Reconstruction of *L. scolopendra*. Illustrations rendered by author using the vector software program Adobe (San Jose, CA, USA, Illustrator CS6).

was then possible to insert the leaves and see how they interplayed three-dimensionally. The results were surprising. The hook-like leaf bases of *L. scolopendra* formed a shield-like barricade of erect fan-shaped leaf bases. However, even more surprisingly, the spine-like leaf segments formed an overlapping labyrinth of spines projecting in all directions (Illustration 11.2). Only by rendering the leaves as vectors digitally was it possible to see how they would precisely interlock and interact with each other along the branch.

This was an important point, because the extent to which the leaves overlapped would have been difficult to anticipate if only using one's imagination and sketching from hand. Any minor inaccuracy in the angle, attachment or proportionality of the leaves would yield an inaccurate, more sparsely arranged, structure. This was a barrier in artistic expression, because no human had ever seen a plant structurally arranged quite like *L. scolopendra*. After all, this plant has been extinct for over 375 million years and may have been easy to overlook as a fossil.

Reconstructing extinct life forms can be hampered by expectations that a fossil form would resemble something already known in the modern world. There are limitations to our imagination, and historic realities obscured by the fossil record can, do, and invariably will turn preconceived notions upside down with additional discoveries.

In the case of *L. scolopendra*, hand-drawn draft reconstructions looked more like modern clubmosses with cilate (hair-tipped) leaves. However, the digitally calibrated leaves and stem taken together made the final rendition look superficially more akin to a marine invertebrate than a plant – specifically a deep-sea glass sponge (a hexactinellid). In this particular case, using scientific data to inform the artmaking process resulted in a reconstruction that may have captured how a branch fragment of this extinct species would have appeared when alive 375 million years ago.

The final touch to artistically representing *L. scolopendra* came down to choices in colouration of the plant. As no fragments of early trees (e.g. *Archaeopteris*) were found in the Washington fossil beds, it seems likely that *L. scolopendra* was not growing near closed-canopy forests but instead occupied open, sunlit environments. Since many modern clubmosses continue to grow in open, sunlit environments, their colouration was used as a conservative guide. Moreover, after developing the first successful cultivation techniques for a wide range of terrestrial clubmosses,[39] it was possible to use clubmosses of several genera grown side-by-side under controlled greenhouse conditions as the references for developing the colour pallet of *L. scolopendra*. In extant Lycopodiaceae, most species occupying open habitats range in colouration between dark apple green to bright golden, yellow-green (Illustration 11.1). *L. scolopendra* was therefore depicted at a midpoint of the spectra of modern Lycopodiaceae from sunlit habitats (Illustration 11.2D).

Sometimes the reconstruction of an organism can surface entirely new lines of scientific inquiry about an extinct organism. In the case of *L. scolopendra*, why the leaves formed such a complex three-dimensional cloud of spine-like segments was uncertain. Perhaps the shield-like leaf bases and projecting leaf segments were a form of protection to the developing sporangia from desiccation or herbivory (like the shielding leaves or bracts of developing cones).[40] However, such ideas of adaptive significance in structure are difficult to test or verify in the fossil record. Furthermore, no known stem fragment of *Leclercqia* or its closest relatives – members of the extinct lycopsid order Protolepidodendrales – are attached to rooting organs or rhizomes to date. It is therefore unclear whether stem fragments of *Leclercqia* come from a herbaceous, vining plant resembling modern clubmosses – as they have been traditionally envisioned[41] – or branch tips from canopies of telephone-pole-like trees.[42] In the case of *Leclercqia*, more complete fossils are needed to confidently undertake whole-plant reconstructions.

Conclusion

Accurate and conservative palaeobotanical reconstructions most often accompany scientific studies that can be difficult for the public to access. However, these works serve as indispensable guides for a growing number of palaeoartists undertaking more holistic ecosystem reconstructions that can, in turn, be presented to the public. The extinction rate of land plants is now up to five hundred times pre-Anthropocene background extinction rates.[43] It is therefore vital that the next generation of thinkers, scientists, activists and conservationists see plants for the incredible and charismatic organisms that they are.

Jeffrey P. Benca is a research associate at the University of Washington's Burke Museum of Natural History and Culture, as well as a horticulturist for Amazon's Horticulture Program and Seattle Spheres in Washington state, USA. Receiving his doctorate in integrative biology from the University of California, Berkeley, his Devonian lycopsid reconstruction was featured as the *American Journal of Botany* centennial cover. He has also contributed to modern lycopsid research and conservation through pioneering cultivation techniques for terrestrial clubmosses. He led the first experimental study to test a proposed driver of Earth's largest mass extinction using modern plants, discovering a non-lethal mechanism for global ecosystem collapse under stratospheric ozone weakening, published in *Science Advances*.

Notes

1. Bar-On, Phillips and Milo, 'The Biomass Distribution on Earth'.
2. Manchester, Calvillo-Canadell and Cevallos-Ferriz, 'Assembling Extinct Plants'.
3. Johnson, 'Using Paleobotany to Make Better Reconstructions'.
4. Wandersee and Schussler, 'Preventing Plant Blindness'.
5. Knapp, 'Are Humans Really Blind to Plants?'
6. Wilson, *The Diversity of Life*; Larsen et al., 'Inordinate Fondness Multiplied and Redistributed'.
7. Berenguer, 'The Effect of Empathy'; Knapp, 'Are Humans Really Blind to Plants?'
8. Wandersee and Schussler, 'Toward a Theory of Plant Blindness'.
9. Kahneman, *Thinking, Fast and Slow*; Knapp, 'Are Humans Really Blind to Plants?'
10. Kahneman, *Thinking, Fast and Slow*.
11. Ibid.; Knapp, 'Are Humans Really Blind to Plants?'
12. Balas and Momsen, 'Attention "Blinks" Differently'.
13. Ibid.
14. Berenguer, 'The Effect of Empathy'.
15. Hutton, *Theory of the Earth*.
16. McPhee, *Basin and Range*.
17. Gomez et al., '*Montsechia*, an Ancient Aquatic Angiosperm'.
18. Knapp, 'Are Humans Really Blind to Plants?'
19. Phillips and DiMichele, 'Comparative Ecology and Life-History Biology'.
20. Shear and Selden, 'Rustling in the Undergrowth'.
21. Phillips and DiMichele, 'Comparative Ecology and Life-History Biology'.
22. Kahneman, *Thinking, Fast and Slow*.
23. Ibid.
24. Niklas, 'The Evolution of Plant Body Plans'.
25. Ibid.
26. Manchester, Calvillo-Canadell and Cevallos-Ferriz, 'Assembling Extinct Plants'.
27. Ibid.
28. Ambrose, 'The Morphology and Development of Lycophytes'.
29. Hetherington, Berry and Dolan, 'Networks of Highly Branched Stigmarian Rootlets'.
30. Brown, Gehrles and Valencia, 'Chilliwack Composite Terrane'.
31. Benca et al., 'Applying Morphometrics'.
32. Gensel and Kasper Jr., 'A New Species of the Devonian Lycopod Genus'.
33. Benca et al., 'Applying Morphometrics'.
34. Benca, 'Cultivation Techniques'.
35. Benca et al., 'Applying Morphometrics'.
36. Ibid.
37. Banks, Bonamo and Grierson, '*Leclercqia Complexa* Gen. Et Sp. Nov.'; Grierson and Bonamo, '*Leclercqia Complexa*: Earliest Ligulate Lycopod'; Bonamo, Banks and Grierson, '*Leclercqia, Haskinsia*, and the Role of Leaves'.
38. Benca et al., 'Applying Morphometrics'.

39. Benca, 'Cultivation Techniques'.
40. Benca et al., 'Applying Morphometrics'.
41. Liu et al., 'Reinvestigation of the Lycopsid *Minarodendron Cathaysiense*'.
42. Stein et al., 'Mid-Devonian *Archaeopteris*'.
43. De Vos et al., 'Estimating the Normal Background Rate of Species Extinction'; Nic Lughadha et al., 'Extinction Risk and Threats to Plants and Fungi'.

Bibliography

Ambrose, Barbara A. 'The Morphology and Development of Lycophytes'. *Annual Plant Reviews* 45 (2013): 91–114.

Balas, Benjamin, and Jennifer L. Momsen. 'Attention "Blinks" Differently for Plants and Animals'. *CBE – Life Sciences Education* 13(3) (2014): 437–43.

Banks, Harlan P., Patricia M. Bonamo and James D. Grierson. '*Leclercqia complexa* gen. et sp. nov.: A New Lycopod from the Late Middle Devonian of Eastern New York'. *Review of Palaeobotany and Palynology* 14(1–2) (1972): 19–40.

Bar-On, Yinon M., Rob Phillips and Ron Milo. 'The Biomass Distribution on Earth'. *Proceedings of the National Academy of Sciences of the United States of America* 115(25) (2018): 6506–11.

Benca, Jeffrey P. 'Cultivation Techniques for Terrestrial Clubmosses (Lycopodiaceae): Conservation, Research, and Horticultural Opportunities for an Early Diverging Plant Lineage'. *American Fern Journal* 104 (2014): 25–48.

Benca, Jeffrey P., et al. 'Applying Morphometrics to Early Land Plant Systematics: A New *Leclercqia* (Lycopsida) Species from Washington State, USA'. *American Journal of Botany* 101(3) (2014): 510–20.

Berenguer, Jaime. 'The Effect of Empathy in Proenvironmental Attitudes and Behaviors'. *Environment and Behavior* 39(2) (2007): 269–83.

Bonamo, P.M., H.P. Banks and J.D. Grierson. '*Leclercqia, Haskinsia*, and the Role of Leaves in Delineation of Devonian Lycopod Genera'. *Botanical Gazette* 149(2) (1988): 222–39.

Brown, E.H., G.E. Gehrles and V.E. Valencia. 'Chilliwack Composite Terrane in Northwest Washington: Neoproterozoic-Silurian Passive Margin Basement, Ordovician-Silurian Arc Inception'. *Canadian Journal of Earth Sciences* 47(10) (2010): 1347–66.

De Vos, Jurriaan M., et al. 'Estimating the Normal Background Rate of Species Extinction'. *Conservation Biology* 29(2) (2015): 452–62.

Gensel, Patricia G., and Andrew E. Kasper Jr. 'A New Species of the Devonian Lycopod Genus, *Leclercqia*, from the Emsian of New Brunswick, Canada'. *Review of Palaeobotany and Palynology* 137(3–4) (2005): 105–23.

Gomez, Bernard, et al. '*Montsechia*, an Ancient Aquatic Angiosperm'. *Proceedings of the National Academy of Sciences* 112(35) (2015): 10985–88.

Grierson, James D., and Patricia M. Bonamo. '*Leclercqia Complexa*: Earliest Ligulate Lycopod (Middle Devonian)'. *American Journal of Botany* 66(4) (1979): 474–76.

Hetherington, Alexander J., Christopher M. Berry and Liam Dolan. 'Networks of Highly Branched Stigmarian Rootlets Developed on the First Giant Trees'. *Proceedings of the National Academy of Sciences* 113(24) (2016): 6695–6700.

Hutton, James. *Theory of the Earth: With Proofs and Illustrations*, Vol. 1. Edinburgh: Cadell, Junior and Davies, 1795.

Johnson, Kirk R. 'Using Paleobotany to Make Better Reconstructions of Ancient Plants, Landscapes, and Biomes'. 12th International Palynological Congress (IPC-XII), 8th

International Organization of Paleobotany Conference (IOPC-VIII), Abstract Volume, Terra Nostra 2008/2: 134.

Kahneman, Daniel. *Thinking, Fast and Slow*. New York: Farrar, Straus and Giroux, 2011.

Knapp, Sandra. 'Are Humans Really Blind to Plants?' *Plants, People, Planet* 1(25) (2019): 164–68.

Larsen, Brendan B., et al. 'Inordinate Fondness Multiplied and Redistributed: The Number of Species on Earth and the New Pie of Life'. *The Quarterly Review of Biology* 92(3) (2017): 229–65.

Liu, Le, et al. 'Reinvestigation of the Lycopsid *Minarodendron Cathaysiense* from the Middle Devonian of South China'. *Neues Jahrbuch für Geologie und Paläontologie-Abhandlungen* 268(3) (2013): 325–39.

Manchester, Steven R., Laura Calvillo-Canadell and Sergio R. Cevallos-Ferriz. 'Assembling Extinct Plants from their Isolated Parts'. *Boletín de la Sociedad Geológica Mexicana* 66(1) (2014): 53–63.

McPhee, John. *Basin and Range*. New York: Farrar, Straus and Giroux, 1981.

Nic Lughadha, Eimear, et al. 'Extinction Risk and Threats to Plants and Fungi'. *Plants, People, Planet* 2(5) (2020): 389–408.

Niklas, Karl J. 'The Evolution of Plant Body Plans: A Biomechanical Perspective'. *Annals of Botany* 85(4) (2000): 411–38.

Niklas, Karl J., and Ulrich Kutschera. 'The Evolutionary Development of Plant Body Plans'. *Functional Plant Biology* 36(8) (2009): 682–95.

Phillips, Tom L., and William A. DiMichele. 'Comparative Ecology and Life-History Biology of Arborescent Lycopsids in Late Carboniferous Swamps of Euramerica'. *Annals of the Missouri Botanical Garden* 79(3) (1992): 560–88.

Shear, William A., and Paul A. Selden. 'Rustling in the Undergrowth: Animals in Early Terrestrial Ecosystems', in Patricia G. Gensel and Dianne Edwards (eds), *Plants Invade the Land: Evolutionary and Environmental Perspectives* (New York: Columbia University Press, 2001), 29–51.

Stein, William E., et al. 'Mid-Devonian *Archaeopteris* Roots Signal Revolutionary Change in Earliest Fossil Forests'. *Current Biology* 30(3) (2020): 421–31.

Wandersee, James H., and Elisabeth E. Schussler. 'Preventing Plant Blindness'. *The American Biology Teacher* 61(2) (1999): 82–86.

———. 'Toward a Theory of Plant Blindness'. *Plant Science Bulletin* 47(1) (2001): 2–9.

Wilson, E.O. *The Diversity of Life*. Cambridge, MA: Harvard University Press, 1992.

Chapter 12

Ellis Rowan, Extinction and the Politics of Flower Painting

Jeanette Hoorn

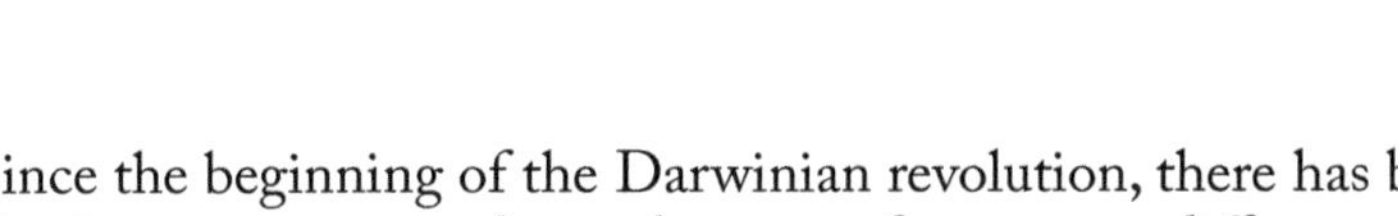

Since the beginning of the Darwinian revolution, there has been a particular interest among educated women from many different countries and a diverse range of communities, in botanical illustration. The recording of plants and flowers is part of the process of naming and identifying characteristic features of 'unknown' species, as well as preserving the memory of plants through the capture and safe storage of seeds, in response to the increasing possibility of future extinctions. There has always been an aesthetic dimension to botanical illustration that transcends the purely scientific, and places the making of visual images among the highest form of artistic production in which art and science are entangled.

In colonial settings, the classification of plants has played a significant role in the exercise of power. Naming means owning, and although male botanists did most of that, the female artists who recorded the features of plants and flowers understood that they were contributing to Western science and to the 'empire' in a significant way.[1] Most of the women who became involved in this enterprise stayed close to home. Not so with the Australian artist Marian Ellis Rowan, who like her mentor Marianne North travelled vast distances on many expeditions to find rare plants, many of which today are either endangered or extinct. Ellis travelled to some of the remotest of locations in Australia, including Far North Queensland, Cape York in the Torres Strait and Albany in Western Australia, where the rates of extinction are now among the highest in the world. It is estimated that she executed more than three thousand paintings and drawings over her lifetime.[2] The National Library of Australia's holdings of her work is the largest of any artist in their collection, placing Rowan in a similar class to Marianne North, whose work is held in a special gallery named for her, The Marianne North Gallery at Kew.[3] The two met in Australia, North having apparently made the journey on the advice of Charles Darwin, to whom she sent a drawing on her return.[4] Rowan painted in New Zealand, Papua New Guinea, the Himalayas and the Caribbean. She also spent four years seeking out rare species in the southern states of the US with the American botanist, Alice Lounsberry. In the preface to their jointly authored *Southern Wild Flowers and Trees*, Lounsberry wrote:

> To see rare ones growing in their natural surroundings, Mrs. Rowan and I travelled in many parts of the south … Through the mountainous region we drove from cabin to cabin, and nowhere could we have met with greater kindness and hospitality … [the] variety and beautiful, luxuriant growth the southern field is perhaps unrivalled.[5]

The Flower Hunter

In Australia, Ellis Rowan made daring forays into country which were not without political baggage. She was exceptionally well connected, always moving in the most elevated circles, bringing into play her extensive contacts. I will endeavour to reveal here the precise circumstances that made Marian Ellis Rowan one of the most powerful members of the Anglo-Australian elite, making it possible to undertake the tasks that enabled her to forward so much plant and other information to the Crown. Wherever she went, Rowan relied upon her contacts, be they scientists, government and vice-regal officials, British naval officers, pastoralists or the leaders of Indigenous communities. Her letters make clear the extent of her personal networks. She 'talked snakes and flowers' with a doctor while travelling with the Bishop of Northern Queensland to the Herbert River.[6] Sir Ferdinand von Mueller wrote to her in 1894 when she was in London, hoping that she might use her influence to convince the Colonial Office to open Antarctica for exploration: 'Now dear madam, can you with your intellectual power and lady's grace give us also some help'.[7] Sir Walter Baldwin Spencer, the foundation professor of biology at the University of Melbourne, included her in newspaper articles in which he recognized the role that women played in collecting material for science.[8] In England, she received many commissions to paint floral panels in the homes and landed estates of patrons.[9]

During her expeditions into tropical Queensland, she accepted the hospitality of frontier pastoralists, such as the Caseys and the Jardines, in order to gain access to the rare specimens she was looking for. She confided in a letter to her husband that she was 'ready for everything' and that her 'visit to Mr Casey's station at Normanby had made "such a man of me"'. The Mr Casey to whom Rowan refers was Reginald Gardiner Casey, a pastoralist and miner who was a member of the Queensland Legislative Assembly for Warrego (1888–93). His son, Sir Richard Gavin Gardiner Casey, Baron Casey (29 August 1890 – 17 June 1976) was the eminent Australian statesman who served as the sixteenth governor-general of Australia from 1965 to 1969. He was a distinguished army officer, long-serving cabinet minister, ambassador to the United States, member of Churchill's War Cabinet, and viceroy of Bengal. In 1926, he married Ellis Rowan's niece Ethel Marian Sumner (Maie) Ryan – later Baroness Casey. The RG *Casey Building* in Canberra, headquarters of the Australian Department of Foreign Affairs and Trade, is named in his honour. Her line could hardly be more notable or more closely aligned to high public office.[10]

Rowan had lengthy stays and undertook long journeys with Francis and Sana Solia Jardine, whose family held extensive pastoral landholdings set up by the Scottish-born patriarch John Jardine when he was appointed by the Queensland Government to establish a British outpost in Cape York in 1864. The resistance of local people to the drive into Far North Queensland over the following two decades met with violent retribution from Jardine, who as a police magistrate appointed colonial law enforcement agents, including native police, to remove Aboriginal owners from what the state considered to be his family's property. In 1896 in a report to the government of Queensland, it was estimated that in the 1870s the Indigenous population between the County of Newcastle and Cape York was over three thousand. By Ellis's arrival the population was well on its way to falling to less than three hundred. Whether it was through the introduction of disease, the exclusion of Aboriginal owners from traditional hunting grounds, or the day-to-day killing of resistance fighters as well as large-scale massacres, the numbers of First Nations people that Rowan encountered during her trip to the Torres Strait in 1991 were greatly reduced.[11] Several groups are believed to have occupied this region prior to the arrival of the British, including the Atambaya, Gudang, Yadhaykenu, Ankamuthi, Wuthathi and the Kaurareg people.[12]

Rowan was introduced to Frank Jardine on her first trip to Thursday Island in 1891. On her arrival, she was met by the governor of Queensland, Sir Henry Norman, and by her brother-in-law, Admiral Lord Charles Scott, who at the time was commander in charge of what was termed 'The Australian Station'. Admiral Scott was the fourth son of the fifth Duke of Buccleuch, the second largest landholder in the United Kingdom. Ellis's sister, Ada Mary (Lady Charles Scott), became his wife in 1883. As a member of one of the best-bred families in Britain, Rowan was able to exert her authority, giving her access to material she would not otherwise have been able to reach. This included the labour of local people who collected plants for her to paint. On trips to Jervis Island, for example, Rowan was able to organize children from the community to collect flowers for her.[13] She was guided to inaccessible spots by local trackers while working on Chillagoe Cattle Station:

> I spent one day with a little native boy on a flower-hunting expedition. We went in search of a particular plant that grows only on a certain ridge of these limestone rocks ... Round the base of the rocks I found innumerable pods, berries and flowers that were new to me. My little guide, with his bare feet, skipped over the rocks like a goat familiar with every feature; while I, left behind, toiled wearily over [them], half-baked with the scorching heat.[14]

Rowan put in tremendously long hours herself, and her labours are truly worthy of admiration. There are frequent accounts of the exhaustion she suffered. From Rockhampton she wrote: 'I sat up painting until well into daylight next morning – how shocked you would have been at my burning the candle at both ends!'[15] On her departure she noted: '[M]y cabin was already filled with flowers, and there and then, between my qualms of sea-sickness, I painted them in'. Speaking of two weeks spent in Cooktown: 'My eyes wore out sometimes painting from daylight till dark, for I couldn't keep pace with the flowers that came, and while I spent hours with wet bandages on them, Mrs Bauer [her host] sat beside me and read aloud'.[16]

Not only did she 'paint in' plants, she dried specimens on the spot, which she carefully enclosed in sheets of paper and brought back to Melbourne. We do not know, however, if she tried to establish the native names of the plants she collected or whether she inquired into their medicinal uses. This information was not included in the naming and classification that Sir Ferdinand von Mueller, the government botanist for Victoria (1853–73), and Frederick Manson Bailey, the colonial botanist for Queensland (1881–1915), undertook following her delivery of specimens.[17]

Her prose is full of rapture as she described the physical features of the landscapes through which she travelled, and engaging anecdotes of life on the properties of the pastoralists with whom she stayed. Accounts of First Nations communities she encounters are redolent with spectacle. Rowan comes upon large and well-organized communities, and an abundant and fertile nature, which today no longer exists. Her descriptions of Murray Island in her letters are ravishing:

> We were awakened at six next morning, when a buxom maiden came up the hill with gourds of water for our baths. I had mine in a very primitive dish. The sun had just risen but the landscape was still drowned in

vapours, while the heavens above were roofed with a sapphire blue; as the mists rolled away the view each moment grew more beautiful, and how often I wished that some fairy godmother, by reason of her wand, could have wafted you to me. Each leaf sparkled with dewdrops, the sea without a ripple, lightly spreading over shallow sands, was of that peculiar shade of green that is only seen in tropical waters.[18]

But the majority of remaining members of Aboriginal communities were, in the following decades, removed from their lands and placed into missions. Rowan described the South Sea Islander communities, who were brought in at this time under conditions of slavery for labour in the sugar and pastoral industries, under the heading of 'happy darkies', assuring the reader that '[t]hey have a very happy time of it on these plantations, and all seem very jolly with the wives, children and those belongings most precious to them, including fowls and pigs'.[19] Alongside such troubling descriptions, her prose has sublime passages in which she is overcome by the beauty of what she sees before her. At many points, however, she reveals her experience of abjection as she encounters people (both Indigenous and non-Indigenous) whom she considers to be inferior, backward and at times dangerous.[20] In one such instance, Ellis informs the reader that she wears gloves when shaking hands with local people, describing her relief that she has a pair at hand. Her accounts of the Australian wilderness are full of awe and wonder, but are also peppered with tales of head-hunting, cannibalism and other manifestations of White fantasy.[21]

Rowan's career as a botanical illustrator was intimately bound up with the project of Empire and the damage it inflicted on Australia's First Nations peoples and its flora and fauna. This destruction is invisible in her botanical drawings. The works of twenty-first-century artist Lucienne Rickard, which I will discuss later, are important in this context, forming a powerful counterpoint as they foreground the violence of erasure.

Disgust and disdain are nevertheless tempered by enthusiasm, as Rowan's fascination for the beauty of the plants, birds and fish that come across her path outweigh her fears. No effort is too great for the intrepid Ellis Rowan to capture on paper the material generously made available to her. With the help of local pastoralists and Aboriginal assistants, Rowan contributed raw material from this new frontier for botany and the science of Empire. In doing so, centuries of Indigenous knowledge and practice went unrecorded.

Her ability to engender support at the highest levels and to make friendships with like-minded people, gave her entrée into powerful circles and the support needed to travel over large distances, and in doing so, she was able to bring vast landscapes under 'scientific' control, making her a link in the chain through which First Nations Australians were dispossessed of their country.[22]

Extinction, botanical illustration and flower painting are entwined in colonial settings, bound together in the process of capturing territory for Empire and for science. Botanical illustration and flower painting are usually seen to be distinct but related minor genres that are not broadly thought about in relationship to conquest. But in the circumstances that confront us today, in which many plants are threatened with extinction, attempts by art historians to continue to argue for a separation of the two genres and to attempt to exclude artists, predominantly women, who mix genres from the canon, are unhelpful. Nor is it useful to read their endeavours as something that had nothing to do with the politics of dispossession. Rowan's primary aim was to record new specimens in remote places, but when the governor of New South Wales, Sir Walter Davidson, opened an exhibition of her paintings in 1920, he noted that her paintings provided a record that was 'instructive to future generations ... before many of the beautiful birds become extinct ... and the flowers disappear beneath the heels of civilisation'.[23] While Rowan was recording plants against possible extinction, she was part of the process that facilitated a completely new regime of land management, which, in the years to come, resulted in the destruction of whole economies, human, plant and animal, creating extinctions in far greater numbers than she could have envisaged.

Chrysanthemums in the Picture

Recently, the representation of plants and flowers has emerged in dramatic form in a celebration and remembrance of the plant world that so many of us have been unable to access – a reminder of what life could be like without our immersion in the natural world. One of the most striking has been a particular performance of *Crisantemi, Elegy for String Quartet*, written by the great Italian composer, Giacomo Puccini.

This enduring piece of music is regularly performed today; but never before like this. On 24 June 2020, the UceLi Quartet, led by conceptual artist Eugenio Ampudia, performed *Crisantemi* as part of his 'Concert for the Biocene' to celebrate the opening of the city's Gran Teatre del Liceu, which had been closed since mid-March due to COVID-19, with a performance of Puccini's hauntingly beautiful piece to an audience made up only of plants – 2,292 of them. Speaking in the context of COVID lockdowns, Ampudia made a plea for humanity to reconnect with nature, writing: 'I heard many more birds singing, and the plants in my garden and outside [were] growing faster and without a doubt. I thought that maybe I could now relate in a much more intimate way with people and nature'.[24] In his article, 'Cybercene or Biocene: Which Future Will We Choose?', Mark

Sommer argues that the human species is now living in the 'twilight years of the Anthropocene' and in the age of the planet's human-made sixth mass extinction. He calls for a Biocene future, one where we reject the values of the post-human for our humanity and nature. Ampudia has these values in mind when he performs 'Concert for the Biocene', choosing for it the Chrysanthemum, whose classic varieties are now threatened.[25]

Chrysanthemums are flowers of commemoration and mourning. Today many species traditionally grown in the UK are threatened with extinction, due to the encroachment of human populations in areas where they used to grow wild, and to their susceptibility to disease through neglect.[26] Puccini wrote his piece in memory of his friend, Prince Amadeo, Duke of Savoy. The Barcelona performance of *Crisantemi* memoralized the deaths of thousands of Catalans caused by COVID-19, and the plants, which were the audience, were given to Barcelona's medical workers as a recognition for their work during the first wave of the pandemic.[27] Puccini's choice of the chrysanthemum took place at a time when academic artists from all over Europe had joined in the rage for chrysanthemums. In Australia, many of the country's most celebrated colonial painters, such as Tom Roberts, Arthur Streeton, Charles Condor and Jane Sutherland, all known for their landscapes and portraits, began to honour the chrysanthemum.

While Ellis Rowan is best known for her wild flowers, she holds a special place in Melbourne's history for the controversy that arose surrounding her *Chrysanthemums* (Fig. 12.1), the picture for which she won the Melbourne International Exhibition of 1888, the most sought-after accolade of the day by Melbourne's fledging community of academic artists, against whom she competed and over whom she triumphed.[28] Ellis's choice of chrysanthemums as a subject, two years before Puccini named his piece for his friend Amadeo, created a debate that lasted several months and bears comparison with some of the most notable scandals in the Australian art world, such as the one that surrounded Emmanuel Frémiet's gift of his bronze *Gorilla carrying off a woman* to the National Gallery of Victoria in 1906, and the purchase of Jackson Pollock's *Blue Poles* for the National Gallery in Canberra in 1973.

When Rowan won at the Melbourne International Centennial Exhibition of 1888–89, she was selected over both local and international artists.[29] The Victorian Artists' Society led by senior artists Julian Ashton and John Mather, lodged an appeal against Rowan's award. Most of the Australian artists had trained at the National Gallery Art School in Melbourne, and some were also returned from popular French and British academies. Ellis had not attended an art school, taking lessons privately in Melbourne and in London.[30] Rowan was the only artist to win gold for the academy's high art category, outclassing everyone in every class, including a vast array of European artists. Ellis won the competition partly on account of her very large painting of a blazing bouquet of Chrysanthemums painted in a style quite unlike those she had submitted elsewhere in the competition. In doing so, she joined a group of elite artists then painting chrysanthemums. Claude Monet and Vincent Van Gogh painted literally dozens of the Japanese blossoms, each work larger in size and more breathtaking than the preceding one. Rowan's large oils do not look an unworthy offering in their company. She also entered two folios of wild flowers into the Queensland and New Zealand galleries, for which she received orders of merit. While most of her paintings are watercolours of modest size, her two known works of *Chrysanthemums* were substantial oil paintings whose dimensions are much larger than her standard work, and also much bigger than the bush pictures, genre paintings or portraits submitted by her fellow artists into the competition.[31] Pictures of this scale, usually thought to be suitable for subjects of serious import, were most often paintings about national life, such as Tom Roberts's *Shearing the Rams* (1890), which today has a two-page entry devoted to it in Wikipedia.

Little discussed is the fact that among the eight paintings offered for the competition by the then upcoming artist Tom Roberts was his own flower painting, *The Glory of Chrysanthemums*.[32] Roberts and fellow artists submitted paintings that today are considered to constitute the nation's patrimony, among them Frederick McCubbin's *Lost*, and Roberts's *Summer Morning Tiff* and *Reconciliation*.[33] All were overlooked. The judges had works before them that documented the rise of British colonial culture in Australia – now priceless. But they were all disregarded in favour of *Chrysanthemums*!

Although Roberts, throughout his career, also painted several works of roses, poppies and pelargoniums, his paintings of flowers are relegated into the background of his oeuvre. Andrew Montana is the only one who has suggested that Roberts's paintings of Chrysanthemums are of interest, albeit within the context of decorative art, noting that the artist 'exploited the flower's potential as a fashionable subject' using 'vigorous brushwork and deep rich colour to achieve his glowing autumnal effect'.[34] Like Rowan's splendid pictures of chrysanthemums, albeit of smaller dimensions, Roberts's works compare well to those painted by other Belle Époque artists, decorative and fashionable or not.

In 2022, I would argue that these paintings of chrysanthemums by Roberts and Rowan, and indeed those painted by so many artists of that age, gain new significance and are no longer usefully relegated to a minor genre of painting. Chrysanthemums were one of hundreds of non-native species introduced to Australia by settler-colonists in the nineteenth century. Some of these species have, over time, damaged the natural ecosystem and threatened native-plant diversity, contributing to the extinction of indigenous flora. While Rowan was carefully recording the rich native plant life of Australia, her statement piece at the centennial exhibition portrays a non-native ornamental species popular in Britain, Europe and America as an exotic, then linked to orientalist fantasies of East Asia.[35] Rowan's work, along with Roberts's painting of the same subject, indexes how the inter-

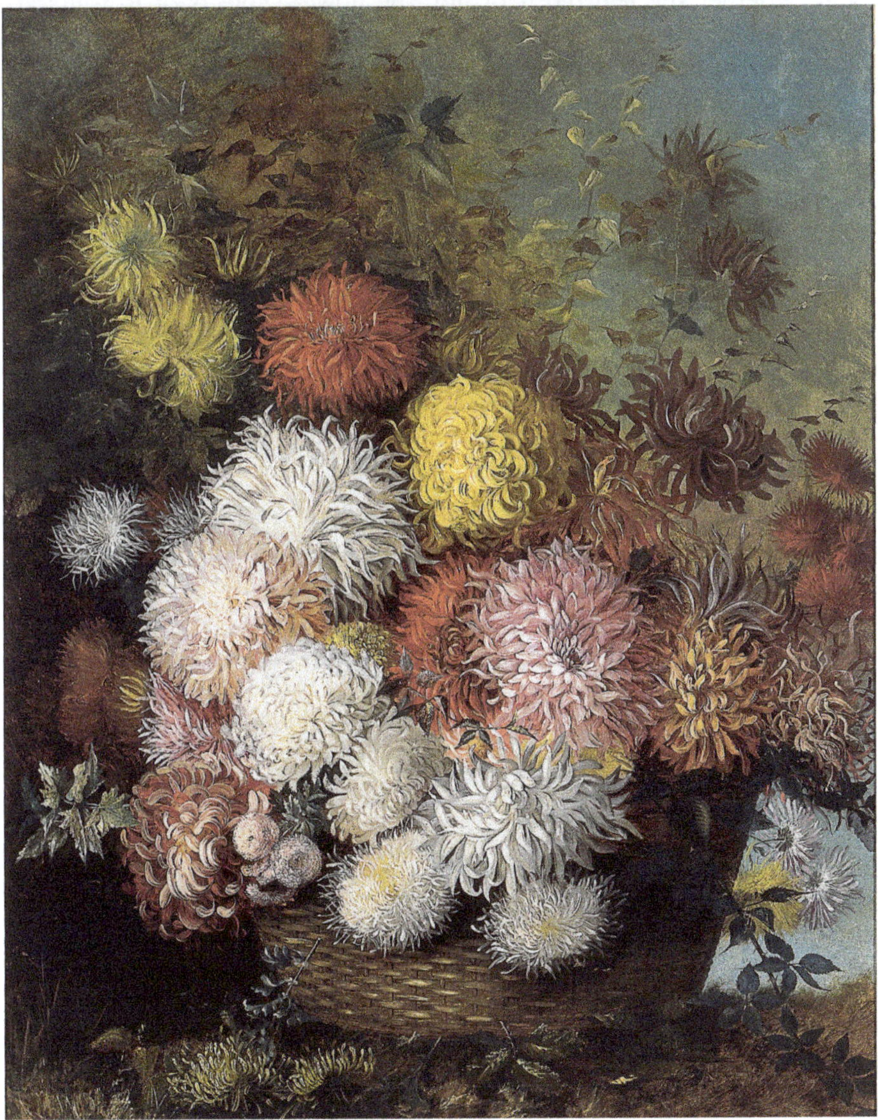

Illustration 12.1 Marian Ellis Rowan, *Chrysanthemums*, oil on canvas, 119.5 × 91.5 cm, private collection, 1888.

national horticultural trade encompassed Australia. This trade contributed, and continues to contribute, to many species extinctions through competitive exclusion, niche displacement, and the spread of disease.

Chrysanthemums indicates that Ellis was able to work outside of the framework of botanical illustration, and that the boundaries between science and art may not have always been as fiercely observed as we believe today; whether she realized it or not, the flower subjects of Rowan's art were intimately bound up with fate of the native flora she recorded. The boundaries of botanical illustration and flower painting were crossed by Rowan and North, both of whom developed a style that integrated elements of art and science, creating a new genre of painting. Ellis Rowan had an international reputation and was more visible than that of any other Australian artist at the time. Travelling extensively, she successfully took part in numerous exhibitions all over the world, and her flowers entered many private and public collections, including the Royal collection in London following the purchase by Queen Victoria of three of her works.[36] Art historians have failed to grasp just how elevated a role Rowan occupied in both the national and international art world. Bernard Smith did not believe her to be sufficiently important to include her in any of his survey histories of Australian art. In general, art historians have also failed to understand how mainstream flower painting actually was, or how widely it was practised by both male and female artists. The remarks of the critic of *The Argus* in reviewing the Victorian Artists' Society spring exhibition of 1888 are indicative:

> Mrs Ellis Rowan's basket of fruit blooms and Mr Tom Roberts' fine group of roses … A panel of maiden's blush roses by Miss Williamson; Mr A Streeton's chrysanthemums; a cluster of daffodils *in situ*, by Mrs Anderson; two panels of irises, by Miss R.C. Atterley; some yellow and crimson chrysanthemums by Miss Jane Sutherland; and some cloth of gold rosebuds by Miss F.A. Fuller, serve to heighten the general attractiveness of the exhibition.[37]

It is worth a thought that Marion Ellis Rowan may have actually won the Melbourne international competition, not because she was so well connected and knew some of the members of the panel of judges – only one of whom was an artist and which included scientists and prominent businessmen in their ranks – but because it was the best picture of the show. Perhaps one of the reasons that Roberts was so offended was that the judges preferred her Chrysanthemums to his? In any case, three years later they were painting on the beach together on Murray (Mer) Island in the Torres Strait. Rowan, clearly a good sport, showed concern over a portrait of a Meriam man completed by Roberts *en plein air*, in which she provides us with the only account in existence of the making of Roberts's extremely important cycle of portraits from Cape York:

> He made a splendid study of his model, who had an ornament of feathers stuck in his woolly head and a big scarlet hibiscus in his ear-ring . . . Just as the forty-guinea finishing touches were being put on the picture, the whole thing fell, butter-side downwards, into the sand. It looked hopeless but the next day the oils had dried, and we carefully wiped off the sand, after which it was re-touched, and looked as well as ever.[38]

Conclusion: The Importance of Plants

In this concluding section, I want to come back to Rowan's botanical illustrations, before turning to the work of the contemporary Australian artist Lucienne Rickard. As noted earlier, a number of the plants Rowan drew are now endangered or extinct. The Cooktown orchid (*Dendrobium bigibbum*), which she painted several times, is classed as 'vulnerable'. This orchid is now the Queensland State floral emblem. In a work from 1891, which combines a depiction of the Cooktown orchid with one of the native bee orchid (*Dendrobium bifalce*), it is clear from the white spot in the centre of the labellum that the variety she depicts is native to the Cape York Peninsula.[39] Rowan's attentiveness to the individual characteristics of the Cooktown orchid, her careful recording of the specifics of the orchid's labellum, speaks to her engaging with plants as subjects. Her way of seeing plants was inspirational, especially at a time when we are largely 'plant blind' when it comes to their conservation. The performance of Puccini's *Elegy* reminds us that diverse plant species are now under threat of disappearing for ever, through extinctions brought on by changes to the climate, the intrusion of introduced plants into native landscapes, and the pressure of human and exotic animal habitation. Performing to an audience of plants, like Rowan's careful and sensitive illustrations, works to affirm plants as subjects, and to counter a tendency to see plants simply as greenery, in the background and unimportant.

Seeing and not seeing are central to Tasmanian graphic artist Lucienne Rickard's concern with species extinction. Her drawings barely emerge before they vanish, becoming afterimages, rubbed out for ever (see Illustration 12.2). Her project, *Extinction Studies*, is an installation of twelve-months duration, which at the time

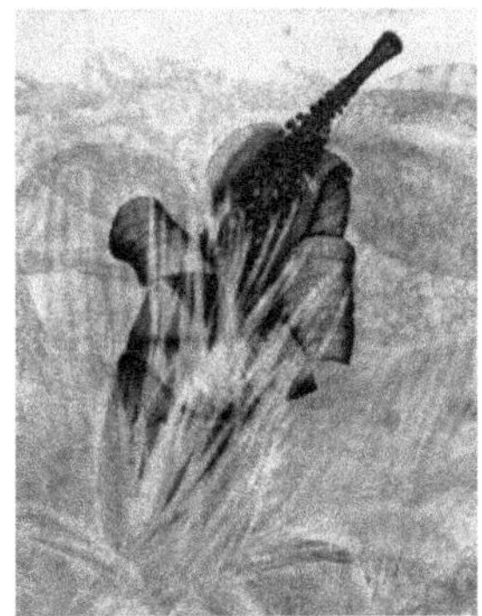

Illustration 12.2 Lucienne Rickard, *Hibiscus*, Extinction Studies project, Hobart, 2020.

of writing is being performed at the Tasmanian Gallery's Link Foyer in Hobart, stretching from September 2019 to December 2020 and now extended to run until 24 January 2021. So far, she has completed thirty-seven drawings of animals and plants, now gone forever. Her performance consists of representing on paper, with an HB pencil, plants and animals and insects from all over the world that are now extinct. When she has completed a drawing, which usually takes at least a week to finish and sometimes several months, she then erases the drawing and begins to draw a different extinct species on the same sheet of paper.

It has taken Rickard months to record in painstaking detail, for example, some of the scales on the wings of the Xerces blue (*Glaucopsyche xerces*) butterfly of San Francisco, which number some 112,000. Her works are exquisite. When the drawing is complete, she rubs it out. There is always ghosting remaining on the surface – an uncanny glimpse of an afterlife. This process of drawing and erasure, of evolution and extinction, is repeated in full knowledge that the paper will deteriorate and eraser shavings will accumulate.[40] Many of those who watch her work, beg her not to erase the completed picture. Rickard is herself a force of nature as she demonstrates, with a visceral power, the terrifying reality of extinction.

Acknowledgements

I would like to thank Michael Morton-Evans, Geoffrey Smith, Nick Chare, Lucienne Rickard, Warren Joel and the staff of the Pictorial Collection, Australian National Library, Canberra, for assistance with the research and writing of this chapter.

Jeanette Hoorn is an honorary professorial fellow in the Faculty of Arts at the University of Melbourne. Her research examines racial and gender issues surrounding the production and reception of modern art and the cinema. Her publications have included studies of the pastoral in Australian art, the impact of Darwinian theory, and the special circumstances surrounding the art and culture of First Nations Australians. She has written on Hilda Rix Nicholas in Morocco; Joseph Lycett and the doctrine of *Terra Nullius*; the portraits of Julie Dowling and Tom Roberts and the place of the yam in Emily Kame Kngwarreye's paintings. She has also written on the Australian films of Michael Powell and Charles Chauvel, on the photographs of Tracey Moffatt and Destiny Deacon, and on the representation of the *mission civilisatrice* in French Colonial cinema. She is working to find ways of revealing and challenging the operation of apartheid in Australia through her writing.

Notes

1. Gates, *Kindred Nature*.
2. Morton-Evans, *Ellis Rowan: A Life in Pictures*, 3.
3. 970 works by Rowan are held in the National Library of Australia in Canberra and 108 works are housed in the collection of the Royal Botanic Gardens at Kew. There are 832 works by North in the Marianne North Gallery at Kew.
4. A letter written to North by Darwin thanked her for her drawing of the shrub, *Raoulia eximia* 'Australian Sheep'. Letter from Charles Darwin to Marianne North, 2 August 1881. Letter no. DCP-LETT-13269A. Darwin Correspondence Project, Cambridge University. Retrieved 6 January 2021 from: https://www.darwinproject.ac.uk/letter/DCP-LETT-13269A.xml.
5. Lounsberry and Rowan, *Southern Wild Flowers and Trees*, ix–x.
6. Rowan, *A Flower Hunter*, 11.
7. Morton-Evans and Morton-Evans, *The Flower Hunter*, 206.
8. In an article in *The Argus*, Baldwin Spencer named Rowan among a number of Australian women, including Emily Peloe and Louisa Meredith, as women 'who have notably tried with pen and pencil to reveal to the ordinary nature lover the beauty and interest of Australian wild flowers'. Quoted in Moore, *The Story of Australian Art*, xviii.
9. Montana, 'Constance Roth and Ellis Rowan', 213.
10. Rowan's writings are replete with references to her friends, family and professional contacts. See for example, *A Flower Hunter*.
11. Sharp, *Footprints*, 55–58.
12. Ibid., 85. See Liddell, *The Savage Frontier*, 98; Richards, *The Secret War*.
13. Clarke, cited in Maroske, 'A Taste for Botanic Science', 75.
14. Rowan, *A Flower Hunter*, 65–66. For a discussion of the often overlooked but crucial role of Aboriginal Australians in the nineteenth-century botanical 'discovery' of Australia's plant life, see Clarke, *Aboriginal Plant Collectors*.
15. Rowan, *A Flower Hunter*, 6.
16. Morton-Evans and Morton-Evans, *The Flower Hunter*, 98.
17. Rowan was a prominent member of a group of over two hundred female artists and collectors of specimens who collected for von Mueller. Maroske, 'A Taste for Botanic Science', 73.
18. Rowan, *A Flower Hunter*, 157.
19. Ibid., 17.
20. Ibid., 84.
21. Ibid., 161.
22. Her travels brought her into direct contact with frontier violence such as that which took place during the same trip to the Torres Strait, when she was hosted by the government resident, John Douglas, on board the steamer *Albatross*. On this occasion, she was brought into public controversy when a newspaper report from Thursday Island complained that the murder of Indigenous people at the Seven Rivers had not been properly investigated. According to the report this was 'because the *Albatross* was absent, escorting Mrs. Rowan, the artist, around the islands'. The matter was taken up in the Queensland parliament by the member for Ipswich, AH Barlow, who sought a parliamentary enquiry into the government resident's conduct. Queensland Parliamentary Debates of the Legislative Assembly, Vol. 64, 30 September 1891, p. 1295, quoted in McKay, *Ellis Rowan, a Flower-Hunter*, 16.
23. Morton-Evans, *Ellis Rowan: A Life in Pictures*, 5.

24. Neira, 'Barcelona Opera House'.

25. Sommer, 'Cybercene or Biocene'.

26. Louise Gray, 'Unfashionable flowers in danger of dying out'. *The Telegraph*, 20 March 2009. Retrieved 6 January 2021 from: https://www.telegraph.co.uk/gardening/502 3101/Unfashionable-flowers-in-danger-of-dying-out.html.

27. Agence France-Presse. 'As Lockdown Lifts, Barcelona Plants Enjoy a Day at the Opera'. *Yahoo! News*, 22 June 2020. Retrieved 6 January 2021 from: https://news .yahoo.com/lockdown-lifts-barcelona-plants-enjoy-day-opera-174357136.html.

28. This victory in 1888 followed on from another major public protest over her success at the Melbourne International Exhibition of 1880.

29. Jordan, 'Tom Roberts, Ellis Rowan'.

30. She did, however, have a family background in scientific illustration. Her maternal grandfather, John Cotton, published two books on British birds: *Resident Song Birds of Great Britain* and *The Song Birds of Great Britain*. He was one of the first ornithologists to record the birds of Victoria. Cotton's manuscript is found in the 'Sketchbook Belonging to John Cotton, with Sketches and Descriptions of Birds of the Port Phillip District of New South Wales [1844–49]'. State Library Victoria. Retrieved 6 January 2021 from: http://handle.slv.vic.gov.au/10381/210587.

31. Dimensions for both are 119.5 x 91.5 cm. Morton-Evans, *Ellis Rowan: A Life in Pictures*, 61, 170.

32. Ibid., 121.

33. Centennial International Exhibition, 119–21.

34. Montana, 'The Chrysanthemum', 114–17.

35. For a discussion of chrysanthemum culture in Victorian Britain, see Chang, 'Chrysanthemums and Cultivated Visions', 178–203.

36. Ellis won many medals in international competitions, including at the World's Columbian Exposition at the World Fair in Chicago in 1893. Rowan painted two large oils of Chrysanthemums at around this time, and Roberts three or four smaller works; we cannot be absolutely sure that those claimed to have been the ones exhibited in 1888 were indeed the pictures the audiences came to see.

37. *The Argus*, 16 November 1888, 6.

38. Rowan, *A Flower Hunter*, 159 cited in McQueen, *Tom Roberts*, 363.

39. This painting is reproduced in Clements, *The Allure of Orchids*, 115.

40. Each extinct species is sourced from the International Union for Conservation of Nature's (IUCN) *Red List of Threatened Species*, widely recognized as the authoritative list of extinct and threatened species used by scientists globally. See 'Exhibitions: Extinction Studies', TMAG, Retrieved 6 January 2021 from: http://www.tmag.tas.gov .au/whats_on/exhibitions/current_upcoming/info/extinction_studies.

Bibliography

Centennial International Exhibition, 1888–1889. Official Guide to the Picture Galleries and Catalogue of Fine Arts. Melbourne: Mason, Firth & M'Cutcheon, 1888.

Chang, Elizabeth. 'Chrysanthemums and Cultivated Visions of the Victorian Garden', in Petra ten-Doesschate Chu and Jennifer Milam (eds), *Beyond Chinoiserie: Artistic Exchange between China and the West during the Late Qing Dynasty (1796–1911)* (Leiden: Brill, 2018), 178–203.

Clarke, Philip A. *Aboriginal Plant Collectors: Botanists and Australian Aboriginal People in the Nineteenth Century*. Kenthurst: Rosenberg Publishing, 2008.

Clements, Mark A. *The Allure of Orchids*. Canberra: National Library of Australia, 2013.

Cotton, John. *The Resident Song Birds of Great Britain*. London: [publisher not identified], 1835.

———. *The Song Birds of Great Britain*. London: Samuel Bentley, 1836.

Gates, Barbara. *Kindred Nature: Victorian and Edwardian Women Embrace the Living World*. Chicago: University of Chicago Press, 1999.

Jordan, Caroline. 'Tom Roberts, Ellis Rowan, and the Struggle for Australian Art at the Great Exhibitions of 1880 and 1888', in Kate Darian-Smith (ed.), *Seize the Day: Exhibitions, Australia and the World* (Clayton, Vic: Monash University Press, 2008), 15.1–15.16.

Liddell, Rodney. *The Savage Frontier*. Redbank, Qld: Rodney Liddell, 1996.

Lounsberry, Alice, and Ellis Rowan. *Southern Wild Flowers and Trees*. New York: Frederick A. Stokes Company, 1901.

Maroske, Sara. 'A Taste for Botanic Science: Ferdinand von Mueller's Female Collectors and the History of Australian Botany'. *Muelleria* 32 (2014): 72–91.

McKay, Judith. *Ellis Rowan, a Flower-Hunter in Queensland*. South Brisbane: Queensland Museum, 1990.

McQueen, Humphrey. *Tom Roberts*. Sydney: Pan Macmillan, 1996.

Montana, Andrew. 'The Chrysanthemum: A Forgotten Flower'. *World of Antiques and Art* 56 (Dec. 1998 – Jun. 1999): 114–17.

———. 'Constance Roth and Ellis Rowan: Lady Artists and Matters of Influence'. *Fabrications* 23(2) (2013): 208–14.

Moore, William. *The Story of Australian Art, Vol. 1*. Sydney: Angus and Robertson, 1980.

Morton-Evans, Christine. *Ellis Rowan: A Life in Pictures*. Canberra: National Library of Australia Press, 2020.

Morton-Evans, Christine, and Michael Morton-Evans. *The Flower Hunter: The Remarkable Life of Ellis Rowan*. Pymble, NSW: Simon & Schuster (Australia), 2008.

Neira, Juliana. 'Barcelona Opera House Reopens with a Concert for more than 2,000 Plants'. *Designboom*, 23 June 2020. Retrieved 6 January 2021 from https://www.design boom.com/art/barcelona-opera-house-concert-for-plants-06-23-2020/.

Richards, Jonathan. *The Secret War: A True History of Queensland's Native Police*. Brisbane: University of Queensland Press, 2008.

Rowan, Marian Ellis. *A Flower Hunter in Queensland and New Zealand*. London: John Murray, 1898.

Sharp, Nonie. *Footprints along the Cape York Sand Beaches*. Canberra: Aboriginal Studies Press, 1992.

Sommer, Mark. 'Cybercene or Biocene: Which Future Will We Choose?' *Data Driven Investor*, 7 June 2019. Retrieved 6 January 2021 from https://medium.com/datadriven investor/cybercene-or-biocene-which-future-will-we-choose-13ca3ef010cb.

Chapter 13

Towards Extinction

Mapping the Vulnerable, Threatened and
Critically Endangered Plant in 'Moments of Friction'

Dawn Sanders

... and the maps of Spring always have to be redrawn
again, in undared forms.
—K. McKittrick, *Sylvia Wynter*

Introduction

Deepening our relationships to plants confronts educators with challenging questions in an age of extinction. As Rose, van Dooren and Chrulew have noted, 'there is no singular phenomenon of extinction: rather extinction is experienced, resisted, measured, enunciated, performed and narrated in a variety of ways to which we must attend'.[1] Bringing art-based approaches to these interstitial spaces of attention can mesh the biological and the cultural into plant encounters that provoke 'human curiosity towards plant-based narratives, and give humans agentic space to experience being with, thinking through, and understanding something genuinely moving of the lives of plants'.[2] Critical animal studies scholars have done much to give agency to non-human animals in these encounters, and in so doing have asked vital questions of educational institutions such as: 'what does education become when humans are not regarded as the only subjects?'[3] In these contested territories of teaching and learning, how can we articulate potential spaces for what Broglio terms 'moments of friction'[4] in the context of 'what is', 'what is not' and 'what might be', when plants are revealed through encounters with art works?

In recent years an interdisciplinary nexus has been generated around what it means to experience life as a plant. From the science of plant behaviours, plant language and meaning-making to plant-based philosophy, plant enquiries are crossing disciplinary and conceptual boundaries. The everyday life of a plant can appear to be static and silent to human perception. And yet, as modern science narratives tell their stories, we are realizing that plants live in complex and often social worlds. Removing plants from the human view makes it easier for us to exploit them, and appears accordingly to reduce our ability to see into their worlds. In our research study 'Beyond Plant Blindness: Seeing the importance of plants for a sustainable world', we asked how, by taking a different view through an interdisciplinary lens, we might improve our understanding and sensitivity to the lives of plants.[5]

In this chapter I draw on the installations created by the artists Snæbjörnsdóttir and Wilson in the aforementioned research project to assert the view that contemporary education needs to be spatially and temporally responsive to 'Life as Plant', particularly when such plants are vulnerable, threatened and endangered. Art, in this space, can reveal aspects of 'plantness'[6] hitherto underexposed to human view, and in so doing, provoke unsettling 'contact zones'[7] between plant and human subjects. In an age rampant with extinction,[8] there is a critical need to resist 'the temptation to simplify a relationship that has too long an abusive history to be mended easily'.[9]

Art and 'Plantness': Albrecht Dürer and *The Large Turf*

Albrecht Dürer produced his watercolour *The Large Turf* in 1503. A relatively small painting (41 × 32 cm), it is seen by many modern commentators as a defining moment in vibrant portraits of plant life. Lubbock, for example, notes that Dürer's painting

> doesn't visually isolate or distinguish its various plants. It presents them in a state of natural disarray, confused, interleaved, entangled. Though each growth is clearly identifiable, the picture is far from being a biological ID parade. It is a slice of living, chaotic undergrowth.[10]

The contemporary British artist Mat Collishaw has wondered why, in an era filled with religious imagery, Dürer chose to focus his artistic attention on a 'seemingly insignificant clump of weeds'.[11] Richard Mabey in his book *Weeds* considers it to be 'painting's discovery of ecology'.[12]

In making this painting, Dürer creates a realistic window on a very ordinary group of living plants, but in taking the 'worm's eye view',[13] he brings us, the human viewers, deep into the world of plants, and allows us to view the structural complexity of each blade and stem, and makes public a subtle palette of greens. In this realism we can identify many of the species portrayed, such as cock's-foot (*Dactylis glomerata*), dandelion (*Taraxacum officinale*), smooth meadow grass (*Poa pratensis*) and greater plantain (*Plantago major*). Thus, this is not generic greenery, but a carefully executed visual record of different types of plant. Each plant, through his painterly rendition, is accorded its singularity. Dürer's excursion into the aesthetics of commonplace plants opens a window onto a little-known world. In making this ordinary world public, Dürer creates a contact zone between human and plant, and, I believe, offers us a critical juncture in the history of artistic representations of 'Life as Plant', albeit that, perhaps, this was not his foremost intention.

Ecologies of Existence

Biologically, '[f]or the vast majority of organisms, the world is the product of plant life, the product of the colonization of the planet by plants'.[14] Moreover, plant biomass on Earth is substantial.[15] And yet, in recent years, we see that many plant populations are struggling to survive, and individual species are increasingly exposed to the stark possibilities of extinction.[16] One of the resounding echoes in contemporary society is that of a world in which plants are viewed as 'Other'.[17] This 'flattening' of plant worlds into a 'thin layer'[18] of otherness has implications for the contact zones between humans and plants: 'In the absence of respect and real curiosity, attentiveness falters. Complex systems become reduced to green blurs, with dangerous consequences for both us and for individual species'.[19]

The root of the word 'ecology' is the Greek word 'oikos', meaning 'home'. But where is 'home' to be found for plants and animals in an era dominated by ecologies of existence of ever-increasing vulnerability, both within habitats and across populations. Contemporary narratives of conservation biology appear to be dominated by extensive loss,[20] mirrored in titles such as: '*Extinction in Our Times*', '*Requiem for Nature*', '*Silence of the Songbirds*'. Furthermore, as Kolbert notes, it is 'not just species vanishing. Entire features of the earth are disappearing'.[21] Critically, for this chapter, she observes: 'Hope and its doleful twin, hopelessness, might be thought of as the co-muses of the modern eco-narrative. Such is the world we've created – a world of wounds – that loss is, almost invariably, the nature writer's subject. The question is how we relate to that loss'.[22]

Working with plant education in a highly urbanized twenty-first century is to engage with Kolbert's question on multiple levels, for the species loss she speaks of is but one absent presence among many; another is the reduction of a highly complex living world into 'an amorphous set of words and a collection of fleeting images',[23] which fail to engage human attention; such perceptual redaction can result in people neither registering the importance of plants and animals, nor mourning their loss. Education can be considered a significant cultural actor in such relational processes. However, much of modern education appears to be mired in competitive measurement and performativity structures,[24] operating in time-constrained spaces that function as inhibitors of lingering encounters with the more-than-human world. Affording plants space and time to grow in school and university settings necessitates active participation from teachers and students, and environments conducive to plant life. In addition, these experiences require sustained human witness beyond the young emergent seedling of an annual cycle. Moreover, relationships with older, mature plants are necessary to understand that many plant lives have timescales beyond most humans, and in this longevity they offer shelter and nourishment to a broad range of living beings.[25]

In the following section I reflect on three particular art installations developed by the artists Snæbjörnsdóttir and Wilson in the aforementioned interdisciplinary research project. They are situated in relation to mapping a vulnerable plant, *Stipa pennata*, whose existence is threatened, in a specifically Swedish context. I intend to discuss potential 'contact zones' between plant and human in order to identify a role for art in the context of making 'Life as Plant' public in an age when the threat of extinction is, for many, imminent.

Searching for Stipa

A vast and highly detailed enlargement of a *Stipa pennata* (European feather grass) plume-like awn, hangs in a barn-like building (Stolpboden) in Gothenburg Botanic Garden, Sweden, in 2017. As visitors enter a central doorway they are confronted by the horizontal, fourteen-metre-long paper print; a seamless unification of twenty-nine separate Scanning Electron Microscope (SEM) scans lit by lamps specifically installed for the work:

> Often with natural objects depicted in art there is an underlying scientific narrative to tell. The artwork *Searching for Stipa* is no exception.

> When looking at the 14 m long print in Stolpboden, one is impressed
> by its sheer size and beauty, but what is it that is shown? Why does the
> plant possess such a structure? In the rain shelter we look at the remnants
> of a Swedish Stipa population, and are moved by the fact that a species
> is disappearing from our flora; but why, in Sweden, is the plant so rare?[26]

The following year, in House B, Pedagogen, University of Gothenburg,
the same image, this time conceived as a woven wool tapestry, is hung
vertically outside the stairwell (Fig. 13.1). In conversation with the artists,
provoked by questions from the editors of this collection, we have reflected
on both the material of the tapestry and significance of the tapestry form.
Tapestries have traditionally been made of cotton or wool, and have utilized symbols of heraldry and pageantry to convey both power and community. Therefore, their cultural role is not only to commemorate military
prowess, ancestors and strength or to elicit religious respect, but also to
join particular communities together – for example, trades unions – and
to symbolize that communal bond in celebratory parades. Students and
teachers entering House B are confronted by a specific aspect of 'plantness'
(the awn of a seed) made statuesque in a union of animal (domesticated
sheep) and plant (*Stipa pennata*). Thus, the artists suggest this amalgama-
tion of accommodating animal and plant, the cultural history of tapestries,
and the technological work in constructing the work can be considered as
an entanglement of political ecology.

A symposium event planned to coincide with the public opening of the
artwork asked: '*Where can a single plant take you?*', mirroring the provocation of the artwork, hanging in the foyer, of the building in which the sem-
inar took place. In its scale, Snæbjörnsdóttir and Wilson's tapestry renders
the miniature awn monumental and confronts with its enormity, whereas
Dürer's watercolour is relatively small and quietly invites the viewer to
wander into a small clump of common plants. Thus, the tapestry, as a representation of 'plantness', offers artistic contrast to the Dürer representation as contact zones between plant and human.

The artworks that make up *Seaching for Stipa #1*, *Searching for Stipa #2*
and *Searching for Stipa Tapestry* intimately engage with one species (*Stipa
pennata*) and its competitive struggles to reproduce. These plants are living in two 'moments of friction'; one biological, as a wind-pollinated and
wind-dispersed plant, the other cultural, as a plant whose Swedish habitats
have been altered and diminished, leaving them with few places to survive.
In their biological world:

> Stipa pennata is wind-pollinated. In order for seeds to develop, pollen
> must be carried through the air by the wind to reach the stigma of an-
> other plant. When populations are small, the mere chance to reach the

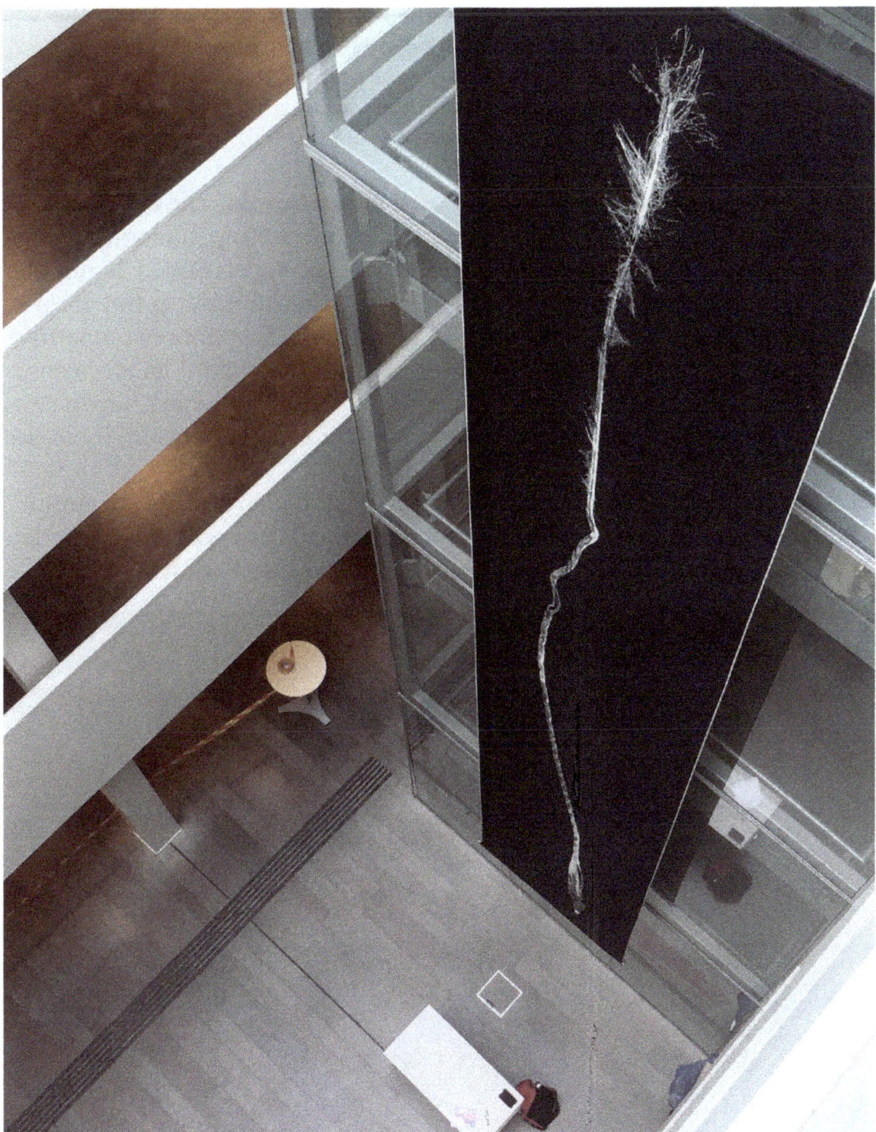

Illustration 13.1 Snæbjörnsdóttir and Wilson, *Searching for Stipa*, 2018, Tapestry, 14 m × 1.5 m, installed in Hus B, Pedagogen, University of Gothenburg.

right partner is low. Wind will also be saturated by pollen from other grass species, soon clogging stigmas, and leaving no room for pollen of the right kind to interact. Consequently, the seed set is lowered. The seeds are also distributed by wind. As a means [of] dispersal it has a 20 cm long plume-like awn attached to its fruit. This structure is what we see clearly in the Stolpboden print and later in the artist's tapestry [Fig. 13.1]. The awn not only allows the fruit to become airborne, but once it lands, helps drill the fruit, with the seed inside, into the ground. Repeated changes in humidity between night and day cause the awn to twist and untwist, and in the process, the pointed fruit is pushed into the soil.[27]

Problems arise when this biological seed structure joins the cultural habitat context. 'This self-burial mechanism is interesting biologically, but requires patches of open soil to work. In dense vegetation it is not the most reliable system … In the absence of cattle to break up the ground, the European feathergrass is at risk of being overrun by more competitive plants in the area',[28] and so a cultural strategy of fencing off an area causes stress to the plant, as the cows – and their hooves – no longer have access to the site, and so vegetation, and thus biological competition, increases in density.

Searching for Stipa #2 (Fig. 13.2) shows an entangled meadow of competition within which the remaining populations of *Stipa pennata* in Sweden strive to survive. Hence, in the narrative of the art installations, one specific

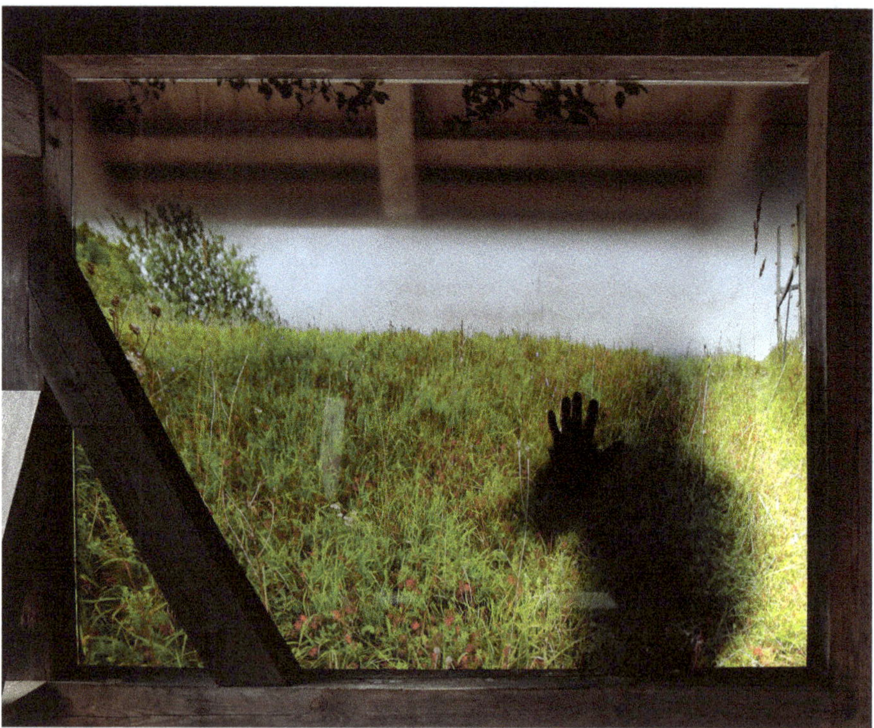

Illustration 13.2 Snæbjörnsdóttir and Wilson, *Searching for Stipa #2*, 2017, installation in the rain shelter, Botaniska (installation detail).

seed, *Stipa pennata*, is made large in *Searching for Stipa #1* only to vanish amongst a throng of competing plants in *Searching for Stipa #2*. As Bente Eriksen asks: 'The long, silver-white plumes of the Stipa make an impressive spectacle when they move in the wind; the question remains whether we will see this sight in Sweden for much longer'.[29]

In our study of viewers' impressions, these artworks provoked complex responses. Of particular interest was the impact of the 'scaling-up' of microscopic seed parts in *Searching for Stipa #1*.

> [This] appeared to create an intersectional zone between plant and animal identities ... As a result, these participants appeared to question, examine and recalibrate the taxonomies of their impressions. For example, several participants moved between thinking they were looking at an earthworm, or even a one-celled animal, to deciding it might be part of a plant ... [T]he installation provoked reflections on what it is to be an animal or plant at all, in terms of visible characteristics.[30]

Indeed, the 'scale-jumping' from microscopic awn to seed-dispersal mechanism made large seemed to provoke fascination, curiosity and confusion. Several participants agreed the work gave them a 'new perspective' on plants. For some viewers, *Searching for Stipa #2* evoked emotions mainly related to aesthetic qualities of beauty and colour. Beauty was seen, in this context, as associated both to 'the beauty of nature' and to the beauty of the artwork itself. Furthermore, it appeared to provoke a deep engagement with the act of looking at plants.

Art and Botanical Attention

Finding the narratives that make public the 'Life as Plant' story without negating the attributes that make plants and their lives 'other' in the eyes of human audiences is a challenge for science communicators and educators. In our interdisciplinary research project 'Beyond Plant Blindness', we found that integrating art, science and education as a shared praxis generated new possibilities for humans to see into the world of plants; art in such contexts did not merely represent plants, but placed them in 'contact zones' in which humans were provoked to think about, and reflect on, what they were seeing – and to consider if, indeed, it even was a plant, or part of a plant, that they were witnessing. In this setting, Snæbjörnsdóttir and Wilson, through changes of scale (*Seaching for Stipa #1, Searching for Stipa Tapestry*), and immersions of place (*Searching for Stipa #2*), render the familiar strange and encourage elements of perceptual disorientation in order to bring a specific grass narrative to our attention. In this way the biography of *Stipa pennata*, and its specific seed physiology and habitat, is foregrounded and cannot be redacted from human view.

In drawing our attention to vulnerable and threatened plants through the lens of one single species, the artworks of Snæbjörnsdóttir and Wilson focus our botanical gaze and do appear to provoke the question: 'where can a single plant take you'? If we choose to follow their subject, we might then question what it means to live 'Life as Plant' at this point in time. Perhaps, in these encounters with *Stipa*, we can begin to comprehend the enormity of the notion that 'extinction is an absolute term, meaning that no individual of a species remains alive'.[31] *Stipa pennata* is *not* going extinct in its wider European range, but the possibility is very real in the Swedish context. Thereby, by taking one plant from Sweden, living such a precarious existence, the artworks *Searching for Stipa #1*, the *Tapestry* and *Searching for Stipa #2* map the contemporary challenges of 'Being Plant' in one singular biocultural context.

There is another perspective to this encounter; by representing the plant on different scales (i.e. a close-up of a specifically adapted seed distribution mechanism and the remaining plants lost in an entangled clamour of competitors), Snæbjörnsdóttir and Wilson make public the unique story of 'what is', 'what is not' and 'what might be' for *Stipa pennata* in Sweden. In so doing, they mirror the vibrancy of Dürer's intensely present turf portrait, and affirm Fortey's assertion that '[e]very species has a narrative of its own, a biography. The loss of a species is not just one lower point on a graph of biodiversity, it is also the loss of a unique story'.[32]

In an era dominated by ecologies of existence that perch on the edge of life, represented by science as specific categories of vulnerability, art that utilizes individual plant narratives can make public what it means to *be* 'plant', both structurally and contextually, in a modern world where extinction proliferates. Perhaps, by installing art-based research in educational contexts, such as botanic gardens and universities, such works might provoke contemporary spectators to view plants as individual subjects that draw their attention. For, as noted by the artists, 'In *Searching for Stipa*, through scale, medium *and* genre, the form of one part of the plant, an airborne seed awn, is made the subject of celebration and wonder – exactly and strategically made strange and de-familiarised, in order to provoke new readings and respect' (Snæbjörnsdóttir and Wilson, personal communication, 22 October 2020).

In the face of exponential levels of extinction, there is a pressing need for all of us to engage, both conceptually and responsibly, with other organisms possessing entirely different physiologies and behaviours. In our engagement with such difference, it is vital that we are not diverted into human registers and 'terms of resemblance',[33] but rather to participate in the particular subjectivities of 'plantness'. In this context we have found that:

> Art not only offers visual beauty, but can unsettle and disrupt human perspectives on 'plantness'. It may connect to one's emotions through the interpretation of what one sees, and it can also challenge the long-held view … Our research participants saw beauty, but also stopped to think about biological diversity and how different plants are to humans.[34]

In providing intimate contact with those we consider to be 'other' in art, our project appears to provoke curiosity, confusion and emotion in the presence of plants. In viewing *Stipa pennata* as a single species struggling to persist in a changing Swedish landscape, Snæbjörnsdóttir and Wilson's works engage with an individual narrative; an approach that might foster greater understanding and recognition of 'Life *as* Plant'. In the looming presence of extinction, such singular spaces of encounter are vital contact zones between plant and human.

> Slowly every kind of one comes into ordered recognition. More and more, then, it is wonderful in living the subtle variations coming clear into ordered recognition, coming to make everyone a part of some kind of them.[35]

Dawn Sanders is an associate professor in the Faculty of Education at the University of Gothenburg, Sweden. Her education combines training in fine art with further studies in ecology, culminating in a doctoral study that considered the educational role of botanic gardens (Geography Department, Sussex University, UK, 2005). She led the interdisciplinary research team in the Swedish Research Council funded project 'Beyond Plant Blindness: Seeing the importance of plants for a sustainable world', resulting in related publications in research journals and in a book of the same name published in 2020 by The Green Box in Berlin. She continues to work with artists on questions concerning human attention to plants.

Notes

1. Rose, van Dooren and Chrulew, *Extinction Studies*.
2. Sanders, 'On Trying to Understand', 48.
3. Dinker and Pedersen, 'Critical Animal Pedagogies'.
4. Broglio, *Surface Encounters*.

5. Snæbjörnsdóttir, Wilson and Sanders, *Beyond Plant Blindness*.
6. Darley, 'The Essence of "Plantness"'.
7. Broglio, *Surface Encounters*.
8. Willis, 'State of the World's Plants 2017'.
9. Nitzke, 'Listening to What Trees Have to Say'.
10. Tom Lubbock's article 'Dürer, Albrecht: *The Large Turf* (1503)', *The Independent*, 18 January 2008. Retrieved 28 January 2021 from: https://www.independent.co.uk/arts-entertainment/art/great-works/durer-albrecht-the-large-turf-1503-770976.html.
11. Collishaw, 'Burning Flowers'.
12. Mabey, *Weeds*.
13. Pavord, *The Naming of Names*.
14. Coccia, *The Life of Plants*.
15. Bar-On, Phillips and Milo, 'The Biomass Distribution on Earth'.
16. See, for example, Willis, 'State of the World's Plants 2017'.
17. Eriksen and Sanders, 'Seeing Significance in Plants'.
18. Broglio, *Surface Encounters*.
19. Mabey, *The Cabaret of Plants*, 5.
20. Elizabeth Kolbert, 'How to Write about a Vanishing World'. *The New Yorker*, 15 October 2018. Retrieved 1 October 2020 from: https://www.newyorker.com/magazine/2018/10/15/how-to-write-about-a-vanishing-world.
21. Ibid.
22. Ibid.
23. Hatley, 'Walking with Ōkami', 32.
24. Biesta, *Educational Research*.
25. Affifi, 'Plant Blindness Leads to Extinction Blindness'.
26. Eriksen, 'The European Feathergrass', 92.
27. Ibid.
28. Ibid.
29. Ibid.
30. Sanders, 'On Trying to Understand', 47.
31. Willis, 'State of the World's Plants 2017'.
32. Fortey, 'Island Life'.
33. Houle, 'Animal, Vegetable, Mineral'.
34. Eriksen and Sanders, 'Seeing Significance in Plants', 18.
35. Stein, *The Making of Americans*.

Bibliography

Affifi, R. 'Plant Blindness Leads to Extinction Blindness: Lessons from *Kok Mai Bak*', in B. Snæbjörnsdóttir, M. Wilson and D. Sanders (eds), *Beyond Plant Blindness: Seeing the Importance of Plants for a Sustainable World* (Berlin: The Green Box, 2020), 140–45.

Bar-On, M.Y., R. Phillips and R. Milo. 'The Biomass Distribution on Earth'. *Proceedings of the National Academy of Sciences* 115(25) (2018): 6506–11.

Biesta, G. *Educational Research: An Unorthodox Introduction*. London: Bloomsbury, 2020.

Broglio, R. *Surface Encounters: Thinking with Animals and Art*. Minneapolis: University of Minnesota Press, 2011.

Coccia, E. *The Life of Plants: A Metaphysics of Mixture*. Cambridge: Polity Press, 2019.

Collishaw, M. 'Burning Flowers: Interview with Mat Collishaw', in Giovanni Aloi (ed.), *Why Look at Plants?: The Botanical Emergence in Contemporary Art* (Rotterdam: Brill/Rodopi, 2019), 255–59.

Darley, W.M. 'The Essence of "Plantness"'. *The American Biology Teacher* 52(6) (1990): 354–57.

Dinker, K.G., and H. Pedersen. 'Critical Animal Pedagogies: Re-learning Our Relations with Animal Others', in H. Lees and N. Noddings (eds), *The Palgrave International Handbook of Alternative Education* (London: Palgrave Macmillan, 2016), 415–30.

Eriksen, B. 'The European Feathergrass, *Stipa pennata L.*', in B. Snæbjörnsdóttir, M. Wilson and D. Sanders (eds), *Beyond Plant Blindness: Seeing the Importance of Plants for a Sustainable World* (Berlin: The Green Box, 2020), 92–93.

Eriksen, B., and D. Sanders. 'Seeing Significance in Plants: A Role for Art in Botanic Gardens', in B. Snæbjörnsdóttir, M. Wilson and D. Sanders (eds), *Beyond Plant Blindness: Seeing the Importance of Plants for a Sustainable World* (Berlin: The Green Box, 2020), 14–19.

Fortey, R. 'Island Life', in Bergit Arends and Siân Ede (eds), *Galápagos* (Lisbon: Calouste Gulbenkian Foundation, 2012), 16–21.

Hatley, J. 'Walking with Ōkami, the Large-Mouthed Pure God', in D.B. Rose, T. van Dooren and M. Chrulew (eds), *Extinction Studies: Stories of Time, Death and Generations* (New York: Columbia University Press, 2017), 19–48.

Houle, K. 'Animal, Vegetable, Mineral: Ethics as Extension or Becoming? The Case of Becoming-Plant'. *Journal for Critical Animal Studies* IX(1/2) (2011): 89–116.

Mabey, R. *The Cabaret of Plants: Botany and the Imagination*. London: Profile Books, 2015.

———. *Weeds: How Vagabond Plants Gatecrashed Civilisation and Changed the Way We Think about Nature*. London: Profile Books, 2010.

McKittrick, K., ed. *Sylvia Wynter: On Being Human as Praxis*. Durham, NC: Duke University Press, 2015.

Nitzke, S. 'Listening to What Trees Have to Say'. *Edge Effects*, 29 September 2020. Retrieved 28 January from https://edgeeffects.net/tree-story/.

Pavord, A. *The Naming of Names: The Search for Order in the World of Plants*. London: Bloomsbury, 2005.

Rose, D.B., T. van Dooren and M. Chrulew. *Extinction Studies: Stories of Time, Death and Generations*. New York: Columbia University Press, 2017.

Sanders, D. 'On Trying to Understand "Life as Plant": Fielding Impressions of Art-Based Research Installation in a Botanic Garden', in B. Snæbjörnsdóttir, M. Wilson and D. Sanders (eds), *Beyond Plant Blindness: Seeing the Importance of Plants for a Sustainable World* (Berlin: The Green Box, 2020), 40–49.

Stein, G. *The Making of Americans: Being a History of a Family's Progress*. New York: Something Else Press, (1925) 1966.

Willis, K.J. (ed.). 'State of the World's Plants 2017. Report'. London: Royal Botanic Gardens, Kew, 2017.

Chapter 14

Sweetness, Power, Yeasts and Entomo-terroir

Robert R. Dunn, Monica C. Sanchez and Matthew Morse Booker

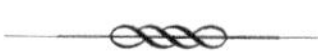

Yeasts are single-celled fungi. They are essential components of the modern world, ours and theirs, and yet their stories are hidden. Their consequences often take centre stage, but they themselves are rarely featured, whether in history, art, or even science. This absence has taken on new importance as it is realized that some yeast species may have begun to go extinct. These yeasts are threatened, of all things, by the declines in insect populations and, of course, indirectly and directly, by us. But this is the end of the story – let us start a little earlier.

Before

In the beginning, before humans evolved, before farms were planted, yeasts lived quiet lives immersed in the small patches of sweet things that can be found in nature – the nectar in the cups of flowers, the sap that leaks from oak trees, the honeydew that pours from aphids and scale insects, the flecks of sugar on oak leaves. In each of these bits of sweetness, the yeast consumed sugars. As waste, they exhaled carbon dioxide and excreted alcohol. Yeasts thrive on sugar. For most species, it is their ancient necessity. Yet, this necessity poses a challenge. The yeasts must find the sugar in the first place. They must find sugar even though they lack legs and wings, and even though, unlike bacteria, they do not readily become airborne. From the perspective of yeasts, all of the patches of sugar in the world, from those on jungle leaves to those inside the most delicate tundra flowers, are remote. They are perfect and delectable and yet hard to come by; islands in a sugarless sea.[1]

Eventually, some yeast species evolved a way around this problem of travel. They evolved the ability to produce, when eating sugar, aromas that attract insects. Different yeasts produce different aromas and attract different insects. Some yeasts call to bees. Others put out their aromas, like a hitchhiker's thumb, to call for wasps – but not just any wasp, specific wasps. The yeast *Saccharomyces cerevisiae*, for instance, appears particularly predisposed to call to social wasps of the genus *Polistes*, paper wasps.[2] Each group of sugar-feeding insects, and in some cases even each species of sugar-feeding insect, seems to carry its own characteristic portfolio of yeasts. They take these yeasts with them as they pursue sweetness; they have carried them from one delicacy to another across months, seasons, years, and millennia.[3]

Most yeasts thrive on the sugars produced by plants – in nectar, in fruits, and on their leaves – but not all. Some yeasts grow instead from the sugars of the dead, the sweetness of animal bodies. But death, like flowers, is not everywhere. To find the dead, these yeasts ride in and on the bodies of carrion beetles. The carrion beetles carry the yeasts to the dead, whether it be dead mice or dead people. The beetles then strip the dead of their fur, roll their muscle into a ball, inoculate the ball with yeast and antibacterial compounds, bury the ball and lay their eggs near to the ball. Once planted by the beetles, the yeast begins to slowly eat the buried flesh, but, as they do, they also ward off other microbes, especially bacteria. As a result, when the beetles' babies hatch from their eggs and crawl to the ball of flesh, it has been fermented in such a way as to make it safe and delicious.[4]

Beetles benefit from their relationships with yeasts. But they are not the only such beneficiary. Wasps and the bees sometimes, and perhaps often, use the aromas of yeasts to find patches of sugars that they might otherwise have missed, patches of sugar in which (thanks to the alcohol produced by the yeast) pathogens are unlikely to be present. Like barflies they wing their way towards the rewarding aroma of fermentation.

Sex and Winter

Yeasts have needs, of course, beyond transportation and food. They need a way to survive in the times when food is scarce (such as winter), and they need a place to have sex. Here too, the insects can oblige. In the winter, some yeasts survive inside the bodies of the subset of insects that live more than one year. They wait there, patiently, until spring. As for sex, it appears to be especially likely to occur inside those same insects (or, in some cases, after having passed through insects). This sex inside insects is, recent studies have shown, particularly likely to be promiscuous. Different strains mate and so too, even, do different species. Inside a wasp, hybrids are made.[5]

Humans

Ancient humans evolved around 1.9 million years ago. At some point those ancient humans began to leave out some of their food in containers.[6] Wasps and bees came to the food and left, inadvertently, their yeasts. The food rotted. The food bubbled. The food fermented. At first in Africa and then elsewhere, ancient humans and then modern humans would come to control this rotting, bubbling, fermentation. They could use it to their advantage. Fruits, when fermented, yielded alcohol and new flavours. Roots, when fermented, became more palatable and flavourful. Meat, when fermented, could be stored, for weeks, months or even longer. No one knew that the agent of this fermentation was a fungus, much less a yeast, and yet the biological consequences of this yeast's metabolism were known. They could be smelled and tasted and remembered. They were not depicted and yet the foods and drinks they helped to engender were a kind of art, an art that in some cultures came to be holy. In Christianity, for example, the body of Christ is represented as leavened communion bread, the blood of Christ as fermented communion wine.

No precise tally exists, but it appears that hundreds and perhaps thousands of ancient and modern human cultures independently found ways to ferment foods. Each of these foods attracted different insects and were colonized by different yeasts. Some meat, for instance, was colonized by the yeast used by carrion beetles;[7] the yeast protected the meat gathered by people, just as it had protected meat gathered by the beetles. Fermented fruits attracted sugar-loving yeasts. But there was a little more to these twisting, tangling, rotting relationships.

Before humans, the story of yeasts and insects had unfolded in different places independently. It began tens of millions of years ago, and so evolved in separate directions on each continent or island, and even in one habitat relative to another. There are yeasts, for example, found riding in only one kind of insect on one Hawaiian island.[8] Similarly unique local yeasts and yeast–insect relationships exist or existed all around the world. As a result, when the first human fermenters began to rely on yeasts, they relied on and began to shape the evolution of different yeasts in different places.

Just Before Agriculture

At some point before agriculture, people in and around China and the Fertile Crescent began to rely more and more on fermentation.[9] As they did, they began to feed the yeasts on which they relied with ever larger quantities of sugars. They did this by gathering grains in baskets. They offered these baskets of food to the yeasts, the way one would offer gifts to the gods. Where necessary, they even treated the grains so that they would be easier for the yeasts to eat. They mashed them. They malted them. They added dates to them. They worked hard to prepare better meals for the yeasts and the other microbes in their fermentations so as to facilitate the transformation, a kind of rebirth, that they hoped would occur.

During this period, something new was happening. In China and the Fertile Crescent, and probably also in other places, humans were beginning to more fully take over the role of the insect. The humans carried the yeasts on their hands, in their bodies and in their vessels, from one place to another. They carried them while riding horses. They carried them while in ships. They carried them while walking in groups from one village to another. As this was occurring, humans were no more aware of their role as vectors of the yeasts than were the wasps.

With Agriculture

The benefits yeasts provide to insects are varied and depend on the context in which a particular yeast is found, as well as on the details of the biology of the insect itself. The benefits yeasts were providing to humans would also come to be varied. Yeasts helped to store food (by preventing food spoilage bacteria from growing) and to make unsafe drinking water safe (by killing the pathogens in the water with alcohol). Yeasts made unpalatable foods, such as some roots, not only palatable but delicious. Yeasts leavened bread. Yeasts, for humans, enriched the culinary world. But they also did something else – they produced alcohol. Thanks to obscure details of the machinery of their brains, humans are pleased by alcohol. They can become addicted to alcohol. In drinking alcohol, they want more alcohol. To produce more alcohol, they need to grow more grains or fruits to feed more yeasts. Agriculture, it has been repeatedly argued, emerged in part in order to produce enough food not for humans, but instead for yeasts; enough food for the yeasts so that the yeasts might provide something in return: alcohol.[10]

In the context of the scaling up of the production of alcohol, a subset of kinds of yeasts was disproportionately likely to survive. The subset included yeasts that grew quickly and yeasts that were tolerant of high concentrations of alcohol, and also produced high concentrations of alcohol. Such yeasts included the yeast *Saccharomyces cerevisiae*, brewer's yeast. Brewer's yeast is from a group of yeasts now thought to be native to China. Recent data suggest, and we hypothesize, that these yeasts from China spread, one fermentation at a time, to the Fertile Crescent. Human bodies and vessels carried them there, like wasp bodies but even more effectively.[11] Brewer's yeasts can live on bakers' hands.[12] Brewer's yeasts can live in beer drinkers' guts. From the Fertile Crescent, brewer's yeast began to spread with bakers, brewers, drinkers and eaters around the world. As it did, it became ever better at thriving in the conditions humans created, and ever better at outcompeting other yeasts. It became, inarguably, the single species on which humans were most predictably dependent: not wheat, not barley, not grapes, not rice, but brewer's yeast.

The Columbian Exchange

In 1492, Columbus set sail with three ships full of fermented foods and drinks. His ships carried fermented milk curds in the form of cheese. The cheese was alive with bacteria and fungi. His ships carried vinegar; the vinegar was alive with acetic acid bacteria and nematode worms (vinegar eels). His ships also carried salt-cured meat. That meat was covered with the bacteria and yeasts used in curing, such as the carrion beetle yeasts. And his ships included extraordinary quantities of wine, and hence the yeast and bacteria still living in the wine. The microbes from these foods and drinks appear to have spread everywhere in the ships, on the sailors and in the sailors, on the wood and in the wood. The ships were described as smelling 'frightfully of mules',[13] but must also have borne the aromas of yeasts and bacteria. This would have been true not only of Columbus's ships, but also the ships of each later wave of colonists. Consider the *Mayflower*, the ship that carried the pilgrims to New England. The *Mayflower* had been used, prior to its journey to North America, to carry wool from England to France, and wine from France to England. Ships that made this journey were called 'sweet ships', because the wine that spilled during the transit from France to England soaked into the wooden hull, making it smell sweet, or at least sweeter than it might otherwise have smelled.[14]

When Columbus set sail, the microbes he brought with him to the Americas are likely to have included European strains of brewer's yeasts. The same is likely to have been true of later colonists, including even the puritanical pilgrims on the *Mayflower*. Before Columbus arrived, there were thousands of kinds of fermentations being carried out in the Americas: fermented corn, fermented cactus, fermented cassava, fermented cacao. Those fermentations would have been carried out by different yeasts in different places. However, we do not fully know which yeasts were involved in those fermentations. This is in part because many of these fermentations have been poorly studied. But it may also be because the yeast that the colonists brought with them were so successful in their spread that they overtook the native yeasts. The story is muddy, and yet it appears as though European yeast strains spread to the Americas and around the world like cows and smallpox, like sugarcane and rum, like wheat and the conquistadors. At least that is our understanding as of today. It is certainly the case in New Zealand, where all the strains of *Saccharomyces* that are present in the country stem from colonization events from Europe.[15]

Isolation

Most of the people in the story of yeasts have become, thanks to the passage of time, anonymous, just as have most of the yeast species – most, but not all. In France, Louis Pasteur discovered germ theory not by studying pathogens (at least not initially) but instead by studying alcoholic fermentations, first beet juice and then beer and wine. Building on the work of other scientists, Pasteur provided the most unambiguous evidence that fermentation was due to living organisms in general, and to yeasts in particular. He discovered that if he sterilized grape juice that it would not ferment. Meanwhile, if he added in the organisms found in grape juice, it would. He then isolated and identified those organisms, including yeasts, which he then depicted. Pasteur's first germ theory, inspired by these yeasts, was that the germ of microbial life was required to make fermented foods. It was only later that he focused on 'germs' as the cause of disease, and even then his initial consideration was on 'diseases' of wine caused by organisms that competed with the yeasts. Ever the good caretaker of his yeasts, he focused first on sick fermentations and then only much later on sick humans.

In Copenhagen, inspired by Louis Pasteur, Emil Hansen at the Carlsberg Brewery in Denmark cultured and observed the yeast that he found in lager beer that Carlsberg was brewing (using cultures originally from the Spat brewery in Munich). He then developed technology for growing those yeasts in pure culture, on their own. As a result, Carlsberg was able to make beer that relied not just on a single yeast, but on a single strain of a single kind of lager yeast, *Saccharomyces pastorianus*. Lager yeast is a hybrid perhaps originally formed in the body of an insect, though no one knows. It is a mix of brewer's yeast and a yeast, *Saccharomyces eubayanus*, from Patagonia, and it appears to have travelled back to Europe on the colonizers' ships. Its existence may be an artefact of both conquest and a German beer law that prevented brewers in southern Germany from brewing during the summer and so, inadvertently, favoured a cold tolerant yeast hybrid. Its isolation brought Carlsberg great wealth, but it also, as it turns out, began a process that would precipitate loss.

The strains of hybrid lager yeast cultured by Hansen spread, brewery to brewery, around the world. Something similar would come to happen with the few strains of *Saccharomyces cerevisiae*, brewer's yeast, which came to be used to make ale beers around the world, but also bread and wine and much else. Before these strains were isolated, the lager yeasts and brewer's yeasts in different places were slightly different, reflecting unique histories. The scientists at the Carlsberg brewery and elsewhere in Europe helped to obscure those differences. In doing so, they helped to favour two of the most successful eukaryotic organisms ever to live. A single brewery contains more cells of the most popular strain of brewer's yeast, for instance, than the number of humans ever to have lived. By any broad perspective, brewer's yeast and its hybrid, lager yeast, domesticated humans and led us to create a world in which it most thrives at the expense of other yeasts, and in which the foods it needs, the plant foods, thrive at the expense of other plants. The area now planted in grapes for wine is bigger than some countries. The area planted in barley for beer-making is bigger still – all to feed the yeast to make the alcohol that pleases us, sometimes at our own expense.

Visual History

This chapter is printed in a book featuring stories about loss of species, and visualizations of those species. With yeasts, there have been losses that we will shortly consider. But these losses are hidden from the visual story of yeasts.

The visual story of yeasts is largely the story of yeasts depicted in the form of the foods and drinks they create, yeasts depicted as a function of their actions. In every culture that is known to have relied heavily on yeasts, the products of yeasts are featured in art. The Egyptians depicted bread and beer, again and again. The Etruscans sculpted bowls of fermented drink onto tombs, so that the dead might sip from them. Medieval art, to the extent to which it featured any food, often included yeast-fermented wine. In the Renaissance, fermentation was everywhere. In Leonardo da Vinci's *Last Supper*, the scene is rich with gestures but also with wine glasses and leavened bread. Later, when Dutch painters began (radically) to replace religious figures and kings with still life at the centre stage of paintings, they did so with a focus on fermented foods, and nearly always included wine or beer, and implicitly its yeasts.

Nearly every art museum in the world features the products of yeasts in one way or another. One painting, however, of which we are especially fond, sits in the Danish National Art Museum. It is a painting by the Danish artist Carl Bloch (Fig. 14.1). The painting features a busy scene in an osteria in Rome in which two women are drinking with a man. The story

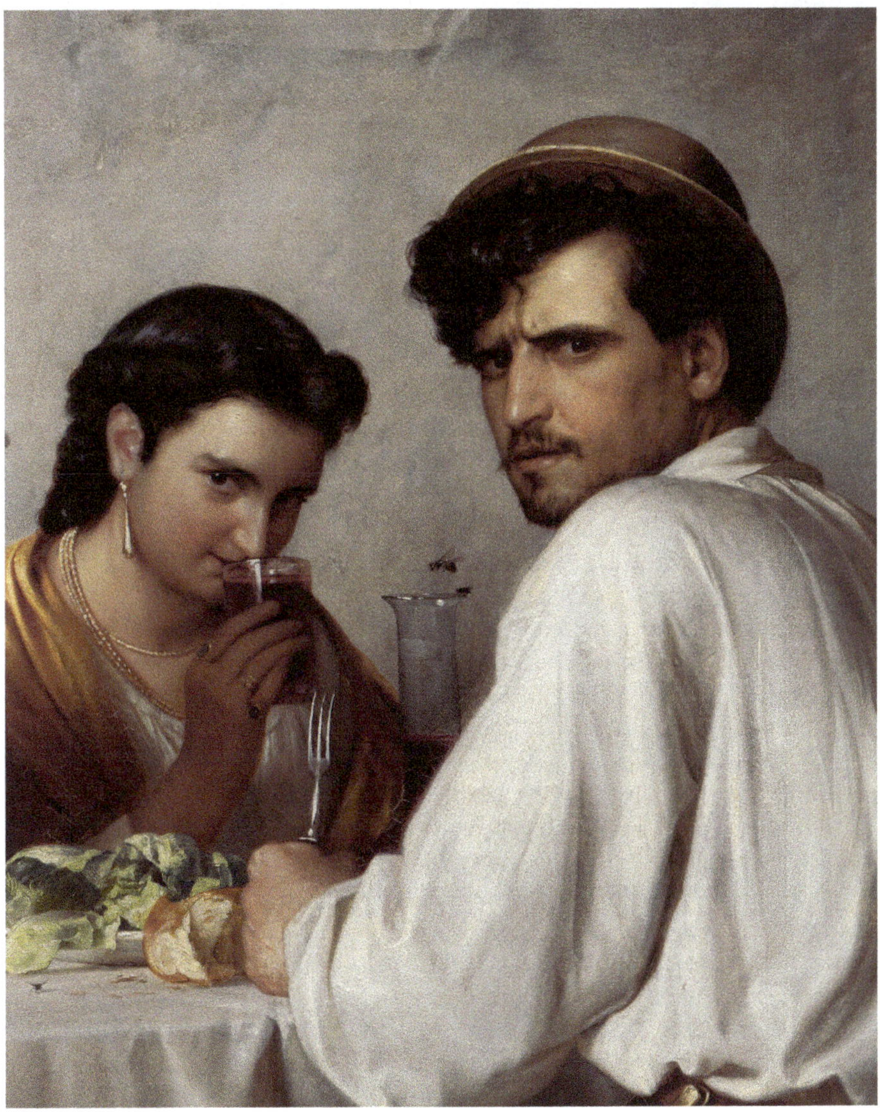

Illustration 14.1 Carl Bloch, *In a Roman Osteria* (detail), 1866. Photo by Robert Dunn.

of the painting is about the human protagonists in the scene, and also the interaction between the people and the painting and you, the viewer. But in the photo of the painting that we have shared here, we have zoomed in a little. In doing so, we reveal what is, to the painting, a subplot; and yet, as we have suggested here, a very important one. Right in the middle of the painting are glasses of wine, filled with yeasts. And hovering not far from one of those glasses is a wasp or a bee. This is one of the few pieces of art of which we are aware in which the ancient relationship between wasps, bees and yeasts, and the modern relationship between humans and yeasts, are both depicted. As for the yeasts themselves, they are hidden. Like the interaction between the man and the women or between the viewer, the man and the women, their power is implied.

What Was Lost

For as much as the products of yeast are often featured in art, the yeasts themselves rarely are. Of course, yeasts do not really care if we paint them. To the extent to which they know pleasure, theirs is simple. They enjoy being fed, cared for and ushered into the future. But because we have not depicted the yeast cells, we are relatively unable to tell which types of yeast cells might, over time, have gone missing. Meanwhile, the reason we have not depicted these yeast cells is simple – their beauty and differences are not visually conspicuous. Yeasts are visually boring. They are round. Each one is like a kind of planet, its roundness interrupted only by the umbilical scar marking the point at which, in birth, it tore away from another, identical such planet. An example here makes the point. One of us (Rob) recently asked the world expert on yeasts whether a yeast cell discovered in an ancient bread was a *Saccharomyces* cell or a yeast of some other kind. He could not tell. 'They nearly all', he said, 'look like that. Round.' The trick here is that the wondrous differences among yeasts are not with regard to their appearance; they are with regard to their consequences.

Yet, when yeasts are painted just in the form of their consequences, the foods they create, their details are hidden. In a wine glass, from a distance, no matter how well it might be painted, a rare yeast from the Amazon and the most common brewer's yeast both look the same. Their differences, even when represented in foods and drinks, are not visual. They are instead in the form of aromas and flavours.

Human noses are sophisticated. They are central to how we learn and learn to love. And yet, they are not as central to how we describe and depict the world as our eyes. Were social insects to have invented art, masterworks to grace their societies, things might have been different. Beehives might feature cells dedicated to the pleasure of aromas and the ways in which they can be mixed and set against each other to evoke responses. 'This,' they might say, 'this combination makes the queen weep. This, this triggers her awe.' But maybe we do not need to look so far, because in our own societies, are not such cells just what culinary artists have created; is not the kitchen the place in which aromas have come to be manipulated, along with tastes, to alter us, not as the viewer, but instead as the taster, as the smeller, as the one before the food. Is not a great kitchen somehow an ephemeral sort of Louvre?

The art of the kitchen features yeasts. What is more, we have words to describe the ways in which different yeasts offer us different experiences. During the same years in which Pasteur was revealing the consequences of the invisible world of microbes, the word *terroir* was brought from French into English to refer to the flavour in wine offered by the details of the land on which the grape was grown and the ways in which the wine was made. The word terroir was and is a way to capture unnameable details. Food writers have come, with time, to talk about the terroir of not only wine, but also cheese, bread and even kimchi. In the years since Pasteur, we have learned that a big contributor to terroir of a particular fermentation is the mix of microbes involved in that fermentation, including the strains of yeasts. In recent comparisons, wines made from the same grapes but different yeast strains have been shown to be more different in their flavours and aromas than wines made from different grapes but the same yeast strains. In this way, we might say that terroir is partially about yeasts and their differences, and so as we have homogenized yeasts, we have extinguished many unique types of terroir and, so too, whole galleries of culinary art, or at least the potential for such art. But there is a little more here.

Because it is hard to record culinary art, to memorialize it for posterity, we do not know what we have lost as yeasts have become homogenized. We know little with regard to which yeasts were lost as Chinese, Middle Eastern and later European cultures spread brewer's yeasts or lager yeasts in general, or the domesticated strains of those yeasts in particular, around the world. No doubt old strains of brewer's yeast and lager yeast disappeared. Some of those missing yeasts might still be found, lurking in old vessels, buried alive beneath a crumb of malt. But the bigger loss is likely to have been strains of other species and genera of yeasts, domesticated here and there around the world. It does not seem a stretch to imagine that there were ten thousand human cultures that relied on yeasts for fermentation a thousand years ago. In all likelihood, each of those ten thousand cultures relied on a subtly (or even very) different yeast or set of yeasts for its fermentation. But they also knew how best to use those yeasts, how to favour them, how to call to them for their services. No one has gone to look for these yeasts with any earnestness. Which were the precolonial yeasts of the Americas? We do not know. Which were the precolonial yeasts of Africa? We do not know. When did brewer's yeasts arrive in the Fertile Crescent, and what was there before? We do not know. These missing microbes are implied in the ancient art of diverse cultures, implied in the depictions of the products of yeasts the way that a baker is implied in a painting of bread, a cook in a depiction of a stew, a fisherman in the fish hanging on a hook. They are implied and yet unknown, lost ingredients of our collective culinary art.

We could attempt to recreate lost yeasts by finding them in nature and culturing them, and then making new foods and drinks from them. We could look for them in the little islands of sugars out in the wild. Or we could search among the insects. Recently, Anne Madden has undertaken the latter approach. In doing so, she was able to isolate a yeast from wasps that can be used to make, on its own, sour beers, which normally require the use of a mixture of yeasts and bacteria. Recently, Anne, along with collaborators, has also found tens of yeasts in insects capable of making new kinds of breads.

Anne's work is exciting. But it depends upon the insects. Anne can only recover yeasts from insect species that still exist. Many no longer do.[16] One recent study found 70 to 80 per cent declines in insect biomass in parks in Germany. No one is sure why. Land use change? Pesticides? Climate change? Maybe all of these things. As these insects have declined in Germany and as insects have declined elsewhere around the world, have their yeasts been lost[17] via coextinction? Coextinction is the process in which one species goes extinct due to the loss of a species on which it depends.[18] What happens to a yeast if it no longer has an insect species to carry it from here to there? Some yeasts, no doubt, find new modes of transport. Others disappear.

As we lose insect species, we lose some of the yeasts that ride them. As we lose the yeasts that ride them, we lose the potential to create new kinds of culinary art, arts that have never been savoured before. If the yeasts in a particular patch of land contribute to the terroir of the wine that can be made on that land, we might also say that the insects in any land contribute to the terroir of that land; in their movements of yeasts and in providing places for yeasts to have sex, they are part of the terroir. Theirs is the entomo-terroir, the wriggling, writhing, many-legged secret to the sauce.

After

Yeasts have existed for more than a hundred million years. And while individual yeast strains and species are threatened by the ways in which we have altered the world, yeasts as a whole will live on. After humans, someday, become extinct (all species do, this is one of nature's true laws), at least some species of yeasts will still be around, taking whatever ride they can get from sweetness to sweetness. Until that time, we have some control over which yeast species we carry with us into our future. We, the three authors of this chapter, would like to think that humans will find reasons to save a rich diversity of wild yeasts and the species they depend on, with which we might make more culinary art. A paper wasp carries wild yeasts from the autumn until the spring. We need to find a way, in the context of our global society, to carry the paper wasps, and millions of other species, from this year to the next. What attracts the wasp to the yeast is the aroma of the yeast. What attracts us to the rest of life is something less concrete. It is hard to describe, and yet ancient fondness for the living world, a fondness that when sated pleases us, is a fondness on which we and many other species ultimately depend.

Acknowledgements

Special thanks to Delphine Siccard and Caiti Heil, who read and commented on this chapter.

Robert R. Dunn is professor of applied ecology at North Carolina State University and in the Center for Evolutionary Hologenomics at the University of Copenhagen. He is the author of numerous books including the book he co-authored (with Monica Sanchez) *Delicious: The Evolution of Flavor and How It Made Us Human* (2021).

Monica C. Sanchez is a medical anthropologist who studies the cultural aspects of health and well-being. Her most recent book (co-authored with Robert Dunn) is *Delicious: The Evolution of Flavor and How It Made Us Human* (2021).

Matthew Morse Booker is associate professor of history at North Carolina State University and vice president for scholarly programs at the National Humanities Center in North Carolina. He publishes in environmental and food history, including *Down by the Bay: San Francisco's History between the Tides* (2013) and *Food Fights: How History Matters to Contemporary Food Debates* (2019).

Notes

1. Madden et al., 'The Ecology of Insect–Yeast Relationships'.
2. Stefanini et al., 'Role of Social Wasps'.
3. Stefanini, 'Yeast–Insect Associations'.
4. Shukla et al., 'Microbiome-Assisted Carrion Preservation'.
5. Stefanini et al., 'Social Wasps Are a Saccharomyces Mating Nest'.
6. Amato, *Current Anthropology*.
7. Patrignani et al., 'Role of Surface-Inoculated *Debaryomyces Hansenii*'.
8. Lachance et al., 'Metschnikowia hamakuensis sp. nov.'.
9. Liu et al., 'Fermented Beverage and Food Storage'.
10. Katz and Voigt, 'Bread and Beer'.
11. Pontes et al., 'Revisiting the Taxonomic Synonyms'.
12. Reese et al., 'Influences of Ingredients and Bakers'.
13. Nuttall, 'Gaspar de Portolá'.
14. Philbrick, *Mayflower*.
15. Gayevskiy, Lee and Goddard, 'European Derived *Saccharomyces cerevisiae*'.
16. Dunn, 'Modern Insect Extinctions'.
17. Hallmann, 'More Than 75 Percent Decline'.
18. Dunn et al., 'The Sixth Mass Coextinction'.

Bibliography

Amato, Katherine R., Elizabeth K. Mallott, Paula D'Almeida Maia and Maria Luisa Savo Sardaro. 'Predigestion as an Evolutionary Impetus for Human Use of Fermented Food'. *Current Anthropology* 62, no. S24 (2021): S207–S219.

Dunn, Robert R. 'Modern Insect Extinctions, the Neglected Majority'. *Conservation Biology* 19(4) (2005): 1030–36.

Dunn, Robert R., et al. 'The Sixth Mass Coextinction: Are Most Endangered Species Parasites and Mutualists?' *Proceedings of the Royal Society B: Biological Sciences* 276(1670) (2009): 3037–45.

Gayevskiy, Velimir, Soon Lee and Matthew R. Goddard. 'European Derived *Saccharomyces cerevisiae* Colonisation of New Zealand Vineyards Aided by Humans'. *FEMS Yeast Research* 16(7) (2016): fow091.

Hallmann, Caspar A., et al. 'More Than 75 Percent Decline over 27 Years in Total Flying Insect Biomass in Protected Areas'. *PloS one* 12(10) (2017): e0185809.

Katz, Solomon H., and Mary M. Voigt. 'Bread and Beer'. *Expedition* 28(2) (1986): 23–34.

Lachance, Marc-Andre, et al. '*Metschnikowia hamakuensis* sp. nov., *Metschnikowia kamakouana* sp. nov. and *Metschnikowia mauinuiana* sp. nov., Three Endemic Yeasts from

Hawaiian Nitidulid Beetles'. *International Journal of Systematic and Evolutionary Microbiology* 55(3) (2005): 1369–77.

Liu, Li, et al. 'Fermented Beverage and Food Storage in 13,000 Y-Old Stone Mortars at Raqefet Cave, Israel: Investigating Natufian Ritual Feasting'. *Journal of Archaeological Science: Reports* 21 (2018): 783–93.

Madden, Anne A., et al. 'The Ecology of Insect–Yeast Relationships and Its Relevance to Human Industry'. *Proceedings of the Royal Society B: Biological Sciences* 285(1875) (2018): 20172733.

Nuttall, Donald A. 'Gaspar de Portolá: Disenchanted Conquistador of Spanish Upper California'. *Southern California Quarterly* 53(3) (1971): 185–98.

Patrignani, Francesca, et al. 'Role of Surface-Inoculated *Debaryomyces Hansenii* and *Yarrowia Lipolytica* Strains in Dried Fermented Sausage Manufacture. Part 1: Evaluation of Their Effects on Microbial Evolution, Lipolytic and Proteolytic Patterns'. *Meat Science* 75(4) (2007): 676–86.

Philbrick, Nathaniel. *Mayflower: A Story of Courage, Community, and War*. New York: Penguin, 2006.

Pontes, Ana, et al. 'Revisiting the Taxonomic Synonyms and Populations of *Saccharomyces Cerevisiae*: Phylogeny, Phenotypes, Ecology and Domestication'. *Microorganisms* 8(6) (2020): 903.

Reese, Aspen T., et al. 'Influences of Ingredients and Bakers on the Bacteria and Fungi in Sourdough Starters and Bread.' *Msphere* 5(1) (2020): e00950-19.

Shukla, Shantanu P., et al. 'Microbiome-Assisted Carrion Preservation Aids Larval Development in a Burying Beetle'. *Proceedings of the National Academy of Sciences* 115(44) (2018): 11274–79.

Stefanini, Irene. 'Yeast–Insect Associations: It Takes Guts'. *Yeast* 35(4) (2018): 315–30.

Stefanini, Irene, et al. 'Role of Social Wasps in *Saccharomyces Cerevisiae* Ecology and Evolution'. *Proceedings of the National Academy of Sciences* 109(33) (2012): 13398–403.

———. 'Social Wasps Are a Saccharomyces Mating Nest'. *Proceedings of the National Academy of Sciences* 113(8) (2016): 2247–51.

Part V

Representing Extinct Mammals

Chapter 15

Animal Extinction, Film and the Death Drive

Barbara Creed

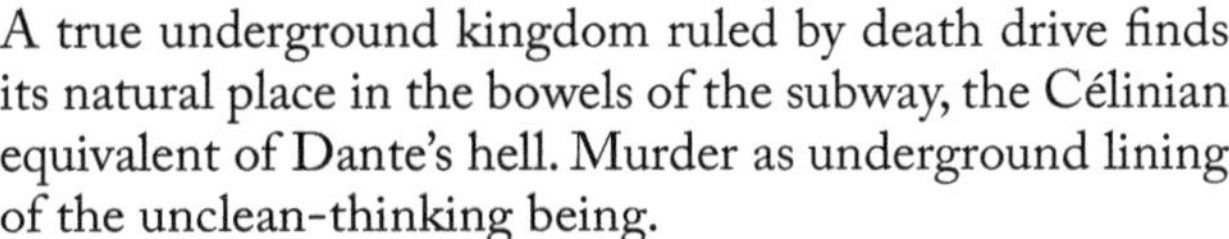

A true underground kingdom ruled by death drive finds its natural place in the bowels of the subway, the Célinian equivalent of Dante's hell. Murder as underground lining of the unclean-thinking being.
—Julia Kristeva, *Powers of Horror*

Faces of elephants, monkeys, pigs, lions, seals, ostriches, alligators and bears stare out of the frame at the viewer in Nicolas Philibert's remarkable documentary of the restoration of the Galerie de Zoologie in Paris. The camera places us in intimate contact with the inhabitants of this strange menagerie. Although glass, their 'animated' eyes endow their stilled bodies with life. These are the myriad of dead animals who have been killed, skinned, tanned, stuffed or mounted and brought back to life with glass eyes, make-up, fur patches and needle and thread in order to create a simulacra of life. Uncanny spectres, they haunt the imagination. Taxidermy, meaning 'arrangement of the skin', is a method used to record a life-like impression of a living creature or species, including those that are extinct and threatened. It is sometimes used to memorialize pets. In his acclaimed film, *Un animal, des animaux*, 1996 (animals, more animals), Nicolas Philibert explores what took place in 1994 during the refurbishment and re-opening of the Galerie de Zoologie of Le Muséum national d'Histoire naturelle (The French National Museum of Natural History), which had been closed since 1965. The museum's new design is no doubt intended to allow spectators to walk amongst the exhibits, secure in the knowledge of the superiority and power of the human species over all others. What prompts humanity's need to control nature and to replace her vast kingdom of marvellous and endlessly diverse species with its own uncanny kingdom of the dead, preserved and taxidermied? The relationship between natural history museums and extinction has in recent years assumed a central place in demonstrations by the global-wide Extinction Rebellion movement. In April 2019, protesters took over the Natural History Museum in London and lay down beneath the vast hanging skeleton of the famous Blue Whale.

This chapter will discuss animal extinction in three animal documentaries directed by Nicolas Philibert in relation to the context of the human death drive, as presented in the psychoanalytic theories of Sabina Spielrein, Sigmund Freud and Julia Kristeva. The death drive has played a little understood role in the extinction of species through hunting, science, and the creation of natural history museums and zoos. The chapter will also consider the artistic practice of Australian artist Janet Laurence, who creates a very different testimony to death and extinction in her many exhibitions of the natural world. Both Philibert's and Laurence's films and installations represent extinction in the context of what has become known, in the Age of the Anthropocene, as 'ecological grief'. It will ask whether or not the human death drive has made the coming of the Anthropocene and the extinction of species inevitable – including the human. Finally, it will explore the connection between art, practices of extinction and personal revolt.

Established in 1635, the French National Museum of Natural History is now composed of over fourteen sites, encompassing three zoos, two museums, four scientific sites and three botanical gardens. The main museum is in Paris. Its aims are to create collections, conduct scientific research and educate the public. The Galerie de Zoologie was established in 1889 with over one million specimens. It is impossible to view Philibert's *Un animal, des animaux* without thinking of the operations of French colonialism that permitted the creation of such a vast empire of dead animals, fossils and plants, taken from so many countries. Filmed in Paris between 1991 and 1994, Philibert's film focuses on the meticulous repairing and restoring of the museum's immense collection. Over a three-year period, taxidermists restored over a thousand animals – mammals, birds, fish and reptiles. Philibert carefully organizes the way in which he films the animals. The opening scene is of various animals (zebras, a polar bear, a group of deer) being transported to their newly renovated home in an open van, so we see them clearly as they journey to their destination – almost as if they were alive and eagerly anticipating their arrival. There is almost something comical, perhaps surreal, about the moment.

As Adrian Danks[1] notes, Philibert's restaging does not follow the principles that informed the strict arrangement of specimens in the Grande Galerie de L'Evolution. Instead, Philibert focuses on individual animals taken from a mix of periods and countries – their faces and expressions preserved through taxidermy to create a sense of their individual uniqueness. His approach is informal, undermining a sense of colonial achievement, order, science and ceremony that underpins the patriarchal Symbolic order – the order that has brought into being what we now call 'civilization'. Throughout, Philibert frames the faces and the gazes of these animals as they are repaired and resurrected, as if they were also watching us; we are encouraged to exchange looks with them as if they were alive. Philibert does not take a position or reveal what he might think about taxidermy; he leaves this to the viewer. Philibert's documentary demonstrates how such a museum offers a way of staring death in the face, the death of non-human species, without having to experience death in the real. Confronting images of dead animals, reconstructed to appear as if they were alive, can also be a very distressing, even abject, experience.

Philibert's documentary is accompanied by a shorter film, *Dans la peau d'un blaireau*, 1994 (In a badger's skin), which shows in intimate detail a taxidermist preparing a badger for display. We see how a dead badger is skinned, stuffed and restored to look as if it were alive. Despite the brutality that drives the whole process of killing animals and restoring them to look lifelike, the film reveals each process in an objective and understated way. The horror of what has happened to the lifeless badger speaks for itself. The animal is simply an object, a thing to be eviscerated and manipulated in order to uphold the scientific and philosophical goals of the museum. The wild animal is, like nature, a life form that serves civilization in the latter's desire to separate itself out from the natural world. An institution of the patriarchal Symbolic world of law and language, natural history museums around the world, which are responsible for these macabre collections of the dead, exist alongside, but in opposition to, the world of nature, which operates according to its own laws, which are aligned with the body and the visible workings of the natural life–death cycle.

The natural history museum is a testament to what the Russian psychoanalyst, Sabina Spielrein, in 1912, named the death instinct, a concept developed later by Sigmund Freud into his theory of the death drive (*Todestrieb*).[2] Spielrein proposed her concept, which she related to reproduction, in a paper entitled 'Destruction as the Cause of Coming into Being'.[3] Spielrein was one of the first women psychoanalysts and is known as a pioneer of the movement. Although largely forgotten, her remarkable achievements have been brought to the fore with the publication of her work, entitled *The Essential Writings of Sabina Spielrein: Pioneer of Psycho-*

analysis (2018). In *Beyond the Pleasure Principle* (1920), Freud stated (in a now-famous footnote)[4] that Spielrein's paper had inspired his thinking that led to his concept of the death drive. Spielrein conceptualized destructiveness as aiding the reproductive instinct rather than being an instinct per se. She proposed that the sex drive comprised both an instinct of destructiveness (with sadistic components) and one of transformation. According to Fátima Caropreso, Spielrein sees destruction as essential to the act of 'coming into being', which is complemented by a 'static' drive and a 'dynamic' drive. It is Spielrein's static drive that Freud develops in his theory of the death drive as one that subsumes the dynamic drive and the desire for 'resurrection' or transformation. It was Freud who developed the concept of the 'static' drive into what Caropreso describes as a 'purely negative drive' that seeks 'the annihilation of life'.[5]

In explaining his concept of the death drive in his 1920 essay 'Beyond the Pleasure Principle', Freud argued that there is a tendency in all organic forms of life to return to an inorganic state. He proposes the hypothesis '[t]hat all instincts tend towards the restoration of an earlier state of things'.[6] He states that 'the aim of all life is death'.[7] This drive is stronger than the instinct for pleasure. The death instinct can be turned inwards, leading to self-harm, or outwards, resulting in harm to others. It is expressed through a range of actions such as self-destructiveness, aggression and repetition compulsion; hence it is kept in check by the superego, but when it imposes itself too harshly on the ego, it can lead to a 'pure culture of the death drive'.[8] In a subsequent essay, 'The Future of an Illusion' (1927), Freud connects the death drive to nature, which man views as the enemy of civilization and his own happiness. Man believes he must engage in a continuous war with nature in the face of her 'majestic, cruel and inexorable' powers such as earthquakes and floods, which destroy his achievements. Nature also mocks man because she is the cause of 'the painful riddle of death'.[9]

In a later essay, 'Civilization and Its Discontents' (1930), Freud explores the death drive in greater detail. Man believes control of nature will bring greater happiness, and this includes the extermination of 'wild and dangerous animals'.[10] With the development of civilization, the totemic animal lost its sacred powers, which were transferred to other gods, and the animal became a thing to be hunted and killed. When turned outwards or 'diverted towards the external world', the death drive becomes 'an instinct of aggressiveness and destructiveness'.[11] Freud argued: 'It is clearly not easy for men to give up the satisfaction of the inclination to aggression. They do not feel comfortable without it'.[12] Man's aggressiveness towards nature and wild animals is an expression of his death drive turned outwards. Freud's conclusion is grim: 'Men have gained control over the forces of nature to such an extent that with their help they would have no difficulty in exterminating one another to the last man'.[13]

Significantly, Freud does not appear to include women in this struggle against nature. Women he says are 'hostile' to civilization. Because the 'work of civilization has become increasingly the business of men',[14] women are confined to the domestic arena. Here Freud is referring to what Jacques Lacan later theorized as the Symbolic order, the realm of law and language, by which civilization is organized and sustained. Freud does not refer to woman's relationship with nature. As I will discuss shortly, the Symbolic however locates woman, along with nature and the animal, outside the Symbolic. Importantly, Freud notes that despite everything, man is not happy; he does not enjoy civilization because of all that he must repress for civilization to function, such as his libido, bisexuality, natural aggression and his desires to inflict death on his enemies. According to Freud, given the degree of repression, 'may we not be justified in reaching the diagnosis that ... some civilizations, or some epochs of civilization – possibly the whole of mankind – have become neurotic?'[15]

The death drive, which Freud argues 'can also be turned outwards', sheds light, I argue, on the unconscious drives and destructive forms of human behaviour that inform the very establishment of the natural history museum. Dedicated to scientific research, the museum is also a mausoleum designed to control all of nature and her myriad of species. The death drive helps to explain the countless acts of human aggression that have led to the deaths of so many animals; the compulsion to repeat (to kill, stuff and display one species after another); the display of dead bodies of animals as if they were alive, thus invoking dread associated with the uncanny; the overwhelming sense of stasis; and the prevailing atmosphere of melancholia in the museum, which arises from the unconscious feeling that no matter how many animals are killed it will never be enough to resolve the unconscious fear of one's own death. It is as if the human species has unconsciously displaced its own death drive onto all non-human species as a way of staving off knowledge of, or the advent of, its own death – for which, Freud argues, man blames nature. The mind, however, substitutes a new aim to explain the need for the mass slaughter of animals – such are the demands of science. Freud points to science and technology as evidence of man's greatest achievements, of his 'omnipotence and omniscience'.[16]

In *Powers of Horror*, Julia Kristeva draws on the concept of abjection to rework Freud's concept of the death drive. Kristeva offers yet another way of thinking about extinction, the role of the animal in nature and its significance for the civilized human world. Kristeva explores further Freud's argument that woman is hostile to civilization, because she is excluded. Kristeva asks what this might mean for woman. How is she excluded? Here

Kristeva proposes her theory of the abject to explain woman's marginal-ization.[17] Kristeva's conclusion is that the Symbolic order actively excludes woman (she does not set herself apart in the way Freud describes) – that is, abjects woman – because of her association with procreation, the body, the animal and nature. Because she is unclean, she also becomes the object of male aggression and the death drive. It could be argued, as Klaus Theweleit does, that in extremely masculine and fascist cultures masculine identity is moulded by their dread and fear of women.[18]

Nature, which threatens to undermine culture, must be separated out from all that is civilized and male. Woman is associated with bodily fluids such as menstrual blood and breast milk, as well as her infant's excreta and vomit, which are regarded as unclean and abject (Freud also refers to the 'taboo on menstruation' and the disgust aroused by infant excreta and other strong bodily smells).[19] By contrast, the clean and proper body of the Symbolic (that is the male body) is one that shows no sign of a debt to, or association with birth, nature or the animal: 'The abject confronts us, on the one hand, with those fragile states where man strays onto the territories of animal'.[20] The child must eventually learn to reject the abject mother and her world, as well as nature and the animal, in order to enter the paternal world of civilized behaviour and values – language, the law, culture and sociality.

> The abject confronts us, on the other hand, and this time within our personal archaeology, with our earliest attempts to release the hold of *maternal* entity even before existing outside of her, thanks to the auton-omy of language. It is a violent, clumsy breaking away, with the constant risk of falling back under the sway of a power as securing as it is stifling.[21]

Kristeva focuses on woman's exclusion from the Symbolic: 'The difficulty a mother has in acknowledging (or being acknowledged by) the symbolic realm – in other words, the problem she has with the phallus that her father or her husband stands for ...'.[22] As the child leaves this early pe-riod, he/she begins to acquire language and enter the paternal Symbolic, a monolithic uniform system that has clear boundaries, particularly between self and other – boundaries that are permanently threatened with collapse.

According to Clifford Davis, Kristeva's theory creates an abject form of the death instinct associated with the maternal: 'It could be seen as a rather horrifying intensification or transmutation of the Freudian concept of the death instinct ... Significantly, it is the very act of exclusion by the super-ego that transforms the maternal object into the subversive, horrifying abject'.[23] In the context of this gender opposition, the death drive, when turned outwards, could be seen as directed towards those others defined by the Symbolic as representing the unclean and improper, such as women, ethnic minorities, and the animal. These are the unclean and improper who must suffer in place of the clean and proper upright (male) subject of the Symbolic, who nonetheless unconsciously harbours a belief that he too is unclean, particularly given the abject nature of his birth; hence femininity and abjection are within, and undermine from within.

Kristeva argues that a journey into abjection can lead the subject to revolt.[24] 'In abjection, revolt is completely within being. Within the being of language. Contrary to hysteria, which brings about, ignores, or seduces the symbolic but does not produce it, the subject of abjection is eminently productive of culture.'[25] What this new or transformed culture may look like remains open to debate; the films discussed in this chapter, however, offer possibilities for thinking about change through their representations of nature and the animal. According to Davis, 'Kristeva identifies the monolithic patriarchal Symbolic with all cultural institutions'.[26] In my view, two of the most significant are the natural history museum and the zoological institution. From the perspective of the Symbolic, the function of these institutions is, I have argued, to assert mastery over nature and the animal (and by extension, woman) in order to curb the threat they offer to the proper functioning of the patriarchal Symbolic order of law and language.

Kristeva's theory of the operations of the death drive offers one way of understanding how the human death drive impacts woman, the natural world and its animals. It makes sense of practices such as big game hunting (murdering the threatening abject animal), taxidermy (replacing the animal's abject insides with clean, bloodless stuffing), classifying and naming the specimens (creating order from the abject chaos of nature), and placing captive animals behind the walls and bars of zoos (controlling nature) to live a bare and often torturous existence. The history of animal abuse in zoos and amusement parks is horrific. The shocking executions of captive and performing elephants, for instance, who killed their often-cruel trainers, attest to the sadistic cruelty of the human death drive when displaced onto animals. Chunee, Topsy and Mary died by firing squad, electrocution and hanging, respectively, for killing their trainers when under extreme duress.[27]

The third film in Philibert's cinematic investigation into human/animal relationships is set in the zoo of the Jardin des Plantes in Paris. *Nénette* (2010) is a documentary about a forty-year-old female orangutan, who was born in 1969 in the forests of Borneo, taken captive and dispatched to the menagerie of the Jardin des Plantes, where she arrived in 1972. She has spent all her adult life in captivity, along with other orangutans. Nénette has given birth to four offspring. In 2019, on her fiftieth birthday, she took up painting. She is a great favourite with zoo visitors, some of whom visit her daily. They know her by name and stand close to the glass wall as they talk, whispering their secrets, while commenting on her life, as they might with any other close relative. She appears to listen intently to their confidences.

Since 2008, the Bornean orangutan, has been listed as critically endangered. Their population has shrunk by 60 per cent as a result of illegal hunting, demand from zoos for exhibits, the outlawed pet trade, and destruction of their habitat by the palm oil and rubber industries. Nénette may be among the last of her kind. She offers an example of a living creature, captured when her species were coming under threat of extinction, and still living as a captive animal in a land very distant from her home.

Philibert's style of filmmaking undermines anthropocentrism – he is focused not on the human but on the animal. His masterstroke, in constructing his documentary, is to show only Nénette on screen. As with *Un animal*, the subject of Philibert's *Nénette* is the animal herself. Her human visitors exist only as opaque reflections on the screen. We do, however, hear what they are saying; it is almost as if Nénette were their personal psychoanalyst. Nénette's visitors speak of Nénette's loneliness and of their own, of her yearning for home and of their own. One says: 'I think she's depressed'. She is shown alternatively eating yoghurt, covering herself with a blanket when she desires privacy, apparently listening intently to the words of her many visitors, even seeming to mimic their expressions from her side of the glass. Philibert says that the visitors come to have fun: 'But after half a minute here, they stop looking. Because they are struck by something more tragic. They start thinking about the situation of these animals in the wild and about what we are doing with our planet'.[28] One keeper says: 'All of us working in zoos share an inner sense of deep-seated guilt'.

In addition, Nénette offers a different threat – that of abjection. She is both included and excluded from the human domain. She is included as a captive, with seemingly endearing behaviours and alluring personality, while simultaneously excluded because of her abject animal appearance and potentially dangerous nature. Her visitors and keepers remark that she is 'enormous', with long red hair and sagging breasts, that she is dangerous, scary, fertile, menstrual, and has a strange pendulous sac hanging beneath her throat. A keeper says that she is 'sweet', but if the glass were to break 'it would be panic stations'. She lives on the border of two worlds, reminding us of our animal origins, our own bodies and our part in the evolutionary process. The spectator's encounter with Nénette and her abject bare existence in both the zoo and the cinema encourages revolt against the cruelty of the Symbolic.

Australian artist Janet Laurence, like Philibert, has worked with zoos and museum collections. In her 'Stilled Lives' exhibition (Melbourne Museum, Australia, 2000) she overturned museological principles to enable the viewer to form their own interpretation of each specimen as a distinctive life form with its own history. Laurence was Australia's representative at the United Nations 'Artists 4 Climate' conference (Paris 2015), at which she focused international attention on the plight of Australia's Great Barrier Reef through her multimedia installation 'Deep Breathing (Resuscitation for the Reef)'. In her exhibition *After Eden* (2012), Laurence draws on the moving image to encourage the spectator to empathize with the plight of animals under the threat of extinction. Her installations evoke a strong sense of ecological grief, a concept akin to mourning. *After Eden* is also on the theme of habitat loss. Different works are organized as a series of tableaux, comprising projected images, with titles such as 'Abandoned', 'Traded', 'Extinction' and 'Anthropocene'. The lighting is very soft. There are video images of elephants, monkeys and tigers – all of which seem to be travelling through a dreamscape from another age. There are also stuffed animals such as a dingo and an owl, as well as jars of marsupials at the embryonic stage. Each tableau is established inside a net or gauze screen in the shape of a cylinder, which is suspended from above. These are constructed in such a way that spectators can walk into some of these hanging habitats or simply look through the latticed surface. The soft screens create an impression of entanglement, and a sense that all lives, human and animal, are interconnected. The various creatures are presented with empathy and deep respect. Laurence explains her motivation: 'I wanted each cellular structure with its semi-transparent veils/membrane to reveal specific components or particular stories in ways that allow the viewer to experience the spatial relationships and to create connections'.[29] She speaks of the relationship between narrative and loss. Laurence believes attitudes to the dingo in Australia as not unlike those that led to the extinction of the Tasmanian 'tiger', or thylacine, whose loss is now 'creating a myth of longing'. The artist's comments indicate the power of stories to generate emotions leading to the creation of new myths about loss and longing. This narrative power helps to explain why films about extinction are able to tap into emotions of empathy and of ecological grief. The moving images appear behind veils, projected onto walls flickering in the darkness, present yet absent, alive yet dead. There is something phantasmatic, even hallucinatory, about Laurence's reconstructed scenes of the dead and extinct animal. These films and exhibits cannot restore or even set out to restore the past; instead, they speak of irretrievable loss – a loss so profound it may inspire the individual spectator to revolt.

The idea of ecological grief refers to a profound sense of loss felt by many at the degradation of the Earth and the extinction of species. While we usually refer to the death of someone close to us as a loss, it is also possible to feel loss for the Earth and its species, which form a crucial part of the human habitat – of the wider human home. In particular this loss affects how we think about the future. Films and exhibits, using the moving image, that explore human–animal relationships in the context of extinction have the power to expose the workings of the death drive, and to question a Symbolic order that seeks to abject woman, the animal and nature. Freud argues that man's aggressive instinct works against civilization.

> But man's natural aggressive instinct, the hostility of each against all and of all against each, opposes this program of civilisation. . . . This struggle is what all life essentially consists of, and the evolution of civilisation may therefore be simply described as the struggle for life of the human species.[30]

Drawing on Kristeva's theory of abjection, I have argued that the death drive, as destructive of others, is not constitutive of all subjects, but rather of the subject whose identity is produced by a violent phallocentric Symbolic order that is too harsh, that crushes those who do not conform. Those elements of the Symbolic, which do not respect the other, open the door to revolt. As Kristeva writes: 'In abjection, revolt is completely within being'.

Given humankind's long history of destructive behaviour, extinction is for many now the crucial issue. Cultural theorist Claire Colebrook, who has written extensively on the Anthropocene, sees the human as no longer a 'rational animal' but 'instead something like a geological event'.[31] 'Literally, the concept of the Anthropocene is that of an irrevocable and inhuman humanity: man is that animal who has detached himself from his putative ecological animality and lived in such a way that his life is destructive of his milieu.'[32] This observation reminds us of Freud's lament about human destructiveness and the possible end of humanity. In discussing man's invention of science and technology, Freud stated that 'Man has, as it were, become a kind of prosthetic God'.[33] If so, then Freud might have noted that man's major act of creation, as a 'fake' God, has been the disastrous advent of the Anthropocene and the mass extinction of species. But as Colebrook crucially points out, not all human beings are caught up with, or responsible for, the Anthropocene.

> Just to take one example that is fairly obvious, it is probably the case that most indigenous forms of existence didn't have the global reach of what called itself Western humanity . . . That's the problem with saying all humans are involved, because of course they're not. This is important, because in looking forward to the future, when we think about the end of *our* world, we have a really impoverished imagination about what other forms of human existence might be viable and which we shouldn't necessarily depict with horror.[34]

Those who lament the extinction of species, and understand the meaning of ecological grief, hope for a very different future – a future based on transformation of the Symbolic order inspired by an anti-anthropocentric ethic and informed by global movements such as Extinction Rebellion, as well as the transformative work of artists and individuals. It is only by looking directly at the tragic face of the consequences of the human death drive, as represented by contemporary filmmakers and artists, that it might become possible for the individual spectator to transform the experience of encountering abjection into revolt.

Barbara Creed publishes in the areas of feminism, psychoanalytic theory, human rights and animal ethics. She is the author of six books, including *The Monstrous-Feminine: Film, Feminism, Psychoanalysis* (1993), now in its ninth edition; *Darwin's Screens: Evolutionary Aesthetics, Time and Sexual Display in the Cinema* (2009); and most recently, *Stray: Human–Animal Ethics in the Anthropocene* (2017). She is the co-founder of the Human Rights and Animal Ethics Research Network (HRAE) in the Arts Faculty at the University of Melbourne, where she is a Redmond Barry distinguished professor emeritus. Her new book, *Return of the Monstrous-Feminine*, will be published by Routledge in 2022.

Notes

1. Danks, 'The Raw and the Cooked'.
2. See de Lauretis, *Freud's Drive*, 93–95, in which she analyses Freud's theory of the death drive, in the context of sexuality, with reference to the great significance of Spielrein's original contribution and her lack of recognition in psychoanalytic circles. De Lauretis argues for the continuing relevance of the Freudian theory of the drives.
3. Spielrein, 'Destruction as the Cause'.
4. Freud, 'Beyond the Pleasure Principle', 269–339, fn 2.
5. Caropreso, 'The Death Drive', 418.
6. Freud, 'Beyond the Pleasure Principle', 310.
7. Ibid., 311.
8. Freud, 'The Ego and the Id', 394.
9. Freud, 'The Future of an Illusion', 195.
10. Freud, 'Civilization and Its Discontents', 281.
11. Ibid., 311.
12. Ibid., 304–5.
13. Ibid., 340.
14. Ibid., 293.
15. Ibid., 338.

16. Ibid., 280.
17. For a discussion of the nature of exclusion, see Creed, 'Kristeva and the Abject Stray'.
18. Theweleit, *Male Fantasies*.
19. Freud, 'Civilization and Its Discontents', 288–89, n1.
20. Kristeva, *Powers of Horror*, 12.
21. Ibid., 13.
22. Ibid.
23. Davis, 'The Abject', 8.
24. Kristeva, *Intimate Revolt*. In *Powers of Horror*, Kristeva allows for individual revolt emerging from an encounter with the abject. In this new book, Kristeva extends this focus. She is particularly interested in the revolt of the individual rather than the group, because, as she argues, power that is constitutive of the Symbolic has become diffuse and is hence difficult for the group to revolt against.
25. Kristeva, *Powers of Horror*, 45.
26. Davis, 'The Abject', 7.
27. For an account of the death of Chunee, see Simons, *The Tiger that Swallowed the Boy*. See Leafe, 'The Town that Hanged an Elephant'. For a detailed discussion of the case of Topsy, see Creed, 'Animal Deaths on Screen', and Doane, *The Emergence of Cinematic Time*.
28. Shoard, 'Nicholas Philibert'.
29. Janet Laurence quoted in Merrillees, 'An Interview with Janet Laurence', 73.
30. Freud, 'Civilization and Its Discontents', 313–14.
31. Colebrook, 'Not Symbiosis', 187–88.
32. Ibid., 207.
33. Freud, 'Civilization and Its Discontents', 280.
34. Adkins, Parkins and Colebrook, 'Victorian Studies in the Anthropocene'.

Bibliography

Adkins, Peter, Wendy Parkins and Claire Colebrook. 'Victorian Studies in the Anthropocene: An Interview with Claire Colebrook'. *19: Interdisciplinary Studies in the Long Nineteenth Century* 26 (2018). Retrieved 28 January 2021 from https://doi.org/10.16995/ntn.819.

Caropreso, Fátima. 'The Death Drive According to Sabina Spielrein'. *Psicologia USP* 27(3) (2016): 414–19.

Colebrook, Claire. 'Not Symbiosis, Not Now: Why Anthropogenic Change is not Really Human'. *The Oxford Literary Review* 34(2) (2012): 185–209.

Creed, Barbara. 'Animal Deaths on Screen: Film and Ethics'. *Relations: Beyond Anthropocentrism* 2(1) (2014): 15–31. Retrieved 28 January 2021 from http://www.ledonline.it/index.php/Relations/issue/view/54.

———. 'Kristeva and the Abject Stray', in *Stray: Human–Animal Ethics in the Anthropocene* (Sydney: Power Publications, 2007), 21–27.

Danks, Adrian. '"The Raw and the Cooked": The Peculiar Poetics of Nicolas Philibert's *Un animal des animaux*'. *Senses of Cinema* 60 (2011). Retrieved 28 January 2021 from https://www.sensesofcinema.com/2011/cteq/the-raw-and-the-cooked-the-peculiar-poetics-of-nicolas-philiberts-un-animal-des-animaux/.

Davis, Clifford. 'The Abject: Kristeva and the Antigone'. *Paroles gelées* 13(1) (1995): 5–23.

De Lauretis, Teresa. *Freud's Drive: Psychoanalysis, Literature and Film*. Basingstoke: Palgrave Macmillan, 2010.

Doane, Mary Ann. *The Emergence of Cinematic Time: Modernity, Contingency, the Archive*. Cambridge, MA: Harvard University Press, 2002.

Freud, Sigmund. 'Beyond the Pleasure Principle', in *On Metapsychology: The Theory of Psychoanalysis*, Vol. 11. Harmondsworth, England: The Pelican Freud Library, 1955.

———. 'Civilization and Its Discontents', in *Civilization, Society and Religion*, Vol. 12. Harmondsworth, England: The Pelican Freud Library, 1985.

———. 'The Ego and the Id', in *On Metapsychology: The Theory of Psychoanalysis*, Vol. 11. Harmondsworth, England: The Pelican Freud Library, 1955.

———. 'The Future of an Illusion', in *Civilization, Society and Religion*, Vol. 12. Harmondsworth, England: The Pelican Freud Library, 1985.

Kristeva, Julia. *Intimate Revolt: The Powers and the Limits of Psychoanalysis*. New York: Columbia University Press, 2003.

———. *Powers of Horror: An Essay on Abjection*. Translated by Leon Samuel Roudiez. New York: Columbia University Press, 1982.

Leafe, David. 'The Town that Hanged an Elephant'. *Daily Mail*, 15 February 2014. Retrieved 28 January 2021 from https://www.pressreader.com/uk/daily-mail/20140215/283549448551541.

Merrillees, Dolla S. 'An Interview with Janet Laurence', in Janet Laurence (ed.), *Janet Laurence: After Eden* (Paddington, NSW: Sherman Contemporary Art Foundation, 2012), 73–89.

Shoard, Catherine. 'Nicholas Philibert: The Future's Orange'. *The Guardian*, 27 January 2011. Retrieved 28 January 2021 from https://www.theguardian.com/film/2011/jan/27/nicolas-philibert-nenette.

Simons, John. *The Tiger that Swallowed the Boy: Exotic Animals in Victorian England*. Faringdon, Oxfordshire: Libri Publishing, 2012.

Spielrein, Sabina. 'Destruction as the Cause of Coming into Being'. *Journal of Analytical Psychology* 39(2) (1994): 155–86.

Theweleit, Klaus. *Male Fantasies*. Cambridge: Polity Press, 1987.

Chapter 16

Tasmanian Tiger

Precious Little Remains

David Maynard

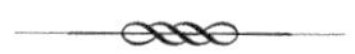

Introduction

The thylacine, or Tasmanian tiger (*Thylacinus cynocephalus*), is integral to Tasmania's historical and cultural identity and is recognized globally as an icon for extinction.[1] Having lived alongside Tasmanian Aboriginal people for over 40,000 years this unique species was driven to extinction just 133 years after European invasion.[2] Today, some people believe that the thylacine still exists, hidden in the remote wilderness of Tasmania's rugged west coast. Why do they continue to believe that the elusive thylacine will one day be bought in front of the cameras? Why do 'true believers' still 'go bush' searching for evidence of its survival? The lack of irrefutable evidence of the species continuing existence, and our scientific understanding of the population dynamics of large carnivores, should be justification enough for the community to accept extinction. These unfounded beliefs appear to stem from a lack of understanding of thylacine phylogeny, biology and ecology. What could we, the Queen Victoria Museum and Art Gallery (QVMAG), do to correct these misconceptions? This communal misunderstanding positioned the museum as the nexus of thylacine education.

The QVMAG was established in 1891 in Launceston, northern Tasmania, in the heart of what was thylacine country. The museum's vision is: 'Our Country, Our People, Our Stories: QVMAG is a place where our community explores, connects and is inspired', which is extremely applicable to the thylacine. The museum's thylacine collection and archives are significant locally, nationally and internationally. It holds five thylacine mounts, eight complete skulls (cranium and mandibles), one cranium, two sets of mandibles and four sets of assorted post-cranial elements, as well as associated documentation. There are also fourteen lots of thylacine cave deposit material in the geology collection. QVMAG has exhibited thylacines since at least 1908, when one (possibly two) mounts appeared in a display case.[3] The next photographic record of thylacines on display is from 1931, with four thylacines appearing in a diorama named 'Carnivore Rock' alongside other 'brute creatures' – a vulgar hyena and a sly fox, all overlooked by a leopard (Illustration 16.1).[4] The thylacines were arranged in a staggered, radiating fashion, which gives the appearance of a 'pack' startled into action, one baring its teeth. This presentation would have done little to placate the community's fear and hatred of the species at a time when it was on the brink of extinction. The three thylacines that are clearly visible in Fig. 16.1 remain in the museum collection today.

The thylacine holds a special place in the hearts and minds of the Tasmanian community. To this day people still report sightings and bring thylacine 'evidence' to the museum. However, over this past decade there has been a steep decline in 'true believers' asking to be heard. Fewer 'old timers' are relating their sightings and historical interactions, while younger Tas-

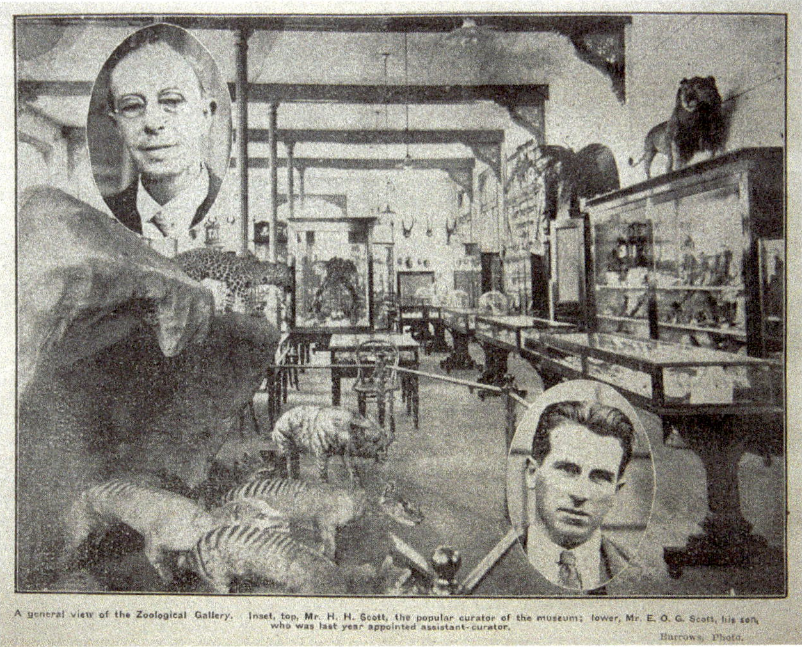

Illustration 16.1 The QVMAG zoology gallery in 1931 showing the 'Carnivore Rock'. *Weekly Courier*, 17 June 1931, p. 4.

manians are producing weather-worn skeletal remains questioning their identity, or rediscovered historical documents that may contribute to our combined knowledge of the species. This change in demographic, and their respective 'evidence', reflects Tasmania's ageing population and their differing positions on the species. We are seeing the replacement of a cohort that has first-hand accounts of the thylacine, or are willing to continue the search, with a younger generation that accepts that the thylacine could not have survived overharvesting, competition and anthropization.

This recent shift in attitudes is a small part of the continuum of changing attitudes to the species since European invasion. The earliest drawing of the thylacine produced by a European settler, the surveyor George Prideaux Robert Harris in 1808, reflected that generation's love of natural history and species discovery;[5] however, the artist delivered a misshapen and dangerous-looking animal, somewhat like a hyena. The 'fierce predator' imagery, and choice of language, influenced attitudes and actions up to, and after, extinction.[6] Beginning in the 1920s, there were growing but futile calls for thylacine conservation; however, these were drowned out by the more vocal 'eradication' lobby. From the death of 'Benjamin', the last captive thylacine, in 1936, through to the 1980s, searches for remnant populations took place. This delusional, head-in-the-sand attitude gave the community false hope that extinction had not occurred at their hands, and that the species persisted in Tasmania's remote south-west.[7] There were even plans hatched to relocate the thylacines to an island *when*, not *if*, they were found,[8] the justification being that they would be better off in some form of captivity than in the wild. Authorities declared the thylacine extinct in 1986, fifty years after 'Benjamin' had died at the Beaumaris Zoo in Hobart. There were still people, mainly older Tasmanians, who insisted that the thylacine was still out there, but by this time the 'fierce predator' image was being replaced by an elusive creature awaiting rediscovery.

At the same time as mortality was overcoming these elderly 'true believers', younger generations were coming through with a better appreciation of the depth and breadth of human impacts on whole ecosystems and landscapes. They understood that parts of Tasmania – once truly remote and considered thylacine refuges, like the forests in the west and south-west – had now been opened up to industrial logging, mining, hydro-power generation and tourism. They also understood that these forests were as inhospitable for the thylacine as they were for humans. These younger Tasmanians were not looking for the thylacine, but for thylacine artefacts and historical information. Fewer and fewer sightings, photographs and footprint casts of the extant but cryptic, more and more relics of the long lost.

That is my experience and interpretation of the recent changes in attitudes in our community, and I credit part of that change to the museum's exhibition about the thylacine, 'Tasmanian Tiger: Precious Little Remains', and the accompanying book by the same name. These were initiated at a time when alleged sightings and the presentation of physical evidence were a common occurrence at the museum, and amateur searches continued in Tasmania's wilderness. We were spending hours with 'true believers' and 'old timers' in a futile attempt to refute their 'evidence', and correct their understanding of biology, ecology and extinction. We felt it was time to put together an exhibition that shared the full story of the thylacine, to effectively draw a line under this extinct species, and to highlight that mistakes could be repeated today, threatening a suite of Tasmanian wildlife with extinction, which was surely where the community's energy should be focused. However, the research needed for this exhibition and book was considerable and, as we progressed, we found that our focus changed.

This chapter will explore the balancing act of educating the public about the tiger's life history and the many drivers for extinction, and the need to dispel myths and factoids, while also tracing the necessity to simultaneously record and celebrate the historical and cultural aspects of the thylacine. Firstly, I will reflect on our background work, making passing comments on the exhibition, before exploring how the thylacine was portrayed in our exhibition.

From inception to completion, this project was a challenging journey in itself; facts were scarce, historical information was rich and misinformation was widespread. Also, one author (me) had no knowledge of this species, while the second author had had a decade of historical research. Yet this mismatch worked. As a thylacine novice and sceptic, I was able to question dogma and put forward alternative views and theories, while my colleague, who was well versed in the social history, focused on addressing the need for evidence, and collating and structuring historical and cultural content. This unlikely pairing first produced the definitive book on the thylacine, stretching back 25 million years, capturing its mystique and role in the environment, and then balancing this against the ignorance and its persecution at the hands of Europeans. In turn, the cross-disciplinary exhibition was developed to reflect the book's content, using a mixture of traditional object-based methods supported by Indigenous content, new technologies, and aural and tactile elements. *Tasmanian Tiger: Precious Little Remains* is a permanent record of the community's love–hate relationship with the thylacine, exploring environmental, economic and cultural aspects of the species demise in the face of unprecedented change and prejudice. It also highlighted the ongoing threat of extinction to other Tasmanian wildlife.

Understanding the Thylacine

In this section I lay out the journey that we took to develop themes and content for the book and exhibition. My original vision was for purely scientific content that would bring to an end any delusions that the thylacine still lived. However, this was just a small component of what made it onto the floor.

Peer Review

It was important to us to produce a permanent record, a peer-reviewed book, of our research, something from which the exhibition could be drawn. All research comes under scrutiny, but thylacine-related content will always attract more than its fair share of critics – the book was our way of dealing with this. It had to present the scientific evidence for extinction and its many drivers, as well as to reflect the historical and cultural importance of the species. The book explored four themes: (1) thylacine biology and ecology, and some of the drivers for extinction; (2) the many government-sponsored and private thylacine searches that occurred after it became a protected species; (3) the QVMAG thylacine collection, and the museum's role in the local and global thylacine trade; and (4) the thylacine as represented in early accounts, oral histories, newspaper articles, and tales.

Some elements of this work and background research were directly translated into exhibition content as they reflected the QVMAG's vision ('Our Country, Our People, Our Stories'). A very specific example was a sketch of an elaborate trapping system of fences, pitfalls and box traps installed by local landowner, Robert Stevenson. This highlighted his familiarity with thylacine and his awareness of its behaviour. The evidence of Stevenson's prowess was displayed with the sketch, a multi-skin thylacine buggy rug. Another element was first-hand recollections of the thylacine, made available as audio recordings – the visitor needed to hear, not read, these to fully appreciate that this species had been alive, experienced and lost within living memory, and was not some temporally distant legend. Another part of the book that was translated directly into the exhibition was 'Dilger's tiger', the museum's best thylacine mount with the best provenance, and the supporting museum correspondence about its acquisition, which related a financial 'scandal' of sorts. Finally, but one of the most important elements from the book, was the Indigenous content. There is a paucity of historical information about the First Tasmanians relationship with the thylacine, yet the exhibition was an opportunity for contemporary Indigenous views and stories to be told through art and language.

We will return to the exhibition content in more detail later.

Science

The priority for me was the science, as this would solve our ongoing problem of dispelling common fallacies and sightings that could never be substantiated. We synthesized what was known – and just as importantly what was not known – about the thylacine, and the many factors that contributed to its extinction.

We explored convergent evolution, pre-European Tasmania and the traditional land management practices of the First Tasmanians, the government bounty scheme enacted to eradicate the thylacine, the wider demand for live and dead specimens as they became rare, and the introduction of European farming methods and sheep. The result? Extinction. The reason? The interplay between multiple factors, all stemming from European invasion.

However, during this process of digesting, distilling and interpreting facts, figures and models, it became clear that we were doing an injustice to the cultural importance of the thylacine. We needed to do more than simply educate the community about the *science* of the species and extinction. In fact, early on in our scientific research it was evident that we first needed to understand the British Empire's contemporary views of the natural world for us to properly interpret the quick, dramatic and disastrous transformations bought about by Europeans in Tasmania, and how this led to the thylacines demise. In short, 'brute creatures', particularly predators, were considered an economic risk and a threat to the safety of explorers and settlers, and thus should be killed.

Language

This was the view of the government of the day and the rural constituency to which they were closely linked. Indifference, fear and ignorance ruled, and is reflected in the language used in reference to the thylacine: primitive, inferior, skulking and stupid; savage, cowardly and treacherous; flexible and powerful predator; vicious, furious and surly; child stealer; undesirable, pest, sheep killer; vermin to be eradicated. Who would want such a beast living near them. Yet there was a love–hate relationship, particularly as the thylacine became rare in the early 1900s. It was referred to as a most singular and novel creature; affectionate, timid, shy and misunderstood. There was community interest in management measures that would preserve the species; official recognition as a protected species, reserves, captive breeding programmes and a prohibition of thylacine exports. Unfortunately, economics outweighed conservation. It was most unfortunate that the thylacine was a valuable commodity alive or dead, with trappers, furriers, bounty and trophy hunters, museums and zoos, all driving demand for the thylacine. Plus, sheep farmers were never interested in making real economic sacrifices in order to preserve the species.

This multitude of views are presented in the exhibition as oversized facsimiles of historical newspaper articles, editorials and advertisements, interspersed with historical images of the thylacine as a trophy or captive beast. Observers of the time drew on their existing knowledge of nature, and specifically in this case of striped carnivores; the thylacine was immediately likened to the then exotic and dangerous tiger as 'carnivorous and voracious'.[9] By the 1880s thylacines were being bought and sold. Advertisements reflecting their value as a commodity were common; 'perfect, large Native Tigers, Devils . . . and Platypus. All these I want dead' advertises a furrier;[10] 'For Sale – Native tiger, very quiet, feeds well', presumably sold as a pet. By 1880, the thylacine was an established threat to sheep farming in the eyes of landholders, and a real risk to personal safety. A boy aged 11 was 'dragged a considerable distance by the savage beast. . . . residents in the vicinity are greatly concerned at the presence of such an undesirable visitor'.[11] And let us not forget the government bounty scheme for the 'destruction of native tigers'.[12]

The public debate weighing up conservation against eradication continued to appear in the newspapers until well after the last known thylacine died. Arguments for the necessity of preservation and the importance of sanctuaries were countered by farmers like the previously mentioned Robert Stevenson, who wrote that he would openly flout laws protecting the species and 'pop a rifle bullet into a tiger . . . right in the neck'.[13] The exhibition reflected the community's attitude at that time, including in the language used, and it was an economic one – more valuable dead or in captivity.

Searching

It was clear that the government and community knew the thylacine was in demise from about 1910, yet they were incapable of action. It was in the 1930s, when the species was probably functionally extinct, that there was real interest in searching for remnant populations, primarily to protect, but sadly also to exploit financially in different ways.

There were many formal searches. Six decades (1930s–1980s) of government-sponsored searches came to nothing. From the 1950s to the 1980s, recognized researchers and research organizations dominated the search effort, using more advanced techniques as technology developed. Private searches continue to this day, although the premise has changed through time. No irrefutable evidence of the species survival has ever been found. Since the 1980s there have been many searches that have been financially 'inspired': some based on large monetary rewards, others attracting corporate sponsorship; some cashing in on media income, and others taking advantage of paying tourists wanting to join the mystique of the search. The only evidence from these latter searches is the ongoing monetary value of the thylacine.

The decades of search effort and the 'evidence' of thylacine survival feature heavily in the exhibition and are discussed below.

Cultural Searches

The final, most important and culturally significant type of search is the informal search by the 'bushy' – the 'true believer' who still 'goes bush' in search of the elusive Tasmanian tiger. It is these people and their tales of strange animal calls and the flash of a striped hide in the undergrowth that keeps the dream alive and feeds the community's ongoing interest and belief. No amount of science will change their views – they heard what they heard and they saw what they saw. These people and their stories hold a certain place in Tasmanian society, and if ever conversation turns to the thylacine, then their tales will be heard and nobody will refute them. It is this ongoing individual belief, along with the community's acceptance of their belief, that needed to be celebrated in the exhibition 'Tasmanian Tiger: Precious Little Remains'.

And so, at this juncture, we had a clear understanding of what the exhibition would contain alongside the science; it was content that reflected the attitudes of the past, and preserved the memory of the thylacine *and* of the community who knew it first-hand, and who continue to believe in its existence.

Exhibition Content

With the above-mentioned understanding clear in our minds, we set to distil this into two overarching objectives: firstly, to provide the audience with information about the thylacine and its extinction, and to empower them to contribute to the protection Tasmania's current list of threatened species from extinction; secondly, to link thylacine objects, images and stories with people and places, and with the attitudes, decisions and actions of the past.

This multidisciplinary exhibition combined artistic, scientific and historical elements. To achieve this, we loosely employed the VARK learning style.[14] In short, we combined visual, aural, reading and kinesthetic (doing/tactile) elements to better engage the individual, rather than broaden engagement. Not every element could be reproduced in each of these learning styles. Instead, the intention was to provide flexible communication of our main messages to 'deepen' the experience. Equally, particular elements were best presented in particular ways. For instance, you cannot *hear* a thylacine skull with a hatchet cut across the sagittal crest, or *hear* a map intended as a visual guide. However, the stories and tales of thylacine interactions are best *heard* from the mouth of a 'true believer'. The combination of VARK-inspired engagement was intended to give the audience a deeper and more meaningful experience and understanding of the thylacine.

Below is a description of the exhibition content and its themes and subthemes, and the messaging embedded within.

Historical Film

The first thing, and one of the last, that the visitor sees in the exhibition is historical film footage of captive thylacines in the London and Hobart zoos. On entry, the visitor views dog-like creatures held in poor conditions, some being taunted to 'perform' for the audience. However, as they leave the exhibition, with their new understanding of the species and persecution, they see a very different creature.

Indigenous Perspective

The first voice be heard was that of the Tasmanian Aboriginal community, which had lived alongside the thylacine for about 43,000 years. The First Tasmanians have an oral history, and so there are no written records of their ancestral relationships with the thylacine. To capture the Indigenous perspective, Aboriginal artist, Vicki West, produced a thylacine sculpture, titled *Luna 2*, from marine bull kelp, a traditional construction material, as well as netting, twine and recycled fabric (see Illustration 16.2). It was a response to the similarities seen between the story of the thylacine and that of the Tasmanian Aboriginal people. The sculpture, to quote West,

> responds to the mythical status of the thylacine, ongoing stories of sightings challenging notions of extinction, and a world in which extinctions have become part of everyday. The history of the 'extinction' of the thylacine closely parallels the extinction myth of the Tasmanian Aboriginal people, with both having a bounty on their heads under colonial rule, and with both being seen as pests in their own environment. The impact of this 'cleansing' of the landscape to conform to colonial pastoral ideals continues to be felt today.

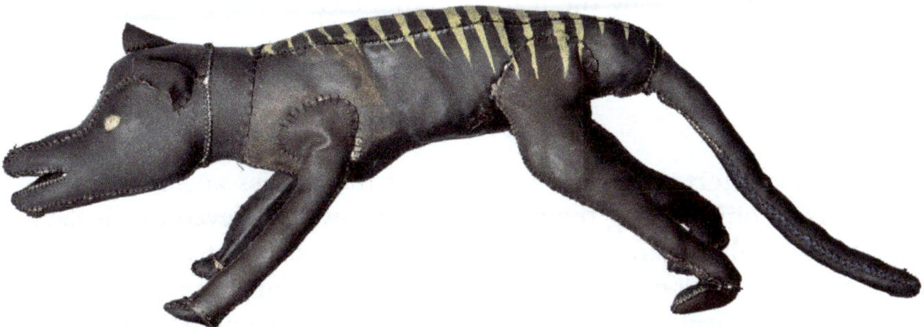

Illustration 16.2 Vicki West, *Luna 2*, 2014. Published with permission.

Alongside West's sculpture is a painting by Indigenous artist Lisa Kennedy. Titled *tebrakunna*, it reflects the concept of long-practised traditional land management practices, mutual hunting grounds, and a continuing culture. Tasmanian Aborigines shaped the landscape and ecology of Tasmania through the deliberate use of fire. They maintained a stable ecosystem that met their needs, and inadvertently those of the thylacine, for thousands of years. Alongside this artwork is a quote by Aboriginal historian, Lyndall Ryan,[15] that describes this traditional practice: 'On the heaths and plains behind their coasts, which they kept open by firing, the men hunted kangaroos, wallabies and emus, and the women hunted possums and other small mammals'. The same benefit was extended to the thylacine.

Europeans recorded Aboriginal language and custom, and two examples are included alongside the Indigenous content. Firstly, there is a selection of traditional thylacine names from across Tasmania. This reflects the wide distribution of both the thylacine and the First Tasmanians, reinforcing Aboriginal ownership of the land and their knowledge of the thylacine.

Finally, one of the earliest European records of Indigenous custom relating to the thylacine is told. George Augustus Robinson, 'conciliator' between settlers and Aborigines, recorded the following account when there was bad weather after his men killed a thylacine: 'The cause of this bad weather is attributed to the circumstance of the carcase of the hyena being left exposed on the ground, and the natives wondered I had not told the white men to have made a hut to cover the bones, which they do themselves, make a little house'.[16]

This mix of contemporary Indigenous art, historical language and the retelling of custom reminds the audience that the thylacine lived alongside the First Tasmanians for millennia. Would Tasmania still have the thylacine today had traditional ownership and land management methods not been pushed aside by the fear and ignorance of Europeans, and their farming practices?

Science

Considerable effort went into interpreting the species and its extinction without overcomplicating the messaging. This was achieved by displaying mounts, skeletal elements, scientific illustrations and photographs, models and a reproduction pelt, all with supporting object labels and text panels. The main case display for this theme featured four elements (Illustration 16.3). Firstly, the mount of an Australian dingo (*Canis lupus dingo*) appears opposite that of a female thylacine. These represent the concept of convergent evolution, the process whereby two unrelated species evolve similar traits as a result of having a similar ecological niche. This particular thylacine mount was displayed because it is very unique – its pouch is open and four everted teats are visible. A strategically placed mirror allows the viewer to see into the pouch. This serves two purposes: it differentiates marsupials (the thylacine, kangaroos, possums, Tasmanian devil, etc.) from placental mammals (the dog, dingo, hyena, etc.); it also identifies this thylacine as a mature female, and suggests that when she was killed, she likely had the care of four joeys that had probably been left in a den while she hunted.

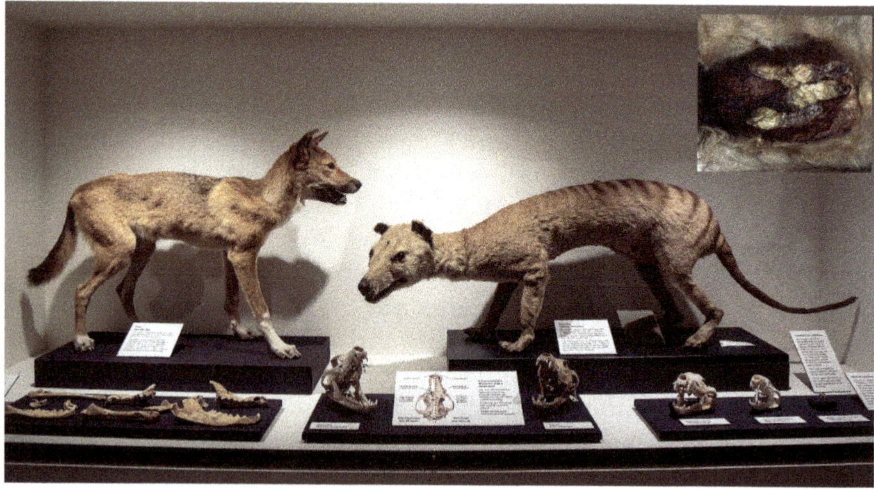

Illustration 16.3 This case display includes mounts of the Australian dingo (left) and a female thylacine (right), highlighting the concept of convergent evolution and the difference between placental and marsupial mammals (marsupial pouch inset). Also featured is a series of skulls of thylacine relatives, a diagnostic photograph, and fossil material. Queen Victoria Museum and Art Gallery.

Secondly, the case includes a comparative photograph of a dog and thylacine skull, highlighting the diagnostic features. This element was included in response to the number of dog skulls being presented by the public for identification, in the hope that it was thylacine. There has been a dramatic reduction in skull identification enquiries since the exhibition opened.

Thirdly, the display included a series of skulls of related Tasmanian carnivore and insectivore marsupials. The intention was to highlight the similarities and differences in size and shape of these species, in turn highlighting the different ecological niches within the Tasmanian environment, and that the thylacine was the largest marsupial carnivore in living memory.

Lastly, there is a display of thylacine fossil remains; these may be around five thousand years old, reinforcing the species presence in Tasmania for millennia prior to the arrival of Europeans and dogs.

Accompanying this display were two text panels that distilled complex topics into just 150 words each. One described the thylacine, its ancestry, its historical distribution, virgin population at European invasion, an outline of European attitudes, and its short time to extinction. The second panel provided an overview of the drivers for extinction.

Adjacent to this were two touchable objects: a commercially produced thylacine model set on the floor for children to handle, and a replica thylacine pelt pinned to the wall like a tanned hide. These tactile elements balanced the dry, clear-cut scientific information, and it is obvious that they are well used, evidenced by the gentle and careful patting by the visitor. One last interactive 'scientific' element was a touchscreen computer featuring a 3D image of a thylacine skull. Developed using photogrammetry, this allowed the visitor to handle and enlarge the skull virtually. The organic surface texture, delicate suture lines, the large sagittal crest that once anchored massive jaw muscles, and impressive tricuspid teeth are fascinating. These features contrast with the handwritten registration number 1962:1:53 on the cranium. For me, this human mark reinforces uniqueness and individualism, but equally transforms it into an inanimate possession for cataloguing and shelving.

Dilger's Tiger

A second thylacine is displayed in an adjacent case. This specimen is set against a natural setting that typifies thylacine habitat – open grasslands bounded by open eucalypt woodland. The case is darkened, silhouetting the thylacine against the backdrop. This is primarily a conservation measure to minimize light exposure to an already light-damaged specimen. However, it also adds an element of mystery. A sensor, triggered by the approaching visitor, slowly illuminates Dilger's tiger, an impressive adult male. The visitor, with their new understanding of the species, is presented with the alpha male. They no longer see a dog-like animal. Instead, the marsupial features are obvious – the stiff tail, the low and rounded haunches, and the head – all clearly, but (genetically) distantly, kangaroo-like.

Dilger's tiger is the transition from the scientific to the historical, and epitomizes our second objective, to link objects, images and stories with people and places. This specimen is the most valuable of the museum's thylacine collection as it has excellent provenance, yet it was purchased well under market value. In short, Alfred Dilger snared this thylacine in 1912 when thylacines were rare and of considerable value. A miscommunication led Dilger to mistakenly sell the dead thylacine to the Queen Victoria Museum and Art Gallery for just 20 shillings plus freight, instead of to Launceston's City Park Zoo for £3 (about three-week's wages). Dilger soon realized his costly error, and wrote to the museum, pleading for better compensation, arguing that he had 'carried him 10 miles, and I tell you he was a fair weight'.[17] He was unsuccessful in his request, as it was rare for a thylacine to be offered to the cash-strapped Tasmanian museums, who could not compete financially with higher-paying wildlife traders.

Dilger's tiger prepares the audience for the next display, 'Hunters and Collectors', a narrative on destruction. This uses historical photography, advertisements and editorials to relate how the thylacine was perceived, persecuted and killed. This sad and confronting display includes: the original 1888 notification of the government bounty scheme promoting the 'Destruction of Native Tigers',[18] competing with advertisements placed by a circus and a zoo for live and dead thylacines; disturbing imagery of zoo-held thylacines caged in substandard conditions by today's standards; proud farmers with gun in hand posing with their trophy thylacine; and a modified photograph of a thylacine, portraying it as evil and dangerous with needle-like teeth, horn-like ears and narrow-set eyes. Lastly, is a quote from respected zoo director and ornithologist William Le Souëf stating, 'Thylacines are now getting scarce as every man's hand is against them ... these animals will probably become extinct before many years'.[19]

Extirpation was achieved by the people of Tasmania. The last-known wild thylacine was killed in 1930 by Wilf Batty (featured) at Mawbanna, north-west Tasmania; and the last captive animal died in Hobart Zoo in 1936 (featured). This triggered a series of searches for remnant populations, which is the focus of the next display.

Sightings and Searches

'Sightings and searches' uses objects, images and text to explore the many attempts to rediscover the thylacine, and material considered to be 'evidence' of its survival. The visitor is provided with a chronology of the many unsuccessful government-sponsored and formalized private searches, and the changing attitudes, methods employed and motives for continuing the hunt. Early searches may have been well intentioned but were futile. With sanctuaries in mind, two-person teams searched for evidence across 1,000–3,000 km^2 of rugged terrain in Tasmania's west and north-west.[20] This was simply a shot in the dark, and was reflected in the results – 'signs of thylacines' and 'footprints suspected to be those of a thylacine'.

More sophisticated searches in the 1960s through to the 1980s are detailed, and supported by objects. Firstly, a change in thinking around how and where to search for the thylacine was adopted by the Thylacine Expeditionary Research Team, and they mapped public records of sightings. An oversized copy of the original map is featured in the exhibition, and its content is revealing – most sightings were in north-east Tasmania and on the central plateau, two areas previously ignored by searchers. Is this evidence of a missed opportunity to have preserved the species in the 1930s? And does the data reflect thylacine distribution, or just human distribution?

This change in thinking around thylacine search effort was accompanied by a change in the technology available for collecting evidence. Snares were still being used right through to the 1980s,[21] and one is displayed with instructions on how to set it; yet one of the first rudimentary motion-activated cameras, used by Guiler in the 1960s, is also displayed.

Modern formalized private searches have morphed into a semi-commercial/commercial business, either producing content for the entertainment industry, or servicing a niche sector of the tourist trade. These searches use modern technology like the motion-activated cameras, but have taken on a more 'experiential' element, 'going bush' like the 'old timers' searching for 'evidence'. However, they are not 'true believers'; instead they are searching for the dollar, not the thylacine. This is reflected in the exhibition in a single, poignant image of a newspaper front-page article featuring foreign cryptozoologists, branded with the logo of a multinational corporation, with the headline 'TIGER HUNT. International team hopes to settle thylacine mystery once and for all'.[22] The world awaits their findings.

The final type of search presented is that of the 'true believer', the old bushman who is convinced that the thylacine persists. They know where (but will never tell anyone) and they have the evidence to prove it. Unfortunately, their 'evidence' never stands up to scrutiny. On display is a selection of 'evidence' donated to the museum – sheep and dog skulls, desiccated dog and Tasmanian devil scats, plaster casts of unidentifiable footprints, a blurry (and probably altered) photograph of wallaby, and the jawbone of a baby fur seal.

The stories of some of these 'true believers' are available in the exhibition as audio files. This aural element provides a personal connection between the visitor and a Tasmanian with a personal link to the thylacine. Some speakers recall sightings and interactions from their youth, at a time when thylacines still lived. Others confidently speak of more recent sightings and strange sounds, and employ a process of elimination to rule out other Tasmanian wildlife as an option. On reflection these are good old yarns, or tiger tales, that feed the imagination, but do not support survival. Finally, and most unhelpful to our modern-day mission to educate the public about the loss of the thylacine, is Robert H. Green, highly respected curator of zoology at QVMAG from the 1970s to the 1990s, who speaks authoritatively about the survival of the thylacine in remote Tasmania, and population growth. Green provided the 'true believers' with oxygen; we are now trying to quench that fire. In all, the audience benefits from hearing from these Tasmanians voicing their passionate beliefs, as it feeds remembrance of the species.

The 'sightings and searches' section poses the question to the audience, 'Are they still out there?' Having explored this far into the exhibition, the viewer is clear that the thylacine has not survived.

Gone but Not Forgotten

The final display highlights how rarity and looming extinction increased global demand from the zoo, museum and university industries, the very entities that today would be integral to the preservation of a threatened species. A selection of images of museum-held skins, skeletons and poor taxidermy from around the world are interspersed with advertisements for the sale of live specimens, the imminent arrival of cashed-up foreign animal traders, expressions of admiration to shooters for having reduced thylacine numbers to record lows, and editorials from Tasmanians decrying proposed measures to protect the species. The thylacine was persecuted and exploited to the end; considered a pest to be exterminated, a commodity to be traded, but never valued as part of Tasmania's unique wildlife. Instead, this exhibition shows that the thylacine's value was as a mount in a public display or an articulated skeleton in the collections of the world's museums, for endless study and reflection.

Conclusion

This brings to an end the exhibition, with the exception of two cases of skeletal remains that can be studied up close. Some bones are easily identifiable (ribs, skull, vertebra), others are harder to place, requiring the viewer to refer to the adjacent scientific illustration of an articulated skeleton. There are partially articulated limb and foot bones delicately wired together by an unknown person in the past for an unknown purpose, a string of vertebra on a length of fencing wire, a skull with an axe mark, another featuring handwritten notes scrawled across the snout identifying the collector, location and date.

Each of these specimens has its registration label attached relating its core data – who, what, when, where. On close inspection, nearly all labels lack data; we do not know where or when many of the specimens were collected, or their sex or age, or the collector. Without this information the question begs, what scientific purpose do these specimens serve, as they tell us nothing directly about thylacine biology, ecology, diet, habitat, distribution, population size, or fecundity. Other than being a potential source of DNA and morphometric data, the thylacine remains have next to no scientific value. So why does the museum retain them? It is for their historical and cultural value. They are objects of remembrance. The story of the thylacine is more about attitudes, places and people, and their decisions and actions.

The exhibition is a window into Tasmania's past, highlighting to this generation and future generations the mistakes of the past, so that we do not repeat them. As the visitor leaves the space, they are faced with a display of some of Tasmania's native wildlife, including a selection of endemic species that are vulnerable to, or threatened with, extinction unless we make environmentally sustainable decisions. There is no excuse for species extinction on our watch, and our exhibition 'Tasmanian Tiger: Precious Little Remains' shows our community that they have the power to ensure that the fate of the thylacine is never repeated.

David Maynard is the senior curator for natural sciences at the Queen Victoria Museum and Art Gallery in Launceston, Tasmania. He took this role in 2012, and quickly came to realize the cultural and social importance of the thylacine to Tasmanians. Since then, he has seen a change in the community's attitude to the Tasmanian tiger, shifting from an unswaying belief that it survived in the remotest parts of Tasmania, to an understanding that the human hand has left no place for the thylacine to hide. He co-authored (with Tammy Gordon) the book *Tasmanian Tiger: Precious Little Remains*, from which the exhibition of the same name was developed.

Notes

1. Kyne and Adams, 'Extinct Flagships'.
2. Department of the Environment, '*Thylacinus cynocephalus*'.
3. 'Launceston's Museum'. *Weekly Courier*, 14 January 1908, 81.
4. This image was reproduced in the article 'Queen Victoria Museum and Art Gallery', *Weekly Courier*, 17 June 1931, 4.
5. See Gates, 'Introduction'.
6. Freeman, 'Figuring Extinction'.
7. Maynard and Gordon, *Tasmanian Tiger*.
8. Guiler, *Thylacine*.
9. *The Sydney Gazette and New South Wales Advertiser*, 21 April 1805, 3.
10. 'Wanted' (advertisement). *The Mercury*, 14 September 1886, 1.
11. 'Country News'. *Launceston Examiner*, 6 April 1899, 7.
12. 'Re Destruction of Native Tigers'. *Launceston Examiner*, 5 March 1888, 1.
13. 'The Native Tiger', *Examiner*, 3 March 1937, 1.
14. Fleming, *VARK*.
15. Ryan, *Tasmanian Aborigines*.
16. George Augustus Robinson, *Friendly Mission*, 921.
17. A.W. Dilger, Letter to H.H. Scott, 4 June 1912. Queen Victoria Museum and Art Gallery.
18. 'Re Destruction of Native Tigers', 1.
19. Le Souëf, *Wild Life in Australia*.
20. 'Move to Preserve Tasmanian Tiger. Sanctuaries Are Suggested'. *The Advocate*, 12 April 1937, 7.
21. Maynard and Gordon, *Tasmanian Tiger*.
22. 'Tiger Hunt'. *The Mercury*, 30 October 2013, 1.

Bibliography

Department of the Environment. '*Thylacinus cynocephalus* in Species Profile and Threats Database'. *Australian Government, Department of Agriculture, Water and the Environment*, n.d. Retrieved 9 June 2020 from http://www.environment.gov.au/sprat.

Fleming, Neil. 'VARK – A Guide to Learning Styles'. Retrieved 11 June 2020 from http://www.vark-learn.com.

Freeman, Carol. 'Figuring Extinction: Visualising the Thylacine in Zoological and Natural History Works, 1808–1936'. Doctoral thesis, University of Tasmania, 2005. Retrieved 9 June 2020 from: https://eprints.utas.edu.au/19821/1/whole_FreemanCarol2005_thesis.pdf.

Gates, Barbara T. 'Introduction: Why Victorian Natural History?' *Victorian Literature and Culture* 35(2) (2007): 539–49.

Guiler, Eric R. *Thylacine: The Tragedy of the Tasmanian Tiger*. Melbourne: Oxford University Press, 1985.

Kyne, Peter M., and Vanessa M. Adams. 'Extinct Flagships: Linking Extinct and Threatened Species'. *Oryx* 51(3) (2017): 471–76.

Le Souëf, W.H. Dudley. *Wild Life in Australia*. Melbourne: Whitcombe & Tombs, 1907.

Maynard, David, and Tammy Gordon. *Tasmanian Tiger: Precious Little Remains*. Launceston, TAS: Queen Victoria Museum and Art Gallery, 2014.
Robinson, George Augustus. *Friendly Mission: The Tasmanian Journals and Papers of George Augustus Robinson, 1829–1934*, edited by N.J.B. Plomley. Hobart: Quintus, 2008.
Ryan, Lyndall. *Tasmanian Aborigines: A History since 1803*. Sydney: Allen & Unwin, 2012.

Chapter 17

From the General to the Particular

Piecing Together the Life and Afterlife of A544,
Louis XVI's Quagga

Valérie Bienvenue

——⣥——

It is with the most precious complexity that the shadow
of death is entangled within the house of life, and we are
always implicated at the threshold.
—Deborah Bird Rose, 'In the Shadow of All this Death'

Introduction: A Quagga Cannot Change Its Stripes

In natural history, the diversity of stripe patterns and colour variations
within the zebra family has historically generated its share of taxonomic
confusion. This confusion was particularly acute in the nineteenth century, when no zebra that was observed or captured ever seemed exactly
like the last one. Individual variations in striping among members of the
same population were erroneously interpreted as evidence for the existence
of distinct species. Camouflaged by varied stripe patterns, the animal appeared to have mastered the art of resisting classification. Amongst these
taxonomically defiant herds of barcoded equines, one type stood out in
particular. Only 'half' striped, with its stripes restricted to the head, neck
and shoulders, it possessed a different colouring from other members of
the zebra family. Its coat was not the usual black and white, but brownish
with cream stripes. Its legs, belly and tail were white. Viewed from an
anthropocentric perspective, it could be described as the leader of a con-
spiracy against typology. The animal was named the quagga.[1] Despite its

stripes, *Equus quagga quagga*, 'elegant in its proportions',[2] was thought for a long time to be a closer relative to the horse than to the zebra. Thanks to genetic testing conducted on fragments of dried flesh collected from taxidermied individuals, the long-standing mystery of the animal's species was solved in the 1980s. The quagga was confirmed to be a subspecies of the Plains Zebra (*Equus quagga*).[3] Its true stripes finally revealed, the quagga is now scientifically accepted as a member of the zebra family.

Because of its distinctive stripes – a feature that may serve as camouflage, as a form of social identification, as an aid for thermoregulation, as a deterrent to flies, or a combination of some or all of these purposes – the quagga, like other zebras, was found aesthetically intriguing by artists. Its unique *zébrure*, to use the French word for a striped pattern, resulted in numerous depictions demonstrating great variation. Although registered extinct in 1883, the quagga is a recurring 'presence' in contemporary visual culture. Such attention, as Carol Freeman has noted in the context of the thylacine, another extinct animal renowned for its stripes or bands, seems motivated by the perceived strangeness of the species.[4] Quaggas or quagga-like creatures have secured a substantial afterlife roaming visual habitats such as logos, video games, artworks, and movies such as *Khumba* (Dir. Anthony Silverston, South Africa, 2013). A sense of the real physical appearance of the quagga can be gleaned from ageing museum specimens. These specimens, which pose major issues for conservation, are mainly located in the collections of European natural history museums.

Photography would also seem a good option for researchers seeking confirmation of the quagga as a phenotype possessing a distinct morphology. Only one individual, however, was photographed while alive. This quagga was a mare held at London Zoo. Five photographs exist of the animal. Three of these were probably taken in 1864 by Frank Haes, and a further two by Frederick Yorke in the summer of 1870.[5] Haes's photographs are taken from outside the quagga's enclosure. The horizontal and vertical bars of the pen mask part of the animal, whose head is turned towards the photographer. In one image, a top-hatted zookeeper is visible looking at her. An artificial habitat constructed of cobbles and brickwork has replaced the grassy, sweeping plains once familiar to the mare. The cobblestones and bricks, cubes and rectangles, provide an abstract backdrop. As repeated, regular geometric forms, straight and hard-edged, the building materials contrast with the quagga's singular curves and stripes, making her appear out of place, too visually unruly for this rigidly ordered environment. Even though these photographs provide a vital record of the physical appearance of a quagga, as they represent a solitary, captive individual their representativeness of the species is highly questionable. Although their value is clear, these images of a single mare cannot come close to giving a sense of the variety that characterized the species as a whole. The mare's uniqueness is a reminder of the singular loss that the death of any animal embodies, but because of this individuality her image cannot be used as the basis for making generalizations about quagga appearance or morphology, and even less about individual psychology.

The artist Ellen Litwiller shares science's curiosity for the living world.[6] In 2014, for her series 'The Last of Their Kind', Litwiller painted the last quagga to die in captivity, a mare held by the Natura Artis Magistra zoo ('Nature is the teacher of art') in Amsterdam, who perished on 12 August 1883. The zoo administrators had not realized that the quagga was the last of her kind until they unsuccessfully sought a replacement. The work by Litwiller recreates the moment of her passing, the instant when the mare finally escapes her captivity, the bars that separate the viewer from the animal seemingly evanescing. Her heavy eyes appear to express both her relief at the end of her incarceration and an unbearable sadness at the disappearance of her kind. The typeface used by Litwiller for the painted plaque at the base of the enclosure that records the quagga's deathday, and also the Latin binomial *equus quagga* that is repeatedly stencilled the length of three of the bars, seems blunt and devoid of feeling. The animal stands atop the metal plaque like an equestrian statue, her pose prefiguring the monument her taxidermied corpse was destined to become. The solidity of the plaque contrasts markedly with the delicate transparency of the skin of her rump, her flesh fading to reveal part of her skeleton. Through this work, Litwiller accomplishes the difficult task of inscribing a crude and unfortunately typical interspecies ecology of death into the context of the quagga's captivity and extinction. The painting acknowledges but also looks beyond the generic, portraying the mare as a specific individual with distinct feelings. To me, the work forces us to confront our lack of hospitality towards living as well as extinct species.

Inspired by Litwiller's case study, I have also chosen to focus on a single quagga, a stallion once in the possession of Louis XVI, the last king of France before the revolution of 1789. This stallion has no name. His only identifier is a catalogue number for the skeleton: A544.[7] I have studied the stallion's physical remains (a taxidermy mount and a skeleton), and also textual and visual records documenting his existence. The quagga in question was imported as a juvenile from southern Africa in 1784. He was corralled for ten years in the royal menagerie at Versailles, far from his natural surroundings of the Karoo, a semi-arid scrubland where species such as gnus and ostriches are common. The quagga was displaced once more in April 1794, moved to the Jardin des Plantes, where he died four years later.[8] In his lifetime, he was prized as a living testament to French colonial power, and viewed as a form of divertissement and as an educational device.

Post-mortem, the quagga's remains and contemporary accounts of his life were divided and secured in different buildings of the Museum of Natural History in Paris. His skin, bones, writings about him and portraits of him assumed independent existences. Because of how these remnants are currently archived, the insights they offer into the quagga are inevitably partial and disconnected. The narrative relating any life is necessarily fragmentary, marked by lacunae and by decisions over what events to include and exclude, but the way the quagga's life is archived amplifies the difficulties involved in telling his personal story. In death, the quagga was eviscerated, his internal organs removed and destroyed, and also quartered, the physical, textual and visual records of his existence despatched to four different locations. This chapter forms an effort to undo something of this violent dismemberment, threading together hitherto discrete archival elements to provide a rounder picture of this singular stallion. Judith Butler has suggested mourning can only occur when a loss of life is recognized, becomes known, becomes valued.[9] She signals the importance of the obituary in this process, a report of a death that also offers a biography, a life story. This chapter can be read as seeking to provide an obituary for the quagga, one that gives elements of his life story, rendering his recognition possible.

Jacques Derrida has observed that although 'we do not know what hospitality is', we human animals have a responsibility 'to open the door' and 'to welcome' the non-human into our home.[10] Here, the home in question is my text and the non-human I am inviting into my words is dead, radically, other to me. Through my prose I present, or represent, a creature that is no longer. My aim is not to possess the quagga but rather to let him roam within my prose. Circling around the archives of his life enables me to draw out details and insights. Gathering these together permits me to bring him back into circulation, detaching him from the abstract violence embodied in the alphanumeric identifier A544.[11] I seek to resurrect something of his personality, his preferences, and the precise state of his physical and mental health. Inspired by Steve Baker's observation that 'not all animals are seen as equally dead',[12] I have sought to reconstruct as many details of the quagga's life story as possible. My aim in writing this chapter has therefore been to open a space in which some of the violence inflicted on this animal, both in life and posthumously, could be, if not undone, at least attested to, and something of the uniqueness of this particular individual brought to the fore, welcomed into my words.

Detached patience is the strongest ally of any humans seeking to forge a connection with the natural world. Waiting is an important premise of interspecies hospitality. Animals, as 'Other' to us, only reveal something of their lives slowly. They do not understand clock time or respect deadlines. Whether they be alive or dead, found in a laboratory, in a museum or in the wild, I argue there is a particular way of seeing, an artful, unmotivated looking, that is particularly favourable to unveiling them. When I approached the quagga in Paris, I did so with a quiet and open mind, adopting a mode of reading his remains that sought to foster the conditions conducive to a holistic encounter. I knew from the outset that my words could not contain the quagga and I could not come to know him. My goal was not to seize his significance, not to impose my will on what endures of him. Rather I have sought to respectfully open a door and invite him to tell his story, letting us in on some of his secrets.

Into the Room of Endangered and Vanished Species

The taxidermy mount of the quagga that forms this case study is displayed in a room that forms part of the Grande Galerie de l'Évolution (Gallery of evolution) at the Muséum national d'Histoire naturelle (National museum of natural history) in Paris. The space, called La salle des espèces menacées et disparues (The room of endangered and vanished species), is reserved for extinct, or near extinct, species of plants and animals. It is a quiet place and, in fact, the very architecture encourages silence. Because of its high ceilings, any sounds are prone to echo, and people therefore instinctively murmur. Loud voices are met with a 'shhhh' or a swift glance of reproach. The immobility of the many species behind glass, their deathliness, also fosters solemnity and silence. There is a sharp contrast with what happens just outside the room. Elsewhere, visitors are also encountering taxidermied animals, but not in the same way.

There are clearly two kinds of death: the medium of taxidermy generating markedly different reactions, dependent upon context. In the adjacent imposing central nave of the gallery, a loud soundtrack simulates different animals calls, including trumpeting, varied roars and a vast cacophony of assorted bird calls. The many animal bodies are organized as if engaging in an immense multispecies parade. Although filled with the remains of dead birds and mammals, it is an intensely lively space. Children scream and laugh, adults talk animatedly. A short distance away, by contrast, there is only silence, punctuated each half hour by the chimes of a monumental clock. Commissioned by the king as a gift to Marie-Antoinette, it was built in 1785 by Robert Robin. Its intrusive marking of time forms an acoustic reminder of the House of Bourbon and the royal menagerie at Versailles where some of these species originated – species ripped from their familiar habitat to serve as exotic spectacles.

The Skin

I approached the mounted skin of the quagga as I would a work of art, studying it in detail, slowly, carefully. I examined the technique of the taxidermist, the quality of their work, and also how well conserved the mount is (Illustration 17.1). As if it were a sculpture on display in a gallery, I studied the context in which the mount was placed, noting its immediate surroundings. In the parquet-floored room, animals and plants are stored in very spacious, even oversized, glass cabinets. The quagga is difficult to locate at first, tucked away on one side of the room. Only the stallion's left side is visible. Compared to more recent taxidermied bodies on display nearby, he shows his age and the hardships he has endured in both life and death. Although his state of preservation is classed as 'good' by the Quagga Project, he is far from 'perfect'.[13] His short hair means old wounds remain in plain sight, and the difficulties encountered by the taxidermist who handled him post-mortem are clearly visible. Mounting the skin on its wooden support was a complicated operation. It has left big scars, which

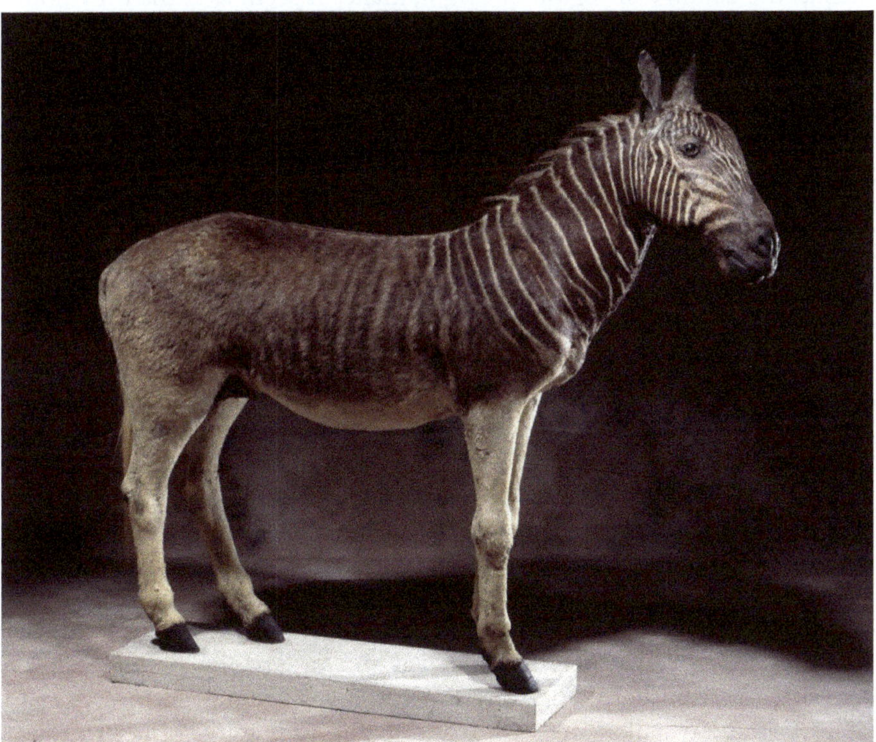

Illustration 17.1 Taxidermy mount of a quagga, Muséum national d'Histoire naturelle, Paris. © MNHN – Laurent Bessol.

are most visible on the inner side of his legs, under his belly, on his neck and around his eyes. Like the rest of his body, his sexual organs were emptied of flesh and then refilled. The skin of his penis and testicles is carefully preserved to show he was a stallion. Human footprints are visible on the rocky surface that coats the base he stands on; at some point a visitor must have left their tracks, whether recently or not is impossible to tell. Time stopped long ago in his glass prison house. Visits are kept to a minimum because clothes moths and *dermestidae* are an ever-present danger, eager to set up home in his remains.[14] The cream-coloured lines on his brown coat are still clearly visible. The harsh artificial lighting enhances the sheen of his fur. Patches of hair are missing here and there. On his head, above his left eye, a hairless patch reveals that the skin itself is also striped.

Labelling is uneven among the endangered and extinct species housed in the room. For those that are endangered, a map or, at the minimum, the name of their country of provenance, is provided. This is not, however, the case for those that are extinct. The mounted skins of vanished birds and mammals list only the name of the species. The lack of basic information, such as the country or region of origin, makes it seem as if the only home these extinct animals have ever known is this habitat of polished wood and glass. For the quagga, the labelling is token. The French word *Couagga* (quagga) is written in white, followed by the word *Éteint* (extinct) scrawled in blue, ending with a small but symbolically weighty final dot. Certain species of plant, either those that are critically endangered, definitively extinct or extinct in nature, such as the Bois de senteur blanc (*Ruizia cordata*) and Gordonia (*Franklinia alatamaha*), also relegated to the side of the room, received the same blue calligraphic treatment regarding their status. Here, it is the species at the centre of the exhibition space that are the unfortunate stars of this spectacle of extinction. Clearly felt to possess more celebrity cachet than the dried flowers and the long-departed quagga, they are the species most likely to be looked at by visitors. Viewable from multiple angles, they also make better material for 'souvenir' pictures.

I stood for hours in front of the quagga's mounted skin, taking copious notes. Now and then, visitors noticed me and therefore took notice of him. Without lifting my eyes from either the animal or my notes, I heard murmured conversations, 'Oh, what is that? A quagga, never heard of that one!' Whenever I do research in a museum, the same phenomenon occurs. My documenting generates a kind of magnetism towards the artwork or other artefact I am focusing on. This time, it is an animal. John Berger makes a direct comparison between zoo visitors, moving from cage to cage, and viewers at an art gallery, moving from picture to picture.[15] In her project 'Zoo World' (ongoing since 2011), artist Julia Lindemalm notes that every year 700 million people from around the globe will visit a zoo, and each of them will stay for an average of forty-six seconds in front of each animal's cage or enclosure.[16] While I was studying the quagga, no one else engaged with him for anything like forty-six seconds. The viewing or dwell time accorded to artworks is also often shown in studies to be on average less than a minute.[17] If an 'exotic' animal at a zoo or a work of art by a celebrated artist in a gallery cannot hold the attention for long, what are the chances that a dead quagga will encourage contemplation in a visitor.

There are twenty-three quagga skins mounted in various museums. These form a 'physical presence' of the animal in the present and are 'the definition of irreplaceable'.[18] Although the quagga in Paris is held in a room of disappeared or vanished species, it persists, visible and substantial. What we see, however, is not a quagga in any simple sense. The treatment of the quagga's ears offers material evidence of this reality. Hairless, erected in a bizarre position and seemingly much too tall, these auricular impostors contribute to rendering the reconstruction of the quagga's head awkward. Brown and dry, they resemble the long-faded flowers and leaves to be found on the other side of the room. Curled up and neglected, like the plants, many of which are taped to cardboard, they are amateurish, redolent of a sham. Displaying animal or vegetal remains of species that are now extinct, often condemned to oblivion through human action, is a delicate task that requires a great measure of deference for the dead.

An examination of the few remaining quagga mounts renders explicit the challenges faced by taxidermists while sewing and preparing them. The young taxidermied quagga known as Sparrman's quagga (after the Swedish naturalist Anders Sparrman), who is held at the Swedish Museum of Natural History, looks like a caricature of the creature.[19] The quagga has been more denatured than naturalized by its fabricant. Positioned in an ultra-rigid pose, with the taxidermist showing no pretention, ability or willingness to make the animal seem like they were once 'alive', the Swedish quagga exhibits what is best described as cruelty through negligence. Their (the quagga's gender is unknown) mouth is coarsely stitched, mummy like, and their thin infant skin is much damaged. Their life cruelly curtailed, they merit more respect than this shoddy workmanship affords.

A544, although also marred, clearly benefited from the hands of a skilled taxidermist capable of implementing the latest techniques.[20] This impressive attention to detail is, in itself, proof that the remains of the 'royal' quagga benefited from the preferential treatment that only a prestigious and wealthy institution could afford to lavish on its dead. Yet even if he was of major importance to the institution's collection when alive, the actual date of his death is unknown. By the length of his coat, I suspect the quagga died in the colder months. He has his winter coat, so probably died then or in the autumn. The worn patches of skin on his nose and also his legs index likely resistance to captivity. These injuries are symptomatic of equine distress. Something of his experience of captivity was therefore conserved with the animal, signs of his psychology and of the effects of human actions upon his mental and physical health inadvertently preserved. Through a close reading of the taxidermy mount, aspects of his life story begin to emerge. I turn to contemporary written accounts of the quagga to see if they lend substance to my suspicions about his suffering.

The Words

The Bibliothèque centrale du Muséum national d'Histoire naturelle (Central library of the national museum of natural history) possesses three copies of the same book dating from 1801 that is dedicated to describing the animals who were living or had lived at the menagerie, and to telling their 'stories'. Promoted as 'the necessary complement to any animal story',[21] the texts were accompanied by 'precise images traced by the pen or engraved by the chisel'.[22] The accounts of thirty-six species were mostly written by Georges Cuvier, with a general introduction provided by Étienne de Lacépède. Fragile, mouldy and incomplete, these old books seem themselves to be on the brink of extinction. Hard to manipulate because of their size and their frail condition, they hold precious textual traces about the quagga. Their content is also of incommensurable value for understanding prevailing attitudes towards non-human animals at the beginning of the nineteenth century in Europe.

In the Introduction, the mandate to educate the public and to participate in elevating science is openly expressed.[23] Writing specifically of the equines exhibited at the Natura Artis Magistra zoo, Rick De Vos explains that '[t]he stable served as a living cabinet, allowing the visitor to distinguish between the equids on display and to discern the similarities and differences between them'.[24] The same pedagogical principle applied to the menagerie. Lacépède and Cuvier were clearly proud of the collection of animals and held good intentions towards them. They were unaware of the degree to which their actions, however well intentioned, would be harmful to the animals they wrote about. A lack of interspecies hospitality at the menagerie was the root cause of the problem – a problem that persists to this day.

> One can compare this immense menagerie to a varied and laughter-filled countryside, where the different species of animals will enjoy all the freedom that it is possible to allow them without endangering the numerous and sometimes imprudent spectators; where they will find a shelter, a way of exhibiting and a form of care tailored to their kind; and where,

living in the midst of the plants and trees of their country, shaded, at the very least, by plants as similar as possible to those of their native land, as if they were returned once more to their birthplace, indulging in their games and their beloved activities, feeling neither their exile nor the loss of their independence, they will present to the eye of the observer the faithful picture of the bounties of living nature in the most remarkable regions of the globe.[25]

I am pretty sure that if it was possible to ask them, the animals in the menagerie would not have agreed with the glowing description that Lacépède provides of their existence. A 'varied and laughter-filled countryside' was not really reflected in the gaggle of animals at the menagerie, uprooted from their home and kin, yoked together from diverse geographies and subject to steady scrutiny by human visitors. Their habitat in Paris was far from a 'faithful picture' of their lands of origin. The presence of a few imported trees would not have prevented the animals from sensing their exile and mourning the loss of their native habitat. The living space of animals such as the quagga, used to roaming and grazing, was drastically reduced. The loss of the ability to move freely across significant distances would clearly have been felt as a loss of independence. These animals, symbolizing uncivilized nature,[26] were not looked upon with any real understanding of their specific needs or modes of existence. They were viewed as curiosities.

Writing specifically of the quagga, Cuvier notes of him that 'despite being locked up at a very young age, captivity had taken almost nothing away from our individual's fierce nature'.[27] The text goes on to acknowledge the fact that, 'in the wild, Quaggas sometimes live in herds of more than a hundred individuals'.[28] The fact that a herd animal such as the quagga was forced to live alone at the menagerie, as most captive quaggas did, did not seem to register as a possible explanation for the 'fierce nature' that Cuvier described. In their natural habitat, taking turns, individual quaggas will doze standing, while the rest of the herd is able to sleep soundly on the ground. The male at the Jardin des Plantes had no other quaggas to watch his back as he slept. Captive, bereft of companionship, deprived of mutual grooming, without opportunities for courtship and mating, these trying conditions would have triggered immense stress and frustration. Recent studies show that 'caging large mammals and putting them on display is undeniably cruel from a neural perspective. It causes brain damage'.[29] Cuvier goes on to observe of the quagga that

> he sometimes allowed himself to be approached and even caressed; but as soon as he was bothered, he would kick, and when [someone] wanted to move him from one park to another, or to change his location in any way, he would become furious; he would try to bite, throw himself on his

knees, and seize everything he came across with his teeth, seeking to tear it up or break it.[30]

Even though the quagga seemed to enjoy interacting with humans at times, he clearly did not like being manhandled. When he was moved, he became violent, biting and kicking. This is also the case with so-called 'wild' horses. When they are finally caught, the halter is never removed, facilitating control of the animal in the future. The scraped skin on the nose of the taxidermied quagga is proof that he received the same treatment. The wounds on his knees and forelegs had already led me to believe he had fallen repeatedly on hard surfaces. The biography confirmed my suspicions, he would 'throw himself on his knees'. Here two different archives, two different parts of the quagga's story, inflect and confirm each other.

Many animals still live on the original grounds of the menagerie. Walking around, I reflected on the quagga's lodging quarters. From the buildings, the iron fences and the mature vegetation, it seems not much has changed since the late eighteenth century. Endangered Przewalski's horses were introduced to the menagerie between 1902 and 1906. The perimeter of their present enclosure, made of large stones slabs, is a sad relic of the material commonly used for the flooring of pens at the time. As was the case with the compound depicted in the photographs of the quagga held at London Zoo that I discussed earlier, the ground of the enclosure in Paris was composed of cobbles. Even though this is a surface that is convenient for humans, making it easy to pick up excrement, it is alien to the quagga. The sensation of sleeping on stone, and the pain of falling on it, is markedly dissimilar to the comfort afforded by the grasses of their native plains.

The Art

The motivation behind the practice of producing drawings on vellum at the menagerie was initially fuelled not by a love of animals but of the botanic. Gaston d'Orléans, the brother of Louis XIII, wished to preserve the beloved flowers of his garden by having their 'portraits' painted by plant specialist Nicolas Robert.[31] When the collection of plant drawings was transferred to the museum on 10 June 1793, its significance shifted and less emphasis was placed on memorialization. There was a growing recognition of the value of draughtsmanship as a scientific tool that could be used for teaching purposes. On 3 December 1793, the artist Nicolas Maréchal was commissioned by the museum to enrich the collection of drawings with representations in watercolour of the mammals resident at the Jardin des Plantes. Since the quagga's death is registered as having occurred in 1798, and working on the assumption that the representation of the animal was made from life, it must have been created sometime in this four year period. Maréchal continued to work for the institution, offering his services to science as an artist until 1802, just a year prior to his death.

Maréchal's illustration of the quagga (Illustration 17.2) is one of seven thousand drawings that form the museum's precious vellums collection, a collection that includes representations of mammals, birds and plants. The museum's Central Library has equipped itself with secure means to protect these priceless treasures. The 107 unbound portfolios, all measuring 46 x 33 centimetres, are stored flat in custom-made locked cabinets and are only exposed to cold lighting. Access to the air-conditioned storeroom is accompanied by extreme restrictions. The museum's collection of vellums is little known, almost never seen and even less frequently touched.[32] Their only chance of survival lies in darkness and solitude. The medium itself is extremely fragile. In the case of the depiction of an extinct species such as the quagga on vellum, the material support in itself attests to immense rarity and precarious existence. We failed to save the quagga but have acted in time to save a precious and endangered likeness of him. This is no consolation.

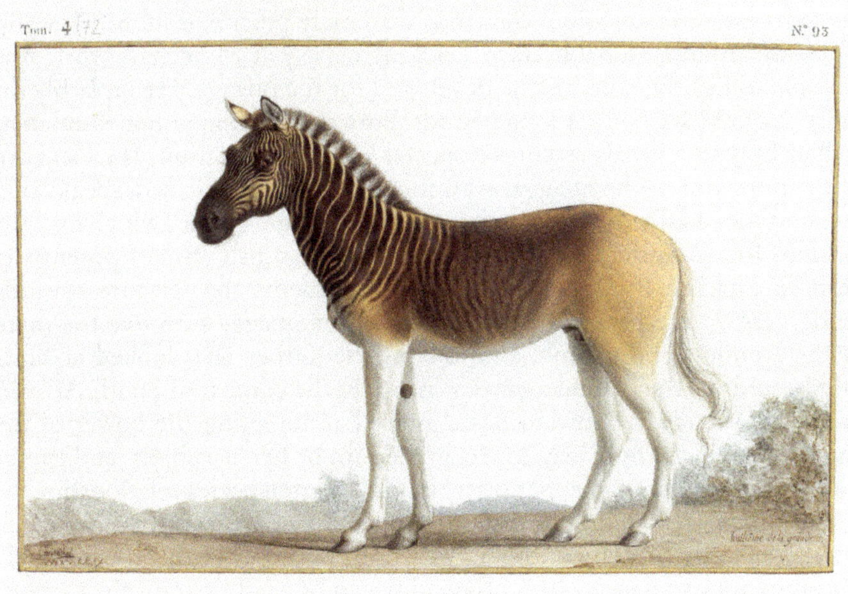

Illustration 17.2 Nicolas Maréchal, *Le couagga*, c. 1794. Watercolour on vellum. 460 × 330 mm. © RMN-Grand Palais / Art Resource, NY.

Monique Ducreux explains that the word vellum comes from the old French '*veel*' or '*veau*' (veal). It is a parchment prepared from the skin of a stillborn calf or a calf slaughtered before it was more than fifteen days old. Ducreux observes of vellum: 'Very white, fine and transparent, uniform, supple and light, the blood system of the calf is still seen by transparency. The skin is of small dimension, the trace of the spinal column hardly perceptible'.[33] To our contemporary eyes, vellum can appear an eccentric, even unethical, medium to use for making scientific records. But in the context of A544, vellum also embodies an extremely powerful material metaphor. Death is inscribed in the medium itself. Additionally, the pigments used by Maréchal would have been of mineral, vegetable and animal origin. Ox gall (*fiel de boeuf*) was often used as a wetting agent when making watercolours. Maréchal's paintbrush would have been made of sable. The drawing therefore took life through animal death. The vellum stands as a memorial not just to the extinct quagga but also to other animals, anonymous casualties killed in the name of art. These animals helped to bring the quagga's likeness into being, their bodies now forever intertwined with his. His flesh is their blood and skin.

Watercolour is a capricious medium. For Maréchal to achieve his fine-detailed portrait, he would have had to 'caress the skin with great meticulousness'.[34] Watercolour demands extremely precise gestures. Any slip will leave an unpardonable trace. Looking closely at Maréchal's work, it is possible to feel the empathy he developed for the quagga. He probably sat in front of the stallion for long periods. Portrait painting of non-human as well as human animals requires a special kind of dedication. The elegance of the portrayal of the quagga, even though it is romanticized, speaks of a certain type of affection for another species, for the 'Other', which exceeds a desire for scientific objectivity.[35] The delicate brush strokes potentially bespeak attachment. They are delicate movements that gesture towards tenderness on the part of their maker. Did the quagga turn like the mare in the London photograph, and look at the 'Other' that looked at him? Did he address the painter's gaze? Writing in the context of South African literature, Wendy Woodward has argued that accepting the power of the animal gaze is the only way to recognize him or her as subject, and therefore to question human superiority within an interspecies relationship.[36]

On vellum, the stallion feels almost weightless. There is shadow, which grants the body substance. The hooves, however, rest lightly on bare earth. The animal lacks density. This ethereality can be read figuratively as evoking both the animal's imminent death and a willingness to raise the animal to the status of an ideal, an idea, that transcends the quagga in his captive reality, his distress and depression. Maréchal grants the quagga nobility – he stands upright, immaculately groomed. The painting is from life but is not realistic. This quagga looks young and healthy. His rump is well rounded in contrast to the stallion's taxidermied self. His coat is shiny, lighter in colour and no scars are shown. The four hooves are clean and look well maintained. The veins of the animal, clearly visible on both his left front and hind legs, are gorged with blood. This quagga is in fine fettle. Maréchal gives him a vigour that was probably illusory.

On vellum, the quagga is revitalized. He is shown at his best; a beautiful animal with a perfectly combed mane, an elegantly swirled tail and a clean coat. This vision jars with the textual description of a difficult animal and the physical traces of distress that mark the taxidermy mount. The quagga is also shown in a landscape, the colours of which are in harmony with his coat. He is depicted as at one with nature. There is no trace of his incarceration. As was tradition, the frame surrounding the portrait is of gold leaf and the name of the species is also written in gold. Overall, there's an aureate glow to the image that signals admiration but also indicates that this likeness is not based on strict observation.

Despite his idealizing, Maréchal has faithfully captured some particulars. The long hairs under the quagga's eyes and on his nose are realistic details that a casual observer might easily overlook. The ears of the animal in the watercolour are 'horse like', very delicate.[37] They are far different from the foreign objects that seem as if they have been stuck onto the head of the taxidermy mount. The eye Maréchal accords the quagga also cannot be confused with the bulging ones of the mount. Glass eyes can never come close to the real thing. There's a vitality to the eye visible in the watercolour that invites recognition. The glass eyes that were transplanted to the quagga post-mortem are too small and give him an unrealistic gaze.[38] They make for an impossible encounter. The eradication of the quagga as a species deprived us of the possibility of meeting its gaze, but Maréchal gives us a glimpse of it, the possibility of a fleeting connection.

The nineteenth-century French painter of animals, Rosa Bonheur, was able to reveal an animal's personality by painting their gaze with almost surgical precision.[39] When treated with that level of attention, the animal gaze in a picture becomes a means to attest to their singularity. The gaze becomes like a signature, a guarantee of individual identity. Maréchal's treatment of the quagga's eye is noteworthy for its finesse. Through it, he acknowledges that his 'sitter' had distinct qualities. He was not just a quagga – not generic but particular. With his head turned three-quarter view, Maréchal makes us believe that the animal posed for his portrait as a human would. This obviously involves a measure of anthropomorphism. I would argue, however, that something of the quagga exceeds Maréchal's anthropomorphizing tendencies, enabling his equine character to show through also.

The only other image left of the quagga is a mouldy black-and-white engraving made by Simon-Charles Miger after Maréchal's watercolour.[40] The reversed image is generally loyal to the original portrait, apart from the animal's left hind hoof, which does not touch the ground completely as it should. Miger's rendering of the quagga is, however, less vital. Produced after the stallion had died, death is inscribed in the image. The quagga is now pictured confronting his own mortality: there is a pile of bones lying in front of him. In accordance with the scientific remit of the institution, these bones were probably his own. Miger would have had access to them so that he could draw them accurately. The stallion is therefore forced to gaze upon his own insides. Now skin without bones, he takes an impossible last stand. The vegetation alongside where the skull and bones are carefully arranged is clearly ornamental. Like Maréchal's watercolour, Miger's engraving does not seek to record the quagga's natural habitat. The background in the engraving, however, is more extensively worked. In the book, no other species depicted by Miger receives anything like the same careful treatment of their surrounds.

There were representations contemporary with Miger's that provide truer portrayals of the quagga's habitat. The hand-coloured aquatint engraving of a quagga from Samuel Daniell's book *African Scenery and Animals* (1804–5), for example, shows the quagga in a more naturalistic environment. In the background of the engraving, Daniell even goes so far as to depict an animal, most likely another quagga, being captured. He therefore alludes to colonization, and to how European attitudes towards animals were based on a logic of domination. Maréchal's and Miger's fantastical backdrops distract from the lived reality of the quagga's plight. Both artists have erased all traces of the quagga's uprooting and granted him a specious liberty. Yet Maréchal's watercolour, in particular, also adds to our knowledge of the Paris stallion, correcting misrepresentations such as the false ears of the mount, even as it in turn misrepresents.

The Bones

Globally, no more than seven complete quagga skeletons still exist. None remain in countries that overlap the quagga's former natural habitat. At the Grant Museum of Zoology in London, the now restored skeleton of the quagga known as Z581, safe behind glass, is one of the most popular exhibits amongst visitors. As part of their 'Bone Idols' project, Nigel Larkin and his staff carefully restored this 'iconic skeleton'. Once cleaned with appropriate chemicals, it was reframed in a more anatomically correct position. This particular quagga was renowned for its missing fourth leg.

Thanks to cutting-edge technology, a CT-scan of the quagga's existing right hind leg was made. The data was then 'mirrored' and 3D printed in black to become its left hind leg, hence remedying an imbalance that had lasted for over a century. In the process, the quagga was also given a right scapula. For Jack Ashby, manager of the Grant Museum of Zoology, the creation of the prosthesis permitted the animal 'to ride again'.[41] This demonstration of commitment towards the well-being of a dead 'Other' gestures towards what could be called a kind of posthumous hospitality.

The skeleton of A544 is a permanent resident of the impressive Galerie de Paléontologie et d'Anatomie comparée (Palaeontology and comparative anatomy gallery). There, he is exhibited in the front row of what looks like an army of skeletons marching towards the exit door. He is housed in a glass case which he shares with an Okapi, an endangered species that is partially striped, and also a Syrian Wild Ass. The latter, a subspecies of onager, became extinct about 1930. Entering the gallery, visitors are 'welcomed' by a flayed man or écorché; the figure was cast by the artist Jean-Pancrace Chastel from a cadaver in 1758. Displayed atop a plinth, the man's penis is concealed by a leaf. His left index finger points to the sky. The way the reddish plaster flesh of this muscled figure contrasts with the sea of fleshless, cream-grey-coloured bones of non-human bodies exhibited behind him is troubling. The *écorché* seems to signal human hegemony. From his pedestal he points heavenwards, indicating his divine origins, while the quagga and the other animals remain earthbound. At the very back of the 80-metre-long gallery, in a frieze inspired by the parietal art at Lascaux, there are representations of animals that potentially evoke a different, more balanced, type of interspecies relationship.[42]

The label for the quagga provides a bit of general history of his species and also gives the death date of the last of his kind, the female who perished at the Natura Artis Magistra zoo. Nothing is written about the stallion, but his skeleton reveals hardships he probably endured before he died. He is missing at least eight teeth. One of his remaining molars is cracked, which would likely have been a source of terrible pain if this damage occurred while he was alive. In 1916, the circus elephant Mary was accused of murder and hanged. Her autopsy revealed a toothache that may have explained her immense irritability. In the case of the quagga, it is possible his propensity to 'seize with his teeth everything he came across, to tear it up or break it'[43] was a symptom of dental distress, of a physical ailment. His 'fierce nature'[44] could be explained by physical as well as psychological trauma.

A piece of his skull is missing, evidence of a necropsy. The cranium was sawn open for his brain to be extracted and dissected. It was then sewn back together carelessly, using a metal spindle. The same kind of metal thread now keeps his rib cage from falling apart. His catalogue number of

A544 is plainly visible, inked on his right shoulder blade. This act of branding objectifies his remains. Numbering the stallion renders him abstract and makes it easier to forget he was once a living creature. Like the quagga held at the Grant Museum, the pose of A544 requires anatomical correction. The alignment of the first and second phalange of his left anterior leg leaves him hanging in a precarious, if not impossible, pose. Holding this counter-intuitive stance, which would be tremendously uncomfortable, the quagga is not in any position to 'ride again'.

Studying the stallion's skeleton, I was convinced that his bones had much more to tell me about the hardship he had endured and I would have liked to consult someone equipped with knowledge of forensic anthropology about the skeleton. Together we would have been better able to read the bones and decipher the stories they hold; however, even without the benefit of scientific expertise, the bones have contributed to the quagga's story. Taken cumulatively, the skeleton, the mount, the textual accounts and the artworks have enabled something of the quagga's singular qualities and experiences to emerge once more.

Conclusion: *In Memoriam* A544

We can only properly learn to say 'adieu' to the quagga once we have come to recognize him. This endeavour requires a collaborative effort that draws on the resources of art and science; only in this way can A544 and others like him become an individual once more. My own efforts here have been directed at weaving together separate archives – physical, textual and visual – each of which reveals important facets of the singular story of this quagga. These archives demonstrate how art and science in the museum are always intertwined. The taxidermy mount displays artistry as well as anatomical knowledge. Maréchal's watercolour had a scientific aim, to record and preserve the appearance of a particular species, but his artwork transcends this intention. Through close readings of these varied documents – bones, skin, paint, words – I have contributed a more rounded picture of the quagga, inviting a recognition of his singularity and welcoming back what was particular to him and to his life. Through making representations of this kind, we can begin to mourn this quagga for who he once was and, in a sense, continues to be.

Acknowledgements

I am grateful to the staff at the Bibliothèque centrale du Muséum national d'Histoire naturelle for their help. I am also particularly indebted to

Mégane Pulby and Cindy Lim of the museum for their assistance in my quest to get to know this individual quagga. Additionally, I want to thank Nicholas Chare for sharing his insightful comments relating to an earlier draft of this chapter.

Valérie Bienvenue is a doctoral candidate in the Department of History of Art and Film Studies at the Université de Montréal. Her thesis critically examines human–equine relations through the prism of modern art and visual culture. Prior to her academic career, she worked for ten years in equestrian circles, including teaching bareback riding and rehabilitating horses suffering from physical and psychological trauma. She is the author of several articles and book chapters.

Notes

1. De Vos points out that the term quagga was used indifferently by both the Khoekhoe and Dutch settlers to refer to all zebras. This caused confusion, and contributed to the disappearance of the 'real' quagga, which passed unnoticed by British hunters and administrators. De Vos, 'Stripes Faded, Barking Silenced', 31. The Khoekhoe (or Khoikhoi, 'men of men') are the traditional nomadic pastoralist Indigenous population of south-western Africa.
2. Cuvier and Lacépède, *La ménagerie du Muséum national d'histoire*, 'Le Couagga', 1.
3. 'The Quagga Project'. Retrieved 20 January 2021 from: https://quaggaproject.org/.
4. Freeman, *Paper Tiger*.
5. The animal was a female bought in 1851 from the animal dealer Carl Jamrach (Edwards, 'The Value of Old Photographs', 142). In the nineteenth century, photographs were perceived as too ephemeral, and were not considered suitable for the task of creating scientific records (ibid., 148). Professional photographers were often motivated by profit rather than a love of science and exactitude (ibid., 150).
6. 'Portfolio'. *Ellen Litwiller*, n.d. Retrieved 10 October 2021 from: https://www.art worksforchange.org/portfolio/ellen-litwiller/
7. Several skeletons of equines are numerically close to the quagga, including an Arab horse (*Cheval arabe*) numbered A538, a Bashkir horse numbered A542 and a Dauw (a zebra) numbered A547. The close grouping of these numbers suggests the cataloguing was carried out by genus some years after the quagga died. These skeletons are all listed in the Nouvelles Archives du Muséum d'histoire, 173.
8. In captivity many large mammals develop chronic stress which a change of location can exacerbate.
9. Butler, *Precarious Life*, 34.
10. Derrida and Dufourmantelle, *Of Hospitality*.
11. In a recent article, 'Sexual dimorphism', Peter Heywood refers to A544 by the name 'Paris quagga' (2760). This posthumous linking of the identity of the quagga to a city where he was held captive is problematic as, for many, 'Paris' connotes romance and

sophistication. If the quagga experienced 'Paris', it was in physical terms as an imposed, stressful, solitary confinement. Paris was a cage, not a city.

12. Baker, 'Dead, dead', 290.
13. Museum national d'Histoire naturelle, Paris, France. 'The Quagga Project', n.d. Retrieved 20 January 2021 from https://quaggaproject.org/skins/museum-nation al-d-histoire-naturelle-paris-france/.
14. Graham, 'Le soin des collections d'histoire naturelle', Figure 22a.
15. Berger, *Pourquoi regarder les animaux?*, 48.
16. For an overview of this powerful project, see 'Zoo World', Julia Lindemalm, n.d. Retrieved 12 October 2021 from: http://kontinent.se/wp-content/uploads/ZOO_J_1 .pdf.
17. Stephanie Rosenbloom, 'The Art of Slowing Down in a Museum', *The New York Times*, 9 October 2014. Retrieved 6 December 2020 from: https://www.nytimes .com/2014/10/12/travel/the-art-of-slowing-down-in-a-museum.html.
18. Poliquin, *The Breathless Zoo*, 4.
19. The quagga, a foetus, was collected by Sparrman in 1875 and presented to the Royal Swedish Academy of Sciences in 1876.
20. A constant improvement in taxidermy techniques now makes it possible to see the animal's teeth through its mouth, as in the case of several zebras installed in the Gallery of Evolution.
21. Cuvier and Lacépède, *La ménagerie du Muséum national d'histoire*, 'Introduction', 9.
22. Ibid.
23. Ibid., 6–7. '[T]o serve public curiosity to diffuse accessible and sustainable education', 'to give naturalists the real means to perfect zoology' and 'to acclimatize the animals demanded by the public economy'.
24. De Vos, 'Stripes Faded, Barking Silenced', 31.
25. Cuvier and Lacépède, *La ménagerie du Muséum national d'histoire naturelle*, 'Introduction', 6.
26. Ritvo, 'Animal Planet', 209.
27. Cuvier and Lacépède, *La ménagerie du Muséum national d'histoire*, 'Le Couagga', 2.
28. Ibid.
29. Jacobs and Marino, 'The Neural Cruelty of Captivity'.
30. Cuvier and Lacépède, *La ménagerie du Muséum national d'histoire*, 'Le Couagga', 2. Inspired by Thomas Pennant's *History of Quadrapeds* (1793), Harriet Ritvo has recently noted that the quagga was historically perceived by some naturalists as morally balanced, as by turns brave and meek. See Ritvo, 'Q is for Quagga', 146.
31. Cardinal, 'Les vélins du Muséum'. Nicolas Robert (1614–1685) once painted a quagga in watercolour and gouache on vellum; see 'Old Master Drawings: 66 Nicolas Robert'. Sotheby's, n.d. Retrieved 20 January 2021 from: https://www.sothebys.com/en/auc tions/ecatalogue/2007/old-master-drawings-n08281/lot.66.html.
32. The exhibition 'Précieux vélins. Trois siècles d'illustration naturaliste' was organized by the museum, and took place from September 2016 to January 2017. Because of the extreme fragility of the vellum only 40 works were exhibited at any one time and these were changed every month. Approximately 150 different vellums were displayed in total over the duration of the show.
33. Ducreux, 'La technique des vélins', 609.
34. Ibid., 609–10.
35. In an important recent reading, Peter Heywood views the watercolour through the prism of morphological truth, finding the painting exaggerated and lacking in veracity. It is too horse-like compared to the taxidermy mount. See Heywood, 'Ways of Seeing

Nonhuman Animals', 7–8. For me, neither of these forms of representation is totalizing and to be privileged. Each offers a valuable perspective on who A544 was.
36. Woodward, *The Animal Gaze*. How species such as plants, which have no eyes, can aspire to recognition under this logic is an unanswered question.
37. Hamilton Smith, *Equus*, 230. See also Cuvier and Lacépède, *La ménagerie du Muséum national d'histoire*, 'Le Couagga' (1) : '[T]his animal shares to a certain extent the beauty of the Zebra's dress, surpasses it by the elegance of the proportions, and looks more like our prettiest Horses.'
38. Even though the official photo of the animal, provided by the museum for this chapter, strives to promote, with the right angle and carefully chosen lighting, an aesthetically pleasing gaze, the encounter I had with the quagga offered a very different experience.
39. See my discussion of Bonheur's treatment of the animal gaze in Bienvenue, 'Au-dela de l'hégémonie humaine', 82–83.
40. Of the three copies of the book consulted, only one still contains the engraving made by Miger. In the other two, the engraving had been removed along with the ones of the zebu (*Bos taurus indicus*), another animal probably sought after for their atypical look.
41. Ashby, 'The World's Rarest Skeleton'.
42. For a discussion of the significance of animals, including horses, in palaeolithic art (including the parietal art of Lascaux), see Guthrie, *The Nature of Palaeolithic Art*.
43. Cuvier and Lacépède, *La ménagerie du Muséum national d'histoire*, 'Le Couagga', 2.
44. Ibid.

Bibliography

Ashby, Jack. 'The World's Rarest Skeleton Rides Again. . . on Four Legs'. *UCL Culture Blog*, 28 July 2015. Retrieved 6 December 2020 from https://blogs.ucl.ac.uk/muse ums/2015/07/28/the-worlds-rarest-skeleton-rides-again-on-four-legs/.

Baker, Steve. 'Dead, Dead, Dead, Dead, Dead', in Garry Marvin and Susan McHugh (eds), *Routledge Handbook of Human–Animal Studies* (New York: Routledge, 2014), 290–304.

Berger, John. *Pourquoi regarder les animaux?* Translated by Katia Berger Andreadakis et al. Geneva: Éditions Héros-Limite, 2011.

Bienvenue, Valérie. 'Au-delà de l'hégémonie humaine. Examen de deux œuvres équestres de Rosa Bonheur: *Marché aux chevaux* (1853–1855) et *Rocky Bear and Chief Red Shirt* (1899)'. Master's thesis, Université de Montréal, 2016.

Butler, Judith. *Precarious Life: The Powers of Mourning and Violence*. London: Verso, 2004.

Cardinal, Catherine. 'Les vélins du Muséum national d'Histoire naturelle: une collection mise en lumière'. *Artefact* 5, 15 Novembre 2017. Retrieved 6 December 2020 from https://journals.openedition.org/artefact/716.

Cuvier, Georges, and Bernardin G.E. de La Ville Lacépède. *La Ménagerie du Muséum national d'histoire naturelle, ou Description et histoire des animaux qui y vivent ou qui y ont vécu*. Paris: Miger, Patris, Gilbert, Grandcher and Dentu, 1801.

Derrida, Jacques, and Anne Dufourmantelle. *Of Hospitality*. Stanford, CA: Stanford University Press, 2000.

De Vos, Rick. 'Stripes Faded, Barking Silenced: Remembering Quagga'. *Animal Studies Journal* 3(1) (2014): 29–45.

Ducreux, Monique. 'La technique des vélins', in Pascale Heurtel and Michelle Lenoir (eds), *Les vélins du muséum national d'histoire Naturelle* (Paris: Citadelles and Mazenod, 2016), 609–10.

Edwards, John C. 'The Value of Old Photographs of Zoological Collections', in R.J. Hoage and William A. Deiss (eds), *New Worlds, New Animals: From Menagerie to Zoological Park in the Nineteenth Century* (Baltimore, MD: The Johns Hopkins University Press, 1996), 141–50.

Freeman, Carol. *Paper Tiger: How Pictures Shaped the Thylacine*. Tasmania: Forty South Publishing, (2010) 2014.

Graham, Fiona. 'Le soin des collections d'histoire naturelle'. *Government of Canada*, 15 July 2020. Retrieved 20 January 2021 from https://www.canada.ca/fr/institut-conservation/services/conservation-preventive/lignes-directrices-collections/histoire-naturelle.html #a4.

Guthrie, R. Dale. *The Nature of Palaeolithic Art*. Chicago: University of Chicago Press, 2005.

Hamilton Smith, Charles. *Equus: Comprising the Natural History of the Horse, Ass, Onager, Quagga and Zebra*. The Equestrian Wisdom and History Series. The Long Riders' Guild Press, (1841) 2009.

Heywood, Peter. 'Sexual Dimorphism of Body Size in Taxidermy Specimens of *Equus quagga quagga* Boddaert (Equidae)', *Journal of Natural History* 53(45–46) (2019): 2757–2761.

————. 'Ways of Seeing Nonhuman Animals: Some Likened Zebras to Horses, Others to Asses', *Society & Animals* (2020). doi:10.1163/15685306-bja10027.

Jacobs, Bob, and Lori Marino. 'The Neural Cruelty of Captivity: Keeping Large Mammals in Zoos and Aquariums Damages Their Brains'. The Conversation, 24 September 2020. Retrieved 6 December 2020 from https://theconversation.com/the-neural-cruelty-of-captivity-keeping-large-mammals-in-zoos-and-aquariums-damages-their-brains-142240.

Nouvelles archives du Muséum d'Histoire naturelle, deuxième série, volume 3. Paris: G. Masson, 1880.

Poliquin, Rachel. *The Breathless Zoo: Taxidermy and the Culture of Longing*. University Park: Pennsylvania State University Press, 2012.

Ritvo, Harriet. 'Animal Planet'. *Environmental History* 9(2) (2004): 204–20.

————. 'Q is for Quagga', in Antoinette Burton and Renisa Mawani (eds), *Animalia: An Anti-Imperial Bestiary for Our Times* (Durham: Duke University Press, 2020), 145–51.

Rose, Deborah Bird. 'In the Shadow of All this Death', in Jay Johnston and Fiona Probyn-Rapsey (eds), *Animal Death* (Sydney: Sydney University Press, 2013), 1–20.

Woodward, Wendy. *The Animal Gaze: Animal Subjectivities in Southern African Narratives*. Johannesburg: Wits University Press, 2008.

Part VI

Exhibiting Extinction

Chapter 18

Three Variations on the Theme of Extinction

Looking Anew at the Art and Science of Mark Dion

Anne-Sophie Miclo

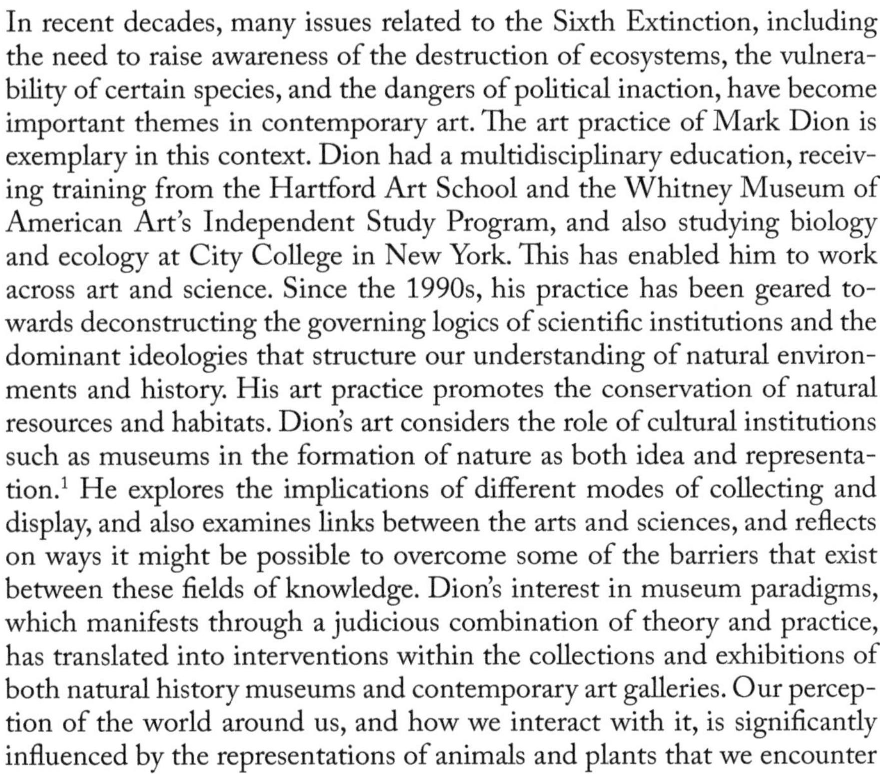

In recent decades, many issues related to the Sixth Extinction, including the need to raise awareness of the destruction of ecosystems, the vulnerability of certain species, and the dangers of political inaction, have become important themes in contemporary art. The art practice of Mark Dion is exemplary in this context. Dion had a multidisciplinary education, receiving training from the Hartford Art School and the Whitney Museum of American Art's Independent Study Program, and also studying biology and ecology at City College in New York. This has enabled him to work across art and science. Since the 1990s, his practice has been geared towards deconstructing the governing logics of scientific institutions and the dominant ideologies that structure our understanding of natural environments and history. His art practice promotes the conservation of natural resources and habitats. Dion's art considers the role of cultural institutions such as museums in the formation of nature as both idea and representation.[1] He explores the implications of different modes of collecting and display, and also examines links between the arts and sciences, and reflects on ways it might be possible to overcome some of the barriers that exist between these fields of knowledge. Dion's interest in museum paradigms, which manifests through a judicious combination of theory and practice, has translated into interventions within the collections and exhibitions of both natural history museums and contemporary art galleries. Our perception of the world around us, and how we interact with it, is significantly influenced by the representations of animals and plants that we encounter in culture.[2] In museology, studies of display practices have revealed the impact that exhibitions can have on a visitor's understanding of a given subject.[3]

In this chapter, Dion's art will be analysed through the prism of species extinctions, with particular emphasis given to how the artist disrupts existing museum practices. Extinction as it relates to the museum raises a number of important questions. How can something that is now absent, such as an extinct species, be displayed? And similarly, how can the process be shown by which something came to be extinct? How should extinction be exhibited to the public? From the very beginning, representing extinction involves a number of contradictions and paradoxes. It is certainly possible to put surviving specimens of an extinct species on display. This approach is commonly adopted in natural history museums. The specimens become a kind of witness attesting to their own vanishment. Is this recourse to a phantom figure, to a 'what once was', however, the only means to showcase a creature's extinction? Is it the most effective way for the public to grasp the magnitude of its loss?

Dion's artworks denounce anthropocentrism and critique systems of classification prevalent in Western thinking that can be traced back to Aristotle's ordering of beings – his *Scala Naturae* or Natural Ladder – and to Christian teachings about the hierarchy of life. It is usual for his corpus to be approached either in an overarching way as a unified project or with emphasis given to a single dimension (such as Dion's use of taxidermy).[4] My own reading, which focuses on the theme of extinction, falls somewhere between these modes of analysis. Species extinction (in the past and yet to come) forms a key theme in Dion's efforts to foreground and subvert the pretence of objectivity that characterizes institutional discourses. In studying Dion's works, I have come to realize that Dion's treatment of extinction usually falls into one of three categories: extinct species; critically endangered species; and species (including the human) that might be held responsible for a given extinction.

The three categories are porous, with the same species potentially assuming different roles in the triad, dependent on context. By appropriating the scientific procedures of collecting and classifying, and borrowing and subverting modes of display employing taxidermy such as the diorama, Dion brings notions of scientific objectivity into question. He returns repeatedly to particular strategies to make extinction visible, including isolation, absence and the use of animal skins. In this chapter, I begin with a consideration of institutional modes of exhibition. I then move on to consider my three identified categories (extinct, endangered, culpable), using specific case studies. These case studies have been selected for their complementariness, for their being emblematic of the artist's practice, and finally, for how well they showcase Dion's strategies for subverting standard approaches to display. The artist's critical interrogation of display practices enables him to address how systems of categorization and classification shape our perception and treatment of species, and to reflect on what kind of changes to such practices might be of benefit.

On Display

How can extinction be made visible? What models or paradigms of display should be used? For the last few decades, the boundaries between artworks and their exhibition have become increasingly blurred, with artists giving as much attention to the display of their works as to their production.[5] Dion's practice, for example, pays close heed both to the artwork and to how it is exhibited. Artwork and exhibition are, in fact, often inseparable and indistinguishable from each other. Dion's works combine and disrupt exhibition strategies. They also critically examine forms of classification used in the natural sciences and, concomitantly, in botanical gardens, museums and zoos. As a proponent of institutional critique, one of Dion's major interests is in the storytelling practices adopted by museums and zoos. Through museography (the methods of classification and display used by museums), museums shape the reception of their collections. Their display practices embody specific viewpoints (a reality which is frequently disavowed).[6] Mieke Bal has deftly exposed what might be called the 'ventriloquism' of the museum, the pretence it maintains that it simply shares the truth of things when, in fact, the museum is an institution of power, and the knowledge it communicates is always partial.[7]

Contrary to the museum, which organizes, produces and disseminates knowledge without openly acknowledging the ideological underpinnings those processes, Dion renders ideology visible, the better to unsettle it. To achieve this, he seizes control of aspects of institutional discourse (such as dioramas and forms of classification, including taxonomy) that traditionally produce knowledge, and shape our understanding of the world. These methods of classification and display contribute as much as the knowledge they ostensibly simply vehicle to the production of the broader narrative of science and nature offered by the museum. It is this narrative that is at the heart of Dion's work and that he seeks to disrupt. As Marie Fraser has noted, natural history museums and zoos share the same system of classifying and ordering the world as art galleries, one that is embodied in the collection, how it is catalogued and how it is displayed.[8] It is this shared system that shapes our understanding of both gallery and museum collections. Museums and zoos tend to focus on living animals and their stories rather than featuring narratives of extinction. The collections of natural history museums, however, usually possess materials related to the five previous mass extinctions, as well as artefacts related to past climate change events.

Zoos are often tasked with the conservation of rare and endangered living species. As John Berger has observed, through enclosing and displaying the living creatures they possess in particular ways, zoos encourage them to be seen in a specific manner and to transform their very nature. Berger suggests that zoos be understood as sites of mourning both for the animals themselves and for animal–human relations. In a sense, Berger observes, the animals form a living monument to their own extinction.[9] Dion's works bring the spectacular instrumentalization of the natural world identified by Berger to the fore. He also encourages us to question the paradoxical discourse articulated by institutions such as zoos. Filipa Ramos emphasizes this paradox when she states:

> Zoos juxtapose incompatible states, presenting a sampled, condensed fauna: Bengal tigers and African lions together, Florida bottlenose dolphins and Alaskan bald eagles, South American boa constrictors, and polar bears. Providing an experience of hallucinated observation, zoos offer the illusion of education via entertainment, perpetuating the positivist tradition that to see is to learn with an added twist, for how much can you learn if what you see is an assemblage of simulacra?[10]

As an artist who is not beholden to scientific institutions, Dion is able to manipulate display practices with greater freedom and to critique them more effectively. His relation to science is nuanced: he appreciates its factual dimensions, but his artworks are never uncritical; he always questions the use to which facts are put, and the significance that is accorded to them. In the past, the fields of art and science were closely related, both spiritually and practically. By the seventeenth century, however, they had begun to assert their independence and affirm a relative indifference to each other. Contrary to scientific discourse, which relies on deploying facts to advance 'truths', Dion's work (and, indeed, art in general) has recourse to allegory, humour or irony in relation to a given subject. Such rhetorical modes free narrative from the familiar aims of education and from denunciation regarding biodiversity loss. Through embracing a more critical and disruptive approach, Dion does more than simply state how things are. In a delicate balancing act, he addresses the climate emergency and the collapse of ecosystems, taking his audience to task but refusing to condemn them outright. The exhibition of Dion's works in contemporary art galleries, as often as in natural history museums, contributes significantly to the bridges the artist is able to build between the art and science.

Now That They Are Gone

Although of considerable interest, I will not be examining Dion's treatment of past extinctions in depth. I will, however, provide a brief overview of this aspect of the artist's practice. To remind us of vanished animals and plants, Dion uses specimens drawn from museum collections, or produces representations of them. Working across artmaking and curating, Dion has reflected on some major extinctions from the deep past. For example, the 1995 installation *When Dinosaurs Ruled the Earth (Toys 'R' U.S.)*, which recreates a child's bedroom, combines a number of real and imagined representations of dinosaurs in different media, including plastic play toys, duvet covers, wallpaper, decals and videos. The work immerses the visitor in how a child's vision of the age of dinosaurs is formed, with dinosaur species of different periods and habitats combined pell-mell. The commodification of dinosaurs is also brought to the fore, as these extinct species are now big business. Some projects make direct reference to extinct species through their titles, such as the 1995 solo exhibition *DODO* held at the Tanya Rumpff gallery in Amsterdam.

In the same spirit, the 2014 installation *Harbingers of the Fifth Season* takes the form of a workroom, a desk and swivel chair shielded from view on one side by a wooden screen. On the desk a box of paints lies open alongside an enamel tray containing a magnifying glass and other items. A number of books, with titles such as *Animal Invaders* and *Pests and Diseases*, are lined up at the back of the desktop. The implication is that this is the desk of a scientific illustrator. It is possible to walk around the installation and view the screen from both sides. On one side, the side facing inwards towards the workspace, a map of the world is tacked to a corkboard, alongside watercolours of animal and plant species adjudged to be invasive (such as ants, ticks and several rodents). The side facing outwards comprises a blackboard on which the names of extinct species, such as the dodo, the passenger pigeon, the pink-headed duck, the quagga and the thylacine, are scrawled in chalk. Although the list looks beyond 'canonical' extinct species, listing lesser-known examples such as Schomburgk's deer and the Indefatigable Galápagos mouse, it is still highly selective. Dion also includes no likenesses of the extinct animals that are referenced.[11] In the work, they are only alluded to in the abstract, their past existence solely attested to by way of their common names. The names would be invisible to anyone seated at the desk, a powerful metaphor for their literal disappearance.

Through a work such as *Harbingers of the Fifth Season*, Dion therefore exploits the power dynamics of exhibition practices linked to the representation and understanding of species for critical purposes. The work refuses to simply serve as a memorial, commemorating extinct species and listing the extinct and/or endangered. Rather, Dion's installation encourages visitors to reflect on the potential interrelationships that exist between the extinct and the invasive species referenced on opposing sides of the screen, drawing links between them. The empty chair in the workshop might be read as standing for the artist/scientist. He has created what we see, but takes his leave to enable us to draw our own connections. The work can be read as a commentary on extinction and on how it might responsibly be represented.

They Are the Last

In 2019, *The Life of a Dead Tree* was exhibited at the Museum of Contemporary Art in Toronto. For this temporary work, Dion drew on scientific approaches to tackle the issue of endangered species. *The Life of a Dead Tree* is hybrid in nature. It included a scientific laboratory, scientific photographs and dioramas. These might all feature in a science museum but here they appeared in a museum of contemporary art. In the exhibition space, a dead ash tree was displayed on its side. It was substantial, monumental in scale. A post-mortem of the tree revealed that it had formed the habitat for a whole host of different species, including fungi and invertebrates. These lived on and in its roots and bark. One of these species, the emerald ash borer (*Agrilus planipennis*) caused its death. It is a small beetle. The ash tree therefore dwarfed the insect, bringing home the reality that great size is no protector against affliction. The beetle gets its name from its colouring, as its exoskeleton has an emerald green metallic sheen. The smaragdine aspect of the shell of the insect, its bejewelled appearance, renders it aesthetically appealing. It conforms to commonly held notions of beauty. This shocked some visitors, as they found themselves admiring the very insect that killed the tree. The reality that such a small creature could destroy such a giant tree also piqued their curiosity. The borer therefore embodied a double-bind, both attracting and repelling.

Organisms such as insects are usually unwelcomed in galleries, posing a threat to the art on display. Dion's work, however, deliberately introduced them into the gallery space. Among the display panels, there were macroscopic photographs and diagrams providing information about the insects (Illustration 18.1). There was also a film about the how the tree was sourced, and a mock science laboratory. A scientist, Alexandra Ntoukas, was sometimes present conducting research, contributing to the ongoing development of the work. Her participation physically served to break down any simple opposition between art and science. The exhibition involved collat-

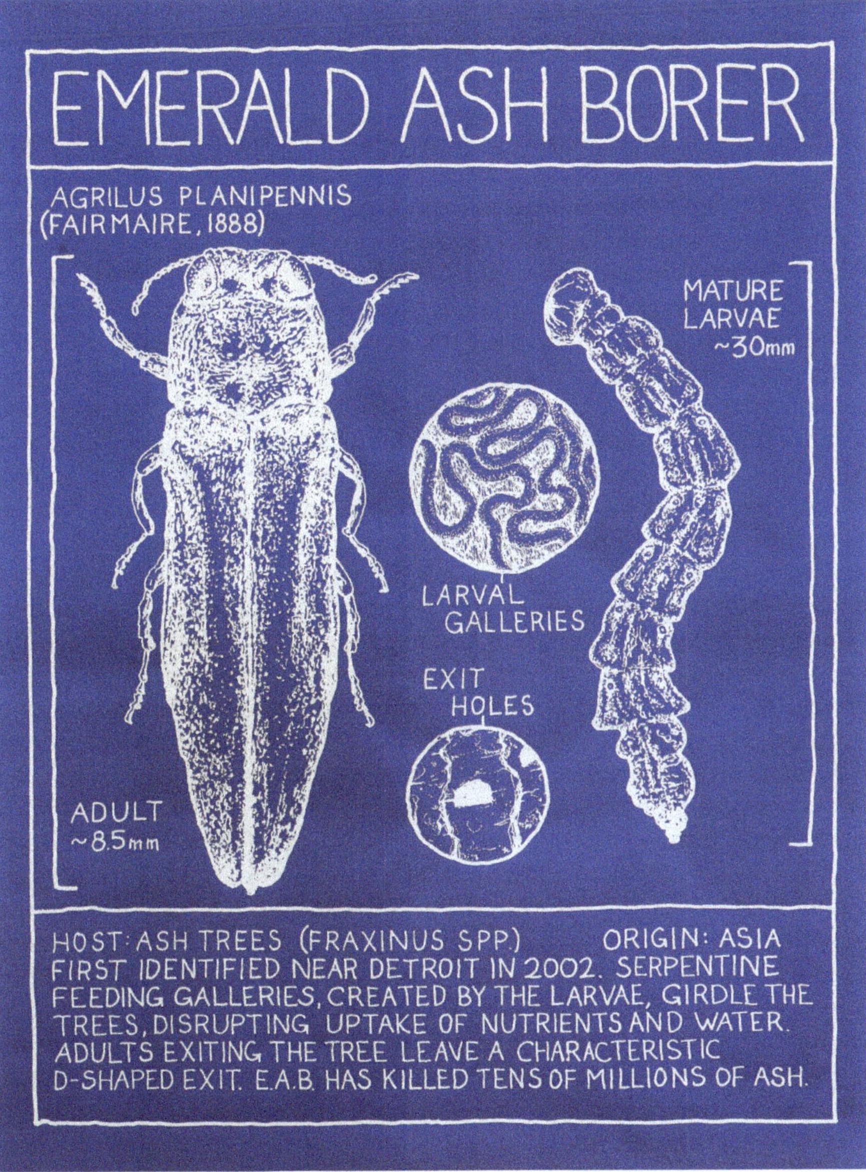

Illustration 18.1 Mark Dion, *The Life of a Dead Tree – Emerald Ash Borer*, 2019. Illustration by Matthew Wells. Silkscreen on paper. 24 × 18 inches. Photo: Tom Arban. Courtesy the artist, Tanya Bonakdar Gallery, New York/Los Angeles.

ing data that was shared with both the Royal Ontario Museum and the University of Toronto's Faculty of Forestry. Sample-taking occurred under the watchful eye of visitors. Some of the more inquisitive among them could approach Ntoukas as she worked, seeing what she did and asking questions. Dion therefore encouraged a sense of science being accessible.

Species of ash tree first appeared 50 million years ago, but the genus is now in danger of disappearing from the Western world because of the emerald ash borer. The beetle is viewed as invasive and highly destructive. Female borers lay their eggs beneath the bark of trees or in crevices in the bark. Once the eggs hatch, the larvae burrow to where the bark meets the wood and feed there, blocking the tree's circulation of nutrients. The borer is already responsible for the deaths of millions of North American trees, ravaging the forests of Canada and the United States.[12] A species native to East Asia, the borer is common in countries such as China, Japan and Mongolia. It was first noted in the United States in May 2002 in southeast Michigan, but has since spread through much of the North American continent. Dion's work drew attention to the impact of the borer not just on the ash tree but on the other organisms for which it provides a habitat. Through the accompanying photographs, *The Life of a Dead Tree* catalogued the various invertebrates and fungi that shelter in the ash. In viewing the exhibition, visitors come to realize that the death of the ash is just the first step in a chain of events that impact many different species.

Another way of displaying critically endangered or endangered species is to focus on an individual animal from a herd as a means to monitor the effects of population decline. In his 2001 work *Park (Mobile Wilderness Unit)*, Dion features a taxidermy mount of a bison (*Bison bison*) enclosed in a narrow glass case, the base of which is covered in earth, stones, twigs and broken branches so as to resemble a forest floor. The case is mounted on wheels, so in some ways it resembles a goods wagon. The resemblance to a form of railway transport provides a visual reminder of the devastating impact of the Transcontinental Railroad on bison in the United States. The cramped confines of the case give a strong sense of encroachment. The display also connotes dioramas of the kind found in natural history museums, and the situation of living bison today, many of which are held in captivity, unable to roam freely. A species that is now massively reduced in numbers compared to the early nineteenth century, corralled in national parks, continues to embody an image of wilderness in the popular imagination. Dion exploits this fantasy of the bison, subverting the notion of the wild through rendering it the subject of his artwork. The stuffed bison in its artificial habitat is shown to be a product of culture rather than nature. In fact, as Rachel Poliquin observes of such mounts: 'As dead and mounted animals, [they] are thoroughly cultural objects: yet as pieces of nature, [they] are thoroughly beyond culture. Animal or object? Animal and object? This is the irresolvable tension that defines taxidermy'.[13]

Dion's critique is articulated through a number of aspects of the artwork: the isolation of the taxidermy mount, its objectification, its constriction, the painted background and the addition of wheels rendering the display mobile. Above all, the work references the diorama, a mode of display that supposedly grants the museum visitor access to an animal's natural environment. As he often does when employing museum display methods, Dion engages in a practice of demystification, here revealing the shaky foundations of the diorama's claim to offer a realistic portrayal of nature. Usually, the epistemological and rhetorical dimensions in operation within a given diorama are veiled.[14] Dion, however, exposes and problematizes them, such that, as Giovanni Aloi explains, the diorama becomes 'a tool through which this very rhetoric can be dismantled and appraised'.[15] This appraisal includes noting how 'the specimen also is a deterritorialized animal body that has acquired the status of species representative through a state of isolation and preservation in the scientific cabinet'.[16] Beyond using a solitary animal to showcase a highly social species, deterritorialization is also suggested by way of the wheels, which indicate a willingness to displace this symbolism of the bison, traditionally an icon of wilderness. The artwork reflects the human footprint that shapes both environment and species.

The work *Extinction Series: Black Rhino with Head* (1989) provides another example of a work that examines a single endangered species: the black rhinoceros (*Diceros bicornis*). It features a wooden shipping crate with the lid removed so as to reveal a rhino head in profile emerging from a bed of straw packing. The scene is reminiscent of how artworks are transported. Dion is exploring legacies of colonialism such as animal trafficking. Rhinoceros horns are sometimes used as an expensive ingredient in traditional Asian medicine, and this has greatly encouraged poaching. The work connotes traffic of this kind, and also the traffic in ideas that leads to such practices. Nearby, further crates, this time sealed, are stacked on a pallet. They are suggestive of fragility and of far-off places. These additional crates display images, a map of Africa in Pan-African colours, and photographs portraying the rhinoceros's natural habitat and the perils it faces. Dion draws on taxidermy here to explore ideas about endangerment and extinction. The taxidermy he employs differs from that employed in natural history museums, which follow long-standing conventions and which idealize animals. Museum taxidermy occurs at the intersection of concerns about aesthetics, science and education. Specimens are often shown frozen in action or at rest, but seldom suffering or dying. In Dion's artwork, the rhino head seems serene, despite the absence of its body. The context, however, changes the reception of the taxidermy mount. It connotes the slaughter and trafficking that the species is subject to.

In a departure from other artworks where Dion presents his viewers with a specimen (such as a taxidermy mount) or foregrounds its absence, the artist also sometimes examines the popular representation of certain species (using drawings or soft toys). In the work *Survival of the Cutest (Who gets on the Ark?)*, which formed part of the 'Wheelbarrows of Progress' series that Dion produced in collaboration with William Schefferine, the duo used humour to explore the issue of the Sixth Extinction. Exploiting the freedom of expression that art allows, Dion and Schefferine employ cutting irony to playfully transgress conventions. Contemporary art provides a conduit to think critically about conservation issues. Through its title, the work refers mockingly to Noah's Ark, and draws attention to the criteria behind decision-making about conservation and the saving of wildlife. It features soft toys in a wheelbarrow, including an elephant, a panda and a polar bear, that have been singled out for saving because of their perceived charisma or cuteness. As well as critiquing the reification of certain species of animal above others, the display condemns the way species are classified and the role of human decision-making in the management and conservation of biodiversity. As Aloi has noted, the use of the wheelbarrow is key, as it helps to foreground Western attitudes towards the natural world, in which nature is viewed as a resource to be managed and exploited.[17] Discussing this work with Miwon Kwon, Dion highlights the issue of charismatic megafauna:

> Generally, in order to raise money for the protection of endangered ecosystems, conservation organizations draw isolated attention to extremely attractive and photogenic animals – tigers, whales, pandas. These are not keystone species, so the system won't collapse if they are taken out. Of course, all members of an ecosystem are important, but these animals are often the least critical ones, usually peripheral animals at the top of the food chain. They're not like the beaver or corals, which produce systems that support other animals.[18]

The use of soft toys of animals that are iconic and widely appreciated (other animals that appear in the wheelbarrow include a killer whale, a moose and a zebra) powerfully brings home the idea of cuteness. As Vincent Lavoie has examined, however, far from being a means of escape from life's horrors, the fascination with cuteness may bind us to it more forcefully.[19] Through *Survival of the Cutest (Who gets on the ark?)*, the discourse used to raise our awareness regarding endangered wildlife – in which cuteness is a key criterion and tool – is critically questioned and ironized.

Hung Out to Die

Lastly, to foster critical reflection regarding vanishing wildlife, Dion discusses issues of guilt. This topic is difficult to address, even through art. Dion does so in ways that are both nuanced and ambivalent, thereby signalling something of the complexity surrounding the issue. Like other similar works, such as *Killer Killed* (1994–2007), *Tar and Feathers* (1996) and *Monument to the Birds of Puffin Island* (2006), in *Monument to the Birds of Guam* (2005) Dion introduces species that, from an anthropocentric perspective, are viewed as 'guilty' of the extinction of others (Illustration 18.2). This guilt is announced as much through the titles of the works as their content. In each of the works, the pest species hangs from trees (in scenes reminiscent of lynching) or are covered in tar (and sometimes feathers). Tarring and feathering was a form of public punishment enacted in Early Modern Europe and later in North America.

Monument to the Birds of Guam calls to mind the fact that on the island of Guam (a territory of the United States), the brown tree snake (*Boiga irregularis*) has caused the disappearance of a number of native species of bird. Guam possesses considerable biodiversity, including small mammals, reptiles and numerous bird species. Snakes, however, have no stakeholding in this broad variety of animal life. The snake was inadvertently introduced

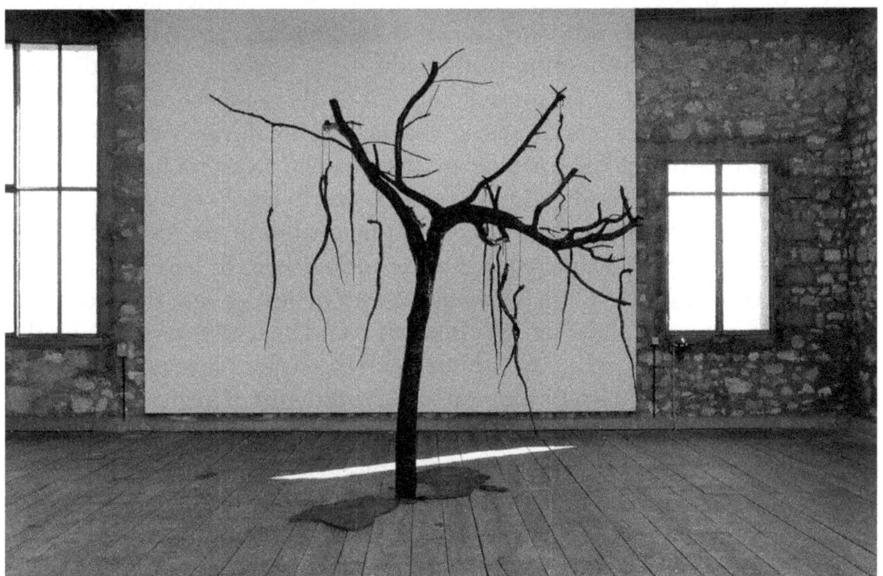

Illustration 18.2 Mark Dion, *Monument to the Birds of Guam*, 2005. 250 × 350 cm. Courtesy of the artist and Tanya Bonakdar Gallery, New York/Los Angeles. (TBG 20104)

to Guam shortly after the Second World War by naval vessels that visited the island. It is native to Australia, Indonesia, Papua New Guinea and the Solomon Islands. The snake's population increased exponentially, and it came to imperil the island's biodiversity because it preys on the nests of birds. It has had a devastating impact on Guam's native wildlife, particularly its avifauna including the Guam rail (*Hypotaenidia owstoni*) and the Guam flycatcher (*Myiagra freycineti*), but also on small mammals such as the Guam Mariana fruit bat (*Pteropus mariannus mariannus*) and reptiles like geckos and skinks.

As collateral damage, a war between humans therefore caused a fight for survival between other species in which the brown tree snake has been the victor. The United States Congress officially recognized the threat posed by the snake in the Aquatic Nuisance Prevention and Control Act of 1990, which aimed to prevent the introduction of non-indigenous aquatic nuisance species to American waters and to limit the spread of those already present. This was followed in 2004 by the Brown Tree Snake Control and Eradication Act. Any 'guilt' attributed to the snake is, however, relative. The snake may indeed be the cause of the disappearance of indigenous birdlife, yet human actions brought the reptile to the island. As it had no predators, the species then multiplied until its population was so great as to critically upset the ecosystem. The snake has caused the extinction of eight endemic species of bird, leading Guam to become an avian wasteland.[20] Among those species that are not extinct, some are held in captivity pending a potential future reintroduction to the wild if the snake can be eradicated. *Monument to the Birds of Guam* was inspired by this process of eradication. Made of wood, rubber, metal, string and tar, the work takes the form of a tree of death with snakes hanging by their 'necks' from its branches. The leafless tree, an improvised gallows, seems rooted in a pool of tar shaped like Guam. Because the entire tree appears to be coated in tar (connoting the practice of tarring and feathering) and the snakes seem to have been strung up, the work establishes a link between the idea of 'justice' and the process of eradication currently underway in Guam. It could be read as embodying the notion of 'lex talionis', of an eye for an eye. The snakes have caused extinctions and will in their turn be extirpated. Any thought of justice is, however, complicated by the kind of punishment that is displayed. Depending on the historical context and on your point of view, practices such as hanging or tarring and feathering can be viewed as forms of injustice.

This dimension to the work gives it an added critical force. The work's title, *Monument to the Birds of Guam*, suggests it commemorates the birds that have become extinct. The snakes have been held accountable for their 'unjust' actions. Monuments, however, usually ennoble or solemnify their subject. Dion's monument – a tree of death that acts as a scaffold – does not conform to conventions of the genre. It provides a bleak vision of loss, the arboreal snakes put to death in the branches of a tree they might otherwise call home. No birds are visible in this blackened tree. The monument to their extinction is a cycle of further destruction. Through its showing the tarred snakes, the work provides a nuanced perspective on the situation in Guam. Visitors feel empathy for the reptiles in the face of the fate that awaits them. The monument figures the legislation passed by Congress that sanctioned the extermination of the brown tree snake. The artwork therefore foregrounds the complex reality of the situation in Guam, the intertwined fates of the birds and snakes, and the important role played by human action in it. The snakes fall victim to their own adaptability but also to the human activities that brought them there in the first place. It is human action that is at the centre of these major disruptions to the Guamanian ecosystem. This is why, in works such as *Monument to the Birds of Guam*, human intervention always lurks in the background, acting like a watermark which becomes visible when the work is viewed in a certain light. Humans are not shown but, through the snakes hung out to dry, it is their actions that are indexed.

A Conclusion of Sorts

In the everyday, it is not always possible to grasp the scale of the Sixth Extinction that is currently unfolding. Dion's artworks and display practices raise awareness of the extinction, documenting and critically examining it. He reminds them that scientific discourse does not have a monopoly regarding how nature is thought about or understood. Dion tirelessly questions the roles and responsibilities of traditional institutions of knowledge, such as museums. Museums are customarily sites where existing knowledge is preserved and communicated. Dion asks in what ways they are able to describe and engage with contemporary concerns. Is there a willingness, for example, to question the taxonomic system, which positions humans at the apex of the natural world, separate from and superior to all other forms of life? Will scientific institutions go out on a limb and cut the topmost branch of the tree of life that splits the human from the animal?

Through my engagement with Dion's work, I have drawn attention to the important role that contemporary art has to play in critically and effectively addressing the topic of extinction. Dion's works of the 1990s are marked by humour and the use of irony. His more recent projects, however, strike a more serious tone, acting as calls to action through their invitation to critical reflection. Climate change and the accelerating rate of extinction have led Dion to insist on the urgency of doing something tangible. He has observed: 'I am increasingly pessimistic about the future. In each area I care about, I see nothing but discouraging developments. My work in general is right now more fuelled by anger than by hope'.[21] This pessimism is reflected in the increasingly serious tenor of his more recent works. The works are not devoid of humour but appear more sober than his early projects. A different, more caustic, jesting has emerged, a biting satire that accompanies the portrayals of the tragic events that he shares with his audience. Before not too long, Dion seems to be saying, humankind will be joining the long list of extinct species.

Translated by Nicholas Chare.

Anne-Sophie Miclo is a teaching assistant and doctoral candidate in history of art at the Université de Québec à Montréal (UQAM). Her research, which explores the impact of living things on museum practices, is supported by the Quebec Research Funds – Society and Culture (FRQSC). She is a member of the research group Figura, which is dedicated to text and the imaginary, and also of the Convulsive Collections research group (CIÉCO), which focuses on the development and promotion of museum collections. She is the author of numerous articles and exhibition catalogues relating to contemporary art practice.

Notes

1. Dion and Thompson, 'Interview with Mark Dion', 53.
2. See Berger, *About Looking*; Berger, *Why Look at Animals?*; and Aloi, *Why Look at Plants?*
3. Thorsen, Rader and Dodd, *Animals on Display*.
4. For a generalist approach to Dion's work, see Graziose, Kwon and Norman, *Mark Dion*. For a specific consideration of the role of taxidermy, see Aloi, *Speculative Taxidermy*.
5. See, for instance, Fraser, 'L'exposition à la puissance deux'.
6. See Bennett, *The Birth of the Museum*.
7. Bal, *Double Exposures*.
8. Fraser, *Zoo*, 55.
9. Berger, *Why Look at Animals?*, 52.
10. Ramos, 'Looking at Animals', 87–89.
11. The Indefatigable Galápagos mouse (*Nesoryzomys indefessus*) even suffered the ignominy of having its name misspelled as 'Glapagos'.
12. These ravages have symbolic as well as ecological impact, because in Norse mythology the ash tree, *Yggdrasil*, was sacred and revered as the centre of the known world.
13. Poliquin, *The Breathless Zoo*, 5.
14. Bal, *Double Exposures*.

15. Aloi, *Speculative Taxidermy*, 105.
16. Aloi, *Art and Animals*, 35.
17. Ibid., 103.
18. Kwon, 'Miwon Kwon in Conversation with Mark Dion', 18.
19. Lavoie, *Trop mignon!*
20. Fritts and Rodda, 'The Role of Introduced Species'.
21. Aloi, *Art and Animals*, 150.

Bibliography

Aloi, Giovanni. *Art and Animals*. London: I.B.Tauris, 2012.
———. *Speculative Taxidermy: Natural History, Animal Surfaces, and Art in the Anthropocene*. New York: Columbia University Press, 2018.
———. *Why Look at Plants? The Botanical Emergence in Contemporary Art*. Leiden: Brill, 2019.
Bal, Mieke. *Double Exposures: The Subject of Cultural Analysis*. New York: Routledge, 1996.
Bennett, Tony. *The Birth of the Museum: History, Theory, Politics*. London: Routledge, 1995.
Berger, John. *About Looking*. New York: Vintage International, 1991.
———. *Why Look at Animals?* London: Penguin Books, 2009.
Dion, Mark, and Nato Thompson. 'Interview with Mark Dion', in Nato Thompson (ed.), *Becoming Animal: Contemporary Art in the Animal Kingdom* (Cambridge, MA: MIT Press, 2005), 52–53.
Fraser, Marie. 'L'exposition à la puissance deux: Qu'est-ce qui fait exposition dans l'exposition?' *esse arts + opinions* 84 (2015): 4–13.
——— (ed.). *Zoo*. Montreal: Musée d'art contemporain de Montréal, 2012.
Fritts, Thomas H., and Gordon H. Rodda. 'The Role of Introduced Species in the Degradation of Island Ecosystems: A Case History of Guam'. *Annual Review of Ecology and Systematics* 29 (1998): 113–40.
Graziose Corrin, Lisa, Miwon Kwon and Norman Bryson (eds), *Mark Dion*. London: Phaidon, 1997.
Kwon, Miwon. 'Miwon Kwon in Conversation with Mark Dion', in Lisa Graziose Corrin, Miwon Kwon and Norman Bryson (eds), *Mark Dion* (London: Phaidon, 1997), 6–33.
Lavoie, Vincent. *Trop mignon ! Mythologies du cute*. Paris: Presses Universitaires de France, 2020.
Poliquin, Rachel. *The Breathless Zoo: Taxidermy and the Cultures of Longing*. University Park: Pennsylvania State University Press, 2012.
Ramos, Filipa. 'Looking at Animals', in Tristan Garcia and Vincent Normand (eds), *Theater, Garden, Bestiary: A Materialist History of Exhibitions* (Lausanne: Sternberg Press, 2019), 85–92.
Thorsen, Liv Emma, Karen A. Rader and Adam Dodd (eds). *Animals on Display: The Creaturely in Museums, Zoos, and Natural History*. University Park: The Pennsylvania State University Press, 2013.

Chapter 19

The Exhibition of Extinct Species
A Critique

Norman MacLeod

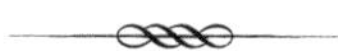

Introduction

Museums and galleries are the places to go if you want to see extinct species. After all, you cannot find any of these intrinsically interesting animals, plants, fungi or protists in zoos, national parks or wildlife refuges, much less in your backyard or local woods. Naturally, you can also meet extinct species on television, radio and the internet, in the movies, in video games and increasingly in theatres via the magic of animatronics, not to mention in books and magazines. But the sense of awe most people get from being in the physical presence of an accurate portrayal of even the humblest extinct species is an aesthetic experience that is often carried in memory for the remainder of their lives.

The relation between extinct species and a wide range of contemporary cultural, socio-economic, political, historical and scientific issues places a considerable burden on museums and galleries to tell their stories in a way that that deals with these complexities and remains truthful to the biological facts. Sadly, most museum exhibitions stick strictly to the latter two of these factors, perhaps with a knee-jerk tug at the emotional heartstrings as a consequence of extinct species' fates. The extent to which these institutions have been willing to meet the challenges that a genuine understanding of extinct species' stories requires, provides an insight into the expectations they have of themselves and of their audiences, in whose hands the fates of living species ultimately reside. Moreover, as all large national museums and galleries, as well as many regional and local institutions, receive all or part of their funding from public sources, the manner in which these institutions decide how to handle controversial subjects, such as extinction, provides insight into the power relations that exist in all human societies.

A Brief History of Extinction

An appreciation that species can become extinct is a surprisingly recent scientific development in Western culture. From the time of Aristotle (ca. 350 BC) to the early 1800s, the very idea of extinction was dismissed out-of-hand by most scholars. Historically, fossils recognizable as vertebrate and common invertebrate animals or plants were regarded as evidence for mythological creatures,[1] extant species or 'sports of nature'. Even the well-documented extinction of the Mauritius Dodo[2] was rejected as evidence for extinction initially, owing both to the rarity of specimens and the often fanciful nature of eyewitness descriptions. The Dodo's current status as a global extinction icon was not gained until well into the nineteenth century, over a hundred years after it had become extinct.

Acceptance of extinction as a fact came about, in part, because of a French insult to the New World, and the response by an entrepreneurial nineteenth-century American painter. In the run up to the French and Indian War (1754–63), French soldiers collected the teeth and the femur of a large unknown animal that were exposed at a locality in the territory of Ohio, referred to as Big Bone Lick, which was well known to local Indians. These specimens were sent to Paris in 1762 and entered into the collections of the Cabinet du Roi, where the femur was identified as belonging to a 'Siberian mammoth' and the teeth to a hippopotamus by Louis Jean-Marie Daubenton, then a museum curator working under the direction of Georges-Louis Leclerc, Comte de Buffon. Buffon was prominent in European biological circles for advocating the theory of racial degeneration. This theory included his 1789 proposition that, because of their 'smaller, weaker and generally inferior character', New World quadrupeds should be regarded as degenerate forms of European ancestors.[3] Many New World intellectuals considered this a thinly veiled political slur, including one Thomas Jefferson. Natural history was prominent among Jefferson's many interests and he set about collecting specimens, stories, anecdotes, myths and legends that would prove Buffon wrong; hence his interest in fossilized specimens of what the French called the *animal de l'Ohio*, but which Americans had dubbed the *American Incognitum*.

By the late 1700s, *American Incognitum* fossils were turning up fairly regularly in the Quaternary terrace deposits along the continent's eastern seaboard, including, in 1801, a spectacular find of what appeared to be an articulated specimen in a quarry near Newberg, New York. When the noted American portrait painter Charles Wilson Peale learned of this discovery, he rushed to Newburg, inspected the site, bought the bones that had been recovered thus far, and acquired the right to make further excavations from the landowner. Peale understood the symbolism this animal had acquired in the (now) United States, and had a hunch it might serve as a compelling centrepiece attraction in his new Philadelphia museum, which he had created to display his own portraits and natural history collections.[4]

Previously, Peale had painted Jefferson's portrait while the latter was the US Secretary of State. Jefferson was aware of Peale's interest in the *American Incognitum* and encouraged his efforts. With the help of the anatomist Caspar Wister, Peale mounted the full *American Incognitum* skeleton, substituting plaster casts for the missing or broken bones. The mount went on display in Peale's museum in 1801, with the tusks curving downwards (incorrectly) in order to increase the perception that this was a ferocious American monster.

Jefferson held out hope that populations of the *American Incognitum* would be found in the interior of North America, a substantial tract of which he acquired from Napoleon in 1803 as the Louisiana Purchase. Quickly thereafter, Jefferson funded the Lewis and Clark expedition (1804–6), primarily to explore the new territory but also to find a practical route for access to the continent's western coast. However, as documented in an 1803 letter to Meriwether Lewis, Jefferson also wanted the party to survey the territory's economic and resource potential. In particular, he charged Lewis to be on the lookout for 'the animals of the country generally, and especially those not known in the U.S., the remains and accounts of any which may be deemed rare or extinct'.[5]

Lewis and Clark did not find the *American Incognitum*, and Jefferson eventually reconciled himself to its absence from the North American landscape. Meanwhile, in Paris, a young Georges Cuvier was applying his newly developed principles of comparative anatomy to the *animal de l'Ohio* and similar fossils collected from the Paris Basin, all of which now resided in Paris's Musée national d'Histoire naturelle. Upon close inspection, Cuvier concluded that: (1) both the femur and the teeth of *animal de l'Ohio* came from the same animal; (2) neither the African elephant nor the Indian elephant bore close morphological similarities to the unknown bones; and (3) neither did the bones of the Siberian Mammoth (which itself eventually came to be identified as an extinct species).[6] Reasoning that it was unlikely such a large animal could have gone unnoticed by previous explorers, Curvier concluded the animal, which he named a *Mastodon* in recognition of the conical character of its molar cusps, was most likely an extinct species of elephant.[7] Although Cuvier spent much of his remaining career convincing sceptics that extinction was a real phenomenon, such was his standing in the biological community that an increasing number of anatomists accepted his pronouncement on the reality of extinction.

And what of Peale's *American Incognitum* mount? Its popularity with the citizens of Philadelphia knew no bounds. Indeed, Peale's mount, in conjunction with Cuvier's identification and interpretation of the *Mastodon*, set the template that has been followed by museum and gallery exhibitions of extinct species ever since, especially once the bones of that other extinct group, dinosaurs, came to be recognized for what they were – the true extinct monsters that Jefferson had hoped for, and Peale had pretended, the *American Incognitum* to be. All depictions of extinct monsters in museums and all media, from the Crystal Palace dinosaur sculptures (1851) to *Jurassic World: Extinction* (2021), can be traced to Peale's excavation and reconstructed mount of the *American Incognitum*.

A Brief History of Natural History Museums and Galleries

In order to understand how museum and galley exhibitions of extinct species function, and the challenges these institutions face in taking a broader view of the extinction issue, an understanding of the history of these institutions is necessary, especially regarding the radical change in their purpose since their inception. A museum is any institution that houses, cares for and exhibits objects of cultural, artistic, historical and/or scientific importance. Museums originated as collections of objects made (usually) by wealthy or important men as part of their work (e.g. collections of medicinal plants), for aesthetic reasons (e.g. collections of pictures and sculpture), or by virtue of their positions in society (e.g. collections of gifts presented to heads of state). In a more general sense though, museums grew out of a deep human need to collect information and organize it into ordered categories.

While private collections had been made and exhibited by a variety of individuals as far back as Neo-Babylonian times (c. 530 BC), most historians trace the origin of modern museums and galleries to western Europe, specifically the opening of Oxford University's Ashmolean Museum in 1683. The Ashmolean's collection was based on the private collection of Elias Ashmole, which was composed of coins and engravings as well as geological and zoological specimens, including a taxidermy mount of the last Dodo seen alive in Europe, all housed in a purpose-designed museum building.[8] The first public museum was the Louvre, which opened a little over a century later, in 1793. Although the heyday of museum building in western Europe and the United States was the Victorian age (1837–1901), the concept and organization of Victorian museums was illustrated picto-rially by Charles Wilson Peale's 1822 self-portrait, *The Artist in His Museum* (Illustration 19.1).

In this painting, we see Peale holding up a curtain in the manner of the impresario he was, beckoning the viewer into his museum where the

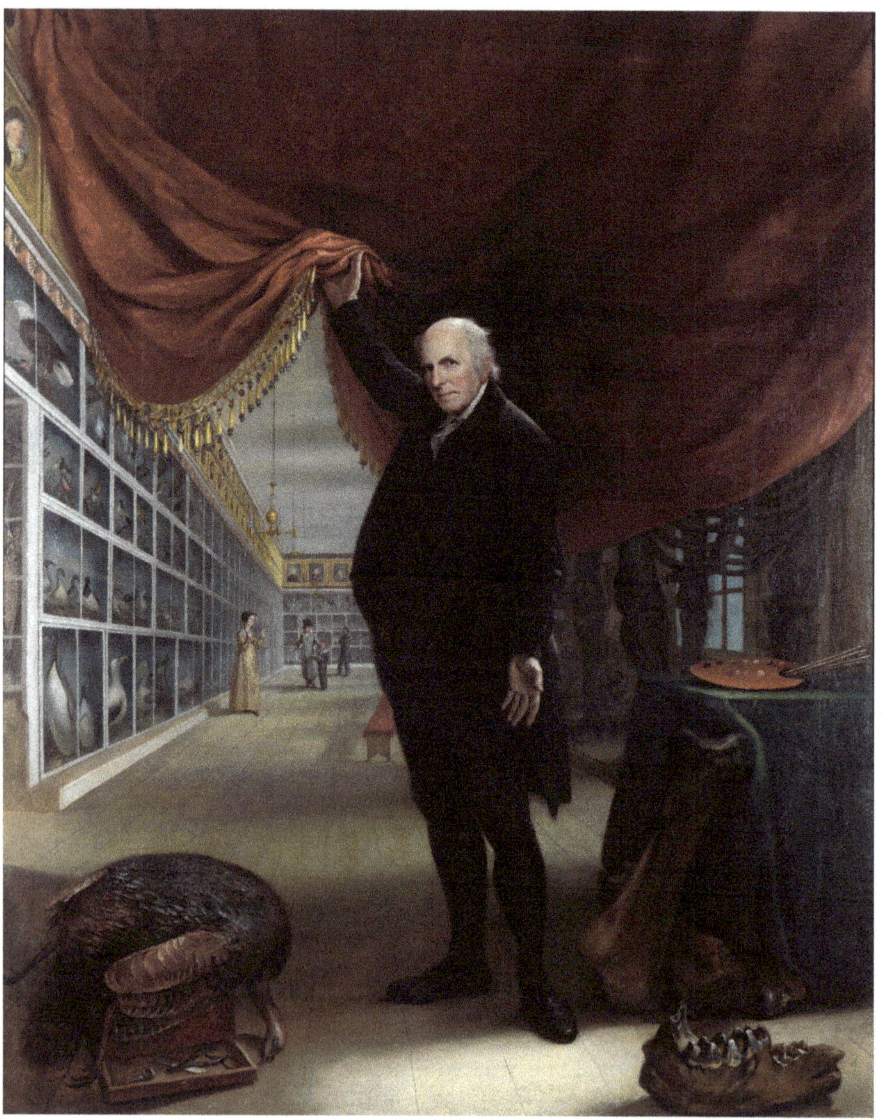

Illustration 19.1 Charles Wilson Peale, *The Artist in His Museum*, 1822. Note a few bones of the *American Incognitum* in the foreground, and the basal parts of the mounted skeleton behind the raised curtain.

specimens have been organized and arranged to tell a story without the need for verbose or complicated labels. Museum historians refer to this design concept as 'object-based epistemology'.[9] As the viewer moved laterally through the exhibit, the specimens changed in a consistent and obvious way (e.g. the zoological transition from sponges to arthropods, or fish to mammals), while, as the viewer's gaze moved up from the lower cases, another dimension of organization was apparent (e.g. simple to highly ornamented, herbivores to carnivores, local species to species from remote regions). In the context of this exhibition-design aesthetic, the specimens are not just the primary focus of the exhibition; they are its only focus. Every aspect of the exhibition's design, and even the gallery design (e.g. large skylights to let natural light in) was present to direct the viewer's gaze to the specimens or objects on display. Furthermore, it was no coincidence that Peale's own portraits of famous and important Americans were placed above the display cabinets. Such placement reinforced the idea that humans (including, in Peale's case, some females) occupied the apex of the natural order, and that the apex of humanity was embodied by rich and noteworthy individuals. Thus, in Victorian museum exhibitions, specimens functioned both as synecdoches and metonyms.

The larger purpose of these institutions was to provide instruction to the general public; but not only in terms of the objects on display. Museums and galleries were also charged with providing, via the examples set by their staff and patrons, standards of the appropriate dress, manners, attitudes and behaviours expected in polite society. In an era before mass entertainment, and when the cities in which most Victorian museums resided played host to an increasingly polyglot population, museum displays were considered one of the most effective means of communicating with the 'common man'. Victorian museums celebrated the accomplishments of the society in which they were embedded, educating the public about those accomplishments, and engaging their visitors in 'civilizing rituals'.[10] At an even more abstract level, Victorian museums were about order, promoting the knowledge that comes about through the understanding of order and, more subtly, emphasizing the importance of preserving the 'natural order' – both scientific and social – if the fruits of human knowledge were to be recognized and developed.

These functions continue to lie at the heart of most museum and gallery exhibitions, and that of the museum/gallery experience in general. Indeed, in our modern, and increasingly postmodern world, when all forms of tradition, authority and order are subject to critical scrutiny, the obsession of museum exhibitions with the promotion of a static order often evokes the sense of tension that arises when one is being subjected to unwanted indoctrination. Indeed, to the casual, modern museum goer, the sight of an old-fashioned Victorian museum gallery – with its long banks of glass cases containing sparsely labelled specimens – is perhaps more likely to elicit a sense of panic rather than pleasure.

Along with their cultural and educational roles, Victorian museums served another, lesser-appreciated purpose; they were centres of intellectual debate and scientific research. This might seem odd to the contemporary reader, for in today's world it seems self-evident that these roles are located primarily in the great research universities. But such was not always the case.

The heroic era of museum and gallery building coincided with a time when natural history specimens, art works, artefacts, manuscripts, and curios of all types were pouring into the urban centres of western Europe and North America owing to the exploration and economic activity that resulted from the Industrial Revolution (c. 1760–1840). In no small measure, the great Victorian museums and galleries were founded in an attempt to deal with these new objects and the new ideas they inspired. Accordingly, a major intellectual challenge of the time involved the cataloguing of these materials, and their placement within a relational classification system that facilitated the identification and/or prediction of properties that could be useful in economic and scientific contexts. Victorian museums and galleries were the institutional foci of this task, which was promoted to the public at large through their exhibitions, lectures, and educational programmes.

Critics who adhere to the views of philosophers such as Marx and Foucault often criticize both Victorian and modern museums for their tendency to use their exhibitions to promote arbitrary and socio-economically selective classification systems aligned with the interests of the wealthy, powerful, and well-educated elite, rather than those of the masses. Indeed, under the rubric of the 'treasure house' (a typical metaphor), museums and galleries can be seen as little more than overbearing signifiers of irresponsible power, in that they both display and legitimate the immoral – and often illegal – expropriation of resources, art works and artefacts from other cultures. While it is certainly the case that all human-devised classification systems are arbitrary, and that almost all of the major museums and galleries include questionably procured specimens, what these critics often overlook is that classification systems are the conduit through which knowledge of the world is gained.[11] Humans have collected specimens and objects of interest and placed them into collections throughout the history of our species, and will continue to do so. When humans went to the Moon, and when they go to Mars, one of the primary purposes of those trips was (and will be) to make collections of the materials they find there. Museums are the places where large collections of such objects are housed and made available for the purposes of education, research and entertainment. The intellectual challenge for museums and galleries, especially in the context of biodiversity conservation and the issue of extinction, is not whether such collections can be used to further the ends of society through their exhibition, but what the ends to which these collections might contribute will, or should, be.

Universities, rather than museums, are now universally regarded as the natural home for high-level education and advanced intellectual inquiry. This change came about around the end of the nineteenth century, when simple documentation and classification were replaced by direct experimentation as a way of gaining knowledge. In science, the intellectual interest in natural objects was replaced by an interest in natural processes, whose investigation did not require access to large collections. Since that time, and with ever-increasing frequency, museum research programmes have been curtailed because museum operating budgets have had to rely on uncertain quantities of public money, either in terms of direct funding for national museums or state/city grants for regional and local museums, augmented by charges levied on visitors, and pleas for donations directed towards the public. As a result, few museums can compete with large research universities in terms of securing the instruments or infrastructure needed for engaging in contemporary scientific research or art collection from their own budgets. A secondary effect of this diminution of institutional expertise has been an increasing trend towards using museums and galleries as destinations for grade-school field trips, when very rudimentary lessons about science, history and culture are taught. The relegation of many once important and proud institutions to the status of 'children's museums' has had a devastating impact on many aspects of museum/gallery culture. Art galleries have managed to escape this cruel fate to a much greater extent than museums insofar as they are patronized/supported by adult members of their communities for their own edification and pleasure.

Extinct and Near-Extinct Species

Darwin's materialist theory of evolution precipitated a scientific and cultural revolution with regard to ideas about how species – including our own – originated, but was surprisingly terse in its treatment of extinction. Darwin accepted that extinct species existed but, aside from regarding extinction as the ultimate consequence of the 'struggle for existence', assigned it little creative role. We now know that extinction plays a major role in promoting biodiversity at all levels by triggering profound changes in extant selective regimes that can abolish the advantages of ecological incumbency (Figure 19.1). Thus, mammals that appeared in the late

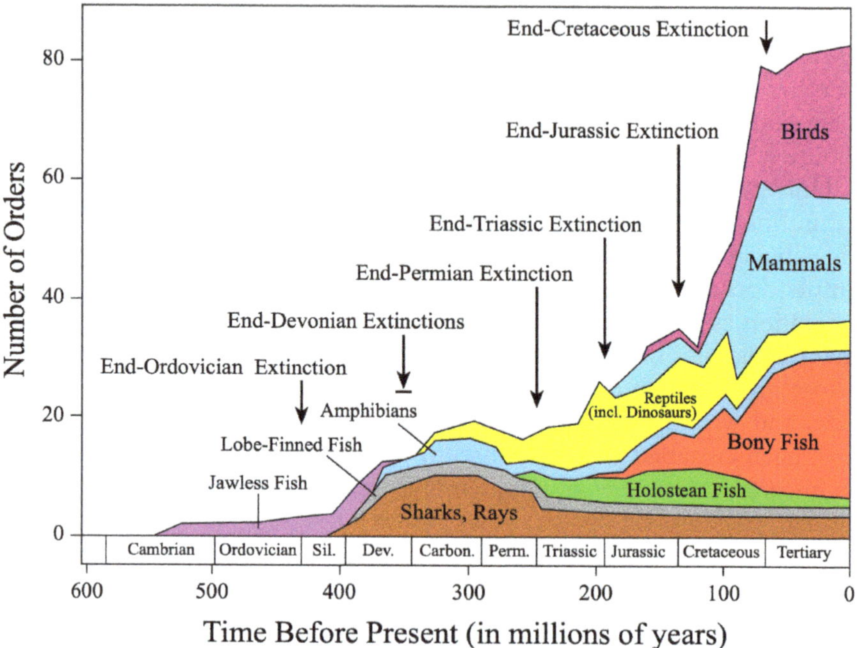

Figure 19.1 Diversity history of the major vertebrate orders. Note how vertebrate diversity has accumulated through time, despite the Earth having suffered major extinction events, as well as the association between the appearance of new orders and the global extinction events that reset selection regimes. © Norman MacLeod.

Triassic Period (c. 200 million years ago), somewhat later than, but about the same time as dinosaurs, were excluded from many ecological roles in the terrestrial landscapes they inhabited because dinosaurs had diversified into those roles first. This situation was reset some 175 million years later by the end-Cretaceous extinction event. The reorganization of local and, in the case of large extinction events, global biotas as a result of extinction events, promotes biodiversity due to evolutionary stacking, as it is rare for all representatives of formerly diverse groups to disappear entirely, even during major extinction intervals. In this way, previous extinction events in Earth's history – especially the so-called mass extinction events – are, in part, responsible for the wonderfully diverse biota we see today, and whose susceptibility to future climate change is a matter of concern to us all.

Ancient Extinctions

The modern scientific interest in extinction research dates from the 1950 publication of Otto Schindewolf's *Grundfragen der Palaontologie* (Handbook of Palaeontology).[12] Schindewolf was an iconoclast among twentieth-century palaeontologists for his rejection of Darwinian natural selection as the primary driver of evolutionary change, his advocacy of directionism and periodicity in earth history, and his catastrophist theory that mass extinctions had taken place in the Earth's geological past driven by radiation from a nearby supernova. All these ideas have been repudiated by subsequent scientific research, except Schindewolf's general concept of mass extinction. Taken up and given fresh empirical support by Norman Newell,[13] and later by David Raup and John J. (Jack) Sepkoski Jr.,[14] the long-suspected idea that very large extinction events had occurred repeatedly in what has come to be called 'Deep Time' had, by the mid-1980s, been established beyond reasonable doubt, though much debate continues regarding these events' cause(s). Early in this process, Raup and Sepkoski developed a quasi-objective statistical test for identifying truly large geological extinction events, and settled on five intervals of earth history they regarded as being characterized by extraordinarily large, or 'mass', extinctions. It is these five events that are referred to as the 'Big Five' mass extinctions of the geological record, with the modern biodiversity crisis often being referred to as an incipient 'sixth'.

Owing to the number of palaeontological species and the difficulty of placing them accurately into a classification system based on modern organisms, most deep-time surveys of biodiversity and extinction employ taxonomic categories higher than that of the species – usually taxonomic families or genera. This makes estimating the true magnitude of geological extinction events, as well as comparison with data from modern species, difficult, because there is no way to tell whether a taxonomic family or genus is represented by only a single surviving species or by several hundred. David Raup addressed this problem in 1979 using a statistical method termed 'rarefaction' to model the relation between percentage family loss and percentage species loss, under the assumption that species were being eliminated in a random manner. The relation between species-level losses and the losses at higher taxonomic levels can be simulated mathematically, and the results of such simulations expressed graphically in the form of what Raup termed a 'kill curve'. Using such kill curves, anyone can transform an empirically validated number of family- or genus-level extinctions into an estimated number of species-level extinctions.

Raup undertook this exercise as a way of estimating the probable species-level loss for the end-Permian extinction event from family-level data.[15] The result astounded everyone. Based on an approximately 20 per cent loss of marine invertebrate families during the last 8 million years of the Permian Period, Raup estimated the level of species loss could range from a low of 76 per cent to a peak of 96 per cent. By 1979, most geologists knew the end-Permian extinction event was the largest in recorded geological history, but few imagined the loss could have been as great as

96 per cent of all fossilizable life in the oceans. Subsequently, Raup and others have calculated the probable species loss inferred from both family- and genus-level counts for the other 'mass' extinction events in Earth's history (Table 19.1). These estimates provide a sobering benchmark against which to appreciate the level of loss the earth's biota has experienced in the geological past, as well as the magnitude of loss it takes to qualify as a true 'mass extinction'.

Single extinction events tend to be the sole research focus of individual researchers and/or teams. Little direct research has been done on multiple events using the same approaches. Many physical mechanisms have been proposed by palaeontologists and others as having caused each of the geological mass extinction events, some serious (e.g. climate change) others fanciful (e.g. mass psychosis in dinosaurs).[16] Among the most consistently proposed causes are sea-level change, climate change, volcanism, marine anoxia, and asteroid/comet impact. Each mechanism has supporters who advocate it as the sole cause of particular extinction events, while others believe the mass extinctions occurred when the operation of various major environmental disruptor mechanisms coincided in time. No single mechanism is accepted as having been the cause of all mass extinction events.

Modern Extinctions

The most respected multi-group source of information on recently extinct and endangered species is the International Union for the Conservation of Nature (IUCN), which maintains and publishes annual updates to its Red List of Threatened Species.[17] Table 19.2 summarizes the most recent (2021) IUCN data for twelve major organismal groups showing percentages of species considered at low risk of extinction, vulnerable, endangered

Table 19.1 Estimates of species-level extinction loss from genus-level palaeontological data, using Raup's kill-curve approach. These results assume equal genus sizes and equal species-specific extinction probabilities. Data from Jablonski, *Extinctions in the Fossil Record*, 18.

Extinction	Age (million years ago)	Est. Percentage Genus Loss	Est. Percentage Species Loss
End-Ordovician	439	60 ± 4.4	85 ± 3.0
End-Devonian	367	52 ± 3.3	83 ± 4.0
End-Permian	245	69 ± 3.8	95 ± 2.0
End-Jurassic	208	60 ± 4.4	80 ± 4.0
End-Cretaceous	65	47 ± 4.1	76 ± 5.0

Table 19.2 Current (2021) estimates of all extinctions of modern species, tabulated by the IUCN from both contemporary and historical data, along with estimates of species numbers in various endangered, vulnerable and low-risk categories. Note that coverage is only adequate for a small proportion of low-diversity, charismatic groups.

Taxonomic Groups	Extinct		Endangered		Vulnerable		Low Risk		Total Evaluated	Est. Total
	No.	%	No.	%	No.	%	No.	%		
Mammals	115	1.93	772	12.97	555	9.32	3,324	55.83	5,954	5,954
Birds (Aves)	186	1.67	683	6.12	798	7.15	8,460	75.82	11,158	11,158
Reptiles	78	0.85	1,026	11.24	561	6.14	5,715	62.58	9,132	10,700
Amphibians	184	2.55	1,723	23.88	721	9.99	3,129	43.37	7,215	8,500
Sharks/Rays	3	0.25	213	17.42	180	14.72	534	43.66	1,223	1,300
Bony Fish	217	1.03	1,626	7.74	1,241	5.91	13,064	62.19	21,006	35,000
Echinoderms	–	–	8	2.14	9	2.41	111	29.76	373	7,000
Insects	141	1.23	1,109	9.66	850	7.40	5,841	50.88	11,480	900,000
Non-insect Arthropods	70	1.87	558	14.94	420	11.25	1,344	35.99	3,734	270,000
Molluscs	464	5.19	1,297	14.52	1,042	11.67	3,331	37.29	8,932	50,000
Corals	1	0.12	32	3.69	234	26.96	293	33.76	868	9,000
Plants	718	1.28	13,828	24.59	8,675	8.39	25,676	45.65	56,245	391,000
Total	2,177	1.59	22,875	16.66	15,286	11.13	70,822	51.57	137,320	1,699,612

or extinct. These data are based on evaluations of over 1 million species against a consistent set of well-defined criteria. As can be seen from the table, while some groups (e.g. mammals, birds) command so much attention from conservationists that, essentially, all species are being monitored, in most cases (e.g. insects, molluscs, arthropods, echinoderms) only very modest efforts are being made to monitor their extinction states. Thus, for many groups, current assessments of extinction risk may not be accurate.

Setting these caveats aside, as these are the best data available at the moment, on a percentage basis the total number of species known to have become extinct, either globally or just in the wild, over historical times is surprisingly small. It is also largely confined to groups associated with ecologically precarious habitats (e.g. birds endemic to islands). Of course, the elimination of any species by causes that can only be deemed as 'non-natural' must be lamented. Moreover, some groups have obviously suffered more extinctions to date than others. But in terms of the proportions of species scientific experts recognize as being lost, the biosphere is currently well below anything that could conceivably be considered a 'mass extinction' (see the prehistoric 'mass extinctions' detailed in Table 19.1).

Concern over modern extinctions within the scientific community comes not so much from the number of species that have become extinct to date, but rather the proportion of species considered endangered and vulnerable, and the rate at which species are moving from the 'low risk' to higher risk categories. The IUCN believes that, if habitat loss, invasive species, pollution, human population, and over-harvesting (HIPPO) trends continue, all species currently considered endangered will become extinct within the next one hundred years. This estimate is controversial. Some specialists consider it far too low, others far too high. Regardless, these data confirm that a substantial proportion of the known biosphere is currently at risk of extinction.

The need for species conservation strategies to be developed and implemented is obvious. The need to inform the public about these issues in a manner they understand is clear. Fortunately, one hundred years is a long time in terms of scientific understanding, technological innovation and public policy. One hundred years ago the dangers of HIPPO practices, and the threat they posed to the biosphere, were appreciated by a vanishingly small number of naturalists and researchers. More generally, few of the predictions made in 1920 about the world of 2020 have come true. This observation is not made to induce any sense of complacency, but rather to give the reader hope that, with goodwill and hard work, there is time to address the extinction issue and avert the future to which the IUCNs data point. Good scientific evidence indicates the predictions of the IUCN, and those of many other conservation organizations, will certainly come to pass if political establishments, regulatory bodies, corporations, and/or the general public worldwide do nothing. But given the high public profile that ecologically sustainable styles of living, working and voting have achieved in just a few decades, it seems unlikely nothing will be done to address this problem and save threatened species. The need, of course, is to educate the public and assist them in mobilizing effective responses to this challenge. This is the area where museums and galleries can, if they choose, make a substantial contribution to species-conservation efforts and, in so doing, reaffirm the purpose for which they were founded originally.

What Stories Should Museums and Galleries Tell?

The challenge that museums and galleries face when mounting extinct species exhibitions is to do justice not only to the best scientific information, but also the associated economic, social, cultural and historical information about this complex subject, and present it in a way that both attracts and informs visitors. Labelling extinction as a scientific or technological problem is insufficient. Extinction is a social and cultural problem.[18] In this context, the level of public understanding of the extinction issue is quite low, and so the potential for visitors coming away better informed is high. But this advantage is offset by the public's generally negative attitude towards the subject. While there is vast popular interest in charismatic endangered species (e.g. felids, proboscideans, raptors), extinction is seen fundamentally as both evidence of, and a metaphor for, failure; failure on the part of extinct species for not winning their struggle for survival, and failure on the part of humanity for thoughtlessly usurping resources on which extant species depend. To take an obvious example, the novel and little-understood creative role extinction has played in promoting biodiversity is almost always ignored in exhibitions in favour of the awkwardness and guilt evoked by popular perceptions surrounding the topic. Indeed, the general lack of informed discussion and education about the extinction issue in such broader contexts harms the entire contemporary conservation movement. But this need not exhaust the range of expectations for such exhibitions, especially among more mature and/or thoughtful audiences.

Owing to its inherent complexity, the extinction issue should always be presented via reference to multiple levels of historio-conceptual understanding, and involve multiple disciplines/interests. Like the magazine articles, books, television documentaries, and videos/movies from which they take their inspiration, museum and gallery treatments of the extinction issue must transcend the tragic and elegiac expository mode, which, in lamenting the passing of selected charismatic species, implicitly casts the finger of accusation back at the exhibition visitor. Justification for this approach is usually made via reference to 'scientific' evidence, as if science is the only source of authoritative evidence, exists in a cultural vacuum, and is in complete agreement with itself. In some instances, such exhibitions have even included trivializations of the subject in the form of extinction-themed video games,[19] thus further complicating the signals being sent. Since this approach has singularly failed to provide a unified, compelling vision of how to address the current crisis – other than to subordinate all social, cultural, economic and political decisions to a vaguely described, authoritarian, environmental 'agenda' – it is little wonder that such increasingly strident exhortations have had the opposite of their intended effect.

In fact, each of the assertions that underpin this approach to extinction exhibitions are deeply flawed. For example, there is no known habitat that has been occupied continuously by humans that has not undergone extensive change and active management throughout the period of that occupation. Thus, the appeal to a time when humans lived in a supposedly 'harmonious' state of nature is largely fictitious.[20] While there is no question that many modern species are in decline, we actually live in an era of biodiversity *increase* owing to the ongoing discovery of new species, the migration of modern species to new habitats,[21] and the evolution of new species adapted to urban environments.[22] The science of extinction is shot through with intriguing complexities and uncertainties.[23] Moreover, the idea that scientists can, much less do, operate in a cultural or political vacuum has been thoroughly debunked.[24] Of late, some conservationists have even begun to challenge the very idea that 'nature', in the a sense of a place separate from the dominant influence of mankind, exists. To the extent this is true – and it certainly is true for many parts of the world – the pertinent issue is not how nature can be returned to a 'natural' state, but rather how 'natural environments' can be managed to meet the needs of both human and non-human species.

This more inclusive view of the extinction issue is beginning to cause an interesting reconsideration of the extinction issue's nature. In addition to widening the scope for discussion, this reconsideration widens the scope for productive and affirmative institutional engagement with a wide variety of local and remote communities.[25] I suspect museums will have considerable difficulty coping with this new approach, tied, as they are, to the concept of their primary purpose being the advocation of order and education of children. Art galleries, however, are much more used to engaging audiences over a wider range of approaches, and are much more comfortable with the idea of their exhibitions appealing to diverse audiences through thoughtful inspiration rather than pedagogic instruction.

One easy lesson that museums could take from galleries is to incorporate a focus on the professional and social lives of the scientists who made the discoveries and found the specimens on display into their traditional focus on the specimens and artefacts themselves. Art gallery exhibitions typically contain precious little information about the techniques of painting and sculpting or theory of perspective, but always include much information about the lives of the artists themselves and the effect their work has had on the lives of other people. The lives of many scientists are no less eventful and interesting, and the mystery of the creative process is largely the same in both groups.

Take, for example, the complex of social factors that influenced relations between two lions of Victorian natural history, Sir Richard Owen and Thomas Henry Huxley (Illustration 19.2). Both had similar middle-class origins, though Huxley's childhood was blighted by family financial misfortunes. Owen enjoyed the benefits of a formal secondary-school education whereas Huxley managed only two years of primary school. Although neither held a university degree, both managed to master their topics more or less through self-education. By dint of his undeniable talent and hard work, Owen, the older of the two, quickly rose through the ranks of Victorian science, developing, along the way, a secondary talent for cultivating influential patrons. Huxley, on the other hand, had a much more difficult early career. After returning from his reputation-making voyage as the surgeon and marine naturalist on board HMS Rattlesnake, Huxley was

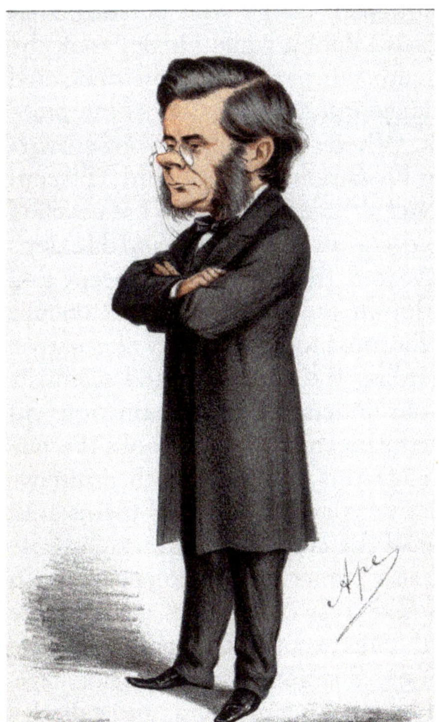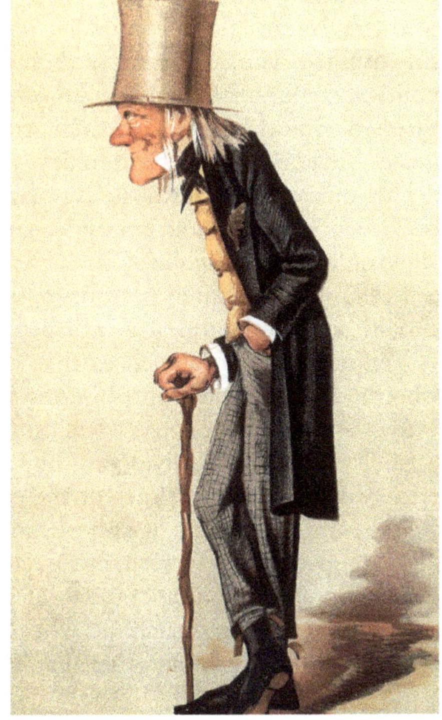

Illustration 19.2 Carlo Pelegrini's caricatures of Sir Richard Owen (left) and Thomas H. Huxley (right) for *Vanity Fair*, c. 1870.

chronically in debt and drifted through a succession of low-paid, tempo-rary positions, despite the efforts of his friend, colleague and most sup-portive patron, Richard Owen. The regard both men had for each other, though, reversed dramatically when they took opposing sides in the most profound scientific and philosophical question of their day – the progres-sion, or evolution, of life.

Owen accepted Cuvier's ideas on extinction, and acknowledged the progressive patterns of change that characterize many fossil lineages. However, in keeping with establishment doctrine, he, like many of his contemporaries, could not bring himself to break with religious dogma, especially over the issue of human origins. His younger colleague, though, had no such qualms. Huxley went well beyond simple explanations of the evidence for evolution, choosing to both reject and deride Owen's personal views, as well as his scientific interpretations, in the most castigating and personal manner, both in print and at the lectern. But rather than a simple disagreement between former associates over a popular debate, the highly acrimonious nature of Owen and Huxley's clash must also be viewed in the context of a much more far-reaching battle for influence and power within the (then) emerging field of modern science. In rejecting evolution, Owen took the side of well-to-do gentlemen clergy who pursued their scientific interests as a somewhat obsessive hobby, while Huxley took the side of those who, as a result of their circumstances as well as their personal politics, favoured the conversion of science into a profession whose prac-titioners would be paid for their work. Whereas the opinions of eyewit-nesses differ as to whether Huxley won his famous 1860 Oxford Museum debate on evolution with the Rev. Samuel Wilberforce – who has coached by Owen[26] – evolution is now accepted almost universally, and Huxley's view of a professionalized science, divorced from religious precept and not reliant on personal patronage, so dominates our view of how science should be organized that it is difficult for most to imagine any alternative.

Telling real stories such as this regarding the naturalists and scientists who made significant contributions to our understanding of extinction and extinct species – not some oversimplified caricature that overlooks the hu-man elements of their lives and the social factors embedded in their discov-eries – alongside the stories of the specimens and/or artefacts themselves, would not only provide a critical aspect of the context within which those discoveries must be viewed, but would also attract a new audience to such exhibitions, and counteract the modern stereotype of scientists as little more than asocial thinking machines who, with rare exceptions, cannot communi-cate with 'normal people' other than through a jargon-rich foreign language.

This brings us, finally, to the issue of how willing museums and galleries are to mount exhibitions of extinct species, and, in so doing, engage with the extinction issue. Here, once again, art galleries will have the advantage insofar as their patron community expects these institutions' exhibitions to be experimental and controversial so as to inspire engagement with diverse audiences. Museums, on the other hand, are far more conservative by nature. Many depend on the support of regional and national governments that prioritize their basic primary and secondary school educational programmes, and view their role as being that of an established and authoritative reference rather than a source of debate, passionate inquiry and inspired engagement. Both these mindsets serve as constraints for both types of institutions in terms of presenting the extinction controversy adequately. But given the importance of this topic, its scope, and the wholly inadequate manner it has been dealt with in the past, the extinction issue embodies an unparalleled opportunity for both museums and galleries to rethink how contemporary issues can be presented to their visitors, what sort of visitors such subject matter might attract, and how, through their exhibitions, they can become a vital player and forum for the wide-ranging discussions that lie at the heart of this debate, rather than being relatively little-known and little-appreciated sideshows.

Norman MacLeod is a distinguished professor at the School of Earth Science and Engineering, Nanjing University, Nanjing, China. He was formerly the Dean of Postgraduate Education and Training (2013–16), Keeper of Palaeontology (2000–13), and Associate Keeper of Palaeontology (1999) at the Natural History Museum, London, in addition to being an honorary professor at University College London and a visiting professor at the Nanjing Institute of Geology and Palaeontology, Chinese Academy of Sciences. He is perhaps best known for his work on the causes of Phanerozoic extinctions, and his theoretical, methodological and applied work in quantitative morphological analysis.

Notes

1. Mayor, *First Fossil Hunters*.
2. Cheke, 'Dodo's Last Island'.
3. Buffon, *Histoire Naturelle*; Dugatkin, 'Buffon, Jefferson and the Theory of New World Degeneracy'.
4. Sellers, *Mr. Peale's Museum*.
5. Rowland, 'Thomas Jefferson'.
6. Cuvier, *Mémoire sur les Espèces d'Éléphans*.
7. In his 1796 monograph, Cuvier not only described the Ohio material, but also noted the living elephants in Africa and India comprised not only separate species, but separate genera. MacLeod, 'The Geological Extinction Record'.

8. This building has survived to the present day, and can be found on Broad Street in Oxford as the History of Science Museum, where it houses a leading collection of scientific instruments.
9. Conn, *Museums and American Intellectual Life*, 24.
10. Ibid., 6.
11. Conn, *Museums and American Intellectual Life*; Kubler, *The Shape of Time*; DiMaggio, 'Classification in Art'; Kopyoff, 'Cultural Biography of Things'.
12. Available in English as Schindewolf and Reif, *Basic Questions in Paleontology*. See also Schindewolf, 'Neokatastophismus?'
13. Newell, 'Revolutions in the History of Life'. See also Newell, 'Crises in the History of Life'.
14. Raup and Sepkoski, 'Mass Extinctions'. See also Raup and Sepkoski, 'Periodicity of Extinctions'; Raup, 'Biological Extinction in Earth History'; Raup, *The Nemesis Affair*.
15. Raup, 'Size of the Permo-Triassic Bottleneck'.
16. See Benton, 'Scientific Methodologies in Collision'.
17. 'Summary Statistics'.
18. Heise, *Imagining Extinction*.
19. Naish, 'Extinction'.
20. See Koch et al., 'Earth System Impacts of the European Arrival'.
21. Sax and Gaines, 'Species Diversity'.
22. Thomas, *Inheritors of the Earth*.
23. Maier, *What's So Good About Biodiversity?*
24. Latour, *Pandora's Hope*; Takacs, *The Idea of Biodiversity*.
25. Carnall et al., 'Natural History Museums'.
26. Desmond, *Huxley: The Devil's Disciple*.

Bibliography

Benton, Michael J. 'Scientific Methodologies in Collision: The History of the Study of the Extinction of the Dinosaurs'. *Evolutionary Biology* 24(37) (1990): 371–400.
Buffon, Georges-Louis Leclerc, comte de. *Histoire Naturelle*. Paris: Imprimerie royale, 1749–1788.
Carnall, Mark, et al. 'Natural History Museums as Provocateurs for Dialogue and Debate'. *Museum Management and Curatorship* 28(1) (2013): 55–71.
Cheke, Anthony. 'Dodo's Last Island: Where Did Volkert Evertsz Meet the Last Wild Dodos?' *Proceedings of the Royal Society of Arts & Sciences of Mauritius* 7 (2004): 7–22.
Conn, Steven. *Museums and American Intellectual Life, 1876–1926*. Chicago: University of Chicago Press, 1998.
Cuvier, Georges. *Mémoire sur les Espèces d'Éléphans Vivants et Fossiles*. Paris: Baudoin, 1796.
Desmond, Adrian. *Huxley: The Devil's Disciple*. London: Michael Joseph, 1994.
DiMaggio, Paul. 'Classification in Art'. *American Sociological Review* 52(4) (1987): 440–55.
Dugatkin, Lee Alan. 'Buffon, Jefferson and the Theory of New World Degeneracy'. *Evolution: Education and Outreach* 12(1) (2019): 1–8.
Heise, Ursula K. *Imagining Extinction*. Chicago: University of Chicago Press, 2016.
Jablonski, David. 'Extinctions in the Fossil Record', in John H. Lawton and Robert M. May (eds), *Extinction Rates* (Oxford: Oxford University Press, 1995), 25–44.
Koch, Alexander, et al. 'Earth System Impacts of the European Arrival and Great Dying in the Americas after 1492'. *Quaternary Science Reviews* 207 (2019): 13–36.

Kopyoff, Igor. 'Cultural Biography of Things', in Arjun Appadurai (ed.), *The Social Life of Things: Commodities in Cultural Perspective* (Cambridge: Cambridge University Press, 1986), 64–91.
Kubler, George. *The Shape of Time: Remarks on the History of Things.* New Haven, CT: Yale University Press, 1962.
Latour, Bruno. *Pandora's Hope: Essays on the Reality of Science Studies.* Cambridge, MA: Harvard University Press, 1999.
MacLeod, Norman. 'The Geological Extinction Record: History, Data, Biases and Testing', in Gerta Keller and Andrew C. Kerr (eds), *Volcanism, Impacts, and Mass Extinctions: Causes and Effects.* Geological Society of America Special Paper 505 (2014), 1–28.
Maier, Donald S. *What's So Good About Biodiversity? A Call for Better Reasoning About Nature's Value.* Dordrecht: Springer, 2012.
Mayor, Adrienne. *First Fossil Hunters: Paleontology in Greek and Roman Times.* Princeton, NJ: Princeton University Press, 1990.
Naish, Darren. 'Extinction: Not the End of the World at London's Natural History Museum'. *Scientific American Blog*, 15 August 2013. Retrieved 11 January 2021 from https:// blogs.scientificamerican.com/tetrapod-zoology/extinction-not-the-end-of-the-world-at-londone28099s-natural-history-museum/.
Newell, Norman D. 'Crises in the History of Life'. *Scientific American* 208(2) (1963): 76–95.
———. 'Revolutions in the History of Life', in Claude C. Albritton Jr. (ed.), *Uniformity and Simplicity: A Symposium on the Principle of the Uniformity of Nature.* Geological Society of America Special Paper 89 (1967), 63–91.
Raup, David M. 'Biological Extinction in Earth History'. *Science* 231(4745) (1986): 1528–33.
———. *The Nemesis Affair: A Story of the Death of Dinosaurs and the Ways of Science.* New York: W.W. Norton, 1986.
———. 'Size of the Permo-Triassic Bottleneck and Its Evolutionary Implications'. *Science* 206(4415) (1979): 217–18.
Raup, David M., and John J. Sepkoski. 'Mass Extinctions in the Marine Fossil Record'. *Science* 215(4539) (1982): 1501–3.
———. 'Periodicity of Extinctions in the Geologic Past'. *Proceedings of the National Academy of Sciences of the United States of America* 81(3) (1984): 801–5.
Rowland, Steven M. 'Thomas Jefferson, Extinction, and the Evolving View of Earth History in the Late Eighteenth and Early Nineteenth Centuries'. *Memoir* 203 (2009): 225–46.
Sax, Dov F., and Steven D. Gaines. 'Species Diversity: From Global Decreases to Local Increases'. *Trends in Ecology & Evolution* 18(11) (2003): 561–66.
Schindewolf, Otto H. 'Neokatastrophismus?' *Zeitschrift der Deutschen Geologischen Gesellschaft* 114 (1963): 430–45.
Schindewolf, Otto H., and Wolf-Ernst Reif. *Basic Questions in Paleontology: Geologic Time, Organic Evolution, and Biological Systematics.* Chicago: University of Chicago Press, 1993.
Sellers, Charles Coleman. *Mr. Peale's Museum: Charles Willson Peale and the First Popular Museum of Natural Science and Art.* New York: Norton, 1980.
'Summary Statistics'. The IUCN Red List of Threatened Species, 10 December 2020. Retrieved 11 January 2021 from https://www.iucnredlist.org/resources/summary-statistics#Summary%20Tables.
Takacs, David. *The Idea of Biodiversity: Philosophies of Paradise.* Baltimore, MD: Johns Hopkins University Press, 1996.
Thomas, Chris D. *Inheritors of the Earth: How Nature Is Thriving in an Age of Extinction.* London: Allen Lane, an imprint of Penguin Books, 2017.

Chapter 20

Exhibiting Extinction

Thylacines in Museum Display

Kathryn Medlock

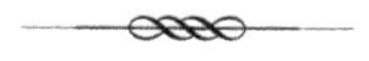

Australia's largest marsupial carnivore, the Tasmanian tiger, wolf or thylacine (*Thylacinus cynocephalus*) became extinct only 133 years, 11 months and 29 days from the date of the European invasion of Tasmania to establish a British penal colony in 1803. From that date, marking the start of European interactions with thylacines, they were consistently regarded as a major pest that hindered the establishment of agriculture and the development of a profitable wool industry across the island. The last-known thylacine died in the Beaumaris Zoo in Hobart on 7 September 1936.[1]

Between 1803 and 1936, the thylacine became increasingly rare, as it failed to survive an organized regime of victimization. The reasons for its extinction are a combination of factors. As European agriculture and sheep farming expanded across Tasmania, thylacines suffered the effects of habitat alteration and loss, as well as a reduction in the availability of prey species when favoured habitats were converted to crops and pasture. Many private landholders, such as those that managed the large Woolnorth property in north-west Tasmania, as well as agricultural societies such as the Buckland and Spring Bay Tiger and Eagle Extermination Society, established in 1884[2] (hereafter referred to as the Buckland Society), willingly paid a reward for dead thylacines.[3] The Buckland Society initially relied on subscriptions from farmers, but it soon ran out of funds to pay the promised £5 reward, which resulted in some of its members, with support from other Tasmanian farmers, campaigning for the government to pay for tiger extermination.[4] Eventually, in 1888 after an extensive campaign, the Tasmanian government passed legislation to pay a bounty of £1 for each adult thylacine killed, and 10 shillings for young animals.[5] To the Tasmanian public, the thylacine was always portrayed as a pest to be destroyed by any means, but the government scheme now gave farmers, bushmen and farm labourers a way to make some money to support their small wages.

Although some of the animals killed under the Buckland scheme were subsequently obtained by the Tasmanian Museum (now the Tasmanian Museum and Art Gallery, TMAG) in Hobart, the rules for the government bounty stipulated that specimens had to be destroyed to prevent a second claim being made, thus cutting off the supply of thylacines for museum purposes. This resulted in the curator of Hobart's Tasmanian Museum, Alexander Morton, proposing that specimens should be sent to the museum for payment of the reward.[6] The strategy was not particularly successful as it was far easier for bounty claimants to transport animals to the nearest police station than send them to Hobart. Therefore, few specimens from the government bounty scheme came to the museum, but 2,209 claims for dead thylacines were paid.[7]

Tasmania's first museum, established by the Royal Society of Van Diemen's Land for Horticulture, Botany and the Advancement of Science (now the Royal Society of Tasmania, RST, but hereinafter 'the society') began operations in 1848.[8] At this time, Tasmania was still a penal settlement, officially known as Van Diemen's Land, and thylacines were still found in the areas where sheep farming had not progressed. This society, the first scientific society in the Southern Hemisphere, was established under the patronage of Sir John and Lady Jane Franklin in 1842, and from the start it announced its desire to establish a museum in the young colony. Opened in 1848, the newly established museum emphasized natural history and ethnography, exotic specimens and curios.[9]

It was not until 1856 that the first thylacine skin was donated to the museum by two businessmen, Lade and Morris, who had caught and killed the thylacine at Falmouth on Tasmania's east coast.[10] The donation, announced via the society's published proceedings as the 'skin of dog opossum (hyena) or tiger',[11] was undoubtedly thought to be an important acquisition; but, on arrival, it was found to be incorrectly prepared for museum purposes because the skull and the leg bones had been removed. However, the society secretary, Joseph Milligan, wrote to Lade and Morris explaining the correct method of preparation for museum taxidermy, as well as asking for a Tasmanian devil (*Sarcophilus harrissii*).[12] As the first thylacine to be acquired, this was a valuable addition for the fledgling museum, and despite the condition of the skin on arrival, it was mounted and placed on exhibition.[13]

I will now provide an overview, the first of its kind, of the exhibition history of the thylacine at the Tasmanian Museum and Art Gallery. This research shows how changing display practices reflected the social values of the time and reveal shifting attitudes towards the thylacine and its cultural significance.

Although Lade and Morris's thylacine was the first for the fledgling museum, this was not the first thylacine to be obtained by a museum. Several specimens had already been sent overseas, by Tasmanian naturalists, many of whom were members of the society. By the 1850s thylacines were held in the collections of Bullock's Museum, London (1812), the Linnean Society of London (1824), the private museum of Joshua Brooks (1827), Leyden, Holland (1827), the Zoological Society of London Museum (1829), the Royal College of Surgeons of England (1846) and the British Museum, London (1843).[14] The first living thylacine arrived at London Zoo in 1850.[15]

Public interest in overseas fauna was at an all-time high in Britain during the early 1800s as new species flooded in from colonial outposts and exploration voyages. For example, kangaroos were highly sought after by London scientists, showmen and menageries, and were viewed with wonder and awe.[16] The thylacine on display in William Bullock's Egyptian Hall – also known as the London Museum and Pantherion, in Piccadilly, London – was claimed, in the museum's official guide book, to be the 'only one known in any collection'.[17] As the first to display a new species, and with visitors flocking to see such spectacles, Bullock was able to get an edge over the competition and ensure a paying audience. Not only was his museum considered groundbreaking for displaying previously unknown species, but it also presented those animals in a suggested natural setting among rocks and vegetation, rather than in individual cases. Bullock's museum became an extremely popular attraction for the London public, but not with the scientists of the day, who wished to establish collections along scientific lines, and complained that the sensationalist displays hindered their studies of valuable new specimens.[18]

It was not until 1858 that the society's museum obtained a second thylacine. Donated by a member of Tasmania's House of Assembly, Charles Shum Henty, this valuable donation was publicly announced in the local press. Noting that the thylacine was 'nearly obsolete', and using language attesting to Henty's bravery and service to the public in obtaining such an animal, it was described as having 'a very formidable appearance, the mouth, like that of the devil, being large, and furnished with long and very strong teeth, as white as ivory, and the jaws extending far into the skull'.[19] This new specimen was clearly valued by the society. It was reported that local taxidermist Mrs Touch was asked to prepare the specimen for display, and a photograph was commissioned. Mrs Touch was actually Mrs Tost, a well-known Hobart taxidermist and naturalist who was frequently contracted to do preparations for the society. The photograph no longer exists, but no doubt the completed specimen accentuated the large teeth and fierce nature of the thylacine in order to support Henty's bravery and public service.

The 1850s also saw the start of a series of international exhibitions spearheaded by the Great Exhibition in London in 1851. International representation such as this gave Tasmania an unprecedented opportunity to display and promote Tasmania's natural and economic resources. Exhibition commissioners, mainly from the Royal Society, oversaw the Tasmanian contributions and organized the construction and shipping of exhibition items.[20] The Tasmanian court primarily promoted economic industries such as timber, whale products, minerals, furs and agricultural products. A taxidermic platypus drew the attention of the *Illustrated London News* of 25 July 1851, noting that it looked like a 'small polecat with a duck's beak sewn on its muzzle'. Only a single tanned thylacine skin described as 'with the hair on' was sent to the original exhibition by William Rout of Hobart, a member of the Royal Society.[21] Displayed with other Tasmanian fur products, the thylacine was not a prominent feature. The Tasmanian exhibition products were selected to boost Tasmania's industry and reputation internationally (even though it was still a penal colony), so perhaps thylacines were not deemed interesting enough (unlike the platypus), not commercially valuable (unlike the possum and wallaby furs on display), or not worth promoting (because of their perceived negative impact on agricultural enterprises).

When the exhibition closed, it was moved to a new site where it became popularly known as the Crystal Palace.[22] A major difference from the earlier 1851 exhibition was that, as well as promoting industry, exhibits could be sold. Therefore, the displays effectively provided a shopfront from which museums, collectors and businesses could purchase products and specimens.[23] When the Crystal Palace exhibition was initially proposed, the Tasmanian commissioners asked the community for gifts of suitable items, but when few put their hands up to provide their products at no cost, the commissioners decided they would pay for items. Predictably, this resulted in a large number of people offering to be part of the exhibition.

Prior to shipment, the entire collection was put on display in Hobart. In the centre of the display hall was a pyramid covered with the skins of Tasmanian animals. The newspaper report of this display describes that 'the two ends of this pyramid are terminated by the antlers of fallow deer, and on the top there is a stuffed specimen of the native tiger. The walls are lined with skins of the opossum, native tiger, tiger cat, and kangaroo'.[24] Clearly, thylacines had somewhat redeemed their reputation as worthy of inclusion because, when the exhibition opened again in the Crystal Palace, it included an entire thylacine family consisting of a male, female and young. Perhaps to encourage a museum purchase of the group, they were displayed in the ethnology department, an area where museum buyers would be more likely to find items of interest. The strategy proved successful, and the whole remarkable group was sold to the Liverpool Museum in 1863 for an unknown amount.[25]

During the second half of the nineteenth century, as museums evolved and increasingly opened their doors to the public, they began to move away from the static scientific displays of previous years. In such displays, specimens were typically arranged in neat rows intended to show the wonders of God's creation, new discoveries, and to encourage an overall appreciation of nature.[26] Alternatively, they were displayed according to their geographical region, which mirrored the illustrations found in natural history books that depicted a range of animals under general headings. Displays of Australasian species might include koalas, kangaroos and thylacines in a single display, creating the (misleading) impression that they lived together in a semi-natural setting.

As museums grew, they increasingly focused more on their audiences, keeping research specimens separated from the public galleries, using methods designed with an educational focus, and integrating popular appeal.[27] Display techniques also started to depict animals in more naturalistic poses and settings. Representing species as a nuclear family, a concept that could easily be understood by nineteenth-century museum visitors, was a way to gain public appeal as well as showing both sexes and young in close proximity to each other.

When the Tasmanian Museum received a female thylacine and her four young from William Turvey, the secretary of the Buckland Society in October 1884, they were immediately prepared for display. Hobart's *Mercury* newspaper reported that, 'Mr Morton has arranged the young ones with graphic ingenuity, and the exhibit is one of which any museum might well be proud'.[28] The group was initially posed with the mother resting on the ground and her young gathered close by, enabling easy movement between the protection of her open pouch and the outside world (Illustration 20.1). Although hailed as a resourceful and imaginative exhibit when it was prepared, in later photographs of the same group, the idealized 'nuclear family' is completed by the addition of a large unrelated male thylacine placed prominently at a higher level than the female and her young, as if protectively guarding his family (Illustration 20.2).

The frequency of portraying male and female specimens in natural history displays was the subject of a study undertaken at the Manchester

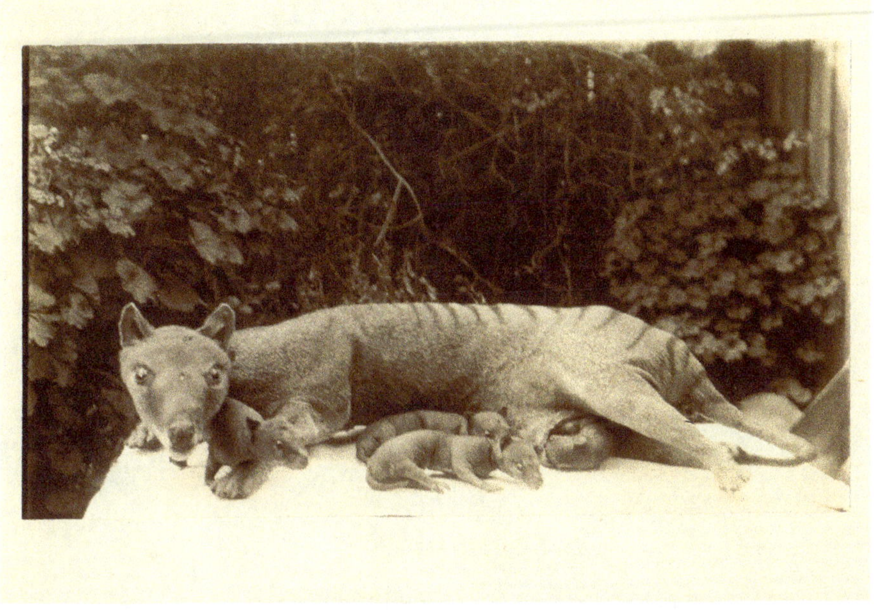

Illustration 20.1 A female thylacine with her four young, presented to the Tasmanian Museum by the Buckland and Spring Bay Tiger and Eagle Extermination Society. Photographer: Edmond Haldane Cotsworth, c. 1885. Collection: Tasmanian Museum and Art Gallery. Q4451.

Museum in England in 2008 examining gender bias in natural history displays. The author found that, although family groups are less common in displays, male specimens significantly outnumber females, and are usually placed at a higher level than females.[29] The Tasmanian Museum family group, always with the added male, remained the mainstay of its thylacine exhibit until at least the late 1920s, when three new thylacines – an adult male collected from the wild in 1926, and an adult female and a younger animal (not related) – were obtained from the Hobart Zoo in 1928.[30] Initially displayed in a case in the museum's foyer, the three unrelated mounts were later used to create a new thylacine family, and this remained on display for approximately forty years.

Of these three family groups, only the most recent survives. The Liverpool specimens were unfortunately destroyed by a Second World War bombing raid on 3 May 1941, and no photographs of it are known to exist.[31] The Buckland group was destroyed in 1935, the year preceding the death of the last captive thylacine.[32] The destruction of these specimens appears to be an example of massive negligence by museum staff. However, closer inspection reveals that the Tasmanian Museum, despite the efforts

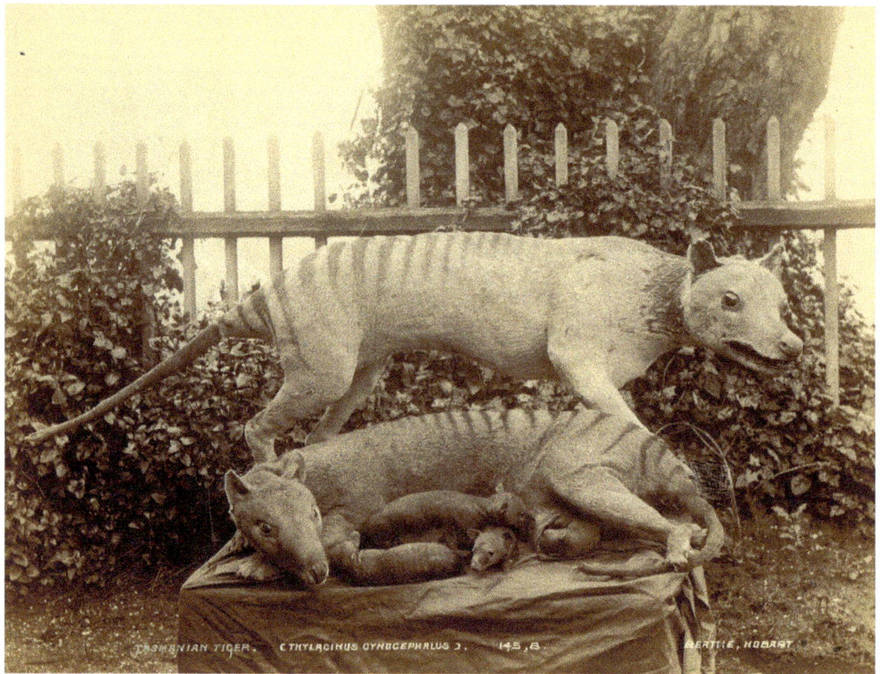

Illustration 20.2 Female thylacine with her four young with the addition of a large unrelated male thylacine. Photographer: John W. Beattie, c. 1890s. Collection: Tasmanian Museum and Art Gallery. Q4450.

of the Board of Trustees, had become extremely run down due to consistently poor funding. When a new director, Joseph Pearson, arrived in 1934, he reported that 'the Zoological Section is of very little value, and is the worst part of the Museum'.[33] The only storage available was in two basement rooms that had become severely overcrowded. In 1935, at the height of the Great Depression, two unemployed men were hired to clean out the stores. The thylacine family became a casualty of this operation, but the annotation recording their destruction in the museum's register is signed with the initials of a full-time staff member.[34] It was not done by mistake, it was done to make space.

Just before Pearson's arrival in Tasmania, the Carnegie Corporation of New York had conducted a study of museums across Australia. The results contained in their report praised some of the Tasmanian Museum displays, but severely criticized display techniques across Australia generally, claiming that Australian museums 'lacked the flair of exhibition'.[35] This report may have influenced the Tasmanian Museum trustees, as a taxidermist and an educator were soon employed, and valuable public spaces were taken over for the establishment of a dedicated education section, a laboratory and a taxidermy room. In 1937, the Carnegie Corporation sponsored a trained taxidermist, Mr Frank Tose from the California Academy of Science, to instruct Australian staff in taxidermy techniques, as well as in the design and construction of habitat dioramas. Whilst in Tasmania, he spent two weeks in Hobart working with museum staff,[36] and gave a lecture to the Royal Society of Tasmania on modern museum methods.

By 1940, four new habitat groups had been installed in TMAG, including one using the three 1920s thylacines placed in a forest setting and posed as a nuclear family[37] (Illustration 20.3). This new diorama was always popular, but demonstrates how human bias can infiltrate a natural history display. Although there are many reports of thylacine in small family groups, the make-up of these groups has not been determined. It is not clear if thylacines actually lived as discreet nuclear family units (male, female and young), or if the display diorama thylacines had been related to each other. Paddle argues for the existence of family groups,[38] but given that so many aspects of thylacine social behaviour were not studied while the species was still alive, the true situation is not known.

As Ashby observes, museums are places made by people, and therefore they are not immune to human stories and biases infiltrating the displays.[39] Clearly the ideal nuclear family group is something that was, and still is,

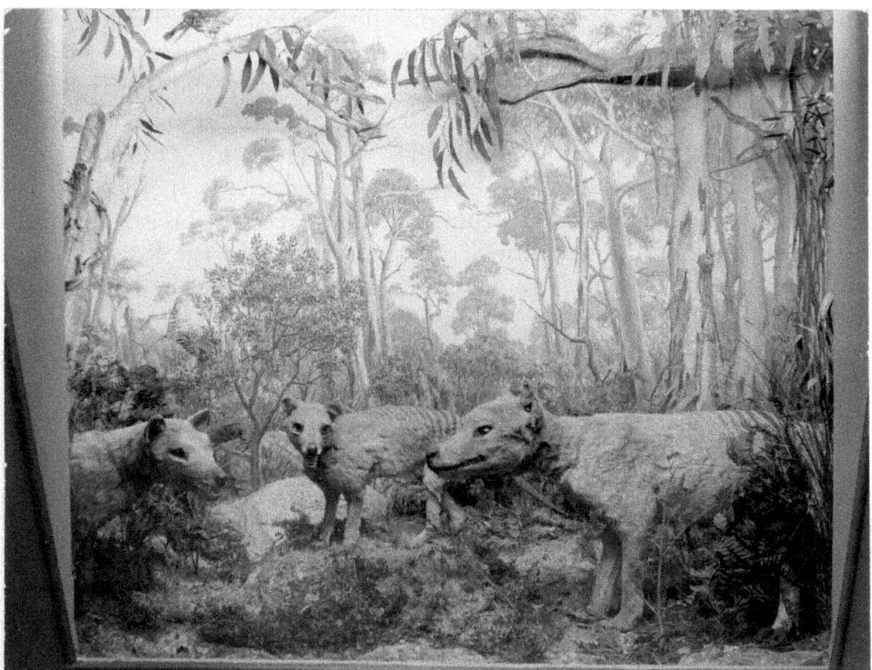

Illustration 20.3 Thylacine Diorama, Tasmanian Museum and Art Gallery. This diorama was on display for approximately forty years. Photographer: Don Stephens, 1965. Collection: Tasmanian Museum and Art Gallery. Q8778.

very familiar with the public, and it certainly conformed with ideas of family composition during the first half of the twentieth century when the diorama was constructed. A second misleading aspect of this diorama is the Tasmanian forest background. Closed forests cover a large proportion of Tasmania, yet do not support large populations of mammals because prey is scarce. Ideal thylacine habitat was relatively open mosaic grassland and woodland habitats that could support an abundance of prey whist at the same time providing the necessary cover. Perhaps the thick forest scene was intended to convey the thylacine as a species that is hidden, yet awaiting rediscovery.

As the threat of extinction became a reality, the desire for museum specimens increased. Warnings of thylacine extinction were well known, and had been promulgated since Professor Richard Owen predicted extinction when addressing a Zoological Society of London meeting on 12 December 1843. Others added their voices, including Tasmanian naturalist Ronald Campbell Gunn, as well as the famous publisher of zoological monographs, John Gould.[40] Paddle identified seven instances from different Tasmanian authors relating to thylacine rarity that specifically mention extinction before 1888.[41]

In 1871, the director of the Australian Museum in Sydney, Gerard Krefft, writing about thylacines, acknowledged extinction yet still encouraged collecting by advising: 'Let us therefore advise our friends to gather their specimens in time, or it may come to pass when the last Thylacine dies, that the scientific men across Bass's Straits will contest as fiercely for its body as they did for that of the last aboriginal man not long ago'.[42] He understood the advantage of having many specimens. In 1866, he had arranged for the Australian Museum's collector, George Masters, to collect in Tasmania, where he was provided with assistance by the Royal Society and the Tasmanian Museum. Masters subsequently returned to Sydney with a total of 297 specimens, representing 23 species of Tasmanian mammals, and the society added a thylacine pouch young to his already large collection. On his return to Sydney, the collection included 25 thylacine specimens.[43]

Despite knowledge of increasing thylacine rarity, the collecting of rare and unusual specimens of species could prove profitable for museums. Excess specimens were regularly exchanged with other museums so that each museum could improve their holdings for display and for research. Many of the specimens collected by Masters were subsequently exchanged with other museums, giving the Australian Museum the opportunity to expand its own collection, the only costs being packing and shipping charges.

Museums wanted thylacines, as much as possible to fulfil the aim of forming an entire species inventory through a comprehensive collection.

Thylacines are also known as a striking example of convergent evolution between placental canids (dogs and wolves) and the marsupial thylacine due to their similar skull and body morphologies, despite evolving separately. So famous is this as an example of convergence that thylacine or dog skulls were often used in university examinations to test students' ability to spot the difference between the skulls of marsupials and those of placental mammals.[44] Many thylacine displays in museums around the world still emphasize this important evolutionary concept through their displays.

After the last-known thylacine died in Hobart Zoo in 1936, Tasmanian fauna authorities immediately implemented efforts to find out where thylacines might still occur. Museum director Pearson offered £50 for a living thylacine in good condition.[45] He was not successful. Over subsequent years many searches have been conducted – none have succeeded. To date, there has been no road kill, no remains found in the bush, and no verified killing or capture. A few films and photographs occasionally appear, but none have proven to be that of a thylacine.

The acquisition of a living thylacine for a zoo or a dead one for a museum, to prove evidence of its continued survival in the wild, never occurred. The International Union for the Conservation of Nature currently deems a species to be extinct when no evidence of its continued existence has been found, despite extensive surveys in known habitat, and there has been a time lapse of fifty years since the last verified evidence. Therefore, on 7 September 1986 the thylacine was officially declared extinct. Many still cling to the hope of finding one in the wild, driven by the understanding that an absence of evidence is not evidence of absence, however, no search has been successful.

The story of the thylacine is one of the best-known extinction stories, capturing the imagination of people around the world as much as those other famous extinctions, the dodo, the passenger pigeon and the great auk. It has become an extinction icon and, although its extinction could have been prevented, it is still commonly used as a symbol to represent Tasmania and all things Tasmanian (in the same way that symbolic lions are associated with the English Test cricket team and the Royal Family).

Thylacines feature on the Tasmanian coat of arms, the current Tasmanian government logo, and branding for sporting teams. The thylacine continues to inspire tourism ventures, artists, furniture makers, writers, filmmakers and poets. If the whole animal is not depicted, the distinctive banding pattern, or even an abstract representation of the pattern, is enough to convey a Tasmanian message. The anniversary of its death, 7 September, is the Australian National Threatened Species Day.

In response to the anomaly of the thylacines' transformation from hated pest to Tasmanian icon, in 1998 the Tasmanian Museum and Art Gallery developed a major touring exhibition on the thylacine. Called 'Tasmanian Tiger: The Mystery of the Thylacine', its themes included biology, history, persecution, extinction, art, design and its use as a Tasmanian symbol. As it travelled to all of Australia's major natural history museums, it inspired talks and seminars on extinction. As the tour coincided with the announcement of a project to clone the thylacine and thus reverse the mistakes of the past, there were also discussions about the ethics of cloning.

In museums with relatively recent displays, thylacine specimens are no longer used only as an example of the largest carnivorous marsupial of modern times alongside other Australian mammals, or as the best example of evolutionary convergence. Instead, they are often depicted in displays that highlight how human behaviour can lead to extinctions. Now they are sometimes included in special galleries devoted to extinction, providing a sober demonstration of how, despite warnings, extinction occurred before anyone cared enough to prevent it happening.

At the Tasmanian Museum and Art Gallery, thylacines now occupy a dedicated gallery. The display consists of a mounted specimen, a male and female skeleton and a flat skin. Films of thylacines are projected on the walls. Framed photographs adorn the space, skulls of thylacine, dog and wolf demonstrate convergent evolution, and artefacts related to some of the searches and historical hunting methods are displayed in cases. Significantly, a panel explains how, despite the accepted 1936 extinction date, in reality, thylacines were functionally or ecologically extinct well before this date. Extremely low population numbers during the early twentieth century almost certainly meant that the species' role as the apex predator in ecosystem regulation was already negligible. The healthy ecological system that existed prior to European colonization had already been significantly disrupted well before 1936.

Today, thylacine collections in museums are in more demand than ever. Researchers increasingly seek access to specimens as they continue to work out aspects of thylacine biology and ecology in order to understand the legacy and impact of extinction. Museums were amongst the players that killed and collected thylacines and transported them around the world. However, they also lobbied for thylacine protection as extinction loomed prior to becoming a reality.

As Anthropocene climate change increasingly alters conditions for all life on earth, and extinction rates increase, museum thylacines, whether on display or stored in collections, continue to be used to educate audiences and facilitate research. They are greatly valued collection items and are frequently named in lists of collection highlights. The specimens have the power to promote feelings of sadness and guilt about a human-induced extinction that could have been prevented, and as a reminder of our ability to disrupt nature with catastrophic results. Although the display of thylacines in museums has altered over time, due to changes in museum philosophy, management and practice, as an example of a recent human-induced extinction their future power is inestimable. Every thylacine specimen is a primary biological and physical reference to a species that, until recently, was integral to the Tasmanian environment. Unfortunately, it is no longer with us.

On 7 September 1936, the Royal Society of Tasmania held its regular monthly meeting in their dedicated space at the Tasmanian Museum and Art Gallery. The subject of this meeting was a special lecture to honour the services of Clive Lord, who had died in 1933. Lord had been the director of the museum and an active member of the Royal Society of Tasmania and had actively pushed for thylacine protection through his role as the museum representative on the Tasmanian Fauna Board. As the society members entered through the museum's foyer, they passed by the case containing the three thylacines obtained between 1926 and 1928. They had no idea that only a few miles away, in Hobart Zoo, the last-known thylacine, locked out of its cage, was to die that night.

Acknowledgements

I wish to thank the following colleagues for their assistance in the preparation of this essay. Robert Paddle read earlier versions of this chapter and provided very useful comments.

Jacqui Ward and Jo Huxley assisted with sourcing the photographs and locating some essential archival records.

Kathryn Medlock spent her career working in curatorial roles in the vertebrate zoology department at the Tasmanian Museum and Art Gallery, where she was responsible for the development, preparation, care and display of the collection. She has actively participated in research projects on reptiles, birds and mammals, including cetaceans, and has curated several museum exhibitions, including one in 1998 on the thylacine. In 2003 she was awarded a Churchill Fellowship to study Tasmanian birds and mammals held in museums in Britain, other parts of Europe, and North America. Currently, her work is focused on the zoological, cultural and historical aspects of thylacine collections around the world, and on the origins of the specimens held in the Tasmanian Museum and Art Gallery. She is passionate about promoting the use of natural history collections for both research and education, as it becomes increasingly urgent to gain an understanding of the implications of habitat loss and extinction.

Notes

1. Hobart City Council Reserves Committee, Minutes of 16 September 1936.
2. 'Tasmanian Telegrams'. *The Mercury*, 25 September 1884, 3.
3. Guiler, *Thylacine*; Guiler and Goddard, *Tasmanian Tiger*; Paddle, *The Last Tasmanian Tiger*.
4. Tasmanian Parliament, 'Eagles and Native Tigers'.
5. Edward Braddon, 'Re: Destruction of Native Tigers'. Lands Department Circular, Hobart, 29 February 1888.
6. Bolton Stafford Bird. 'Re: Native Tigers'. Treasury Memorandum of 15 June 1888.
7. Paddle, 'The Thylacine's Last Straw'.
8. Winter, 'For the Advancement of Science'; Kohlstedt, 'Australian Museums of Natural History'.
9. Ibid. (both).
10. William Lade and James Morris, Letter to Joseph Milligan, 12 November 1856, TMAG Archives.
11. Cooper-Maitland, 'List of Specimens Donated'; Royal Society of Tasmania, *The Courier*, 13 December 1856, 3.
12. Joseph Milligan, Letter to William Lade and James Morris, 18 November 1856. TMAG Archives.
13. Royal Society of Tasmania, 13 December 1856.
14. Renshaw, *More Natural History Essays*; Paddle, *The Last Tasmanian Tiger*.
15. Edwards, '*List of Thylacines*'.
16. Plumb, 'In Fact, One Cannot See'.
17. Bullock, *A Companion to the London Museum and Pantherion*, 130.
18. Wonders, 'Exhibiting Fauna'; Wonders, *Habitat Dioramas*.
19. 'Report of Royal Society Meeting', *Hobart Town Courier*, 10 May 1858, 2–3.
20. 'The Industrial Exhibition'. *The Colonial Times*, 20 August 1850.
21. Royal Commission, 'Official Catalogue of the Great Exhibition'.
22. London Crystal Palace Company (Sydenham), 'The Crystal Palace and Park in 1853'.
23. 'Tasmania in the International Exhibition, 1862'. *The Mercury*, 3 December 1861, 2.
24. Ibid.
25. Liverpool Museum. 'Eleventh Annual Report of the Committee of the Free Public Library, Museum and Gallery of Arts, of the Borough of Liverpool'. Presented to the Town Council, November 1863, 9.
26. Wonders, *Habitat dioramas*.
27. Wonders, 'Exhibiting Fauna'.
28. 'Special Reporter: The Museum: The Progress of a Year'. *The Tasmanian Mail*, 18 July 1885, 4.
29. Machin, 'Gender Representation'.
30. 'Tasmanian Animals: Exhibits at the Museum, Work of Hobart Girl'. *The Mercury*, 29 October 1926, 8.
31. 'History of World Museum'.
32. Tasmanian Museum and Art Gallery, *TM Register 2*, 1890–1913.
33. Joseph Pearson, 'Memorandum on the Tasmanian Museum'. Report presented to the Meeting of Trustees, 6 August 1937, 4.
34. Tasmanian Museum and Art Gallery, *TM Register 2*, 1890–1913.
35. Markham and Richards, *Report on the Museums and Art Galleries*.
36. Tasmanian Museum and Art Gallery, Minutes of Trustees Meeting, 5 August 1937.

37. Tasmanian Museum and Art Gallery, Annual Report for the year ending 30 June 1940.
38. Paddle, *The Last Tasmanian Tiger*.
39. Ashby, 'Four Ways'.
40. Owen, 'On the Rudimental Marsupial Bones'; Gunn, 'Zoology, Section 1'; Gould, *The Mammals of Australia*.
41. Paddle, *The Last Tasmanian Tiger*, 40, note 8.
42. Krefft, *The Mammals of Australia*.
43. Australian Museum, 'Report of the Trustees'.
44. Dawkins, *The Ancestor's Tale*, 229.
45. Tasmanian Museum and Art Gallery, 'Thylacine'. Museum Sub-committee Minutes, 20 April 1939, TMAG Archives.

Bibliography

Ashby, Jack. 'Four Ways Natural History Museums Skew Reality'. *The Conversation*, 13 December 2017. Retrieved 21 January 2021 from https://theconversation.com/four-ways-natural-history-museums-skew-reality-88672.

Australian Museum. 'Report of the Trustees of the Australian Museum, for the Year Ending 31st of December, 1867'. Sydney: Australian Museum, 1868.

Bullock, William. *A Companion to the London Museum and Pantherion: Containing a Brief Description of Upwards of Fifteen Thousand Natural and Foreign Curiosities, Antiquities, and Productions of the Fine Arts; Now Open for the Public in the Egyptian Temple, Piccadilly, London*. London: Printed for the proprietor, 1813.

Cooper-Maitland, Sue. 'List of Specimens Donated to the Royal Society of Tasmania Museum 1849–1886'. Compiled from the Papers and Proceedings of the Royal Society of Tasmania. Hobart: Royal Society of Tasmania, 1968.

Dawkins, Richard. *The Ancestor's Tale: The Dawn of Evolution*. New York: Houghton Mifflin Company, 2004.

Edwards, John. 'List of Thylacines (*Thylacinus cynocephalus*) at London Zoo'. Unpublished manuscript, Zoological Society of London Archives, 2000.

Gould, John. *The Mammals of Australia. Volume 1, Part III*. London: Taylor and Francis, 1851.

Guiler, Eric. *Thylacine: The Tragedy of the Tasmanian Tiger*. Melbourne: Oxford University Press, 1985.

Guiler, Eric, and Phillippe Goddard. *Tasmanian Tiger: A Lesson to Be Learnt*. Perth: Abrolhos Publishing, 1998.

Gunn, Ronald Campbell. 'Zoology, Section 1 – Mammalia', in John West (ed.), *The History of Tasmania, Volume 1* (Launceston: Henry Dowling, 1852), 321–27.

'History of World Museum'. National Museums Liverpool, n.d. Retrieved 1 August 2020 from https://www.liverpoolmuseums.org.uk/history-of-world-museum.

Kohlstedt, Sally Gregory. 'Australian Museums of Natural History: Public Priorities and Scientific Initiatives in the 19th Century'. *Historical Records of Australian Science* 5(4) (1985): 1–29.

Krefft, Gerard. *The Mammals of Australia, Illustrated by Miss Harriett Scott, and Mrs. Helena Forde, for the Council of Education; with a Short Account of All the Species Hitherto Described*. Sydney: The Australian Museum, 1871.

London Crystal Palace Company (Sydenham). 'The Crystal Palace and Park in 1853: What Has Been Done, What Will be Done: Addressed to Intending Exhibitors'. London: William S. Orr and Co., 1853.

Machin, Rebecca. 'Gender Representation in the Natural History Galleries at the Manchester Museum'. *Museum and Society* 6(1) (2008): 54–67.

Markham, S.F., and H.C. Richards. *Report on the Museums and Art Galleries of Australia*. London: The Museums Association, 1933.

Owen, Richard. 'On the Rudimental Marsupial Bones in the *Thylacinus*'. *Proceedings of the Zoological Society of London* 11 (1843): 148–49.

Paddle, Robert. *The Last Tasmanian Tiger: The History and Extinction of the Thylacine*. Cambridge: Cambridge University Press, 2000.

———. 'The Thylacine's Last Straw: Epidemic Disease in a Recent Mammalian Extinction'. *Australian Zoologist* 36(1) (January 2012): 75–92.

Plumb, Christopher. 'In Fact, One Cannot See Without Laughing'. *Museum History Journal* 3(1) (2010): 7–32.

Renshaw, Graham. *More Natural History Essays*. London: Sherratt & Hughes, 1905.

Royal Commission. 'Official Catalogue of the Great Exhibition of the Works of Industry of All Nations, 1851'. London: Spicer Brothers, 1851.

Royal Society of Van Diemen's Land. 'Report of the Royal Society of Van Diemen's Land for Horticulture, Botany and the Advancement of Science'. Hobart: William Gore Elliston, 1846.

Tasmanian Parliament. 'Eagles and Native Tigers: Petition for Powers to Levy Rate for Destruction of, From Municipality of Spring Bay'. *Tasmanian Journals and Printed Papers of the Parliament of Tasmania* 8 (1886).

Winter, Gillian, '"For the Advancement of Science": The Royal Society of Tasmania, 1843–1885'. Honours thesis, University of Tasmania, 1972.

Wonders, Karen. 'Exhibiting Fauna: From Spectacle to Habitat Group'. *Curator* 32(2) (1989): 131–55.

———. *Habitat dioramas: Illusions of Wilderness in Museums of Natural History*. Uppsala: Acta Universitatis Upsaliensis, 1993.

Afterword

After Extinction

Valérie Bienvenue and Nicholas Chare

We Murder to Dissect

Brandon Ballengée's series of artworks 'The Frameworks of Absence' were created using pre-existing illustrations of extinct species.[1] These illustrations, mostly dating from the nineteenth century, have been intentionally marred. The depictions of the animals in them have been cut out, leaving only their silhouettes. Ballengée burned the cut-outs and placed their ashes in glass vials etched with the abbreviation RIP (*Requiescat in pace* or Rest in peace) and the name of the animal whose image has been cremated. He calls the vials funerary urns. Each artwork therefore consists of a despoiled illustration and a cinerary container. The series can be read as a meditation on mourning and loss, as elegiac. This is a common response to extinction, one that has featured in several chapters in this volume. Yet Ballengée destroys not genetic material, not some physical remnant of each species, but a representation of them. He is engaging in acts of iconoclasm, extinguishing *images* of animals, visual remains. The artworks play on a sense of equivalence or substitutability between an animal and their representation.[2] At some level, despoiling the image, ripping into it, seems to repeat the violence each animal was subjected to. Through 'hurting' the illustrations, Ballengée owns anthropogenic extinctions, replicating them in miniature through practices of obliteration, showing the human hand in them. He uses violence to highlight violence.

Are reactions to the works, be they of anger or sadness or bewilderment or something else, generated by the loss of the animal or the loss of the image? Many of the illustrations that Ballengée has damaged beyond repair are dated. They derive from the history of natural history. Something of the power of 'The Frameworks of Absence' therefore stems from it involving acts of vandalism directed towards culturally significant artefacts. The artist's assault on the archive would not possess the same power if all the images were from twenty-first century publications. A spectator's distress potentially derives from their knowledge that the illustrations, which possessed historical importance and financial worth, are now ruined. The illustrations he chooses were often produced roughly at the time the depicted species disappeared. The more ancient the extinction, the greater the possible shock at the mutilation Ballengée has subjected the illustration to.[3] The excision of a Lesser Antillean macaw (*Ara guadeloupensis*) from a 1660 hand-coloured copper plate engraving, for example, is an act that potentially resonates more strongly than the removal of the Rabbs' fringe-limbed treefrog from several photographs that were reproduced in 2008 in the *Journal of Herpetology*.[4] The reaction accorded the images is conceivably bound up more with the age and origin of their material support and less with the species they portray. Ballengée visual attacks are calculated, designed to cause a measured loss. The impact of his slash-and-burns is partly muted by the knowledge that other copies of the images he has used continue to exist. The pictures Ballengée has destroyed are not unique, none of them have become 'extinct'.[5]

The image of the macaw, titled *RIP Lesser Antillean Macaw: After Matthäus Merian* (1660/2014), features a bird labelled in the illustration as an *Araracanga brasiliensibus*, a Brazilian macaw. The name Lesser Antillean macaw, along with the Latin binomial *Ara guadeloupensis*, was not proposed until 1905 by Austin Clark, long after Merian's engraving was made and the bird had disappeared. Ballengée is therefore wishing rest and peace to a macaw that, in a sense, did not exist in the seventeenth century.[6] His decision to adopt Clark's name also reinforces Western dominance over the nomenclature of extinct species. From Jean-Baptiste du Tertre's discussion of the use of tailfeathers as decoration by Indigenous peoples in Guadeloupe, it seems probable the bird had cultural significance.[7] Both Clark's names, however, derive from the Tupi language. The name *Ara* derives from the Tupi word *ará*, which mimics a macaw's vocalization.[8] The name 'macaw' may also derive from a Tupi word, *macavuana*, reputedly the name of a kind of palm tree, the fruit of which is consumed by scarlet macaws (*Ara macao*).[9] Clark's names retain traces of Indigenous terminology yet not from the Kalinago (Carib) groups who lived on Guadeloupe when Europeans colonists first settled there. Their name for the bird may have been *kouléheuc* but records as unclear.[10] As this volume has shown, Indigenous knowledge of more recently extinct species continues to be at risk of disappearing.

'The Frameworks of Absence' works form a predominantly Western vision of natural history. Extinction is figured as loss, as absence. It is also pictured as a modern anthropogenic phenomenon. The nature of the project precludes engaging with previous mass extinctions and background extinctions. As this volume has shown, however, until the Holocene (or the period that many refer to as the Anthropocene) non-anthropogenic extinction was the norm.[11] Extinction is a 'deep time' phenomenon that was present aeons prior to the emergence of archaic humans. The portrayal of extinction offered by 'The Frameworks of Absence' elides this reality. More than that, Ballengée's surgical excisions, the neat-edged voids he creates, imply the loss of a species has no fallout and causes no incidental damage to an ecosystem. Rough cuts that mar the surrounds of each image would better reflect the impact that some extinctions (such as those of keystone species) can have on a given environment.[12] Extinction is often a messy process, with multiple knock-on effects.[13]

Frames of Reference

Ballengée's frameworks as 'frames' perform the function of boundary markers or brackets. Erving Goffman discusses how frames organize experience in *Frame Analysis*, noting that they serve to focus attention and define what is key in a given setting. For Ballengée, loss and mourning are placed front and centre. Certain kinds of extinction – amphibian, avian and mammalian – are also privileged. There is considerable emphasis on North American subspecies, with images of Audubon's bighorn sheep (*Ovis canadensis auduboni*), the California grizzly bear (*Ursus arctos californicus*) the Eastern cougar (*Puma concolor couguar*), the eastern wood bison (*Bison bison pennsylvanicus*) and the southern Rocky Mountain wolf (*Canis lupus youngi*) included. Additionally, some animals seem to require greater mourning than others. There are multiple images, for instance, dedicated to the great auk (*Pinguinus impennis*) and the passenger pigeon (*Ectopistes migratorius*), common extinction pin-ups. Ballengée's personal interests as a biologist specializing in frog deformities have also shaped his vision of extinction, with the number of anurans included probably more than if he had been an expert in carcinology or entomology. Hierarchies of visibility of the kind discussed in this volume therefore register in Ballengée's choice of absences, his inclusions and exclusions.

Plants are notable by their non-appearance as explicit subjects in the series. This may be because extinction rates among plants are relatively low compared to some other kinds of organism. Plant extinction rates, however, vary by region, and are higher in biodiversity hotspots.[14] Ballengée's omission of plants from the current corpus forming 'The Frameworks of Absence' (the work is ongoing) occurs despite the reality that the images he partially destroys are all works on paper and most fibre used in paper pulp derives from trees. Ballengée burns his cut-outs to fuel an ecological message. Meanwhile, forest fires are often being started intentionally to clear land for use in agriculture, causing either total deforestation or significant forest degradation.[15] Conversion of primary or logged forest to plantation crops causes immense loss of biodiversity, and is an extinction driver.[16] This botanic dimension to extinction falls outside Ballengée's frame. As numerous chapters in this volume suggest, plant blindness is an ongoing problem in efforts to address the contemporary extinction crisis.

Cutting out what is the centre of attention in many of the illustrations does encourage spectators to perceive what was previously (in) the background or, at least, tangential, potentially passing unnoticed. In *RIP Lesser Antillean Macaw*, for example, Merian has 'sawn' the branch the bird is perched on to avoid the need to use foreshortening when depicting the tree limb. Often, as with Merian's branch, the putative habitat afforded each animal is minimal, schematic. Occasionally, however, the periphery is of considerable interest, such as in the work *RIP Hare-Indian Dog: After John Woodhouse Audubon*, which portrays the Hare Indian dog (*Canis familiaris lagopus*), also known as the Mackenzie River dog. The canine was used for hunting by the Sahtú or North Slavey (called Hare Indians by settler colonists) people and other First Nations peoples of north-western Canada.[17] Probably a domesticated form of coyote (*Canis latrans*), the Hare Indian dog disappeared sometime after the 1820s.[18] Ballengée's excision of the striking silver-black-striped canine means that the tipis to the left of the illustration gain greater emphasis.[19] One of the tipis has smoke issuing from the flaps at its top, showing it is inhabited.

Usually, natural history illustrations reinforce the idea of *terra nullius*, with animals shown either stark against the snowy paper, or living in an abbreviated setting, a few trees and plants, perhaps some rocks, a land represented as uninhabited by humans. Audubon's hand-coloured lithograph shows an Indigenous presence, one which Ballengée unintentionally brings in from the margins.[20] Michelle Fine and Eve Tuck refer to 'the hegemonic voice-over of colonization' that acts to suppress colonization's ongoing reality.[21] Here that reality is maintained by Ballengée's sidelining of Indigenous epistemologies and animal imageries. 'The Frameworks of Absence' is constructed as if there is only one *natural* history for these extinct animals, absenting Indigenous experiences and perspectives.[22] For some Indigenous peoples, extinction conceived as absence is conceptually aberrant. As Katie

Glaskin notes in the context of Aboriginal Australia, 'the material absence of an entity is not necessarily equated with a corresponding conclusion about its ontological absence'.[23] In Aboriginal cosmologies, the disappearance of a species in embodied form 'does not indicate a final end'.[24]

Matters of Testimony

In making the image of the Hare Indian dog, Audubon worked not from life but using a taxidermy mount. His representation therefore figuratively resuscitates the dog, who is portrayed looking out from a rocky promontory, open-mouthed, in a pose that is loose and relaxed yet alert. From the context, it seems the 'stuffed specimen' that inspired the drawing was sourced for scientific purposes.[25] Audubon's father, John James Audubon, also worked from a 'specimen' to create his depiction of a pair of Great auks.[26] 'Specimens' are examples of species, a typification comprising the whole or part of an organism, which can be used for scientific study. The word 'specimen' enacts distancing, delimiting the animal or other creature as object. Giovanni Aloi describes the 'specimen' as transfixed 'in an atemporal milieu in which any individual history is removed by preparation and decontextualization'.[27] Ballengée's artwork *RIP Endemic Freshwater Mussels of the Americas: After David H. Stansbery* (1971/2015) provides a good example of 'specimens' as Aloi describes them. The work features photographs of mollusc shells derived from various natural history collections, including the Henry Moores collection (Ohio State University Museum of Zoology) and the University of Michigan Museum of Zoology. As Ballengée has decided to focus on mussels, he has left intact two photographs of the shell of an extinct species of snail, the catenoid river snail (*Oxytrema catenoides*), that appear as part of the group.[28] Each of these 'specimens' stands for a unique species of mussel: the shell numbered 1, for example, derives from the sugarspoon mussel, *Dysnomia arcaeformis*; and the shell numbered 3, from the narrow catspaw mussel, *Dysnomia lenior*.[29] Ballengée, however, fails to give the individual names for each of the bivalves; cumulatively they are made to exemplify mollusc extinction in general, the singular sacrificed to the broader picture.

Another work that makes clear use of 'specimens' is *RIP Nelson's Rice Rat: After Edward A. Goldman* (1918/2015) which features skulls from various species of Oryzomys, semi-aquatic rodents. Ballengée has excised the skulls (numbered 1 and 1a) of the Nelson rice rat (*Oryzomys nelsoni*), which was endemic to Isla María Madre in Mexico, from a plate in Edward Goldman's *The Rice Rats of North America (Genus Oryzomys)*.[30] Four individuals from an already small population were collected over a seven-day period in May 1897 in damp thickets near springs at the island's summit. The skins and skulls of the rats are retained by the National Museum of Natural History in Washington DC. The image of the skull cut out and burnt by Ballengée was of USNM 89200, an adult male, the holotype. This cranium is described by C. Hart Merriam as part of his identification of the rice rat as a new species.[31] Merriam's textual sketch of the cranial characteristics is concerned with size and, to a lesser extent, weight: the skull is 'very large' and, when compared to *Oryzomys mexicanus*, 'very much larger and heavier', the interparietal is 'very broad'.[32] The repetition of the adverb of degree 'very' works to intensify physical characteristics of the skull and emphasize its significance. As a concern to accurately record morphology shapes the descriptors and modifiers, with attention to scale the overriding concern, the rat remains an abstract entity, big and weighty but also something of a flat character. Artists such as Audubon, with his portrayal of the Hare Indian dog, put metaphorical flesh on the skins and bones of 'specimens', undoing something of natural history's abstracting and objectifying tendencies and offering a more rounded, characterful sense of an animal.

RIP Nelson's Rice Rat: After Edward A. Goldman leaves the skulls of other river rats such as the subspecies *Oryzomys rostratus megadon* visible. These enduring crania give a good indication of what has been cut out from the image, of the physical remains of the Nelson rice rat. This image and others like it therefore show the lie of absence. Extinction leaves material and textual wreckage in its wake. For the Lesser Antillean macaw we have eyewitness accounts and, recently, the discovery of material remains, albeit a single bone – namely, a terminal phalanx – excavated on Marie-Galante, one of the islands that form Guadeloupe.[33] The bone is roughly ten thousand years old, predating human settlement, and provides material evidence of endemic macaws.[34] Extinction involves the cessation of living examples of a given species, but organic materials from that species often persist, granting them a substantial afterlife. Eye and ear witness accounts also act as afterimages, holding something of the departed species in the fields of vision and audition. An overemphasis on absence risks sidelining material and lay witnesses to 'disappeared' species. Some of the contributors to this volume have gestured to the power of physical memorials, flesh and/or bone archives, as sites of remembrance and sources of information. These are often used in museum displays that explore forms of extinction, as was the case when 'The Frames of Extinction' was exhibited at the Louisiana Art and Science Museum. A monarch butterfly (*Danaus plexippus*) specimen and mounts of a hawksbill sea turtle (*Eretmochelys imbricata*) and a gentoo penguin (*Pygoscelis papua*) were displayed alongside the artworks.

Pictures at an Exhibition

Museum exhibitions frame how any artefacts on display are received, encouraging particular interpretations over others, and are in themselves frames of a kind. As Goffman notes, frames often perform the function of boundary markers or brackets.[35] The art gallery and the natural history museum are spaces that bracket off 'art' and 'nature' respectively from everyday life. 'The Frameworks of Absence' was installed at various venues, including The Armory Show in New York (2015), Expo Chicago (2018) (where the Natural Resources Defense Council partnered with the Ronald Feldman Gallery to present the work at Navy Pier's Festival Hall) and the Louisiana Art and Science Museum (2019). These temporary exhibitions framed how Ballengée's works would be received and understood, bracketing them as art rather than, for instance, vandalism. He also contributed to the framing of the works, such as choosing how they were to be hung and opting to display them against a red background.

Contemporary art is often displayed in a white cube gallery, a seemingly neutral exhibition space. Thomas McEvilley writes that the roots of the white cube are to be found in ritual spaces that are segregated from the outside world, in tombs and places of worship.[36] The white cube aesthetic seeks to 'bleach out the past' and sterilize art, purging it of any links with life, transforming it into something transcendental.[37] Through insisting on a red background, Ballengée refuses pretentions of neutrality. As the artworks have no backing, the red of the wall is visible through those parts of the works that have been cut out. The resultant sanguine silhouettes suggest violence and cruor.[38] Not all the extinct animals that feature in 'The Frameworks of Absence' met bloody ends. The gory backdrop therefore contributes to the project's overgeneralizing tendencies.

One of the potentialities of art that represents animals is to singularize them. The singularizing power of some artistic representations, the capacity to communicate an animal's individuality, bears on ethical questions about how to portray animals, extinct or otherwise. Such questions inflect many of the preceding chapters in this collection. It is clear that some depictions of extinct species, particularly from prehistory, embody 'alternative facts' and bear little relation to contemporary scientific understandings of a creature's appearance and behaviour. Bad palaeoart of this kind, which often appears on the internet, confuses and misleads. Such objectively inaccurate works 'distort proportions measurable from fossils, omit integumentary structures documented from the taxa concerned, include blatant anatomical errors . . . or mix geographically and stratigraphically disparate species'.[39] This is not to say that good palaeoart simply reflects scientific data. There are always gaps in our understanding of the past and the artist must speculatively fill these, employing inference and plausible theorizing. Good palaeoart is therefore credible if not truthful. It is a way of visually thinking through possibilities and probabilities linked to the palaeontological record. Mark Witton emphasizes that in such art notions of accuracy need to be nuanced.[40]

Witton, Naish and Conway suggest that, at its best, 'the relationship between palaeoart and palaentological science is mutualistic, a genuine fusion of artistry and science where both sets of practitioners are inspired by, and learn from, the other'.[41] Their observation presumes that the categories 'art' and 'science' are distinct and stable. One of the aims of this volume has been to demonstrate that art and science are not antithetical or oppositional but imbricated, inseparable. There is an artfulness to all scientific inquiry, a creativity and a capacity for imagining, and equally, through its qualities of observation and experimentation, something scientifical at the heart of art. In this context, Ballengée, rather than being a scientist who is also a practising artist, is an artist-scientist. The scholars from the arts and the scientists who have contributed to this volume, like Ballengée, all embody an awareness of the value of both these 'cultures' of knowing. As the chapters by curators have shown, the museum is frequently the space within which the two 'cultures' meet, conjoin.

Museums, with their display cases and their segregated artefacts, can often contribute to the objectification of animals, plants and other organisms.[42] Images can also objectify. Their potentially objectifying power and their generalizing tendencies raise important ethical questions. Representing non-human animals as if they are objects, for example, encourages viewing them as instruments for human use and pleasure. The reductive focus on form in morphologically motivated depictions of organisms renders a given species into units of data – into colours, shapes, dimensions. 'The Frameworks of Absence' can be read as an exercise in morphology in the negative; form is still privileged but by its absence.

From the General to the Particular

Steve Baker notes that pictures which utilize animals as symbols are inevitably anthropomorphic, and transform them into meaning-making machines serving human ends.[43] Rosi Braidotti has urged a move beyond human exploitation of animals for metaphoric ends and the cultivation instead of a bioegalitarianism, a recognition 'that we humans and animals are in this together'.[44] Portrayals of extinct species are, perhaps, particularly prone to being made into vehicles for political messages. The politicization of a species, their being co-opted to stand as prompts for environmental awareness and conservation, comes at the expense of their singularity. Josephine Donovan views singularizing as crucial to fostering compassion towards animals.[45] Poster species such as the Yangtze River dolphin (*Lipotes vexillifer*), the thylacine (*Thylacinus cynocephalus*) and the passenger pigeon can stand as powerful warnings against inaction regarding the ongoing extinction crisis. As symbols of loss, however, they are evacuated of their particular attributes. Ballengée's artworks in 'The Frameworks of Absence' literalize this voidance. The issue of how images of extinct species are put to work (and the implications of this signifying labour for the ways they are perceived and understood) are themes that unify many of the volume's chapters.

As well as concerns regarding how extinct species are represented, ethical questions arise regarding how representations of extinct species should be received and read. Griselda Pollock has noted the importance of close reading (an imperilled skill as the internet shifts reading habits towards scanning, skimming, speeding and swiping) for appreciating the particularity of a given artwork.[46] Pollock advocates close reading as a means to attend to images as 'singular sites of subjectivity-inflected production of meaning'.[47] She notes of her own reading of an artwork that it 'involves a careful attuning to its otherness, as well as a subtle, always cautious and avowed borrowing from my own experience that might register or resonate with the affective tone, and note the singular turns of a work'.[48] Ballengée's individual readings of the corpus that makes up his 'The Frameworks of Absence' do not seem of this order. All the artworks are treated as substitutable: 'This is a picture of a species that once was'. There are, however, qualitative differences, for example, between a painting by Randy Fehr of the eastern cougar (*Puma concolor couguar*) used as a cover image for the December 1993 issue of *Fur-Fish-Game* and a lithograph by John Gerrard Keulemans of a slender-billed grackle (*Quiscalus palustris*) used in the avian science journal *The Ibis*.

The finely detailed image of the cougar by Fehr, entitled 'Keeper of the Creek' and with the big cat identified by the artist simply as a 'Cougar/Mountain Lion', may have been intended as a homage to the animal but in the context of the magazine for 'practical outdoorsmen', which has a regular feature on 'predator hunting', the large felid becomes a feared enemy, a symbol of wilderness that is untamed and unwelcoming.[49] The cougar, portrayed on the snow-covered bank of a shallow stream, has seemingly been interrupted while fishing (a salmon or similar fish is visible in the foreground, beneath the water's surface).[50] Casting the cat as a purposive freshwater fisher potentially encourages the magazine reader to identify with the predator, given that many hunters are also anglers. The main message, however, as communicated by the bared teeth, seems to be that cougars are aggressive and dangerous. Attributing greater hostility to the cougar enhances the status of a hunter who successfully kills one, as the sense of risk involved is increased. In the context of its appearance in *Fur-Fish-Game*, the image can be read as a stimulus to cougar trophy hunting.[51] Keulemans's grackle is identified as *Quiscalus tenuirostris* by Philip Sclater, the author of the article that the plate accompanies. He states: 'I have little doubt that this is the true *Q. palustris* of Swainson; but it will only create more confusion to supersede the very appropriate name *tenuirostris* given by the same author'.[52] Keulemans depicts a male and female grackle perched on branches. He used 'specimens' or study skins from Sclater's collection.[53] Like Audubon with the Hare Indian dog, Keulemans therefore revives dead animals. His success in such endeavours is evinced by claims that each of his artworks 'hums with life'.[54] Keulemans's birds in *The Ibis* are designed to illustrate Sclater's Latin descriptions, although the ornithologist's careful detailing of the female's colouration seems more muted that the artist's bright yellow-breasted representation. This is a scientific image, whereas Fehr's is a work of drama, consciously seeking to generate an atmosphere, one of fear and awe.

Ballengée encourages seeing both images as on a continuum, their singularity ceded to a general argument about loss. Similarly *RIP California Grizzly Bear: After Felix Octavius Carr Darley* (1888/2015) is presented as of the same order as *RIP Great Auk: After Henry Pawson and Joseph Brailsford* (1896/2014). Darley's drawing, which was made in 1887 and etched by Stephen James Ferris in 1888, is titled *After a Good Day's Sport* and shows a man with a rifle resting against the trunk of a tree, contemplating the corpses of three bears (an adult and two cubs) he has shot. Pawson and Brailsford produced a lithograph of an auk egg (Plate 28) for Henry Seebohm's posthumously published *Coloured Figures of the Eggs of British Birds*. The eggs are linked with national interests, as the great auk is granted 'British citizenship' and because the image of the egg is home grown. In the book's preface, Richard Bowdler Sharpe states with satisfaction that 'while many recent Ornithological illustrations have avowedly been "made in Germany", in this instance all the work in connection with the drawing of the eggs, the lithographing of the plates, the printing of the letterpress and the binding of the volume has been done in Sheffield'.[55] The motivations behind the images of the bears and the egg are markedly different; one celebrates the outdoorsman, the other is an expression of avian national pride. For Ballengée, they are both ultimately concerned with absence.

Despite the etched-glass urns and ashes, despite the ritual of mourning that Ballengée urges enacting, his works overwhelmingly bind extinction to melancholia. Sigmund Freud differentiates mourning from melancholia, with the latter characterized as a failure of perception – the melancholic 'cannot see clearly' (*nicht deutlich erkennen*) what it is that they have lost.[56] In melancholia, that one has lost is recognized, but not what it is that one has lost. Ballengée's artworks, which only allow loss to be known in outline, compellingly figure melancholy. As Ursula Heise notes, melancholy possesses a politically mobilizing power yet it also has its drawbacks. Nostalgic and pessimistic narratives can put off potential supporters of conservation initiatives. Because of this, the challenge Heise identifies for conservationists is to 'enable the imagination not so much of the end of species as of their future'.[57] Drawing attention to extinction can raise awareness but it is also backward looking and risks generating a sense of hopelessness.[58]

Afterimages

All the artworks that form 'The Frameworks of Absence' acknowledge the original creator of the image by way of their name, preceded by the preposition 'After'. This is the case, for instance, with *RIP Maryland Darter: After Aleta Pahl* (1983/2014). The preposition indicates both that the work was made at a later time than Pahl's and that the work is inspired by hers.[59] The linear temporality of the works positions extinction as a thing of the past (every representation has been chosen so its initial production roughly coincides with when the species it portrays died out), even as a contemporary take is provided on each image. This linear conception of temporality is not universally held. Some Indigenous understandings of temporality are 'intergenerational and fold back on themselves'.[60] Non-linear temporalities can lead 'present actions and interactions' to be viewed as 'taking place in a productive dialogue with ancestral pasts to collaboratively establish futures'.[61] Such a concept of time renders the idea of extinction as having an 'after' that is demarcated as strictly separate and set apart from the present questionable; the past is in the present. Our use of the term 'afterimages' in this volume has sought to capture something of this temporal complexity, conceiving extinction as a phenomenon that traverses past and present, undoing any simple partition of the two. This is the case not just for recent extinctions but also ancient ones. Fossil fuels, the use of which contributes to global warming and climate change, derive from extinct organisms. Plastics derived from fossil fuels pollute the environment and can be fatal to animals.[62] Ancient extinctions therefore act upon the present.

Temporal complexity informs artworks by the Tlingit and Unangax̂ artist Nicholas Galanin. His work *Inert* (2009) is composed of a wolf taxidermy mount and a wolf rug which are joined together to appear like a single animal. The *gooch* (wolf) is important to Tlinglit storytelling. Similarly, in Unangax̂ culture, the *alix̂ngix̂* (wolf) has great significance.[63] *Inert* shares similarities with the works in 'The Frameworks of Absence' in that it employs pre-existing materials. Galanin's piece, however, succeeds in being simultaneously backward and forward looking. The wolf can be read as transforming into a rug, yielding to the floor, or as rising up, refusing to be walked all over, inviting respect. Held in suspense between states, either reading is possible, or both can be refused. The wolf is neither a trophy nor at liberty, neither prostrate nor upright. Half-rug, half-animal, *Inert* presents the viewer with a wolf suspended between states and times. A beholder can either embrace the nascent wolf subject, a recognition that manifests care and compassion, or not care less, and see only the 'harvested' pelt, the spreadeagled decoration.

Another of Galanin's works that employs taxidermy is *We Dreamt Deaf* (2017), featuring a polar bear (*Ursus maritimus*) stilled in a state similar to *Inert*: 'fixed in the struggle to survive an unsustainable condition'.[64] The bear was shot by a White hunter as a trophy; he came to Shishmaref in Alaska to 'bag' a bear and then left. He had no connection to the locale. Galanin has described his artistic practice as one that is rooted in connection to land and that pursues freedom and vision for the present and the future.[65] He invites reflection on how people relate to place, to the environment in which they live and the locations they visit. Galanin explains that the title *We Dreamt Deaf* refers to our implication 'in the anthropocentric industrial dream that renders us deaf to our impact on all of our relatives (human and non-human)'. He goes on to emphasize that humans are also animals, but ones that have forgotten their place in the world. We no longer *listen* to the land of which we are a part. Galanin consciously exploits the vulnerable polar bear's iconic status to foreground the ways animals and cultures have been 'decimated through colonial corporate enterprises focused on extraction from land, and the development of capital without care for consequences'. The bear becoming rug is the continuity of this decimation (a decimation that includes many species extinctions), while the rug becoming bear is its refusal.

In *Frame Analysis*, Goffman notes of the 'wooden frame of a picture' that it is 'presumably neither part of the content of [the] activity proper [that is art] nor part of the world outside the activity but rather both inside and outside'.[66] Jacques Derrida similarly calls attention to the undecidability of both material and conceptual frames.[67] In such understandings, the frame of an artwork resists being framed as either 'art' or 'life'. The pictures that feature in Ballengée's 'The Frameworks of Absence' are all carefully framed, literally and figuratively. Galanin's *We Dreamt Deaf* is also 'framed', elevated from the floor by a low white plinth, a metaphorical Arctic region, yet the nature of the work makes it more resistant to bracketing. Our aim in this volume has been to similarly create a space which opens up what representations of extinction might be and might do. The diverse images explored in the volume demonstrate that extinction is not a singular event, one assimilable to a sole frame of reference, but a multiplicity of histories and understandings, some focused on finality, others more equivocal. These differing forms of afterimage each contribute something important to our efforts to feel and think extinction.

Valérie Bienvenue is a doctoral candidate in the Department of History of Art and Film Studies at the Université de Montréal. Her thesis critically examines human–equine relations through the prism of modern art and visual culture. Prior to her academic career, she worked for ten years in equestrian circles, including teaching bareback riding and rehabilitating horses suffering from physical and psychological trauma. She is the author of several articles and book chapters.

Nicholas Chare is professor of art history in the Department of History of Art and Film Studies at the Université de Montréal. He is the author of *After Francis Bacon* (2012). In 2017, with Sébastien Lévesque and Silvestra Mariniello, he founded the baccalaureate (BACCAP) in visual cultures at the Université de Montréal.

Notes

1. The series is sometimes simply called 'Frameworks of Absence' but, following Ballengée, we refer to it as 'The Frameworks of Absence'. http://brandonballengee.com/the-frameworks-of-absence/ (last accessed 9 June 2021). Although our engagement with the project here is sometimes critical, something of the strength and depth of Ballengée's endeavour is revealed in the many reflections that it has prompted in us.
2. To resist objectifying animals, we refer to them as he, she or they (depending on whether their gender is known) rather than 'it'.
3. A 1590 copper engraving identified by Ballengée as the North Atlantic Grey Whale (*Eschrichtius robustus*) seems to be the oldest illustration used. This picture is sourced from Volume IV of Joachim Camerarius's *Symbolorum emblematum centuriae tres*.
4. The journal article is readily accessible whereas the engraving by the Swiss illustrator Matthäus Merian the Elder was sourced from Book 6 of John Jonston's *Historiae naturalis*. A second edition of the *Historiae naturalis* sold at auction for 8,125 USD in 2016. See http://www.bonhams.com/auctions/23255/lot/83/ (last accessed 5 June 2021). Whether the bird that Ballengée has excised is a Lesser Antillean macaw is highly debatable. The bird is identified as *Araracanga brasiliensibus* in the illustration and although described as predominantly coloured an elegant red (*elegantibus rubris*),

Jonston mentions azure (*coeruleis* [coeruleus]) and brown back and tail plumage. The Latin name also links the bird to Brazil. All translations are our own unless otherwise stated. We are grateful to Kristine Tanton for assisting us in translating Jonston's Latin text. See Jonston, *Historiae naturalis*, 201. The section that discusses the macaw in *Historiae naturalis* derives substantially from information sourced from Georg Marcgraf, a naturalist who resided in Brazil. He published *Historia naturalis brasiliae* in 1648 which features an illustration of an *Araracanga brasiliensibus* (206). This looks like a blue-and-yellow macaw (*Ara araruna*) to us. The Lesser Antillean macaw is closer to the scarlet macaw (*Ara macao*), differing in that the former had red tailfeathers. Du Tertre describes the bird from Guadeloupe, for instance, as having a tail that was entirely red. See Du Tertre, *Histoire generale des isles de S. Christophe*, 295. *Araracanga brasiliensibus* would subsequently be identified as the *ara rouge* or scarlet macaw (*Ara macao*) by Buffon in his *Histoire naturelle des oiseaux*, which mentions the Antilles. Buffon questions whether there are a number of distinct species across the Caribbean and neighbouring regions or simply several varities of a single species (239–40). Rochefort suggests in *Histoire naturelle et morale* that the plumage colouring varied and was distinct for each island of the Antilles (154).

5. Some of the lithographs used were hand-coloured and therefore likely differed across each volume of a given publication.

6. Ballengée manifests a tendency to gloss over taxonomic uncertainties, referring with conviction to some depictions as particular species, when things, at least to us, do not appear so clear-cut. He may believe that, as the likeness of the species or subspecies is to be cremated, occasional confusions do not matter. The beholder of each work is meant to mourn the name, not the image.

7. Du Tertre, *Histoire generale des isles de S. Christophe*, 296.

8. The word 'ara' (or the plural 'arras') was also commonly used in European works of natural history to describe macaws in the seventeenth century.

9. The Tupi-Guarani people, the Araweté, use macaw feathers to make arrows and for decoration. Scarlet and blue-throated macaws (*Ara glaucogularis*) are sometimes raised as pets in Araweté villages. See Viveiros de Castro, *From the Enemy's Point of View*, 42.

10. In his Carib vocabulary in *Histoire naturelle et morale des îles Antilles de l'Amérique*, Rochefort gives the name *Kouléhuec* for parrot and *Tónoulou* for bird (525). Although Rochefort divides parrots into three types based on their size, referring only to *perroquets* (173) and not *arras* as *Kouléhuec*, the Carib term may have encompassed the triumvirate without distinction. The term *guacamayo* is sometimes suggested as the Indigenous name for the Lesser Antillean macaw but this is a Taíno word that was used by Christopher Columbus's interpreters to refer to the macaw when he landed on the island. The Spaniards were unable to speak directly to the Caribs. See the entry for *Guacamayo* in Friederici, *Amerikanistisches Wörterbuch*, 267–68. Scientists are increasingly acknowledging the need to engage in retrospective name changes that recognize longstanding Indigenous names for species. See Gillman and Wright, 'Restoring Indigenous Names'.

11. The term Anthropocene fosters a sense of universal human culpability for climate change and, for this reason, we do not adopt it here. In reality, colonialism and the petrochemical industry precipitated climate change. For a useful discussion of the Anthropocene as a term that rightly foregrounds how it violently erases difference, see Davis and Todd, 'On the Importance of a Date'. Kyle Whyte suggests climate change should be understood as intensified colonialism. See Whyte, 'Indigenous Climate Change Studied'.

12. For an analysis of the effect of non-random extinctions on two biosystems (a marine one and a terrestrial one), see Raffaelli, 'How Extinction Patterns Affect Ecosystems'.

13. Van Dooren emphasizes the entanglement of species, their co-shaping coexistence. Extinction forms a loose thread that unravels part of this patchwork of interdependence. See Van Dooren, *Flight Ways*, 42.
14. See Le Roux et al., 'Recent Anthropogenic Plant Extinctions'.
15. See McFarland, *Conservation of Tropical Rainforests*, 36–38.
16. For an analysis of this phenomenon in a South East Asian context, see Wilcove et al., 'Navjot's Nightmare Revisited'. For a discussion of how forest burning can contribute to the reduction of tree diversity, see Tabarelli, Cardoso da Silva and Gascon, 'Forest Fragmentation'.
17. Richardson states that the dog was used 'solely in the chase'. See Richardson, 'Canis f. var. B. Lagopus. *Hare Indian Dog*', 78.
18. Woodhouse, 'The North American Jackal', 148.
19. Tipis are also present in the background of a copper-plate etching of a pair of the dogs created by Thomas Landseer for John Richardson's collaborative project, *Fauna Boreali-Americana*. Landseer's depiction also features two First Nations people with rifles beside the tipis. This portrayal therefore also shows the indirect cause of the dog's extinction: the introduction of rifles rendered the hunting dogs obsolete.
20. Merian's lopped branch also indexes human inhabitants albeit in a more ambiguous way.
21. Tuck and Fine, 'Inner Angles', 147.
22. If Ballengée burnt Indigenous portrayals of extinct species, this would, obviously, raise considerable ethical issues. There is, however, clear scope for dialogue and strategic collaboration with Indigenous stakeholders as a means to acknowledge alternative perspectives and understandings of nature.
23. Glaskin, 'Extinction, Inscription and the Dreaming', 11.
24. Ibid., 15.
25. Audubon and Bachman, *The Viviparous Quadrupeds*, 155.
26. Dry specimens of birds are usually referred to as 'study skins'.
27. Aloi, *Speculative Taxidermy*, 73.
28. Ballengée sourced the images from Stansbery's 'Rare and Endangered Freshwater Mollusks in the United States'.
29. *Epioblasma arcaeformis* and *Dysnomia arcaeformis* are taken to be synonymous.
30. Goldman, *The Rice Rats of North America*.
31. Merriam, 'Oryzomys nelsoni', 15.
32. Ibid.
33. An ulna excavated on the same island and previously attributed to a macaw is now thought to belong to the extinct parrot the Guadeloupe amazon (*Amazona violacea*).
34. Gala and Lenoble, 'Evidence of the Former Existence of an Endemic Macaw'.
35. Goffman, *Frame Analysis*, 251.
36. McEvilley, 'Introduction', 8.
37. Ibid., 11.
38. Boettger notes the importance of the red walls to the tenor of the work. See Boettger, 'Ways of Saying', 259.
39. Witton, Naish and Conway, 'State of the Palaeoart', 3.
40. Witton, *Palaeoartist's Handbook*.
41. Witton, Naish and Conway, 'State of the Palaeoart', 4.
42. For an analysis of the isolation of objects as a stimulus to visitor interest, see Melton, *Problems of Installation in Museums of Art*, 257–60.
43. Baker, *The Postmodern Animal*, 82.
44. Braidotti, 'Animals, Anomalies and Inorganic Others', 528.

45. Donovan, *The Aesthetics of Care*.
46. For an extended analysis of online reading habits, see Herath, 'How Do We Read Online?'
47. Pollock, 'To Play Many Parts', 65. Donovan's call for 'attentiveness' in *The Aesthetics of Care* might also be interpreted as an invitation to close read.
48. Ibid.
49. The artwork is reproduced (currently in reverse to how the image appears on the magazine cover) here: http://www.artcountrycanada.com/images/fehr-randy-keeper-of-the-creek.jpg (last accessed 9 June 2021).
50. Although not a common component of their diet, there are records of cougars consuming fish. The remains of a carp have been found in cougar scat but whether the fish was scavenged or self-caught is unknown. See McClinton, McClinton and Guzman, 'Utilization of Fish'.
51. Hunting cougars as trophies is thought to exacerbate human–cougar conflict. See Teichman, Cristescu and Darimont, 'Hunting as a Management Tool?'
52. Sclater, 'A Review of the Species of the Family *Icteridae*', 158.
53. Ibid.
54. Holmes, 'Exhibition', 1143.
55. Bowdler Sharpe, 'Preface', iii.
56. Freud, 'Trauer und Melancolie', 290.
57. Heise, *Imagining Extinction*, 50.
58. In the 'Introduction' to the edited volume *After Extinction*, Richard Grusin suggests it is possible to view extinction not simply as causing the end of life but as generative of life (ix). In the case of preceding mass extinctions or background extinctions this seems a reasonable assertion. In the case of anthropogenic extinctions, however, some of which involved deliberate efforts at extirpation, it appears perverse.
59. Similarly, the 1888 etching of *After a Good Day's Sport*, which inspired *RIP California Grizzly Bear: After Felix Octavius Carr Darley*, states that it is 'After drawing by F.O.C. Darley'. In this case, Ballengée's work should more accurately be titled '*After Stephen J. Ferris After Felix Octavius Carr Darley*'. Although omitting Ferris's role from the title, Ballengée does include the information that he is the maker of the etching as part of the work. 'After' can also mean 'in pursuit of' and indicate efforts to catch someone or something. Given the creative violence underpinning 'The Frameworks of Absence' it is also possible to hear this sense of the term in Bellangée's titles.
60. Randazzo and Richter, 'The Politics of the Anthropocene', 10.
61. Ibid., 9–10.
62. Virgin plastic is still used in much manufacturing, with recycling of plastics, often an expensive process, being at disappointingly low levels. The negative impact of plastic on the environment has been given considerable visibility in recent years. In a segment of Episode 7 of Blue Planet II (2017), Lucy Quinn, a seabird ecologist working for the British Antarctic Survey, describes the varied plastics found in albatross chick nests on Bird Island in South Georgia; this continues to garner attention. http://www.youtube.com/watch?v=I4QNolP7Khc&t=30s (last accessed 11 June 2021).
63. Wolf is given the alternative spelling of *aliiẋxiẋ* in Black, 'World of the Aleuts', 129.
64. These words are Galanin's, sourced from an interview with Kathleen Wong on behalf of the Honolulu Museum of Art. blog.honoluluacademy.org/nicholas-galanin-the-polar-bear-is-an-iconic-symbol-of-the-struggle-for-survival-of-animals-and-cultu res/ (last accessed 10 June 2021). Carefully crafted, unsettling artworks such as *Inert* and *We Dreamt Deaf*, which are designed to be exhibited in museum spaces, embody a decolonizing impulse that refuses to be reduced to metaphor. The artworks are in-

tended to agitate and act upon the beholder, rather than simply aesthetically please. For a key examination of the importance of not reducing decolonization to a metaphor, see Tuck and Yang, 'Decolonization is Not a Metaphor'. When displayed in a museum context, Galanin's works that employ taxidermy also connote natural history discourse. Their unnatural natures, however, unnerve. They have none of the neutrality associated with science, Dolly Jørgensen has recently emphasized how museums can unite science and affect in this way. See Jørgensen, *Recovering Lost Species in the Modern Age*, 123.

65. http://docs.google.com/document/d/1c7DB1fFxsrjGPA4x41xH5y_lzWwVYVeQ_2D_hoGzGtpY/edit?resourcekey=0-vU5WOau6AxBscHityuKg6g&resourcekey=0-vU5WOau6AxBscHityuKg6g (last accessed 10 June 2021).
66. Goffman, *Frame Analysis*, 252.
67. Derrida, *The Truth in Painting*. For a discussion of aspects of Derrida's vision of animal-human relations, see Bienvenue and Chare, 'The Animal Nude'.

Bibliography

Aloi, Giovanni. *Speculative Taxidermy: Natural History, Animal Surfaces and Art in the Anthropocene*. New York: Columbia University Press, 2018.

Audubon, John James, and John Bachman. *The Viviparous Quadrupeds of North America: Volume III*. New York: V.G. Audubon, 1854.

Baker, Steve. *The Postmodern Animal*. London: Reaktion, 2000.

Bienvenue, Valérie, and Nicholas Chare, 'The Animal Nude: Pony Play and Deborah Bright's *Being & Riding*', in Nicholas Chare and Ersy Contogouris (eds), *On the Nude: Looking Anew at the Naked Body in Art* (New York: Routledge, 2022), 155-172.

Black, Lydia T. 'World of the Aleuts'. *Arctic Anthropology* 35(2) (1998): 126–35.

Boettger, Suzaan. 'Ways of Saying: Thetorical Strategies of Environmentalist Imaging', in T.J. Demos, Emily Eliza Scott and Subhankar Banerjee (eds), *The Routledge Companion to Contemporary Art, Visual Culture, and Climate Change* (New York: Routledge, 2021), 252–62.

Bowdler Sharpe, Richard. 'Preface', in Henry Seebohm, *Coloured Figures of the Eggs of British Birds*, edited by Richard Bowdler Sharpe (Sheffield: Pawson & Brailsford, 1896), iii.

Braidotti, Rosi. 'Animals, Anomalies, and Inorganic Others'. *PMLA* 124(2) (2009): 526–32.

Buffon, George Louis Leclerc, Comte de. *Histoire naturelle des oiseaux : Œuvres complètes : Volume VII*. Paris: Garnier Frères, 1854.

Camerarius, Joachim. *Symbolorum emblematum centuriae tres. Volume IV: Ex aquatilibus & reptilibus*. Nuremberg: Vögelin, 1590–1605.

Davis, Heather, and Zoe Todd. 'On the Importance of a Date, or Decolonizing the Anthropocene'. *ACME: An International Journal for Critical Geographies* 16(4) (2017): 761–80.

Derrida, Jacques. *The Truth in Painting*. Translated by Geoff Bennington and Ian McLeod. Chicago: University of Chicago Press, 1987.

Donovan, Josephine. *The Aesthetics of Care: On the Literary Treatment of Animals*. New York: Bloomsbury, 2016.

Du Tertre, *Histoire generale des isles de S. Christophe, de la Guadeloupe, de la Martinique*. Paris: Jacques Langlois, 1654.

Freud, Sigmund. 'Trauer und Melancolie'. *Internationale Zeitschrift für Ärztliche Psychoanalyse* 4(6) (1917): 288–301.

Friederici, Georg. *Amerikanistisches Wörterbuch und Hilfswörterbuch für den Amerikanisten*. Hamburg: De Gruyter, 1960.

Gala, Monica, and Arnaud Lenoble. 'Evidence of the Former Existence of an Endemic Macaw in Guadeloupe, Lesser Antilles'. *Journal of Ornithology* 156 (2015): 1061–66.

Gillman, Len Norman, and Shane Donald Wright. 'Restoring Indigenous Names in Taxonomy'. *Communications Biology* 3, 609 (2020). DOI: 10.1038/s42003-020-01344-y.

Glaskin, Katie. 'Extinction, Inscription and the Dreaming: Exploring a Thylacine Connection'. *Anthropological Forum* (2021). DOI: 10.1080/00664677.2021.1937513.

Goffman, Erving. *Frame Analysis: An Essay on the Organization of Experience*. Lebanon: Northeastern University Press, (1974) 1986.

Goldman, Edward A. *The Rice Rats of North America (Genus Oryzomys)*. Washington, DC: Government Printing Office, 1918.

Grusin, Richard. 'Introduction', in Richard Grusin (ed.), *After Extinction* (Minneapolis: University of Minnesota Press, 2018), vii–xx.

Heise, Ursula K. *Imagining Extinction: The Cultural Meanings of Endangered Species*. Chicago: University of Chicago Press, 2016.

Herath, Dhammika Channa. 'How Do We Read Online: The Effect of the Internet on Reading Behaviour'. Unpublished MA dissertation. Victoria University of Wellington, 2010.

Holmes, David. 'Exhibition: Art Comes Naturally'. *The Lancet* 377 (2011): 1143.

Jørgensen, Dolly. *Recovering Lost Species in the Modern Age: Histories of Longing and Belonging* (Cambridge, Mass: MIT Press, 2019).

Jonston, John. *Historiae naturalis, De avibus, Libri VI, cum aeneis figuris*. Frankfurt am Main: Impensis haeredum Math. Meriani, 1650.

Le Roux, Johannes J., et al. 'Recent Anthropogenic Plant Extinctions Differ in Biodiversity Hotspots and Coldspots'. *Current Biology* 29(17) (2019): 2912–18, e1–e2.

Marcgraf, Georg. *Historia naturalis Brasiliae*. Leiden: Apud Franciscum Hackium, 1648.

McClinton, Philip L., Susan M. McClinton and Gilbert J. Guzman. 'Utilization of Fish as a Food Item by a Mountain Lion (*Puma concolor*) in the Chihuahuan Desert'. *Texas Journal of Science* 52(3) (2000): 261–63.

McEvilley, Thomas. 'Introduction', in Brian O'Doherty, *Inside the White Cube: The Ideology of the Gallery Space* (Berkeley: University of California Press, [1976] 1999), 7–12.

McFarland, Brian Joseph. *Conservation of Tropical Rainforests: A Review of Financial and Strategic Solutions*. Cham: Palgrave, 2018.

Melton, Arthur W. *Problems of Installation in Museums of Art*. Washington, DC: The American Association of Museums, (1935) 1988.

Merriam, 'Oryzomys nelson sp. nov. Nelson's Rice Rat'. *Proceedings of the Biological Society of Washington* 12 (1898): 15.

Pollock, Griselda. 'To Play Many Parts: Reading between the Lines of Charlotte Salomon's/CS's *Leben? oder Theater?*' *RACAR* 43(1) (2018): 63–80.

Raffaelli, David. 'How Extinction Patterns Affect Ecosystems'. *Science* 306(5699) (2004): 1141–42.

Randazzo, Elisa, and Hannah Richter. 'The Politics of the Anthropocene: Temporality, Ecology, and Indigeneity'. *International Political Sociology*, Advance article, olab006 (2021). https://doi.org/10.1093/ips/olab006.

Richardson, John. 'Canis f. var. B. Lagopus. *Hare Indian Dog*', in John Richardson, *Fauna Boreali-Americana or the Zoology of the Northern Parts of British America*, assisted by William Swainson and William Kirby (London: John Murray, 1829), 78–80.

Rochefort, Charles de. *Histoire naturelle et morale des iles Antilles de l'Amérique*. Rotterdam: Arnould Leers, 1658.

Sclater, Philip Lutley. 'A Review of the Species of the Family *Icteridae*: Part IV Quiscalinae'. *Ibis* 5(6) (1884): 149–67.

Stansbery, David H. 'Rare and Endangered Freshwater Mollusks in the United States', in S.E. Jorgensen and R.W. Sharp (eds), *Proceedings of a Symposium on Rare and Endangered Mollusks (Naiads) of the U.S.* (St. Paul, MN: United States Fish and Wildlife Service, 1971), 5–18.

Tabarelli, Marcelo, José Maria Cardoso da Silva and Claude Gascon. 'Forest Fragmentation, Synergisms and the Impoverishment of Neotropical Forests'. *Biodiversity & Conservation* 13 (2004): 1419–25.

Teichman, Kristine J., Bogdan Cristescu and Chris T. Darimont. 'Hunting as a Management Tool?: Cougar–Human Conflict Is Positively Related to Trophy Hunting'. *BMC Ecology* 16 (2016). https://doi.org/10.1186/s12898-016-0098-4.

Tuck, Eve, and Michelle Fine. 'Inner Angles: A Range of Ethical Responses to/with Indigenous and Decolonizing Theories', in Norman K. Denzin and Michael D. Gardina (eds), *Ethical Futures in Qualitative Research: Decolonizing the Politics of Knowledge* (Walnut Creek, CA: Left Coast Press, 2007), 145–68.

Tuck, Eve, and K. Wayne Yang, 'Decolonization is Not a Metaphor'. *Decolonization: Indigeneity, Education & Society* 1(1) (2012): 1–40.

Van Dooren, Thom. *Flight Ways: Life and Loss at the Edge of Extinction*. New York: Columbia University Press, 2016.

Viveiros de Castro, Eduardo. *From the Enemy's Point of View: Humanity and Divinity in an Amazonian Society*, translated by Catherine V. Howard. Chicago: University of Chicago Press, 1992.

Whyte, Kyle. 'Indigenous Climate Change Studied: Indigenizing Futures, Decolonizing the Anthropocene'. *English Language Notes* 55(1–2) (2017): 153–62.

Wilcove, David S., et al. 'Navjot's Nightmare Revisited: Logging, Agriculture, and Biodiversity in Southeast Asia'. *Trends in Ecology and Evolution* 28(9) (2013): 531–40.

Witton, Mark P. *Palaeoartist's Handbook: Recreating Prehistoric Animals in Art*. Ramsbury, UK: Crowood Press, 2018.

Witton, Mark P., Darren Naish and John Conway. 'State of the Palaeoart'. *Palaeontologica Electronica* 17(3) (2014). Retrieved on 17 October 2021 from http://palaeo-electronica.org/content/pdfs/comment_palaeoart.pdf.

Woodhouse, S.W. 'The North American Jackal: *Canis frustror*'. *Proceedings of the Academy of Natural Sciences of Philadelphia* 5 (1850–51): 147–48.

Contributors

Jeffrey P. Benca is a research associate at the University of Washington's Burke Museum of Natural History and Culture as well as a horticulturist for Amazon's Horticulture Program and Seattle Spheres in Washington state, USA. Receiving his doctorate in integrative biology from the University of California, Berkeley, his Devonian lycopsid reconstruction was featured as the *American Journal of Botany* centennial cover. He has also contributed to modern lycopsid research and conservation through pioneering cultivation techniques for terrestrial clubmosses. He led the first experimental study to test a proposed driver of Earth's largest mass extinction using modern plants, discovering a non-lethal mechanism for global ecosystem collapse under stratospheric ozone weakening, published in *Science Advances*.

Sarah Bezan is postdoctoral research associate in perceptions of biodiversity change at the University of York's Leverhulme Centre for Anthropocene Biodiversity in the UK. Her research focuses on the entangled social and ecological dimensions of species loss and revival in contemporary British, North American and Australian literature and visual culture. She is currently at work on two book projects: *Dead Darwin: Necro-Ecologies in Neo-Victorian Culture* (under advance contract with Manchester University Press), along with a second monograph (in progress) that examines species revivalist representations of the woolly mammoth, great auk, dodo, Steller's sea cow, thylacine, and Pinta Island tortoise.

Valérie Bienvenue is a doctoral candidate in the Department of History of Art and Film Studies at the Université de Montréal. Her thesis critically examines human–equine relations through the prism of modern art and visual culture. Prior to her academic career, she worked for ten years in equestrian circles, including teaching bareback riding and rehabilitating horses suffering from physical and psychological trauma. She is the author of several articles and book chapters.

Matthew Morse Booker is associate professor of history at North Carolina State University and vice president for scholarly programs at the National Humanities Center in North Carolina. He publishes in environmental and food history, including *Down by the Bay: San Francisco's History between the Tides* (2013) and *Food Fights: How History Matters to Contemporary Food Debates* (2019).

Pedro Cardoso is curator at the Finnish Museum of Natural History (Luomus), and adjunct professor in ecology at the University of Helsinki. While heading the Laboratory for Integrative Biodiversity Research (LIBRe, www.biodiversityresearch.org), he is currently mostly interested in understanding global drivers of extinction, and the distribution of species and communities across space and time. To reach such goals he is also developing new statistical and computational tools to quantify extinction risk and biodiversity at all levels: taxonomic, phylogenetic and functional.

Nicholas Chare is professor of art history in the Department of History of Art and Film Studies at the Université de Montréal. He is the author of *After Francis Bacon* (2012). In 2017, with Sébastien Lévesque and Silvestra Mariniello, he founded the baccalaureate (BACCAP) in visual cultures at the Université de Montréal.

Barbara Creed publishes in the areas of feminism, psychoanalytic theory, human rights and animal ethics. She is the author of six books, including *The Monstrous-Feminine: Film, Feminism, Psychoanalysis* (1993), now in its ninth edition; *Darwin's Screens: Evolutionary Aesthetics, Time and Sexual Display in the Cinema* (2009); and, most recently, *Stray: Human–Animal Ethics in the Anthropocene* (2017). She is the co-founder of the Human Rights and Animal Ethics Research Network (HRAE) in the Faculty of Arts at the University of Melbourne, where she is a Redmond Barry distinguished professor emeritus. Her new book, *Return of the Monstrous-Feminine*, will be published by Routledge in 2022.

Robert R. Dunn is professor of applied ecology at North Carolina State University and in the Center for Evolutionary Hologenomics at the University of Copenhagen. He is the author of numerous books including the book he co-authored (with Monica Sanchez) *Delicious: The Evolution of Flavor and How It Made Us Human* (2021).

Jeanette Hoorn is an honorary professorial fellow in the Faculty of Arts at the University of Melbourne. Her research examines racial and gender issues surrounding the production and reception of modern art and the cinema. Her publications have included studies of the pastoral in Australian art, the impact of Darwinian theory, and the special circumstances surrounding the art and culture of First Nations Australians. She has written on Hilda Rix Nicholas in Morocco; Joseph Lycett and the doctrine of *Terra Nullius*; the portraits of Julie Dowling and Tom Roberts; and the place of the yam in Emily Kame Kngwarreye's paintings. She has also written on the Australian films of Michael Powell and Charles Chauvel, on the photographs of Tracey Moffatt and Destiny Deacon, and on the representation of the *mission civilisatrice* in French Colonial cinema. She is working to find ways of revealing and challenging the operation of apartheid in Australia through her writing.

Norman MacLeod is a distinguished professor at the School of Earth Science and Engineering, Nanjing University, China. He was formerly the Dean of Postgraduate Education and Training (2013–16), Keeper of Palaeontology (2000–13), and Associate Keeper of Palaeontology (1999) at the Natural History Museum, London, in addition to being an honorary professor at University College London and a visiting professor at the Nanjing Institute of Geology and Palaeontology, Chinese Academy of Sciences. Professor MacLeod is perhaps best known for his work on the causes of Phanerozoic extinctions, and his theoretical, methodological and applied work in quantitative morphological analysis.

David Maynard is the senior curator for natural sciences at the Queen Victoria Museum and Art Gallery in Launceston, Tasmania. He took on this role in 2012, and quickly came to realize the cultural and social importance of the thylacine to Tasmanians. Since then, he has seen a change in the community's attitude to the Tasmanian tiger, shifting from an unswaying belief that it survived in the remotest parts of Tasmania, to an understanding that the human hand has left no place for the thylacine to hide. He co-authored (with Tammy Gordon) the book *Tasmanian Tiger: Precious Little Remains*, from which the exhibition of the same name was developed.

Kathryn Medlock spent her career working in curatorial roles in the vertebrate zoology department at the Tasmanian Museum and Art Gallery, where she was responsible for the development, preparation, care and display of the collection. She has actively participated in research projects on reptiles, birds and mammals, including cetaceans, and has curated several museum exhibitions, including one in 1998 on the thylacine. In 2003 she was awarded a Churchill Fellowship to study Tasmanian birds and mammals held in museums in Britain, Europe and North America. Currently, her work is focused on the zoological, cultural and historical aspects of thylacine collections around the world, and on the origins of the specimens held in the Tasmanian Museum and Art Gallery. She is passionate about promoting the use of natural history collections for both research and education, as it becomes increasingly urgent to gain an understanding of the implications of habitat loss and extinction.

Charles R. Menzies (hagwil hayetsk) is a member of Gitxaała Nation and a professor of anthropology at the University of British Columbia in Vancouver. His primary research interests are the production of anthropological films, natural resource management (primarily fisheries related), political economy, contemporary First Nations issues, maritime anthropology and the archaeology of north coast BC.

Anne-Sophie Miclo is a teaching assistant and doctoral candidate in history of art at the Université de Québec à Montréal (UQAM). Her research, which explores the impact of living things on museum practices, is supported by the Quebec Research Funds – Society and Culture (FRQSC). She is a member of the research group Figura, which is dedicated to text and the imaginary, and also of the Convulsive Collections research group (CIÉCO), which focuses on the development and promotion of museum collections. She is the author of numerous articles and exhibition catalogues relating to contemporary art practice.

W.J.T. Mitchell is Gaylord Donnelley distinguished service professor of English and art history at the University of Chicago. He is the senior editor of *Critical Inquiry*.

Jingmai O'Connor is an American paleontologist whose research focuses on the dinosaur–bird transition and the Mesozoic evolution of birds and other flying dinosaurs. Her research includes studies of exceptional soft tissues such as lung and ovary traces preserved in specimens from the 130–120 Ma Jehol Biota. She has authored over one hundred papers, including studies published in *Nature, Science, Current Biology*, and *PNAS*. She was the 2019 recipient of the Palaeontological Society's Charles Schuchert Award for outstanding palaeontologist under the age of 40. She is currently serving as the associate curator of fossil reptiles at the Field Museum of Natural History in Chicago.

Stuart Pimm is the Doris Duke chair of conservation ecology at Duke University. He is a world leader in the study of present-day extinctions, and what can be done to prevent them. He received his BSc from Oxford University (1971) and his PhD from New Mexico State University (1974). He is the author of 350 scientific papers and four books. He directs Saving Nature, a 501(c)(3) non-profit, which uses funds for carbon emission offsets to fund local conservation groups to restore degraded lands in areas of exceptional tropical biodiversity. His international honours include the International Cosmos Prize (2019), the Tyler Prize for Environmental Achievement (2010), and the Dr A.H. Heineken Prize for Environmental Sciences from the Royal Netherlands Academy of Arts and Sciences (2006).

Harriet Ritvo is the Arthur J. Conner professor of history emeritus at the Massachusetts Institute of Technology. A past president of the American Society for Environmental History, she is editor of the 'Animals, History, Culture' book series (Johns Hopkins University Press), and co-editor of the 'Flows, Migrations, and Exchanges' book series (University of North Carolina Press). Her research has focused on the history of natural history, the history of human relations with other animals, and environmental history, especially in Britain and the British Empire. Her current work engages issues of wildness and domestication. Her books include *The Animal Estate: The English and Other Creatures in the Victorian Age* (1987), *The Platypus and the Mermaid, and Other Figments of the Classifying Imagination* (1997), *The Dawn of Green: Manchester, Thirlmere, and Modern Environmentalism* (2009), and *Noble Cows and Hybrid Zebras: Essays on Animals and History* (2010).

Monica C. Sanchez is a medical anthropologist who studies the cultural aspects of health and well-being. Her most recent book (co-authored with Robert Dunn) is *Delicious: The Evolution of Flavor and How It Made Us Human* (2021).

Dawn Sanders is an associate professor in the Faculty of Education at the University of Gothenburg, Sweden. Her education combines training in fine art with further studies in ecology, culminating in a doctoral study that considered the educational role of botanic gardens (Geography Department, Sussex University, UK, 2005). She led the interdisciplinary research team in the Swedish Research Council funded project 'Beyond Plant Blindness: Seeing the importance of plants for a sustainable world', resulting in related publications in research journals and in a book of the same name published in 2020 by The Green Box in Berlin. She continues to work with artists on questions concerning human attention to plants.

Gordon M. Sayre is professor of English and folklore at the University of Oregon, and a specialist in colonial history and literature of North America. He is the translator of *The Memoir of Lieutenant Dumont* by Jean-François Benjamin Dumont de Montigny, written in 1747 and first published in French in 2008, and in English in 2012. His work on natural history, species and media has also appeared in the journals *Environmental Humanities* and *The William and Mary Quarterly*.

Samuel T. Turvey is a professor at the Institute of Zoology, Zoological Society of London, where he has worked since 2004. He has previously been based at the University of Canterbury (New Zealand) and the University of Oxford. His research uses 'non-standard' ecological archives, including Indigenous knowledge and historical baselines, to reconstruct human-caused biodiversity loss over time and to inform current-day conservation. He also conducts applied conservation research on highly threatened species to identify effective management and recovery strategies for tiny vulnerable animal populations. He has worked in China for over twenty years and is involved with long-term conservation efforts for Chinese species such as the Hainan gibbon, the Chinese giant salamander and the Yangtze finless porpoise.

Index